Mobile Mapping Technologies

Mobile Mapping Technologies

Special Issue Editors

Pablo Rodríguez-Gonzálvez

Erica Nocerino

Isabella Toschi

MDPI • Basel • Beijing • Wuhan • Barcelona • Belgrade

Special Issue Editors
Pablo Rodríguez-Gonzálvez
Universidad de León
Spain

Erica Nocerino
ETH Zurich
Switzerland

Isabella Toschi
Bruno Kessler Foundation
Italy

Editorial Office
MDPI
St. Alban-Anlage 66
4052 Basel, Switzerland

This is a reprint of articles from the Special Issue published online in the open access journal *Remote Sensing* (ISSN 2072-4292) from 2018 to 2019 (available at: https://www.mdpi.com/journal/remotesensing/special_issues/mobile_mapping).

For citation purposes, cite each article independently as indicated on the article page online and as indicated below:

LastName, A.A.; LastName, B.B.; LastName, C.C. Article Title. *Journal Name* **Year**, *Article Number*, Page Range.

ISBN 978-3-03928-018-6 (Pbk)
ISBN 978-3-03928-019-3 (PDF)

Contents

About the Special Issue Editors

Pablo Rodríguez-Gonzálvez is an assistant professor at the University of León (Spain). He received the B.S. degree in Surveying Engineering from the University of Oviedo, Spain, in 2004, the M.S. degree in geodesy and cartography from the University of Salamanca, Avila, Spain, in 2006, and the PhD degree in Photogrammetry and Computer Vision from the University of Salamanca in 2011. He has authored over 80 peer-reviewed research articles in international journals and conference proceedings, and 10 inventions. His research lines include photogrammetry 3D reconstruction, accuracy assessment, radiometric and geometric calibration of the different geomatic sensors.

Erica Nocerino is currently a Post-Doctoral fellow at ETH Zürich and the University of Marseille (AMU). She holds two PhDs, one in engineering and quality control and the second in geomatics. She serves as secretary of the ISPRS Working Group II/7 in Vision Metrology. She has authored more than 70 scientific articles in international journals and conference proceedings She is editor of different special issues on 3D recording and modelling and acts as a reviewer for international journals and conferences. Her research focuses on the development of procedures and guidelines for the generation of 3D models of known and verified quality, in different application fields such as industrial metrology, cultural heritage (CH), biology and medicine.

Isabella Toschi is a researcher in the 3D Optical Metrology (3DOM) research unit of the Bruno Kessler Foundation (FBK). At the same time she serves as the secretary of the ISPRS Technical Commission II (Photogrammetry), as a member of the O3DM Scientific Committee (ex SPIE Videometrics), and as a reviewer for several international journals. She holds a BSc. and a MSc. degree in Environmental Engineering from the University of Modena and Reggio Emilia, and a PhD from the School of High Mechanics and Automotive Design & Technology of the same university. Her current research interests vary from automation in 3D recording and modelling to sensor integration and geospatial data processing.

Preface to "Mobile Mapping Technologies"

The escalating demand for accurate and dense geo-referenced 3D data is pushing the Geomatics community to develop new tools and algorithms. As a result, Mobile Mapping technologies have experienced a rapidly growing research activity and interest in the last years, due to their ability of acquiring 3D information of large areas dynamically and cost-effectively. This versatility has expanded their application fields from the civil engineering to a broader range (industry, emergency response, cultural heritage...), which is still widening. This increased number of needs, some of them specially challenging, is translating into innovative solutions, ranging from new hardware/open source software approaches, integration of different sensors and devices, up to the adoption and development of artificial intelligence methods for the automatic extraction of salient features and quality assessment for performance verification. In this book, composed of 14 peer-reviewed articles, different trends and experiences are presented, each related to a different aspect of the new paradigm of Mobile Mapping Systems:

- Chapters 1 and 2 present two novel indoor positioning and localization methods based on mobile phone videos and RGB-D (red, green, blue and depth) image database, respectively.
- Chapter 3 presents a novel low-cost visual odometry method for estimating the self-motion of ground vehicles by detecting the changes that motion induces on the images.
- Chapter 4 shows an invesitigation on the feasibility of inferring the room type by using grammars based on geometric maps derived from indoor mapping approaches.
- Chapter 5 proposes an accurate visual-inertial integrated geo-tagging method that can be used to reconstruct crowdsourced trajectory of smartphone users.
- Chapters 6 and 7 show research related to robot navigation; the first proposes an algorithm to estimate the motion of the robot in both dynamic and static environments; the second shows an improved algorithm for a fast-real-time of a mobile robot.
- Chapter 8 proposes a precise and robust segmentation-based localization system designed for high level autonomous driving.
- Chapters 9 and 10 are devoted to the evaluation of Mobile Mapping Systems: the first designs and evaluates the performance of a low-cost backpack indoor mobile mapping system; the second presents a novel extrinsic self-calibration algorithm, fully automatic and completely data-driven.
- Chapter 11 presents a novel framework to extract metro tunnel profiles from 3D point clouds.
- Finally, Chapters 12 to 14 show successful applications of Mobile Mapping Systems in different fields: inspection of thermal and fluid–mechanical facilities (Chapter 12); digitalization and modelling of a complex cultural heritage building (Chapter 13); and 3D digital characterization of vegetation for precision agriculture (Chapter 14).

Pablo Rodríguez-Gonzálvez, Erica Nocerino, Isabella Toschi
Special Issue Editors

Article

Indoor Topological Localization Using a Visual Landmark Sequence

Jiasong Zhu [1], Qing Li [1,2,3,*], Rui Cao [1,3,4], Ke Sun [1], Tao Liu [5], Jonathan M. Garibaldi [2], Qingquan Li [1], Bozhi Liu [3] and Guoping Qiu [2,3]

1 Shenzhen Key Laboratory of Spatial Smart Sensing and Services & Key Laboratory for Geo-Environmental Monitoring of Coastal Zone of the National Administration of Surveying, Mapping and Geoinformation, Shenzhen University, Shenzhen 518060, China; zhujiasong@gmail.com (J.Z.); rui.cao@nottingham.edu.cn (R.C.); sk100.force@gmail.com (K.S.); liqq@szu.edu.cn (Q.L.)
2 School of Computer Science, The University of Nottingham, Nottingham NG8 1BB, UK; jon.garibaldi@nottingham.ac.uk (J.M.G.); qiu@szu.edu.cn (G.Q.)
3 College of Information Engineering & Guangdong Key Laboratory of Intelligent Information Processing, Shenzhen University, Shenzhen 518060, China; bozhi.liu@hotmail.com
4 International Doctoral Innovation Centre & School of Computer Science, The University of Nottingham Ningbo China, Ningbo 315100, China
5 College of Resource and Environment, Henan University of Economics and Law, Zhengzhou 450046, China; liutao@huel.edu.cn
* Correspondence: qing.li@nottingham.ac.uk; Tel.: +86-132-6824-9517

Received: 7 November 2018; Accepted: 31 December 2018; Published: 3 January 2019

Abstract: This paper presents a novel indoor topological localization method based on mobile phone videos. Conventional methods suffer from indoor dynamic environmental changes and scene ambiguity. The proposed Visual Landmark Sequence-based Indoor Localization (VLSIL) method is capable of addressing problems by taking steady indoor objects as landmarks. Unlike many feature or appearance matching-based localization methods, our method utilizes highly abstracted landmark sematic information to represent locations and thus is invariant to illumination changes, temporal variations, and occlusions. We match consistently detected landmarks against the topological map based on the occurrence order in the videos. The proposed approach contains two components: a convolutional neural network (CNN)-based landmark detector and a topological matching algorithm. The proposed detector is capable of reliably and accurately detecting landmarks. The other part is the matching algorithm built on the second order hidden Markov model and it can successfully handle the environmental ambiguity by fusing sematic and connectivity information of landmarks. To evaluate the method, we conduct extensive experiments on the real world dataset collected in two indoor environments, and the results show that our deep neural network-based indoor landmark detector accurately detects all landmarks and is expected to be utilized in similar environments without retraining and that VLSIL can effectively localize indoor landmarks.

Keywords: visual landmark sequence; indoor topological localization; convolutional neural network (CNN); second order hidden Markov model

1. Introduction

Topological localization is a fundamental component for pedestrians and robots localization, navigation, and mobile mapping [1,2]. It is compatible with human understanding as topological maps utilize highly abstracted knowledge of present locations. Represented by a graph, a topological map is a compact and memory-saving way to represent an environment, and thus is suitable for large-scale scene localization [3]. Each node of it indicates a region of the environment, which is associated with a visual feature vector to represent it. The vital problem of the technique is to design robust and distinctive features to represent nodes identically.

Many handcrafted features have been devised based on colors, gradients [3], lines [4], or distinctive points to represent the nodes. Previous work also entails learning the representation of the nodes using machine learning techniques [5]. However, most of them fail in dynamic indoor environments due to camera noise, illumination and perspective changes, or temporal variations. Another serious problem is that there are numbers of visually similar locations in the same environment, which further adds the difficulty of finding the proper visual location representation. Therefore, it still remains a challenging problem to find the robust visual representation for image-based indoor localization.

Exploiting semantic information from videos for localization is more feasible and human-friendly compared to conventional features or appearance matching-based methods. Finding matched features in large scenes is inefficient, and it often fails due to the amount of visually similar locations. In addition, matching multi-modality images is also a problem. Steady elements in the environment are robust representations of locations as they are salient and insensitive to occlusions, illuminations, and view variations. Their ground truth locations are also fixed and known.

In this paper, we propose a novel visual landmark sequence-based approach that exploits the steady objects for indoor topological localization. In the approach, sematic information of steady objects on the wall is used to represent locations, and their occurrence order in the video is used for localization through matching against the topological map. A topological map constructed with the prior of floor plan map of the environment is used to store connectivity information between landmarks. Each node on the map indicates a local region of the environment and is represented by landmark sematic information. To address the environmental ambiguity problem, we extract landmark sequence from a mobile phone video, and match them using the proposed matching algorithm based on their occurrence order. We make the following original contributions:

1. We propose a novel visual landmark sequence-based indoor localization (VLSIL) framework to acquire indoor location through smartphone videos.
2. We propose a novel topological node representation using sematic information of indoor objects.
3. We present a robust landmark detector using a convolutional neural network (CNN) for landmark detection that does not need to retrain for new environments.
4. We present a novel landmark localization system built on a second order hidden Markov model to combine landmark sematic and connectivity information for localization, which is shown to relieve the scene ambiguity problem where traditional methods have failed.

Part of the content is included in our conference paper [6]. Compared to the conference paper, we show the following expansions:

1. We modified the HMM2-based localization algorithm to make it work in the case where part of the multiple-object landmark is detected.
2. We have conducted more comprehensive experiments to demonstrate that our landmark detector outperforms detectors based on handcrafted features.
3. We further tested the algorithm in a new experimental site to verify the generality of the detector and the localization method.
4. We also conducted further analysis over the factors, including landmark sequence length and map complexity, that affect the performance of the algorithm.

The rest of the paper is organized as follows. In Section 2, we review related work on visual landmark representation and image-based localization. In Section 3, we illustrate the basic concept of visual landmark sequence-based indoor localization. Section 4 presents the detail of the CNN-based detector, which detects landmarks from smartphone videos. Section 5 elaborates the proposed matching algorithm based on a second order hidden Markov model. Section 6 presents extensive experimental results, and Section 7 concludes the paper.

2. Related Work

The proposed method is highly related to visual landmark representation and image-based localization methods. We briefly review related works in the two fields.

2.1. Visual Landmark Representation

Visual landmarks can be divided into two categories: artificial landmarks and natural landmarks. Artificial landmarks are purposefully designed to be salient in the environment. Ahn et al. [7] designed a circular coded landmark that is robust with perspective variations. Jang et al. [8] devised landmarks based on color information and recognized them using color distribution. Basiri et al. [9] developed a landmark-based navigation system using QR codes as landmarks, and the user's location was determined and navigated by recognizing the quick response code registered in the landmark's location. Briggs et al. [10] utilized self-similar landmarks based on barcodes and were able to perform localization in real time. Artificial landmarks can be precisely detected since they are manufactured based on prior rules. Such rules allow them to stay robust, facing challenges of the varying illuminations, view points, and scales in images, and help to devise the landmark detectors. Their position can also be coded in the landmark appearance. However, deploying artificial landmarks changes building decorations, which might not be feasible due to economic reasons or the owners' favor. Natural landmarks avoid changing indoor surfaces by exploiting physical objects or scenes in the environment. Common objects such as doors, elevators, and fire extinguishers are good natural landmarks. They remain unchanged over a relatively long period and are common in indoor environments.

Many methods have been proposed to represent locations using natural landmarks [11–13]. Some of them are based on handcrafted features, which make use of color, gradients, or geometric information. Planar and quadrangular objects are viewed as landmarks and are detected based on geometric rules [11,12]. Tian et al. [13] identified indoor objects such as doors, elevators, and cabinets by judging whether detected lines and corners satisfy indoor object shape constraints. SIFT features were chosen to perform natural landmark recognition in [14,15]. Serrão et al. [16] proposed a natural landmark detection approach by leveraging SURF features and line segments. It performed well in detecting doors, stairs, and tags in the environment. Kawaji et al. [17] used omnidirectional panoramic images taken in different positions as landmarks, and PCA-SIFT was applied to perform image matching. Moreover, shape [4,18], light strength [19], or region connection relations [20] have also been exploited to represent landmarks for localizations. Kosmopoulos et al. [21] developed a landmark detection approach based on edges and corners. These methods have achieved good results on specified objects in certain scenes. However, they are likely to fail in other scenes due to the variations in doors and stairs.

In this paper, we propose a robust landmark representation using sematic information. A CNN-based landmark detector is proposed to determine landmark type. Unlike previous approaches using handcrafted features, our detector learns the distinctive features to distinguish target objects and background. Moreover, it can be used for off-the-shelf scenes without changing. The learned features are not derived from a single space but from a combination of color, gradients, and geometric space. With a proper training dataset, it stays robust to landmark variations caused by illumination and other deformations. CNN was selected due to its high performance in image classification [22] and indoor scene recognition [23] and outperforms approaches based on handcrafted features.

2.2. Image-Based Localization

As mobile computing and smartphones are becoming readily accessible, there have been attempts to use smartphone cameras for indoor localization [17,24–36]. These methods exploit computer vision techniques to estimate people's location and mainly fall into two categories: image retrieval-based methods and 3D model-based methods. The former uses images captured by the smartphone camera to search for similar images in the image dataset whose positions and orientations are already known. The pose of the query image is determined with poses of similar images. This approach not only requires significant offline processing but can also easily get stuck in situations where different locations have a similar appearance. The latter approaches estimate location by building corresponding 2D–3D matches. However, they do not work in low texture environments, and they suffer from image blurring caused by camera motion. In addition, environment change significantly decreases the performance of the two types of methods, which frequently occurs in indoor environments.

Many positioning algorithms have introduced landmarks for indoor localization. Basically, landmarks are taken as supporting information to reduce the error drift of dead reckon approaches [37–39]. In this paper, we focus on performing indoor topological localization with only visual landmark information since landmarks play an important role in localizing and navigating pedestrians in an unfamiliar environment [40].

Many approaches perform landmark-based localization under a geometric scheme. Triangle intersection theory is applied to localize users using more than three landmarks [41]. Jang et al. [8] presented an approach with a single landmark. The user's position was estimated based on an affine camera model between the three-dimensional space and the projected image space.

Another type of landmark-based localization utilizes landmark recognition techniques. It assumes that users are close to the detected landmarks. The landmark is identified based on their visual representations [11,12,19]. However, in indoor environments, it is usually not feasible to match landmarks based only on visual features, since locations can have a similar appearance. Additional information is needed to distinguish different landmarks. Tian et al. [13] exploited textual information around doors to address this problem. However, it is not always possible to have tags of text around doors. Contextual information between landmarks was exploited through a hidden Markov model (HMM) to recognize landmarks and achieve good results in [42–44]. However, an HMM model only takes a previous landmark to recognize a current landmark and fails in scenes of high ambiguity.

In this paper, we propose a matching algorithm based on a second order hidden Markov model (HMM2) to utilize landmark connecting information and sematic information for landmark recognition. An HMM2 is able to involve the walking direction in the process of landmark recognition. The walking direction is introduced to constrain the landmark connectivity. In this manner, more contextual information is taken into account for landmark localization, so indoor scene ambiguity is reduced.

3. Visual Landmark Sequence-Based Indoor Localization (VLSIL)

We propose a novel visual landmark sequence-based indoor localization (VLSIL) framework, and we first illustrate its basic idea. Suppose there is an indoor space that has seven locations as shown in Figure 1a. For each location, there is a landmark representing it as shown in Figure 1b and the color indicates the landmark type. Pedestrians can only walk from one location to the others linked by a path. Suppose pedestrians reach the location L(2) without knowing it and observe the red landmark. Their locations cannot be determined since there is more than one location denoted by the red landmark (e.g., LM(5) and LM(7)). Suppose pedestrians observe red, green, and blue landmarks in sequence in their path. They can be sure they start from LM(2), go through LM(4), and arrive at LM(6), because LM(2), LM(4), and LM(6) are the only valid path. The VLSIL achieves localization through taking photos (video) of a location to determine the current position by matching a sequence of previously discovered landmarks against the topological map of the space.

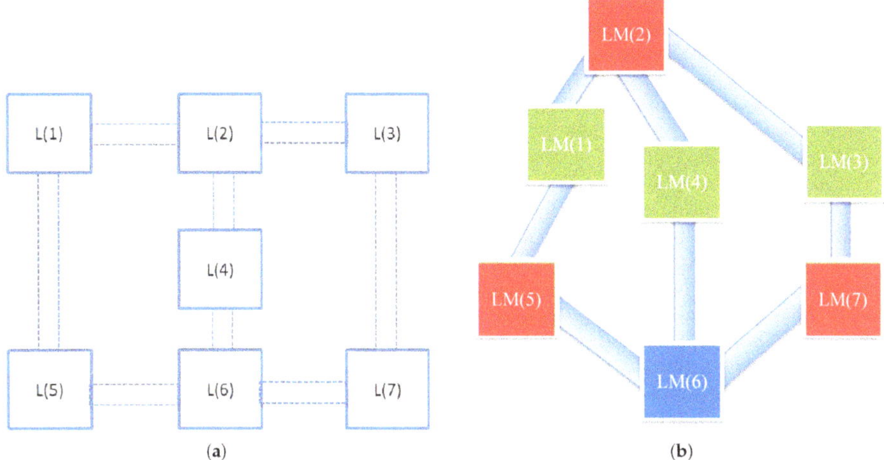

Figure 1. (**a**) Topological map of an indoor space, where there are seven locations. (**b**) In each of the locations of the space, there is a landmark representing it. Landmarks of the same color are identical (e.g., office doors). A person can only walk from one location to the next linked by a path.

4. Landmark Detection

The landmark detection process consists of two phases: the offline phase and the online phase. During the offline phase, landmark types are pre-defined from common indoor objects and scenes, and a CNN is trained to recognize them. The online phase performs the landmark detection from the captured videos. It includes frame extraction, region proposal, and landmark type determination. Figure 2 illustrates the whole process. The offline phase is highlighted with a light blue background, and the rest of it is the online phase.

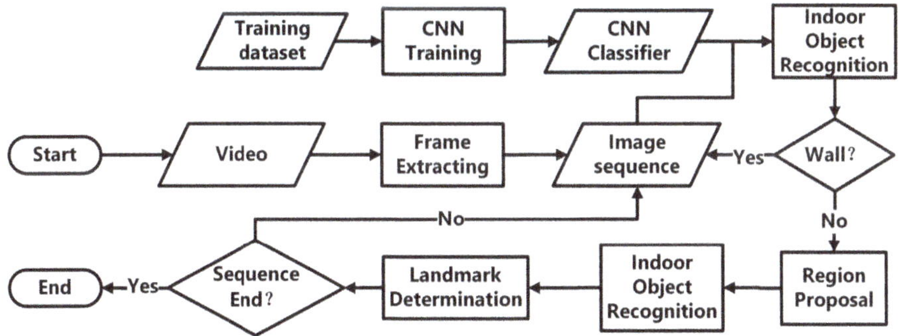

Figure 2. Flowchart of indoor landmark detection.

In the real scene, many of the extracted images only capture the background information, which is usually representative of walls. Applying a selective search to these images is not necessary and decreases the efficiency. Therefore, we first determine whether the extracted image belongs to a wall (the background). If so, the next image is proceeded. If not, a selective search is performed, and proposed patches are identified.

The rest of this section provides a detailed introduction of the offline and online phases of the process.

4.1. Offline Phase

4.1.1. Landmark Definition

In this paper, landmarks are defined using common indoor objects such as doors, fire extinguishers, stairs, and indoor structure locations. Some examples of common objects are shown in Figure 3. Other indoor objects such as chairs and desks are not used because their positions are not fixed.

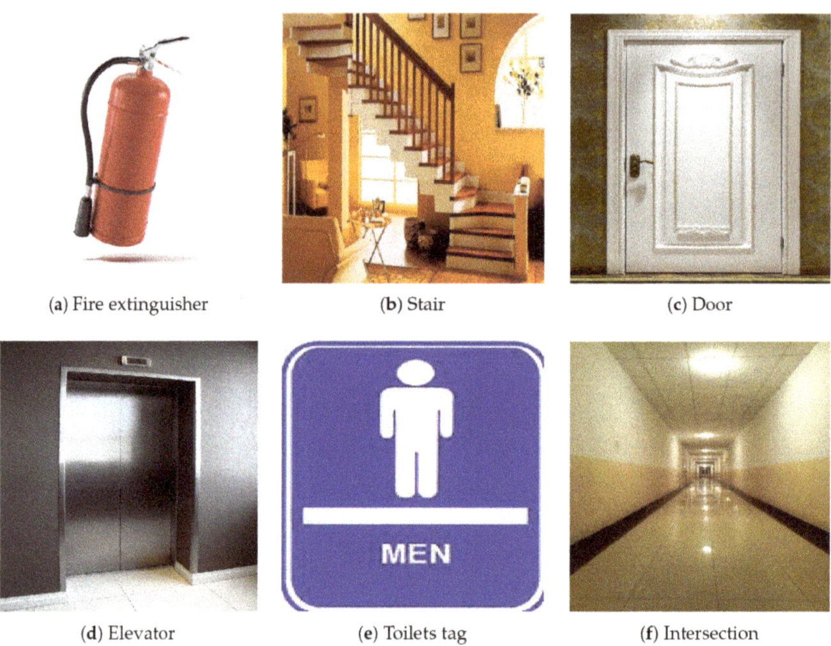

(a) Fire extinguisher (b) Stair (c) Door

(d) Elevator (e) Toilets tag (f) Intersection

Figure 3. Common indoor objects and locations of interest.

Three types of landmarks are defined: single-object landmarks, multiple-object landmarks, and scene landmarks. Single-object landmarks consist of one object such as a fire extinguisher or an elevator. Multiple-object landmarks are defined with more than one object. For instance, office doors are multiple-object landmarks, as they include a doorplate and a door. Combining multiple objects enlarges the landmark distinctiveness and reduces ambiguity of the map. We do not utilize the texts in the doorplate to further distinguish the office doors because motion blur makes text recognition very challenging. Scene landmarks are key locations of the indoor structure such as corners, intersections, or halls that have unique visual patterns.

4.1.2. Training CNN-Based Indoor Object Classifier

Our landmark detection relies on the object detection results of the extracted images. The high accuracy and real-time performance of the CNN on object detection inspired us to use it for our application [45]. In the application, we developed our CNN-based landmark detector by modifying AlexNet [46]. The modified AlexNet contains five convolutional layers and two fully connected layers. Each convolutional layer is tailed by a max pooling layer. Two fully connected layers are used to assemble information from the convolutional layers. AlexNet was selected for two reasons. The first is that its high performance in image classification has been demonstrated in ImageNet competition.

Secondly, it is relatively easy to converge since it has relatively few layers compared to other more complex networks.

Several tricks were applied to train AlexNet for our indoor object detection. Firstly, the output layer was modified to recognize the target indoor objects. AlexNet was originally designed for ImageNet competition, which aims to recognize 1000 types of objects. However, not all indoor objects of our interest were included. We replaced the output layer with new one, in which the number of neurons equals the number of our interesting indoor objects. The softmax function was chosen as the activation function of output layer neurons. Secondly, we retrained AlexNet with a finetuning technique. Only the newly added layer was allowed to retrain, while the weights of the rest of the layers were fixed when fine-tuning. Finally, to eliminate the object variations caused by illuminations, rotations, and movement, we conducted data augmentation by pre-processing the original images. For each original image, we change its brightness by adding 10, 30, -10, and -30 to produce new images. We rotated the original image by $5°$, $10°$, $-5°$, and $-10°$. The movement of pedestrians led to the partial occlusion of targets of interest. We also generated new images by randomly cropping original images to sizes of 224×224. The brightness and rotating images were altered with the original images, and cropping was done in the training stage. In this way, we enlarged the training dataset, and the trained network was robust to those variations.

4.2. Online Phase

The online phase consists of frame extraction, region proposal, indoor object recognition, and landmark type determination. We elaborate the procedures in detail, except the indoor object recognition step, which simply feeds the image patches into the classifier.

4.2.1. Frame Extraction

During the online phase, smartphone videos are sampled at a given rate. The sampling rate is a vital parameter as it impairs landmark detection accuracy and efficiency. Low sampling rate results in low overlap or even no overlap between successive images, which leads to a loss of tracking of certain objects in the image sequence. A high sampling rate leads to large information redundancy, resulting in low landmark detection efficiency, as more images are to be processed. Overlap can be roughly estimated using Equations (1) and (2). They are applied in two scenarios: walking along a line and turning to another direction.

$$Overlap = 1 - \frac{V}{2H\tan(\frac{\theta}{2})Hz} \times 100\% \tag{1}$$

$$Overlap = 1 - \frac{V_{ang}}{Hz\theta} \times 100\% \tag{2}$$

where V represents walking speed, and H is the average distance between camera and surrounding environment. θ is the field of view of camera in each mobile phone. Hz represents sampling rate. V_{ang} is the angular velocity. Empirically, a sampling rate of 3–5 frames per second would work well.

4.2.2. Region Proposal

Cutting target objects out of extracted images is crucial for landmark detection. Feeding images that contain background and target objects directly into the classifier decreases object recognition accuracy. It is because training samples are covered with indoor objects in the majority of image space, while in extracted images target objects may occupy only a small part of it. Therefore, we have to crop the patches with target objects taking up most of the space. Here we choose the selective search algorithm to generate patches of interest from images [47]. Selective search employs a bottom-up strategy to generate patches. The process contains two steps. At first, an over-segmentation algorithm is applied to generate massive initial regions in a variety of color spaces with a range of different parameters. A hierarchical grouping approach is then performed based on diverse similarity measurements including color, texture, shape, and fill, with various starting points. Hundreds or thousands of patches are produced from the algorithm. However, we do not need to process all of them to identify the target objects since there may be too many eligible patches. The first 300 patches are normally used for accuracy and efficiency purposes.

4.2.3. Landmark Type Determination

Landmark type is determined based on the indoor object recognition results. For single-object landmarks and scene landmarks, their types are given with their corresponding indoor objects. Regarding multiple-object landmarks, their types are determined by the combination of detected objects. For instance, the combination of doorplate and door represents an office door landmark.

A sequence of images is used to perform landmark type determination instead of a single image. The main reason is that components of multiple-object landmarks might not appear in the same image. The recognition result of a sequence images can address the problem as the components are sequentially detected. Moreover, it is helpful to eliminate the wrong recognition results. In this paper, indoor objects that are not seen in three successive images are interpreted as false detections. Exploiting image sequences for localization also helps determine the landmark occurrence order when more than one landmarks are observed in a single image. The first landmark detected prior to the current landmark is viewed as the previous landmark of the current detected landmark in the sequence. Sequence image length is set automatically based on the recognition results. A sequence starts from an object, is robustly recognized, and ends at the images (the walls).

5. Visual Landmark Sequence Localization Using the Second Order Hidden Markov Model

Knowing a sequence of landmark types from a video, we match them with the predefined topological map. In this section, we illustrate the defined topological map and the matching algorithm based on the second order hidden Markov model (HMM2) for our applications. We also extend the Viterbi algorithm for our application.

5.1. Topological Map

The topological map provides information of the distribution of landmarks of the indoor environment and indicates the connectivities between landmarks. In our paper, the topological map is a directed graph and is created from the floor plan map of the indoor environment. It consists of two types of elements: nodes and edges. Nodes indicate regions of the environment. Their color represents the landmark type. We use red nodes for fire extinguishers, black for intersections, blue for offices, silver for elevators, yellow for stairs, light green for the disabled toilets, green for men's toilets, and dark green for women's toilets. Edges denote the connecting information between landmarks. An edge starting from node i to node j indicates the sequential direction in which landmark j is detected after landmark i. The arrowed line indicates a one-way connection. In certain situation, two landmarks might be spatially close to each other. They are viewed as two regions and are represented with the corresponding landmarks.

5.2. The Second Order Hidden Markov Model for Indoor Localization

The HMM2 takes context information to perform tasks. It contains five elements: the observations set, the states set, the initial probability, the emission matrix, and the transition matrix. For our application, the observations set includes all landmark types and the states set indicates the landmark locations. Initial probability represents the starting position of a route. In the rest of the section, we detail the emission matrix and transition matrix of the HMM2 in our scenario. We also introduce a new parameter to handle unidentified multiple-object landmarks.

5.2.1. The Emission Matrix of HMM2

The emission matrix represents the state probabilistic distribution over the observation set [48]. Its row count equals the number of states and its column count is the number of observations classes. For our problem, the entry values of the emission matrix indicate the probability of an observed landmark type that belongs to a certain state. We assign the emission matrix value based on the landmark types of a landmark location. The emission matrix is defined as follows: $e_{i,j} = 1$, if landmark type j corresponds to state i; $e_{i,j} = 0$, otherwise.

5.2.2. The Transition Matrix of the HMM2

Unlike the transition matrix of the hidden Markov model which is a two-dimensional matrix, the transition matrix of the HMM2 is three-dimensional [48]. Its value $t_{i,j,k}$ indicates the probability that the next state is k, given the condition that the previous state is i and the current state is j. For the landmark-based indoor localization problem, it represents the probability of going through certain landmark positions given the previous two landmark positions. The matrix is defined as $t_{i,j,k} = 1$ if there is a path from i through j to k; $t_{i,j,k} = 0$, otherwise.

5.2.3. The Probabilistic Matrix of Landmark Type

Ideally, multiple-object landmarks are correctly recognized. However, in some cases, only a component of the landmark is detected. To deal with the problem, a probabilistic matrix, $p_{i,j}$, the probability of landmark type i given detected object j, is defined. This parameter does not affect single-object landmarks or scene landmarks. For them, when the object or scene is detected, its landmark type is determined. It aims to solve the confusion of multiple-object landmark when part of a landmark is observed. This works for situations where an object is detected but its landmark type still remains undetermined. The matrix value $p_{i,j} = 1$ if landmark i is a single-object landmark and j is the object to form it; $p_{i,j} = 0$, otherwise. For multiple-object landmarks, if the detected object cannot be used to recognize landmarks, we split the probability evenly. For example, if a door is detected, its matrix value equals 0.25 since it could belong to either an office or a toilet.

5.3. *The Extended Viterbi Algorithm for Indoor Localization*

Given the modified HMM2 for landmark localization, we extend the Viterbi algorithm to find the landmark sequence corresponding to the sequence of landmark types based on Bayesian theory. The details are below. Assume that the HMM2 has M states for landmarks, and the initial state parameter is π_i, which represents the probability when the process starts from landmark i. The transition matrix value t_{ij} is the transiting probability that the process move from landmark i to landmark j. There are n detected landmarks in the observation sequence, represented by $Y = \{y_1, y_2 \dots y_n\}$. The corresponding locations are represented by $X = \{x_1, x_2 \dots x_n\}$. We aim to find the landmark location sequence X of the maximum probability, given the landmark type sequence Y. Therefore, our objective function is to maximize $P(X|Y)$. From the Bayesian theory,

$$P(X|Y) = \frac{P(Y|X)P(X)}{P(Y)} \tag{3}$$

where $P(Y|X)$ denotes the probability distribution of the landmark type sequence Y, given state sequence X. In the hidden Markov model (HMM), it is represented by the emission matrix. $P(X)$ is the prior probability distribution of state sequence X. $P(Y)$ is the probability distribution of the observation sequence. It is a constant value. Hence, the solution to maximizing $P(X|Y)$ and maximizing $gu(X)$ are the same.

$$gu(X) = P(Y|X)P(X). \tag{4}$$

Taking the logarithm of $gu(X)$, Equation (4) is changed to Equation (5).

$$lgu(X) = log(gu(X)) = \sum_{j=1}^{n} logP(y_i|x_i) + logP(x_1, x_2, \dots x_n). \quad (5)$$

Since the logarithm function is monotonically increasing, $lgu(X)$ and $gu(X)$ share the same solution for the maximization problem. Note that the HMM requires that the next state only depends on the current state. $LogP(x_1, x_2 \dots, x_n)$ can be simplified to Equation (6).

$$logP(x_1, x_2, \dots x_n) = (\sum_{j=2}^{n} logP(x_j|x_{j-1})) + logP(x_1). \quad (6)$$

Equation (5) is transformed to Equation (7).

$$lgu(X) = \sum_{j=1}^{n} logP(y_i|x_i) + (\sum_{j=2}^{n} logP(x_j|x_{j-1})) + logP(x_1). \quad (7)$$

The Viterbi algorithm is used to find the solution to the maximization of $lgu(X)$. It recursively computes the path. Two parameters are updated in the process. At any step t, $V_{t,k}$ is used to record the maximum probability of the landmark sequence ending at landmark k, given t observations. $Ptr(k,t)$ records the previous landmarks before landmark k in the most likely state sequence. The process is as follows.

$$V_{1,k} = e_{y_1,k} \times \pi_k \quad (8)$$

$$V_{t,k} = max(e_{y_t,k} \times t_{x_{t-1},k} \times V_{t-1,x_{t-1}}) \quad (9)$$

$$Ptr(k,t) = \arg\max_{k}(e_{y_t,k} \times t_{x_{t-1},k} \times V_{t-1,x_{t-1}}). \quad (10)$$

The Viterbi algorithm has shown good performance in terms of solving the HMM problem. It has to be modified to solve the HMM2 problem because the HMM2 takes both the previous state and the current state into consideration when predicting the next step. Thus, Equation (6) has to be extended as follows.

$$logP(x_1, \dots, x_n) = \sum_{j=3}^{n} logP(x_j|x_{j-1}, x_{j-2}) + logP(x_2|x_1) + logP(x_1). \quad (11)$$

Another issue is that, during landmark detection, the landmark type might not be clearly recognized. The modified equation is Equation (11). A parameter is added to represent such unclear observations as introduced in Section 5.2.3. The Viterbi algorithm for the HMM2 was initialized by Equations (12) and (13) followed by iteration Equations (14) and (15) and is summarized in Algorithm 1.

$$V_{1,k} = max(p_{y_1,s_1} \times e_{y_1,k} \times \pi_k) \quad (12)$$

$$V_2(x_1,k) = V_{1,x_1} \times t_1(x_1,k) \times max(p_{y_2,s_2}) \times e_{y_2,k} \quad (13)$$

$$V_t(x_{t-1},k) = max(V_{t-1}(x_{t-2}, x_{t-1}) \times t_2(x_{t-2}, x_{t-1}, k)) \times max(p_{y_t,s_t} \times e_{y_t,k}) \quad (14)$$

$$Ptr_t(x_{t-1},k) = \arg\max_{x_{t-2}} (V_{t-1}(x_{t-2}, x_{t-1}) \times t_2(x_{t-2}, x_{t-1}, k)) \quad (15)$$

where S_t is the object type of detected landmark t.

Algorithm 1: Extended Viterbi finds the location sequence of maximum probability.

Input: A sequence of observations Y, transition matrix T_1, T_2, emission matrix E, probabilistic matrix P, initial location π

Output: A sequence of states X

Def: N: number of locations; M: number of landmark type; n: number of observations

Initialization:

$V_1 = T_1 \times \pi \times E \times P$

Recursion:

$V_t = V_{t-1} \times T_2 \times E_t \times P_t$

$Ptr_t = \arg\max\,(V_{t-1} \times T_2)$

Back trace:

$X_K = \arg\max_{col}\,(V_N)$ column index of the V

$X_{K-1} = \arg\max_{row}\,(V_N)$ row index of the V

$X_t = Ptr_{t+1}(X_{t+1}, X_{t+2})$

Return X;

6. Evaluation

6.1. Setup

To evaluate the proposed method, we conducted our experiments on the B floor of the Business South building (BSB) and the B floor of the School of Computer Science building (CSB) at the University of Nottingham, UK. The two sites are typical office environments containing many corridors and office rooms. Floor plan maps of the two sites are shown in Figures 4 and 5, respectively, and their corresponding topological maps are shown in Figures 6 and 7. We selected eight types of landmarks from the two places: office rooms, stairs, elevators, fire extinguishers, men's toilets, women's toilets, disabled toilets, and the intersection (corner). Among them, fire extinguishers, stairs, and elevators are single-object landmarks. Office rooms and toilets are multiple-object landmarks. An intersection is a scene landmark. The BSB is a relatively simple environment, while the CSB is more complex. In the BSB, there are 54 landmarks in total, and there are 65 landmarks in the CSB.

Two female and three male participants were asked to collect videos at both sites using smartphones. Three models of mobile phones were used: an Huawei Honor, a Samsung Note 3, and an iPhone 6s Plus. Each participant wore a mobile phone on their upper arm, with the camera looking sideways. Taking side-viewed videos provides more information about landmarks, as it is orthographic projection on landmarks. Compared to the front view, view variations are relieved. Another reason is that side-view capturing has a narrow field of view, which facilitates the determination of the landmark occurrence order, since the landmarks appear one by one in the video. Participants were asked to walk freely along the corridors in two experimental sites. In our experiments, a real world mobile video dataset of 1.9 h in total was collected for the evaluation of the proposed method.

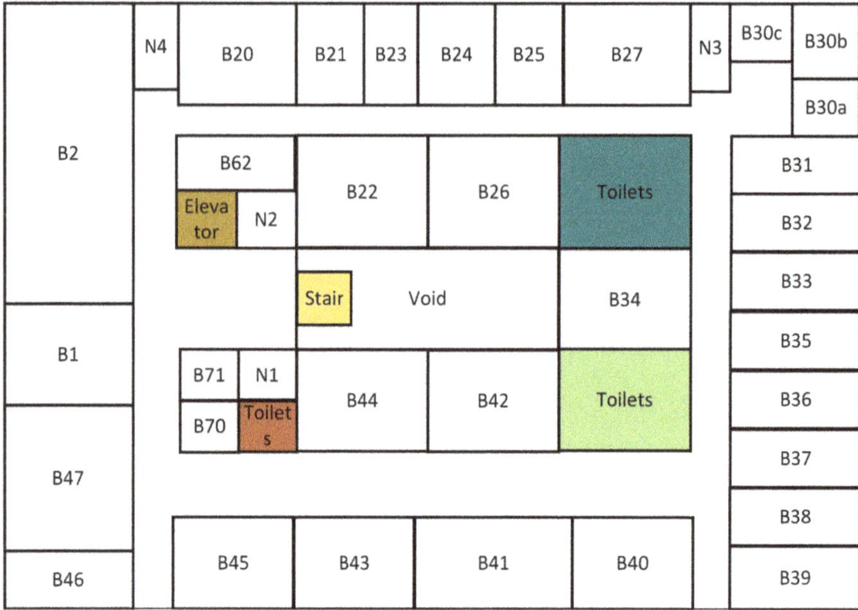

Figure 4. A floor plan map of the B floor in the Business South building (BSB).

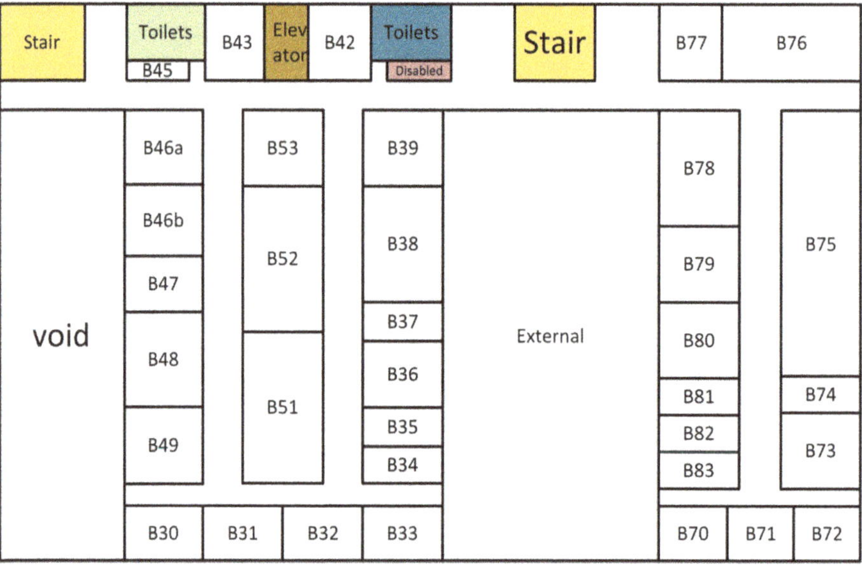

Figure 5. A floor plan map of the B floor in the School of Computer Science building (CSB).

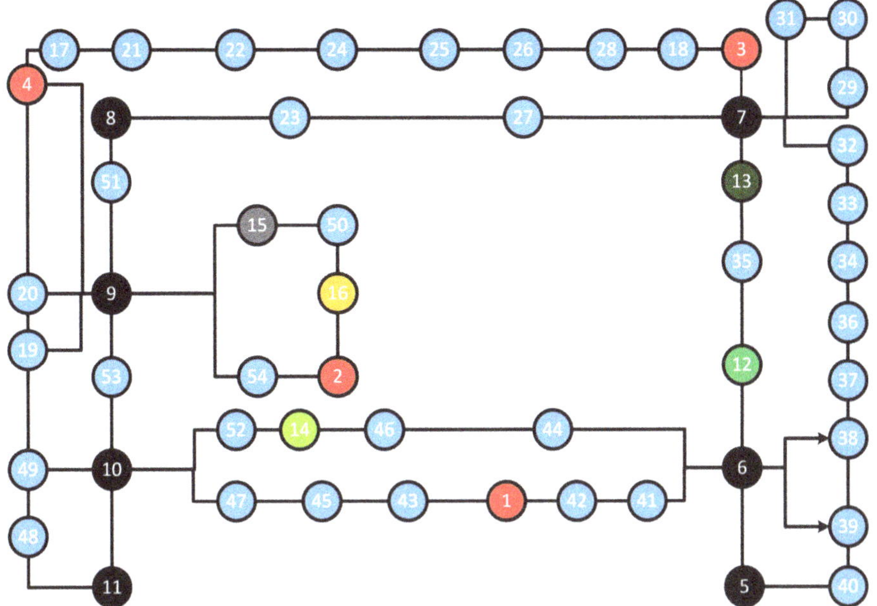

Figure 6. A landmark topological map of the B floor in the BSB.

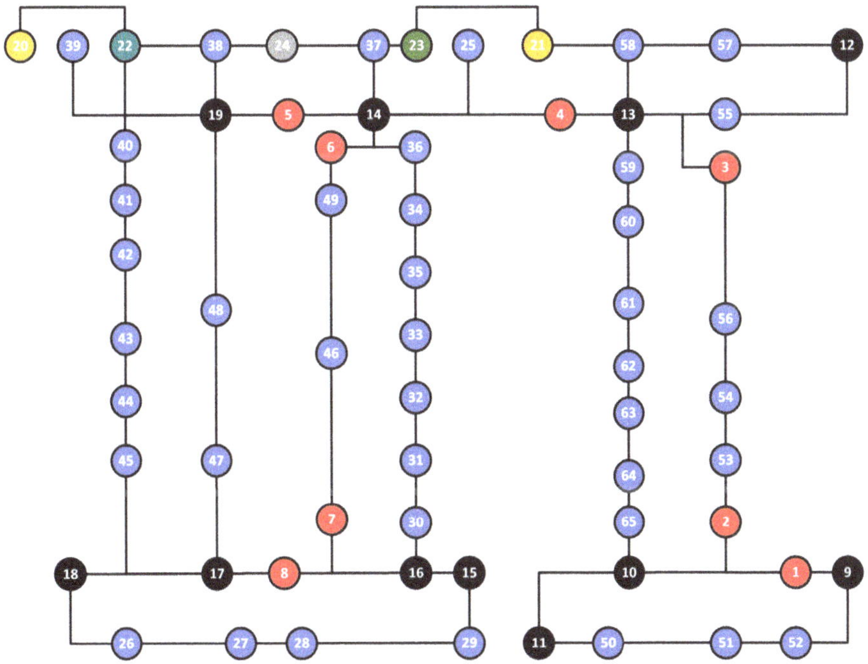

Figure 7. A landmark topological map of the B floor in the CSB.

Seven routes were used as a testing bed to evaluate our method. Two of them were collected in the BSB and five in the CSB. Routes 1–2 are from the BSB and Routes 3–7 are from CSB. The overview of the seven routes are as follows:

Route 1: The route begins at Node 43 and goes through 28 landmarks, ending at Node 47.

Route 2: The route starts from Node 44 and turns left at all four turns before ending at Node 46. There are 16 landmarks in this route.

Route 3: This route goes through 15 landmarks. It starts from an office door (Node 52) and ends in the intersection (Node 14). It walks through a sequence of office doors, containing a corner and a left turn.

Route 4: The route starts from the left stair and goes straight to the end corner of the corridor. In total, 10 landmarks are included in this route.

Route 5: This route contains 14 landmarks. It begins from an intersection (Node 16), goes through a sequence of office doors, turns, and elevators, and finally reaches the left stairs.

Route 6: This route starts from a turn (Node 16) and ends at an office (Node 65), going through three turns, containing 17 landmarks.

Route 7: The route begins from a turn (Node 16) and goes to the end of the corner before turning left. It goes straight until reaching the turn (Node 19). It goes down to the turn (Node 17). There are 22 landmarks in this route.

6.2. *Landmark Detection*

6.2.1. Indoor Object Recognition

The selected landmarks are comprised of nine classes of indoor objects, including eight classes of indoor objects—door (DR), women's toilet tag (WMTT), men's toilet tag (MTT), disabled toilet tag (DTT), fire extinguisher (FE), door plate (DP), elevator (ELV), and stairs (ST)—and one class of scene object (corner or intersection) (CN). Together, they form 8 types of landmarks. We also introduce background as a type of class during the training process, which are uninteresting objects (walls mostly). Uninteresting objects act as negative training samples. This increases the discrimination and generalization ability of the classifier.

We collected about 1300 images containing these 10 types of indoor objects (9 of them are objects of interest and 1 is background). About 1000 of them were used for training (fine-tuning the CNN pre-trained on ImageNet data) and the rest for testing. The distribution of the training and testing datasets are shown in Table 1. These data came from two sources, images on the Internet and video frames from the collected data. We leveraged images from the Internet for two reasons. Firstly, the training dataset could be enlarged and thus the discriminative capacity of the trained classifier over the targeted indoor object was improved. Another reason is that our detector can be used in a new environment without retraining.

Table 1. Distribution of training and testing data.

Type	CN	DTT	DR	DP	ELV	FE	MTT	ST	WLL	WMTT
Training	56	60	155	63	60	250	58	113	104	55
Testing	29	25	33	22	23	36	24	37	31	20

We selected AlexNet as the basic network and fine-tuned it for our application. The output layer was modified by changing the number of neurons from 1000 to 10. Its parameters were initialized with a normal Gaussian distribution. The other layers were initialized with weights that won the Visual Recognition Challenge in 2012. Parameters of the convolutional layers and fully connected layers were kept fixed and only the parameter of the output layer were learned during the training phase. The CNN was implemented using the Caffe framework [49]. The learning rate was 0.05, and the maximum iteration was 40,000. The network was trained in an MSI laptop in GPU mode. The laptop features a

Windows 10 operating system and the processor is Intel i7, and the laptop is fitted with 8 GB of RAM. The graphics processing unit is an Nvidia GTX970M.

We further compare the proposed landmark detection method with traditional handcrafted feature-based methods. Gist [50] is used to represent the visual objects, and the objects are recognized using SVM-based and ANN-based methods. We report the results with the accuracy and the F1 value. F1 value is a measure of classification accuracy, which takes both precision and recall into consideration. Precision represents the number of correct classification results divided by all positive results returned by the classifier. Recall is the number of correct results divided by all the ground true positive samples. The F1 value ranges from 0 to 1, and the higher the value is, the better the performance. F1 can be computed with Equation (16).

$$F1 = 2 \times \frac{precision \times recall}{precision + recall}. \quad (16)$$

The comparison results are shown in Tables 2 and 3, respectively.

Table 2. Indoor object recognition in terms of accuracy.

Methods	CN	DTT	DR	DP	ELV	FE	MTT	ST	WLL	WMTT	Overall
SVM	17.2%	64.0%	90.9%	68.2%	0.0 %	100%	0.0%	56.8%	3.2%	0.0 %	44.3%
ANN	82.8%	80.0%	97.0%	86.4%	73.9%	97.2%	87.5%	70.3%	61.3%	80.0%	81.8%
Ours	100%	96.0%	100%	95.5%	95.7%	100%	100%	100%	100%	95.0%	98.6%

Table 3. Indoor objects recognition in terms of F1 value.

Methods	CN	DTT	DR	DP	ELV	FE	MTT	ST	WLL	WMTT	AVERAGE
SVM	0.29	0.78	0.50	0.77	Nan	0.44	Nan	0.67	0.06	Nan	Nan
ANN	0.89	0.87	0.84	0.90	0.76	0.77	0.88	0.78	0.70	0.86	0.82
Ours	1	0.96	1	0.98	0.98	1	1	1	0.98	0.93	0.98

The results show that our method achieves the best results compared to SVM-based and ANN-based methods on both average accuracy and F1 value. For each type of object, our method outperforms the other two in terms of accuracy and F1 value. The SVM-based method failed to recognize the doorplates and toilets tags. The ANN-based method also obtained high accuracy but it tended to classify the wall as other objects. This affects the localization application as it adds non-existing landmarks to the sequence.

6.2.2. Landmark Detection Performance

All videos of seven routes were empirically sampled at the rate of three frames per second. Some examples of the visual landmark sequences are shown in Figures 8 and 9. Sampled images were processed with the selective search algorithm to generate 300 patches. Landmarks were determined from the classification results according to the strategy described in Section 4.2.3.

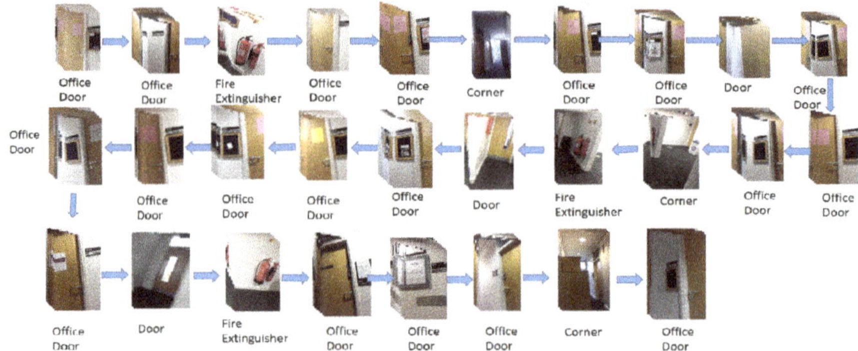

Figure 8. Landmark sequence example of Route 1.

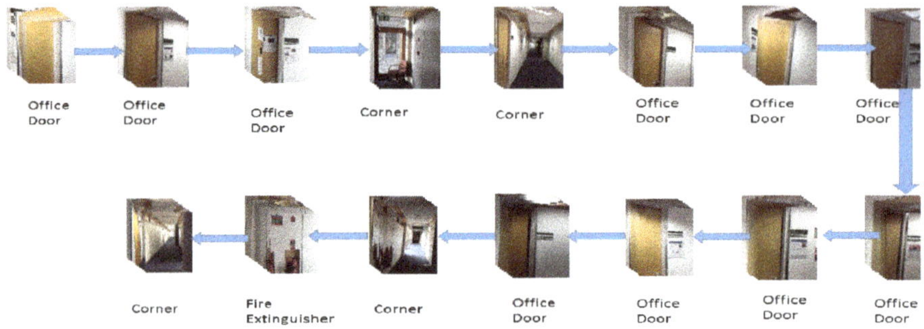

Figure 9. Landmark sequence example of Route 3.

We applied this trained detector and ANN-based detector to the landmark detection on the 1.9 h indoor mobile phone videos. The SVM-based detector is not used due to its low performance on object detection. The results are shown in Table 4.

Table 4. Landmark detection performance in the real data test.

Route	Landmarks Counts	ANN			Ours		
		DL	CDL	WDL	DL	CDL	WDL
1	28	30	25	5	28	28	0
2	16	16	16	0	16	16	0
3	15	20	15	5	15	15	0
4	10	10	10	0	10	10	0
5	14	18	14	4	14	14	0
6	18	26	18	6	18	18	0
7	22	29	22	7	22	22	0

Our method correctly detected all landmarks in all routes. The ANN-based detector correctly detected landmarks in Route 2 and Route 3. Some walls were wrongly detected as doors in Routes 3, 5, 6, and 7. This demonstrates that our detector outperforms the detector using handcrafted features. Currently, the proposed method cannot be achieved in real time. The majority of time is spent on landmark detection. Although the average time of classifying an image is short using our convolutional neural network (about 0.012 s on our machine), the average time to process a landmark image is about

7 s. The process is time-consuming for two reasons. Firstly, we choose an effective selective search algorithm to generate patches from landmark images, which costs about 3–4 s to generate reliable patches. Secondly, we feed 300 patches of a landmark image to the network to correctly detect landmarks, which takes an extra 3 s. It should be noted that the detection process can be optimized with the development of object detection technologies in computer vision.

6.3. Localization

6.3.1. Performance

We match the detected landmarks with topological map on two situations: a known start and an unknown start. The ground truth routes and the predicted routes are shown in Figure 10. The red line indicates the ground truth trajectory. The green line represents the predicted trajectory with an unknown start, while the blue line represents the predicted trajectory with a known start. The route start is represented with a node with a cyan edge, and the route end is denoted as a node with a red edge. For Routes 1, 2, 4, 5, and 7, predictions of both known and unknown starts are correctly localized since the blue and green lines are in accordance with the red line. For Routes 3 and 6, the two blue lines are in accordance with the red lines, indicating that they are accurately localized under a known start condition. For the unknown start case, Route 3 has two predictions: one starts from Node 27 and ends at Node 13, and the other one starts from Node 52 and ends at Node 14. The latter is the correct path. Route 6 also has two preditions: one starts from Node 10 and ends at Node 30, and the other one begins at Node 16 and stops at Node 65, the latter of which is correct. This shows that the two routes cannot be localized with current observations and further observations are eventually required to be localized. This problem can be solved with the start positions since all seven routes are correctly localized under a known start condition. The results demonstrate that our method is capable of localizing users accurately with a known start and it also works well in some cases with an unknown start. Compared to the landmark detection, the localization process barely costs time. We spend about 0.043 s on average to localize each route.

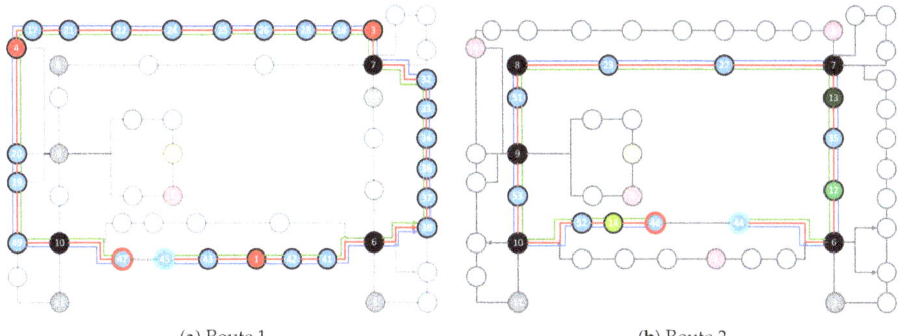

(**a**) Route 1 (**b**) Route 2

Figure 10. *Cont.*

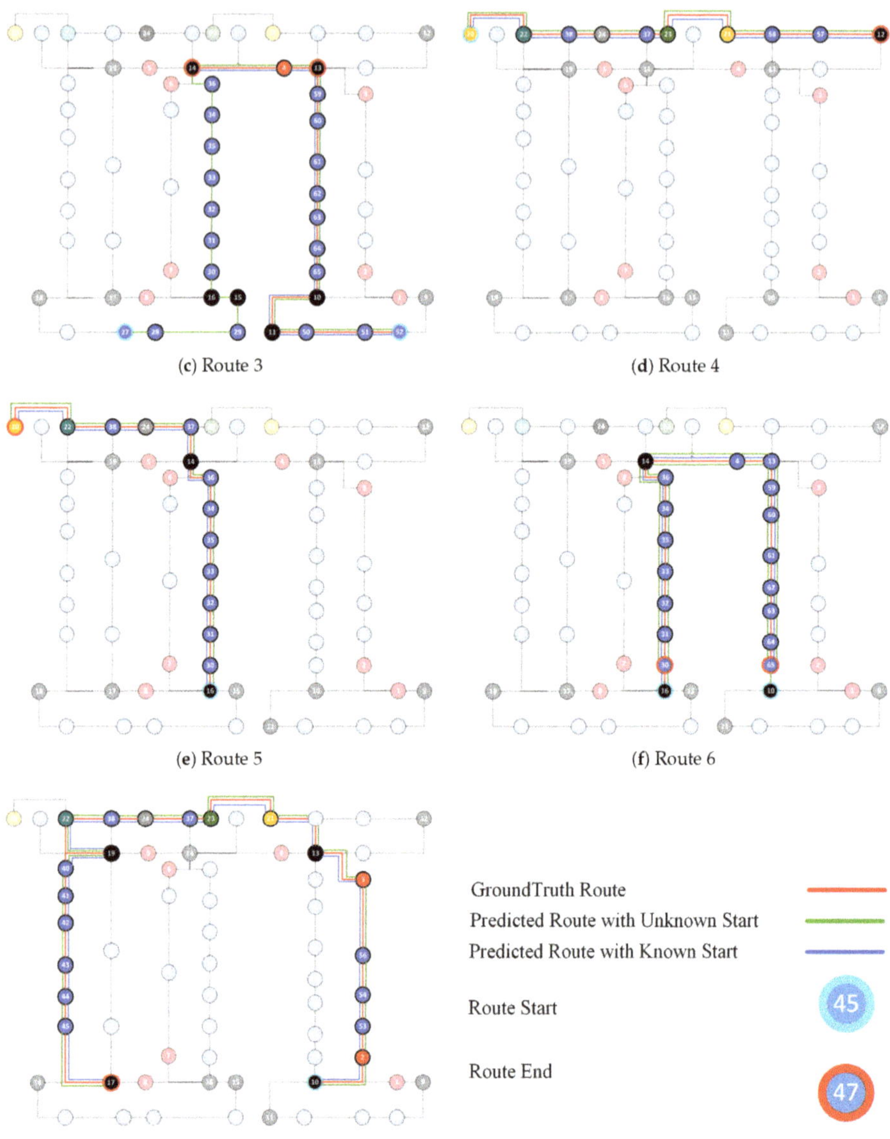

Figure 10. The localization results of seven routes.

We further draw comparisons with the HMM-based method in two situations, and the statistical results are shown in Table 5. The number of possible paths is used to report the comparison result. It is notable that the HMM fails to localize all landmark sequences without a known start and only Route 5 is accurately localized given the start position. In addition, our method outperforms the HMM-based method in seven routes with the same conditions.

Table 5. A statistical comparison landmark sequence localization results of seven routes.

Route	HMM		HMM2	
	Without	With	Without	With
1	18	9	1	1
2	8	2	1	1
3	1137	82	2	1
4	2	2	1	1
5	12	1	1	1
6	18,346	5556	2	1
7	4	2	1	1

6.3.2. Analysis

In this section, we evaluated the localization performance of the proposed method regarding the number of observed landmarks. A number of possible paths was used to report performance. We performed experiments in two scenes using Route 1 and Route 7 along with the number of observed landmarks under unknown start conditions. The performance is shown in Figure 11. Route 1 was localized with six landmarks and Route 7 is localized at the ninth landmark. This is because the CSB is more complex compared with the BSB.

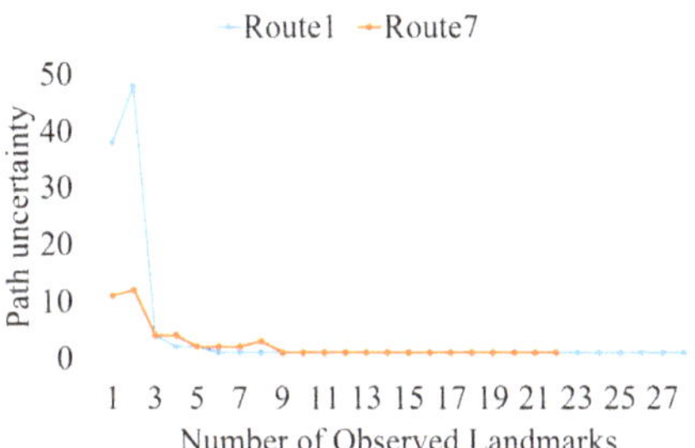

Figure 11. Localization performance with the number of observed landmarks in two scenes.

We also conducted experiments to analyze the effects of given route starts regarding the number of observed landmarks. Route 1 from the BSB and Routes 3 and 7 from the CSB were used to perform experiments. It can be seen from Figure 12 that Route 1 is localized from the third landmark with a known start and from the sixth landmark with an unknown start. Route 7 was localized given nine landmarks with an unknown start and three landmarks with a known start. The proposed method was not able to localize Route 3 given unknown starts but localized a route from the second landmark with a known start. This demonstrates that a known start significantly improves the localization performance in two scenes.

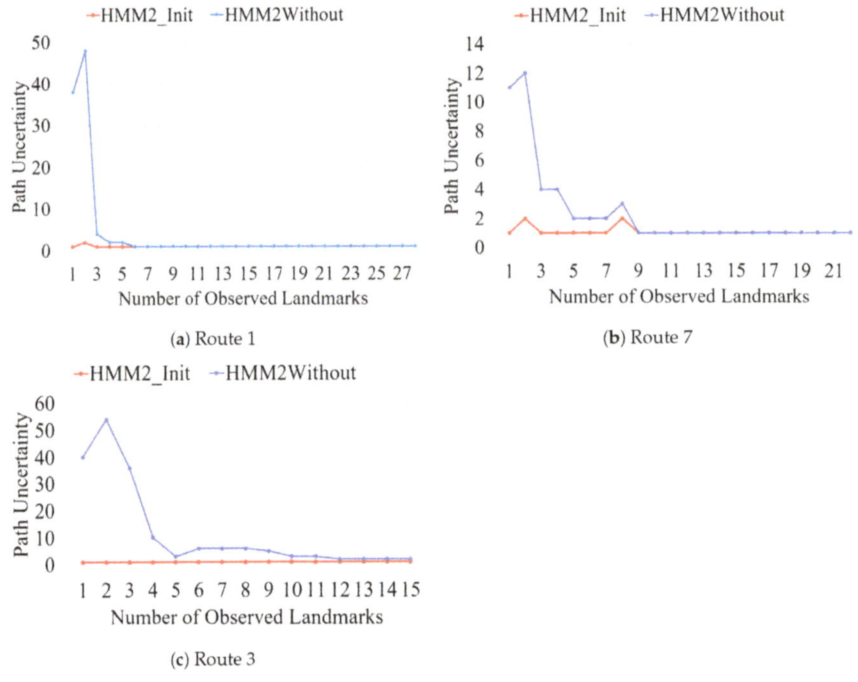

Figure 12. Influence of a known start on localization results.

7. Conclusions

In this paper, we present a visual landmark sequence-based indoor topological localization method. We propose an innovative representation of landmarks on a topological map, a robust landmark detector, and an effective sequence matching algorithm for localization. Sematic information of stable indoor elements is exploited to represent environmental locations. Compared to the traditional landmark represented by local key point features, combined geometric elements or text information, our representation is able to stay robust facing dynamic environmental change caused by moving pedestrians, illumination, and view changes as well as image blurring. This high-level representation reduces the storage requirement and can be extended to large indoor environments. We present a robust CNN-based landmark detector for landmark detection. Previous landmark detecting methods are devised based on the predefined rules or on color and gradient information. Slight environment changes can significantly influence the landmark detection performance. The background also has a significant influence on detection accuracy. We developed the novel landmark detector using a deep learning technique. Instead of designing the feature with a landmark prior, it learns a deep feature representation for landmarks. Experimental results demonstrate that the previous design feature is confused with the background, while our detector is capable of reliably detecting landmarks from the background.

Our matching algorithm achieves good performance to handle indoor scene ambiguity, as it involves more contextual information. Taking object types as landmark representation saves the storage demand but discards landmark details. This further increases scene ambiguity. Methods depending on feature matching fails to work with the scene ambiguity problem. The HMM helps relieve it to a certain degree but still does not solve it. The experiments show that our methods provide better solutions to the problem than the HMM does.

For future work, we plan to investigate the fusion of low-level visual features with semantic features and with geometric features. This would decrease the scene ambiguity and require fewer landmarks for localization. We also plan to extend the proposed method by utilizing all landmarks in both sides of corridors. The current method adopts side-view capturing, which ignores landmarks on the other side of the corridor and results in the loss of information. Another direction to pursue is the automatic construction of the topological map. Currently, we build our topological map manually based on the floor plan map. When there are no floor plan maps of the scenes, a map needs to be constructed from videos. A localization approach is not able to handle a situation in which a camera stops working for a while as we rely on a landmark occurrence sequence to perform localization. If the camera stops working for a period of time, there will be two video segments. The approach will treat the two video segments as independent videos to perform localization. Two landmark sequences are not able to constrain each other because any number of landmarks and any type of landmarks can be observed during the break.

Author Contributions: Conceptualization, Q.L. (Qing Li) and J.Z.; Data curation, Q.L. (Qing Li), J.G. and G.Q.; Formal analysis, Q.L. (Qing Li), J.Z., B.L. and G.Q.; Funding acquisition, J.Z. and Q.L. (Qingquan Li); Investigation, J.Z. and J.G.; Methodology, Q.L. (Qing Li) and J.Z.; Project administration, Q.L. (Qingquan Li) and G.Q.; Resources, R.C., K.S., T.L. and B.L.; Software, Q.L. (Qing Li) and T.L.; Supervision, J.G., Q.L. (Qingquan Li) and G.Q.; Visualization, R.C., K.S. and B.L.; Writing—original draft, Q.L. (Qing Li) and G.Q.; Writing—review & editing, Q.L. (Qing Li), J.Z., R.C., K.S. and Q.L. (Qingquan Li).

Funding: This work was supported in part by the National Natural Science Foundation of China under Grant 41871329, in part by the Shenzhen Future Industry Development Funding Program under Grant 201607281039561400, in part by the Shenzhen Scientific Research and Development Funding Program under Grant JCYJ20170818092931604, and in part by the Horizon Centre for Doctoral Training at the University of Nottingham (RCUK Grant No. EP/L015463/1).

Acknowledgments: In this section, you can acknowledge any support given which is not covered by the author's contribution or funding sections. This may include administrative and technical support, or donations in kind (e.g., materials used for experiments).

Conflicts of Interest: The authors declare that there is no conflict of interest.

References

1. Ranganathan, P.; Hayet, J.B.; Devy, M.; Hutchinson, S.; Lerasle, F. Topological navigation and qualitative localization for indoor environment using multi-sensory perception. *Robot. Auton. Syst.* **2002**, *41*, 137–144. [CrossRef]
2. Cheng, H.; Chen, H.; Liu, Y. Topological Indoor Localization and Navigation for Autonomous Mobile Robot. *IEEE Trans. Autom. Sci. Eng.* **2015**, *12*, 729–738. [CrossRef]
3. Bradley, D.M.; Patel, R.; Vandapel, N.; Thayer, S.M. Real-time image-based topological localization in large outdoor environments. In Proceedings of the IEEE/RSJ International Conference on Intelligent Robots and Systems, Edmonton, AB, Canada, 2–6 August 2005; pp. 3670–3677.
4. Becker, C.; Salas, J.; Tokusei, K.; Latombe, J.C. Reliable navigation using landmarks. In Proceedings of the 1995 IEEE International Conference on Robotics and Automation, Nagoya, Japan, 21–27 May 1995; Volume 1, pp. 401–406.
5. Kosecka, J.; Zhou, L.; Barber, P.; Duric, Z. Qualitative image based localization in indoors environments. In Proceedings of the 2003 IEEE Computer Society Conference on Computer Vision and Pattern Recognition, Madison, WI, USA, 18–20 June 2003; Volume 2, pp. 3–8.
6. Li, Q.; Zhu, J.; Liu, T.; Garibaldi, J.; Li, Q.; Qiu, G. Visual landmark sequence-based indoor localization. In Proceedings of the 1st Workshop on Artificial Intelligence and Deep Learning for Geographic Knowledge Discovery, Los Angeles, CA, USA, 7–10 November 2017; pp. 14–23.
7. Ahn, S.J.; Rauh, W.; Recknagel, M. Circular coded landmark for optical 3D-measurement and robot vision. In Proceedings of the 1999 IEEE/RSJ International Conference on Intelligent Robots and Systems, Kyongju, Korea, 17–21 October 1999; Volume 2, pp. 1128–1133.

8. Jang, G.; Lee, S.; Kweon, I. Color landmark based self-localization for indoor mobile robots. In Proceedings of the 2002 IEEE International Conference on Robotics and Automation, Washington, DC, USA, 11–15 May 2002; Volume 1, pp. 1037–1042.
9. Basiri, A.; Amirian, P.; Winstanley, A. The use of quick response (qr) codes in landmark-based pedestrian navigation. *Int. J. Navig. Obs.* **2014**, *2014*, 897103. [CrossRef]
10. Briggs, A.J.; Scharstein, D.; Braziunas, D.; Dima, C.; Wall, P. Mobile robot navigation using self-similar landmarks. In Proceedings of the IEEE International Conference on Robotics and Automation, San Francisco, CA, USA, 24–28 April 2000; Volume 2, pp. 1428–1434.
11. Hayet, J.B.; Lerasle, F.; Devy, M. A visual landmark framework for indoor mobile robot navigation. In Proceedings of the 2002 IEEE International Conference on Robotics and Automation, Washington, DC, USA, 11–15 May 2002; Volume 4, pp. 3942–3947.
12. Ayala, V.; Hayet, J.B.; Lerasle, F.; Devy, M. Visual localization of a mobile robot in indoor environments using planar landmarks. In Proceedings of the 2000 IEEE/RSJ International Conference on Intelligent Robots and Systems, Takamatsu, Japan, 31 October–5 November 2000; Volume 1, pp. 275–280.
13. Tian, Y.; Yang, X.; Yi, C.; Arditi, A. Toward a computer vision-based wayfinding aid for blind persons to access unfamiliar indoor environments. *Mach. Vis. Appl.* **2013**, *24*, 521–535. [CrossRef] [PubMed]
14. Chen, K.C.; Tsai, W.H. Vision-based autonomous vehicle guidance for indoor security patrolling by a SIFT-based vehicle-localization technique. *IEEE Trans. Veh. Technol.* **2010**, *59*, 3261–3271. [CrossRef]
15. Bai, Y.; Jia, W.; Zhang, H.; Mao, Z.H.; Sun, M. Landmark-based indoor positioning for visually impaired individuals. In Proceedings of the 2014 12th International Conference on Signal Processing, Hangzhou, China, 19–23 October 2014; pp. 668–671.
16. Serrão, M.; Rodrigues, J.M.; Rodrigues, J.; du Buf, J.H. Indoor localization and navigation for blind persons using visual landmarks and a GIS. *Procedia Comput. Sci.* **2012**, *14*, 65–73. [CrossRef]
17. Kawaji, H.; Hatada, K.; Yamasaki, T.; Aizawa, K. Image-based indoor positioning system: Fast image matching using omnidirectional panoramic images. In Proceedings of the 1st ACM International Workshop on Multimodal Pervasive Video Analysis, Firenze, Italy, 29 October 2010; pp. 1–4.
18. Zitová, B.; Flusser, J. Landmark recognition using invariant features. *Pattern Recognit. Lett.* **1999**, *20*, 541–547. [CrossRef]
19. Pinto, A.M.G.; Moreira, A.P.; Costa, P.G. Indoor localization system based on artificial landmarks and monocular vision. *TELKOMNIKA Telecommun. Comput. Electron. Control* **2012**, *10*, 609–620. [CrossRef]
20. Lin, G.; Chen, X. A Robot Indoor Position and Orientation Method based on 2D Barcode Landmark. *JCP* **2011**, *6*, 1191–1197. [CrossRef]
21. Kosmopoulos, D.I.; Chandrinos, K.V. *Definition and Extraction of Visual Landmarks for Indoor Robot Navigation*; Springer: Berlin/Heidelberg, Germany, 2002; pp. 401–412.
22. Girshick, R.; Donahue, J.; Darrell, T.; Malik, J. Rich feature hierarchies for accurate object detection and semantic segmentation. In Proceedings of the IEEE Conference on Computer Vision and Pattern Recognition, Columbus, OH, USA, 23–28 June 2014; pp. 580–587.
23. Zhou, B.; Lapedriza, A.; Xiao, J.; Torralba, A.; Oliva, A. Learning deep features for scene recognition using places database. In *Advances in Neural Information Processing Systems*; 2014; pp. 487–495. Available online: http://places.csail.mit.edu/places_NIPS14.pdf (accessed on 3 January 2019).
24. Werner, M.; Kessel, M.; Marouane, C. Indoor positioning using smartphone camera. In Proceedings of the 2011 International Conference on Indoor Positioning and Indoor Navigation, Guimaraes, Portugal, 21–23 September 2011; pp. 1–6.
25. Liang, J.Z.; Corso, N.; Turner, E.; Zakhor, A. Image based localization in indoor environments. In Proceedings of the 2013 Fourth International Conference on Computing for Geospatial Research and Application, San Jose, CA, USA, 22–24 July 2013; pp. 70–75.
26. Chen, C.; Yang, B.; Song, S.; Tian, M.; Li, J.; Dai, W.; Fang, L. Calibrate Multiple Consumer RGB-D Cameras for Low-Cost and Efficient 3D Indoor Mapping. *Remote Sens.* **2018**, *10*, 328. [CrossRef]
27. Zhao, P.; Hu, Q.; Wang, S.; Ai, M.; Mao, Q. Panoramic Image and Three-Axis Laser Scanner Integrated Approach for Indoor 3D Mapping. *Remote Sens.* **2018**, *10*, 1269. [CrossRef]

28. Lu, G.; Kambhamettu, C. Image-based indoor localization system based on 3d sfm model. In *IS&T/SPIE Electronic Imaging*; International Society for Optics and Photonics, 2014; p. 90250H. Available online: https://www.researchgate.net/publication/269323831_Image-based_indoor_localization_system_based_on_3D_SfM_model (accessed on 3 January 2019).
29. Van Opdenbosch, D.; Schroth, G.; Huitl, R.; Hilsenbeck, S.; Garcea, A.; Steinbach, E. Camera-based indoor positioning using scalable streaming of compressed binary image signatures. In Proceedings of the 2014 IEEE International Conference on Image Processing (ICIP), Paris, France, 27–30 October 2014; pp. 2804–2808.
30. Hile, H.; Borriello, G. Positioning and orientation in indoor environments using camera phones. *IEEE Comput. Gr. Appl.* **2008**, *28*. [CrossRef]
31. Mulloni, A.; Wagner, D.; Barakonyi, I.; Schmalstieg, D. Indoor positioning and navigation with camera phones. *IEEE Pervasive Comput.* **2009**, *8*, 22–31. [CrossRef]
32. Lu, G.; Yan, Y.; Sebe, N.; Kambhamettu, C. Indoor localization via multi-view images and videos. *Comput. Vis. Image Understand.* **2017**, *161*, 145–160. [CrossRef]
33. Lu, G.; Yan, Y.; Ren, L.; Saponaro, P.; Sebe, N.; Kambhamettu, C. Where am i in the dark: Exploring active transfer learning on the use of indoor localization based on thermal imaging. *Neurocomputing* **2016**, *173*, 83–92. [CrossRef]
34. Piciarelli, C. Visual indoor localization in known environments. *IEEE Signal Process. Lett.* **2016**, *23*, 1330–1334. [CrossRef]
35. Vedadi, F.; Valaee, S. Automatic Visual Fingerprinting for Indoor Image-Based Localization Applications. *IEEE Trans. Syst. Man Cybern. Syst.* **2017**. [CrossRef]
36. Lee, N.; Kim, C.; Choi, W.; Pyeon, M.; Kim, Y. Development of indoor localization system using a mobile data acquisition platform and BoW image matching. *KSCE J. Civ. Eng.* **2017**, *21*, 418–430. [CrossRef]
37. Chen, Z.; Zou, H.; Jiang, H.; Zhu, Q.; Soh, Y.C.; Xie, L. Fusion of WiFi, smartphone sensors and landmarks using the Kalman filter for indoor localization. *Sensors* **2015**, *15*, 715–732. [CrossRef]
38. Deng, Z.A.; Wang, G.; Qin, D.; Na, Z.; Cui, Y.; Chen, J. Continuous indoor positioning fusing WiFi, smartphone sensors and landmarks. *Sensors* **2016**, *16*, 1427. [CrossRef]
39. Gu, F.; Khoshelham, K.; Shang, J.; Yu, F. Sensory landmarks for indoor localization. In Proceedings of the 2016 Fourth International Conference on Ubiquitous Positioning, Indoor Navigation and Location Based Services (UPINLBS), Shanghai, China, 2–4 November 2016; pp. 201–206.
40. Millonig, A.; Schechtner, K. Developing landmark-based pedestrian-navigation systems. *IEEE Trans. Intell. Transp. Syst.* **2007**, *8*, 43–49. [CrossRef]
41. Betke, M.; Gurvits, L. Mobile robot localization using landmarks. *IEEE Trans. Robot. Autom.* **1997**, *13*, 251–263. [CrossRef]
42. Boada, B.L.; Blanco, D.; Moreno, L. Symbolic place recognition in voronoi-based maps by using hidden markov models. *J. Intell. Robot. Syst.* **2004**, *39*, 173–197. [CrossRef]
43. Zhou, B.; Li, Q.; Mao, Q.; Tu, W.; Zhang, X. Activity sequence-based indoor pedestrian localization using smartphones. *IEEE Trans. Hum.-Mach. Syst.* **2015**, *45*, 562–574. [CrossRef]
44. Kosecká, J.; Li, F. Vision based topological Markov localization. In Proceedings of the IEEE International Conference on Robotics and Automation, New Orleans, LA, USA, 26 April–1 May 2004; Volume 2, pp. 1481–1486.
45. Ren, S.; He, K.; Girshick, R.; Sun, J. Faster r-cnn: Towards real-time object detection with region proposal networks. In *Advances in Neural Information Processing Systems*; 2015; pp. 91–99. Available online: https://arxiv.org/abs/1506.01497 (accessed on 3 January 2019).
46. Krizhevsky, A.; Sutskever, I.; Hinton, G.E. Imagenet classification with deep convolutional neural networks. In *Advances in Neural Information Processing Systems*; 2012; pp. 1097–1105. Available online: https://papers.nips.cc/paper/4824-imagenet-classification-with-deep-convolutional-neural-networks.pdf (accessed on 3 January 2019).
47. Uijlings, J.R.; Van De Sande, K.E.; Gevers, T.; Smeulders, A.W. Selective search for object recognition. *Int. J. Comput. Vis.* **2013**, *104*, 154–171. [CrossRef]
48. Thede, S.M.; Harper, M.P. A second-order hidden Markov model for part-of-speech tagging. In Proceedings of the the 37th Annual Meeting of the Association for Computational Linguistics on Computational Linguistics, College Park, MD, USA, 20–26 June 1999; pp. 175–182.

49. Jia, Y.; Shelhamer, E.; Donahue, J.; Karayev, S.; Long, J.; Girshick, R.; Guadarrama, S.; Darrell, T. Caffe: Convolutional architecture for fast feature embedding. In Proceedings of the 22nd ACM International Conference on Multimedia, Orlando, FL, USA, 3–7 November 2014; pp. 675–678.
50. Oliva, A.; Torralba, A. Modeling the shape of the scene: A holistic representation of the spatial envelope. *Int. J. Comput. Vis.* **2001**, *42*, 145–175. [CrossRef]

Article

A High-Accuracy Indoor-Positioning Method with Automated RGB-D Image Database Construction

Runzhi Wang [1,2], Wenhui Wan [1,*], Kaichang Di [1], Ruilin Chen [1] and Xiaoxue Feng [3]

1 State Key Laboratory of Remote Sensing Science, Aerospace Information Research Institute, Chinese Academy of Sciences, No. 20A, Datun Road, Chaoyang District, Beijing 100101, China; wangrz@radi.ac.cn (R.W.); dikc@radi.ac.cn (K.D.); reline.chen@myzygroup.com (R.C.)

2 College of Resources and Environment, University of Chinese Academy of Sciences, Beijing 100049, China

3 Institute of Remote Sensing and GIS, School of Earth and Space Sciences, Peking University, Beijing 100871, China; fengxx@pku.edu.cn

* Correspondence: wanwh@radi.ac.cn; Tel.: +86-10-64807987

Received: 27 August 2019; Accepted: 30 October 2019; Published: 1 November 2019

Abstract: High-accuracy indoor positioning is a prerequisite to satisfy the increasing demands of position-based services in complex indoor scenes. Current indoor visual-positioning methods mainly include image retrieval-based methods, visual landmarks-based methods, and learning-based methods. To better overcome the limitations of traditional methods such as them being labor-intensive, of poor accuracy, and time-consuming, this paper proposes a novel indoor-positioning method with automated red, green, blue and depth (RGB-D) image database construction. First, strategies for automated database construction are developed to reduce the workload of manually selecting database images and ensure the requirements of high-accuracy indoor positioning. The database is automatically constructed according to the rules, which is more objective and improves the efficiency of the image-retrieval process. Second, by combining the automated database construction module, convolutional neural network (CNN)-based image-retrieval module, and strict geometric relations-based pose estimation module, we obtain a high-accuracy indoor-positioning system. Furthermore, in order to verify the proposed method, we conducted extensive experiments on the public indoor environment dataset. The detailed experimental results demonstrated the effectiveness and efficiency of our indoor-positioning method.

Keywords: visual positioning; indoor scenes; automated database construction; image retrieval

1. Introduction

Nowadays, position information has become key information in people's daily lives. This has inspired position-based services, which aim to provide personalized services to mobile users whose positions are changing [1]. Therefore, obtaining a precise position is a prerequisite for these services. The most commonly used positioning method in the outdoor environment is the Global Navigation Satellite System (GNSS). In most cases, however, people spend more than 70% of their time indoors [2]. Therefore, accurate indoor positioning has important practical significance. Although GNSS is a good choice for outdoor positioning, due to signal occlusion and attenuations, it is often useless in indoor environments. Thus, positioning people accurately in indoor scenes remains a challenge and it has stimulated a large number of indoor-positioning methods in recent years [3]. Among these methods, fingerprint-based algorithms are widely used. Their fingerprint databases include Wi-Fi [4–8], Bluetooth [9,10], and magnetic field strengths [11,12]. Although these methods are easy to implement, construction of a fingerprint database is usually labor-intensive and time-consuming. Moreover, it is difficult for their results to meet the needs of high-accuracy indoor positioning.

Given that humans use their eyes to see where they are, mobile platforms can also do this with cameras. A number of visual positioning methods have been proposed in recent years. These positioning methods are divided into three categories: image retrieval based methods, visual landmarks-based methods, and learning-based methods.

Image retrieval based methods treat the positioning task as an image retrieval or recognition process [13–15]. They usually have a database that are augmented with geospatial information, and every image in the database is described through the same specific features. These methods perform a first step to retrieve candidate images from the database according to a similarity search, and the coarse position information of the query image is then obtained based on the geospatial information of these candidate images. So the first step, similar image retrieval process, is critical. The brute-force approach, which is a distance comparison between feature descriptor vectors, is often used for similarity search. Some positioning methods based on feature descriptors [16–18] adopt brute-force comparison for the similarity search process of image retrieval. However, it is computationally intensive when the images of a database are described with high-dimensional features, limiting its scope of applications. Azzi et al. [19] use a global feature-based system to reduce the search space and find candidate images in the database, then the local feature scale-invariant feature transform (SIFT) [20] is adopted for points matching in pose estimation. Some researchers try to trade accuracy for rapidity by using approximate nearest neighbor search, such as quantization [21] and vocabulary tree [22]. Another common way to save time and memory of similarity search is principal component analysis (PCA), which has been used to reduce the size of feature vectors and descriptors [23,24]. Some works use correlation algorithms, such as sum of absolute difference (SAD), for computing similarity between query image and database images [25,26]. In recent studies, deep learning-based algorithms are an alternative to aforementioned methods. Razavian et al. [27] use features extracted from a network as an image representation for image retrieval in a diverse set of datasets. Yandex et al. [28] propose a method that aggregates local deep features to product descriptors for image retrieval. After a set of candidate images are retrieved, the position information of the query image is calculated according to the geospatial information of these candidate images through a weighting scheme or linear combination. However, because this position result is not calculated by strict geometric relations, it is rough in most cases and difficult to meet the requirement of high-accuracy positioning.

Visual landmarks-based positioning methods aim to provide a six degrees of freedom (DoF) pose of the query image. Generally, visual landmarks in the indoor environments includes natural landmarks and artificial landmarks. The natural landmarks refer to the geo-tagged 3D database, which is represented by feature descriptors or images with poses. This database could have been built thanks to the mapping module of simultaneous localization and mapping (SLAM) [29,30]. Then the pose of query image is estimated by means of re-localization module and feature correspondence [31–35]. Although the results of these methods are of good accuracy, it takes a long time to match the features of query image with geo-tagged 3D database, especially when the indoor scenes are large. In addition to natural landmarks, there are also positioning methods based on artificial landmarks, e.g., Degol et al. [36] proposed a fiducial marker and detection algorithm. In reference [37], the authors proposed a method to simultaneously solve the problems of positioning from a set of squared planar markers. However, positioning from a planar marker suffers from the ambiguity problem [38]. Since these methods require posting markers in the environments, they are not suitable for places such as shopping malls that maintain a clean appearance.

In addition to the traditional visual-positioning method based on strict geometric relations, with the rapid development of deep learning in recent years scholars have proposed many learning based visual-positioning methods [39–41]. The process of these methods are broken down into two steps: model training and pose prediction. They train models through given images with known pose information, and the indoor environments are expressed as the trained models. The pose of a query image is then regressed through the trained models. Some methods even learn the pose directly [42,43]. These methods, which are based entirely on learning, have better performance in weak-textured indoor scenes, but are less accurate or have lower generalization ability to large indoor environments than traditional visual-positioning methods [44]. Therefore, some methods use trained models to replace modules of traditional visual-positioning methods, such as depth estimation [45–47], loop detection [48], and re-localization [49]. The method proposed by Chen et al. [50] uses a pre-trained network for image recognition. It retrieves two geo-tagged red-green-blue (RGB) images from database, and then use traditional visual positioning method for pose estimation. This method performs well on public dataset, but its database images are hand-picked, which increases the workload of database construction. Moreover, the two retrieved geo-tagged RGB images should have favorable geometric configuration (e.g., sufficient intersection angle) for high-accuracy depth estimation. However, this favorable configuration is not guaranteed by the existing two-image methods. This is a potential disadvantage of these methods. Our RGB-D database method directly provides high accuracy depth information from only one image, this not only ensures high accuracy of positioning, but also improves the efficiency of image retrieval.

To overcome the limitations of the aforementioned methods, in this paper, a high-accuracy indoor visual-positioning method with automated RGB-D image database construction is presented. Firstly, we propose an automated database construction process, making the constructed database more objective than a hand-picked database and thus reducing the workload. The database is automatically constructed according to the rules, which reduces the redundancy of database and improves the efficiency of the image-retrieval process. Secondly, considering the requirement of real-time positioning, we introduce a convolutional neural network (CNN) model for a robust and efficient retrieving candidate images. Thirdly, different from aforementioned image retrieval based positioning methods, we replace rough combination of geospatial information with strict geometric relations to calculate the position of query image for high-accuracy positioning. Finally and most importantly, by combining the above three components into a complete indoor-positioning method, we obtain high-accuracy results in an indoor environment and the whole process is time efficient.

2. Methodology

In this section, the proposed indoor-positioning method consists of three major components: (1) RGB-D indoor-positioning database construction; (2) image retrieval based on the CNN feature vector; (3) position and attitude estimation. Detailed processes in each component are given in the following sub-sections.

2.1. RGB-D Indoor-Positioning Database Construction

In the proposed indoor visual positioning method, RGB-D images are used to build positioning database in an indoor scene. Since most RGB-D image acquisition devices, such as Microsoft Kinect sensor, can provide a frame rate of 30 Hz, images acquired over a period of time have redundancies. Note that a large number of database images need a lot of memory in storage, it takes longer for the image retrieval and positioning. However, if the images in the database are too sparse, it may not achieve high positioning accuracy. In order to meet the requirements of precise and real-time indoor positioning, an automated RGB-D image database construction process is proposed.

Our strategy for RGB-D image database construction is based on the relationships between pose error (i.e., position error and attitude error), number of matching points and pose difference (i.e., position difference and attitude difference). To determine their relationships, we selected several RGB-D image as the database images and more than 1000 other RGB-D images as the query images. These images all come from the Technical University of Munich (TUM) RGB-D dataset [51], which provides ground truth of pose. Figure 1a,b show an example of RGB-D images. The positions of database images and ground truth of trajectory are shown in Figure 1c.

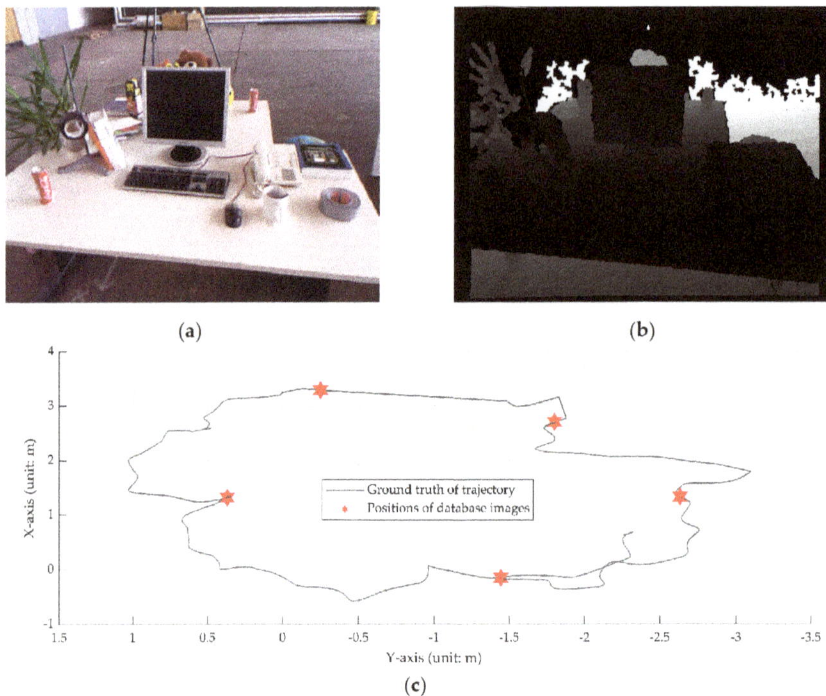

Figure 1. An example of RGB-D image and the positions of database images. (**a**) RGB image; (**b**) corresponding depth image of this RGB image; (**c**) positions of database images (shown by red hexagon) and ground truth of trajectory (shown by gray line).

First, the relationship between pose error and number of matching points is a key criterion of the proposed process. The pose of each query image was calculated by means of the visual-positioning process mentioned in Section 2.3. The number of matching points was recorded in this positioning process. The pose error was obtained by comparing the calculated pose with its ground truth. After testing all the query images, pose errors and corresponding number of matching points for each query image were collected and analyzed to determine their relationship (Figure 2). It is found from Figure 2a,b that both the position error and attitude error fluctuate greatly when the number of matching points is less than 50. However, when the number of matching points is more than 50, the pose errors are basically stable at a small value. In other words, our visual-positioning process can obtain precise and stable results when the number of matching points is more than 50. So we set 50 as the threshold T_{match} for the minimum number of matching points.

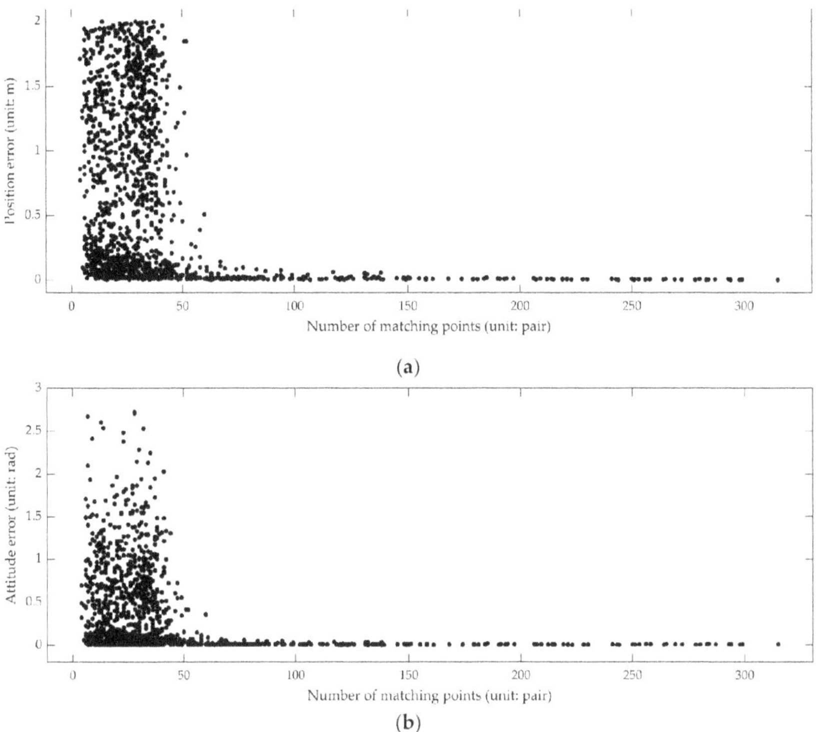

Figure 2. Relationship between pose error and corresponding number of matching points. (**a**) Position error vs. number of matching points; (**b**) attitude error vs. number of matching points.

Second, the relationship between number of matching points and pose difference is also an important criterion in the RGB-D image database construction process. The pose difference was calculated by comparing the ground truth pose of each query image with corresponding database image. Then the pose differences of some images were combined with their number of matching points to fit their relationship. The green fitted curve in Figure 3a shows the fitted relationship between the number of matching points and position difference. Its expression is described as follows:

$$f_p(x) = 19.43 \times {x_p}^{-0.5435}. \quad (1)$$

Here x_p is position difference, $f_p(x)$ is the number of matching points. The blue fitted curve in Figure 3b shows the fitted relationship between the number of matching points and attitude difference. Its expression is described as Equation (2):

$$f_a(x) = 8.846 \times {x_a}^{-0.8503}. \quad (2)$$

Here x_a is attitude difference, $f_a(x)$ is the number of matching points. Then we used some other pose differences and number of matching points of more than seventy images to validate Equations (1) and (2). As shown in Figure 3a,b, the validation data are distributed near the fitted curve. The root mean square errors (RMSE) of fit and validation are shown in Table 1.

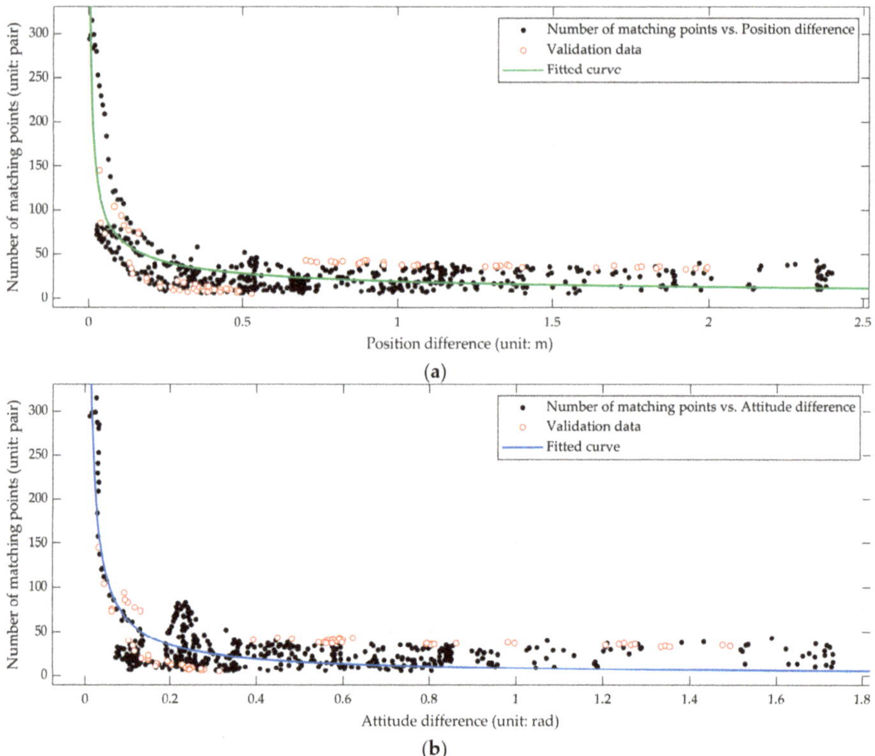

Figure 3. Relationship between number of matching points and corresponding pose difference. (**a**) Fitted curve of number of matching points vs. position difference; (**b**) fitted curve of number of matching points vs. attitude difference.

Table 1. The root mean square errors (RMSEs) of fitted curves and validation data in Figure 3.

Fitted Curves	RMSE of Fit	RMSE of Validation
Figure 3a	23.73	21.88
Figure 3b	23.96	24.08

We can see that the RMSE of validation data is close to the RMSE of fitted curve, which indicates that the fitted curves described in Equations (1) and (2) are applicable to different image data in the same scene. The empirical models are not sensitive to the selection of the query image, the established relationships are reliable to apply in the same scene.

From the trends of the fitted curves in Figure 3, the number of matched points decreases as both of the position difference and attitude difference increase. According to the threshold T_{match} for the number of matching points from Figure 2, the threshold of position difference $T_{\Delta position}$ (i.e., the x_p in Equation (1)), and the threshold of attitude difference $T_{\Delta attitude}$ (i.e., the x_a in Equation (2)), were obtained by substituting $f_p(x)$ and $f_a(x)$ with T_{match}. The results are as follows:

$$\begin{cases} T_{\Delta position} = 0.1757, \text{ unit : m} \\ T_{\Delta attitude} = 0.1304, \text{ unit : rad} \end{cases}. \quad (3)$$

Based on these three thresholds T_{match}, $T_{\Delta position}$ and $T_{\Delta attitude}$, the RGB-D image database construction process was proposed for indoor visual positioning (Figure 4).

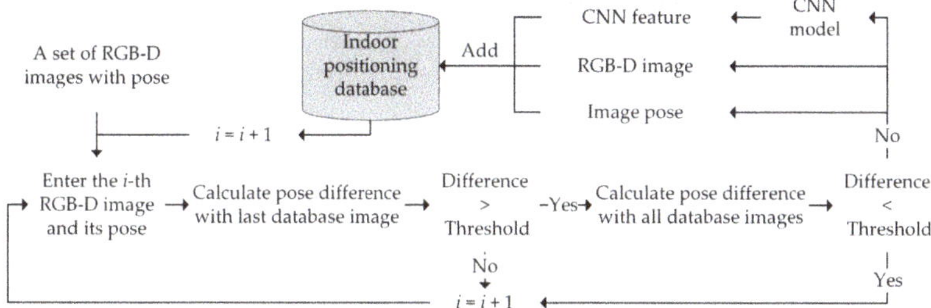

Figure 4. Flowchart of the proposed indoor positioning database construction process.

As shown in Figure 4, first we need to input a set of RGB-D images with known poses in the scene where conducting indoor positioning. The RGB-D images can be captured using a Microsoft Kinect sensor and the ground truth trajectory of camera pose can be obtained from a high-accuracy motion-capture system with high-speed tracking cameras as in reference [51]. In the database construction process, the first RGB-D image is considered as a database image and the other images will be input successively to determine whether they are required to join the database. From Figure 4, the i-th RGB-D image is compared with the last database image joining the database and then compared with all the existing database images to calculate the differences in position and attitude. If the differences meet the preset threshold condition (i.e., $T_{\Delta position}$ and $T_{\Delta attitude}$), the i-th image will be determined as a new database image. It will be also input into the CNN model mentioned in Section 2.2 to calculate its CNN feature vector for subsequent step of image retrieval. Finally, we add the three components of the eligible image into the indoor-positioning database, and then move on to the next RGB-D image until all images are judged.

The three components of the indoor positioning database $B = \{I, Pose, F\}$ are listed as follows. The first one is the RGB-D database images $I = \{RGB\text{-}D_1, \ldots, RGB\text{-}D_i, \ldots, RGB\text{-}D_n\}$ that meet the requirements. The second one is the corresponding pose information $Pose = \{Pose_1, \ldots, Pose_i, \ldots, Pose_n\}$ of database images. The $Pose_i$ here includes 3D position $\{x_i, y_i, z_i\}$ and quaternion form of attitude $\{qx_i, qy_i, qz_i, qw_i\}$. The last one is the CNN feature vector set $F = \{F_1, \ldots, F_i, \ldots, F_n\}$ of database images.

2.2. Image Retrieval Based on Convolutional Neural Network (CNN) Feature Vector

After building the indoor positioning database, it is important to know which RGB-D image in the database is the most similar to the input query RGB image acquiring by the mobile platform. The query RGB image and its most similar RGB-D image will be combined for conducting the subsequent visual positioning. In this sub-section, the CNN model and CNN feature vector-based image-retrieval algorithm were used in our indoor positioning method. We adopted the CNN architecture proposed in reference [52], the main component of which is a generalized vector of locally aggregated descriptors (NetVLAD) layer and it is readily pluggable into standard CNN architecture. The best performing network they trained was adopted to extract image deep features, i.e., CNN feature vector, for image retrieval in this study.

Figure 5 shows the process of image retrieval based on CNN feature vector. With this process, the RGB-D database image, which is the most similar to the input query image, and its pose information were retrieved. First, in Section 2.1, we have calculated and saved the database CNN feature vector set $F = \{F_1, \ldots, F_i, \ldots, F_n\}$ in the indoor positioning database. When a query color image C_q with the same size as the database images is input, the same CNN model is used to calculate its CNN feature vector. This output CNN feature vector F_q of query image has the same length with the feature vector F_i of the database image. In this research, the size of CNN feature vector is 4096×1.Then the distance between

F_q and each feature vector of $F = \{F_1, \ldots, F_i, \ldots, F_n\}$ is calculated to represent their similarity, which is defined as follows:

$$D_{iq} = (F_i - F_q)^{\mathrm{T}} \cdot (F_i - F_q). \tag{4}$$

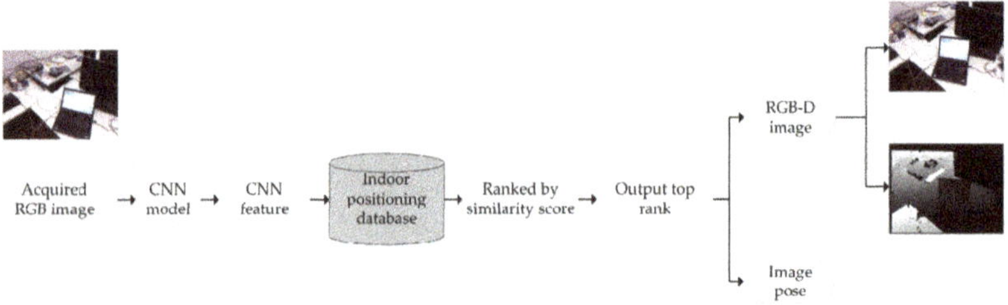

Figure 5. Process of the convolutional neural network (CNN) feature-based image retrieval.

The set of distances is $D = \{D_{1q}, \ldots, D_{iq}, \ldots, D_{nq}\}$. Finally, we output a retrieved RGB-D database image *RGB-D*$_r$ and its pose information *Pose*$_r$, which has the minimum distance D_{rq} with the query color image.

2.3. Position and Attitude Estimation

After retrieving the most similar RGB-D database image with an acquired query RGB image, the visual positioning was achieved by estimation of the position and attitude of the query image based on the retrieved database image and its pose information (Figure 6).

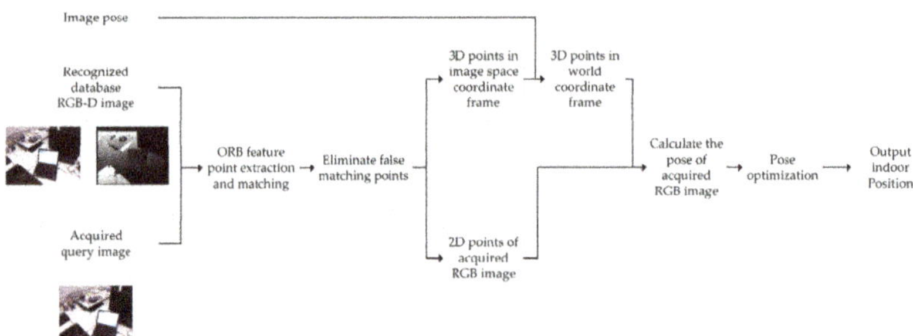

Figure 6. Process of the pose estimation.

In Figure 6, feature point extraction and matching between the acquired query image and the retrieved database image is the first step in the visual-positioning process. The ORB algorithm was adopted in our method to extract 2D feature points and calculate binary descriptors for feature matching. Then, the fundamental matrix constraint [53] and random sample consensus (RANSAC) algorithm [54] were used to eliminate some false matching points. After that, two sets of good matching points pts_q and pts_r in the pixel coordinate frame were obtained. Figure 7 shows the result of good matching points between the acquired query image and the retrieved database image.

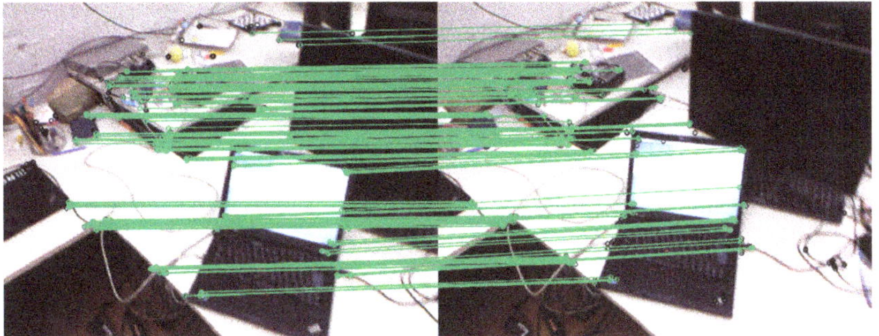

Figure 7. Good matching points of the acquired query image and the retrieved database image.

Second, 3D information in the world coordinate frame of the matching points was obtained by the retrieved RGB-D database image and its image pose. A feature point $p_r(u, v)$ belonging to pts_r in the retrieved database image is a 2D point in the pixel coordinate frame. Its form in the image plane coordinate frame $p_r(x, y)$ is obtained by Equation (5):

$$\begin{cases} x = (u - c_x)/f_x \\ y = (v - c_y)/f_y \end{cases}. \tag{5}$$

Here f_x, f_y, c_x and c_y belong to the intrinsic parameters K of camera, which was calculated through camera calibration process. Through the depth image of the retrieved RGB-D database image, we obtained the depth value $d_{(x,y)}$ of $p_r(x, y)$. Therefore, the 3D point $P_r(X', Y', Z')$ in the image space coordinate frame is obtained by Equation (6):

$$\begin{cases} X' = x \times d_{(x,y)} \\ Y' = y \times d_{(x,y)} \\ Z' = d_{(x,y)} \end{cases}. \tag{6}$$

Next, the input pose information $Pose_r$ of the retrieved image is used to translate $P_r(X', Y', Z')$ into the world coordinate frame. The $Pose_r$ here includes 3D position $\{x_r, y_r, z_r\}$ and quaternion form of attitude $\{qx_r, qy_r, qz_r, qw_r\}$. Usually $Pose_r$ is expressed as a transformation matrix T_{wr} from image space coordinate frame to world coordinate frame by Equation (7):

$$T_{wr} = \begin{bmatrix} R_{wr} & t_{wr} \\ 0 & 1 \end{bmatrix}, \tag{7}$$

where R_{wr} and $t_{wr} = [x_r, y_r, z_r]^{\mathrm{T}}$ are the rotation and translation parts of T_{wr} respectively. R_{wr} is defined as follows:

$$R_{wr} = \begin{bmatrix} 1 - 2{qy_r}^2 - 2{qz_r}^2 & 2qx_r \times qy_r - 2qw_r \times qz_r & 2qx_r \times qz_r + 2qw_r \times qy_r \\ 2qx_r \times qy_r + 2qw_r \times qz_r & 1 - 2{qx_r}^2 - 2{qz_r}^2 & 2qy_r \times qz_r - 2qw_r \times qx_r \\ 2qx_r \times qz_r - 2qw_r \times qy_r & 2qy_r \times qz_r + 2qw_r \times qx_r & 1 - 2{qx_r}^2 - 2{qy_r}^2 \end{bmatrix}. \tag{8}$$

Then T_{wr} is used to transform $P_r(X', Y', Z')$ into the world coordinate frame $P_w(X, Y, Z)$:

$$\begin{bmatrix} X \\ Y \\ Z \\ 1 \end{bmatrix} = T_{wr} \cdot \begin{bmatrix} X' \\ Y' \\ Z' \\ 1 \end{bmatrix} = \begin{bmatrix} R_{wr} & T_{wr} \\ 0 & 1 \end{bmatrix} \cdot \begin{bmatrix} X' \\ Y' \\ Z' \\ 1 \end{bmatrix}. \tag{9}$$

So the relationship between a 3D point $P_w(X,Y,Z)$ in the world coordinate frame and its 2D point $p_r(u,v)$ in the pixel coordinate frame is expressed as follows:

$$\lambda\begin{bmatrix} u \\ v \\ 1 \end{bmatrix} = K\cdot T_{rw}\cdot P_w = \begin{bmatrix} f_x & 0 & c_x \\ 0 & f_y & c_y \\ 0 & 0 & 1 \end{bmatrix}\cdot\left(\begin{bmatrix} R_{rw} & t_{rw} \\ 0 & 1 \end{bmatrix}\cdot\begin{bmatrix} X \\ Y \\ Z \\ 1 \end{bmatrix}\right)_{(1:3)}. \tag{10}$$

Here, matrices T_{rw}, R_{rw} and t_{rw} are the inverse of matrices T_{wr}, R_{wr} and t_{wr} respectively. By using the relationship described in Equation (10), we calculated a set of 3D points Pts_w in the world coordinate frame corresponding to the set of 2D points pts_r in the retrieved image. Because pts_r and pts_q are two sets of matching points, Pts_w is also corresponding to pts_q.

Third, according to the 2D matching points and their 3D points in the query image, the efficient perspective-n-point (EPnP) method [55] was adopted to estimate the initial pose T_{qw} of the query image. The Levenberg–Marquardt algorithm implemented in g2o [56] was then used to optimize the camera pose iteratively. This process can be described as follows:

$$\{R_{qw},t_{qw}\} = \underset{R_{qw},t_{qw}}{\operatorname{argmin}}\sum_{i=1}^{n}\left(\|p_i-\frac{1}{\lambda_i}\cdot K\cdot(T_{qw}\cdot P_i)_{(1:3)}\|^2\right). \tag{11}$$

Here p_i is the i-th 2D point of pts_q and P_i is the i-th 3D point of Pts_w. The number n is the length of pts_q. The poses got from the EPnP method were used as the initial values of T_{qw}.

Through iteration, an optimized pose result T_{qw} of the query image was obtained. And we inverted T_{qw} to get T_{wq}, because T_{wq} is more intuitive, from which we can directly get the pose of the query image. Finally, T_{wq} was saved in the form of $Pose_q=\left\{x_q,y_q,z_q,qx_q,qy_q,qz_q,qw_q\right\}$, where $\left\{x_q,y_q,z_q\right\}$ and $\left\{qx_q,qy_q,qz_q,qw_q\right\}$ are the position and attitude of the query image respectively.

With this process, the precise and real-time position and attitude of the acquired query color image were estimated. In the following experimental section, we performed abundant experiments to verify the accuracy of our indoor-positioning method in common indoor scenes.

3. Experimental Results

We have conducted a series of experiments to evaluate the effectiveness of the proposed indoor positioning method. The first sub-section describes the test data and computer configuration we used in the experiments. In the second sub-section, we evaluate qualitatively the proposed RGB-D image database construction strategy of our indoor positioning method. And the results are reported in the third sub-section. For a complete comparative analysis, the results of our indoor positioning method are also compared with an existing method in reference [50].

3.1. Test Data and Computer Configuration

In order to better evaluate the proposed indoor positioning method, six sequences of the public dataset TUM RGB-D were adopted as the test data. Every sequence of the dataset contains RGB-D images, i.e., RGB and depth images, captured by a Microsoft Kinect sensor at a frame rate of 30 Hz. The size of the RGB-D images was 640×480.

Figure 8 shows the six sequences of TUM RGB-D dataset. These six sequences can well represent the common indoor scenes in daily life. And the intrinsic parameters of the Microsoft Kinect sensor were found in reference [51].

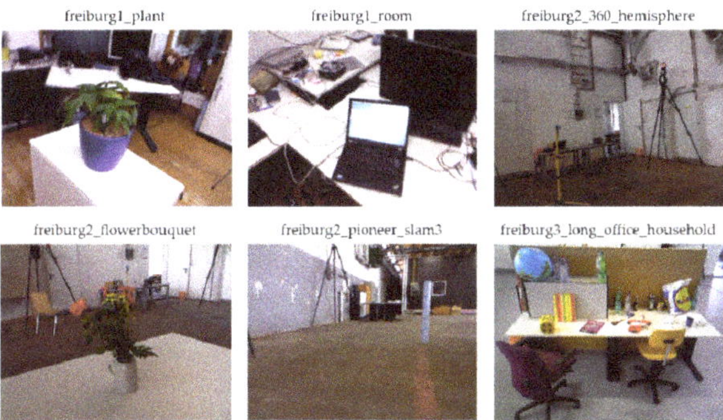

Figure 8. Six sequences of the TUM RGB-D dataset

Before the procedure of RGB-D image database construction, it was important to determine the availability of these sequences. If the associated depth image of a RGB image was missing, then the RGB image was discarded. After that, the remaining test images were used in the experiments of database construction and pose estimation. Then the database images corresponding to the query images with large pose errors were checked manually. If they were motion blur or poorly illuminated images, they were removed from the database. The number of test images in each of these six sequences is shown in Table 2.

Table 2. The number of test images in each of these six sequences of TUM RGB-D dataset.

Six Sequences of TUM RGB-D Dataset	The Number of Test RGB-D Images
freiburg1_plant	1126
freiburg1_room	1352
freiburg2_360_hemisphere	2244
freiburg2_flowerbouquet	2851
freiburg2_pioneer_slam3	2238
freiburg3_long_office_household	2488

We employed a computer with an Intel Core i7-6820HQ CPU @ 2.7 GHz and 16 GB RAM to conduct all the experiments. The procedure of image retrieval based on CNN feature vector was accelerated by a NVIDIA Quadro M2000M GPU. The details of the experimental results are described below.

3.2. Results of RGB-D Database Construction

According to the RGB-D database construction process described in Section 2.2, we got the constructed databases of six sequences shown in Figure 9. The gray lines represent the ground truth of trajectories when recording RGB-D images. The red hexagonal points on the gray lines are the positions of indoor-positioning database images. As can be seen from Figure 9, more database images are selected using the proposed database construction method at the corners where the position and attitude differences between neighboring recorded images change greatly. In these areas with smooth motion, the database images selected by our method are evenly distributed. After selecting the database images from the test images of six sequences, the remaining images were used as query images to conduct the subsequent visual-positioning experiments.

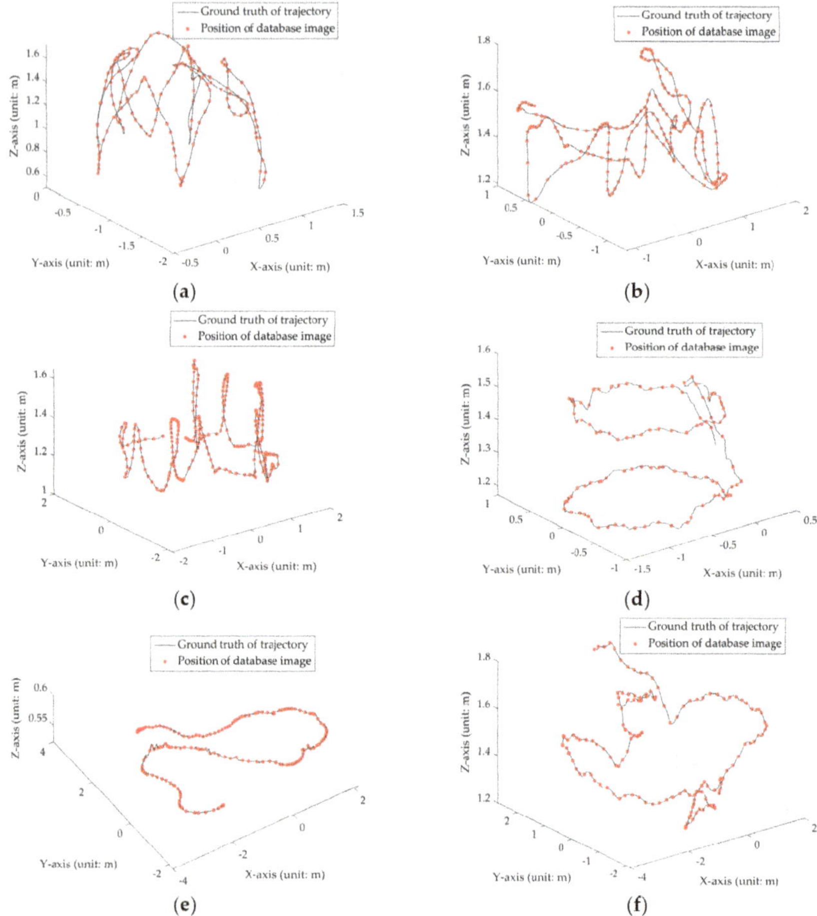

Figure 9. The results of RGB-D database construction for six sequences: (**a**) freibuigl_plant, (**b**) freiburg1_room, (**c**) freiburg2_360_hemisphere, (**d**) freiburg2_flowerbouquet, (**e**) freiburg2_pioneer_slam3 and (**f**) freiburg3_long_office_household.

Considering the redundancy of the RGB-D test images and the efficiency of the visual positioning process, most methods are implemented by selecting representative database images manually, as in reference [50]. The method of hand-pick is subjective and time-consuming. The quality of the database images depends largely on experience. When the number of captured images is large, the workload increases. In contrast, the database images selected by the proposed RGB-D image database construction process are not too redundant or sparse. The database image directly provides high accuracy depth information, this not only ensures high accuracy of positioning, but also improves the efficiency of image retrieval. Therefore, the proposed method can reduce the workload of selecting representative database images and meet the requirements of highly accurate and real-time indoor positioning.

The hand-picked database images of reference method [50] are used for comparison. The numbers of selected database images and used query images in the reference method and the proposed method are shown in Table 3. It can be seen from Table 3 that the number of query images used in our method is about twice the number of query images used in the reference method. This is because the proposed method automatically selects the database images and then takes the remaining images as query images, eliminating the workload of manually selecting images.

Table 3. The numbers of selected database images and query images in the reference method and the proposed method.

Six Sequences of TUM RGB-D Dataset	Images in the Reference Method		Images in the Proposed Method	
	Database	Query	Database	Query
freiburg1_plant	115	456	158	968
freiburg1_room	91	454	193	1159
freiburg2_360_hemisphere	273	1092	253	1991
freiburg2_flowerbouquet	149	1188	104	2747
freiburg2_pioneer_slam3	128	1017	173	2065
freiburg3_long_office_household	130	1034	152	2336

After building the RGB-D indoor-positioning database, we input query images of each sequence in turn. Through the process of image retrieval based on the CNN feature vector, we selected a database RGB-D image with pose information that is most similar to the input query image. Then we performed the visual-positioning process to estimate the position and attitude of each input query image.

3.3. *Quantitative Analysis of Positioning Accuracy*

In this part, the performance of pose estimation in the proposed indoor-positioning method was evaluated by comparing it with the reference method mentioned in the previous section. The estimated pose results of each sequence was saved in a file. The mean pose error and median pose error of these two methods were obtained by comparing the estimated poses with the ground truth trajectory. In addition, both position error and attitude error were calculated as an evaluation of six DoFs.

The pose estimation results of each sequence using the reference method and proposed method are shown in Table 4. As can be seen from the results of the proposed method in Table 4, the mean values and median values of position errors are both at the half-meter level. As for attitude errors, the mean errors are within 5 degrees and the median errors are within 1 degree. These results demonstrate that the database we built in Section 3.2 meets the requirements of high-accuracy visual positioning well. By comparing the results of the reference method and the proposed method, we can see that most of the mean and median pose errors of our method are smaller than those of the reference method. This also indicates that the database constructed by our method can achieve or surpass the accuracy of the hand-picked database to some extent.

Table 4. The results of position error and attitude error of the six sequences using the reference method and the proposed method. Each term contains a mean error and a median error. The numbers in bold indicate that these terms are better than those of the other method.

Six Sequences of TUM RGB-D Dataset	Pose Error of the Reference Method		Pose Error of the Proposed Method	
	Mean	Median	Mean	Median
freiburg1_plant	0.38 m 3.37°	0.12 m **0.01°**	**0.02 m 0.61°**	**0.01 m** 0.44°
freiburg1_room	0.43 m 4.82°	0.17 m 0.54°	**0.02 m 1.14°**	**0.01 m 0.50°**
freiburg2_360_hemisphere	0.38 m 6.55°	0.05 m **0.16°**	**0.22 m 2.77°**	**0.03 m** 0.36°
freiburg2_flowerbouquet	0.15 m 5.32°	0.07 m **0.12°**	**0.04 m 1.57°**	**0.02 m** 0.51°
freiburg2_pioneer_slam3	0.34 m 8.80°	0.13 m **0.13°**	**0.18 m 4.10°**	**0.02 m** 0.43°
freiburg3_long_office_household	0.36 m 3.00°	0.15 m **0.21°**	**0.01 m 0.37°**	**0.01 m** 0.31°

In order to demonstrate the pose estimation results of all the query images intuitively, cumulative distribution function (CDF) was adopted to analyze the positioning accuracy of the proposed method. Figure 10 shows the CDF of position error. From these CDF curves we can see that nearly 95% of the query images have a position error within 0.5 m. Furthermore, the position errors of all query images in sequence freiburg3_long_office_household are within 0.1 m, as shown in Figure 10f. These results show that the proposed method is able to localize the position of a query image well and achieve a high accuracy of better than 1 m in most cases.

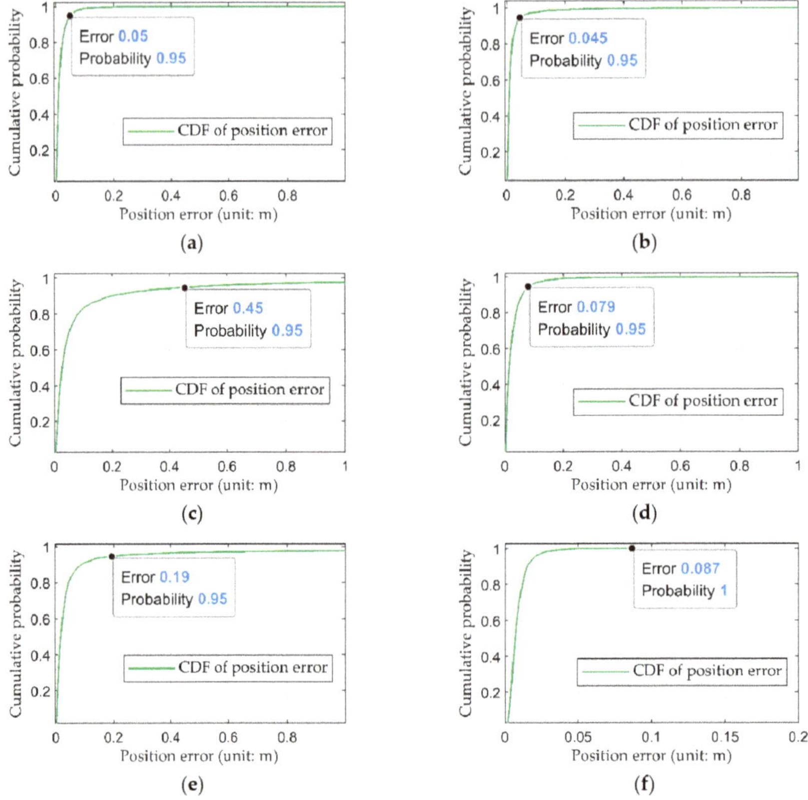

Figure 10. Cumulative distribution function (CDF) of position error in the (**a**) freibuig1_plant, (**b**) freiburg1_room, (**c**) freiburg2_360_hemisphere, (**d**) freiburg2_flowerbouquet, (**e**) freiburg2_pioneer_slam3 and (**f**) freiburg3_long_office_household sequence. The green line of each sub-figure represents the CDF curve. The vertical coordinate represents the cumulative probability of position error.

In addition to CDF of position error, CDF of attitude error was also calculated as an evaluation of six DoFs. Figure 11 shows the CDF of attitude error. The blue line of each sub-figure represents the CDF curve. Similarly, we can see that the attitude errors of all query images in sequence freiburg3_long_office_household are within 3 degrees, as shown in Figure 11f. For the rest of the sequences, almost 95% of the query images have an attitude error within 5 degrees. These results show that the proposed method is able to calculate the attitude of a query image well with an accuracy of better than 5 degrees in most cases.

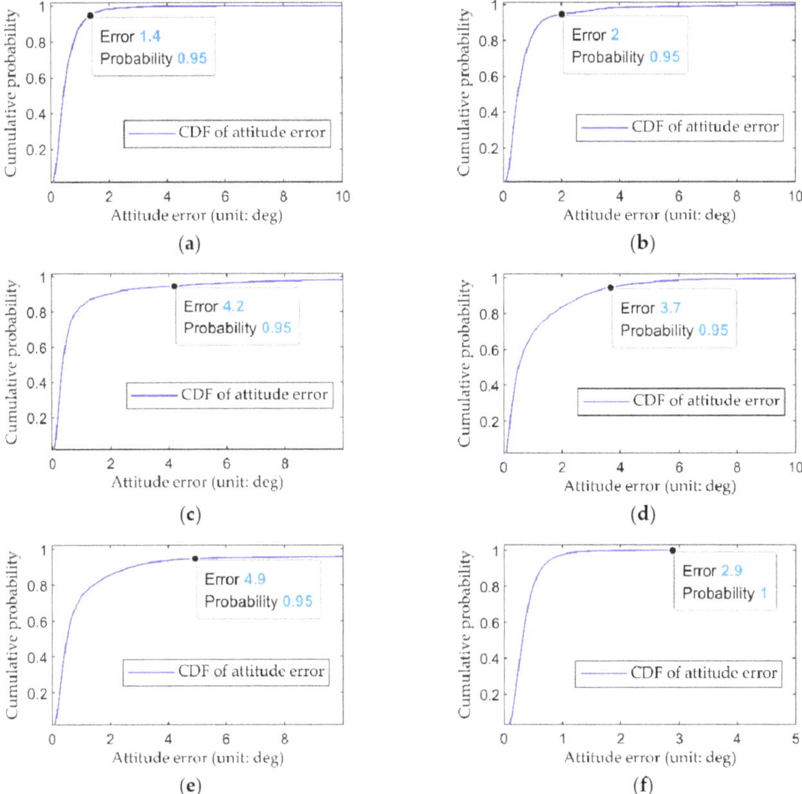

Figure 11. Cumulative distribution function (CDF) of attitude error in the (**a**) freibuig1_plant, (**b**) freiburg1_room, (**c**) freiburg2_360_hemisphere, (**d**) freiburg2_flowerbouquet, (**e**) freiburg2_pioneer_slam3 and (**f**) freiburg3_long_office_household sequence. The blue line of each sub-figure represents the CDF curve. The vertical coordinate represents the cumulative probability of attitude error.

The cumulative pose errors of the reference method and the proposed method were compared, as shown in Table 5. It was found that most of the results using our method outperformed those using the reference method. As can be seen from the cumulative accuracy of the reference method, 90% of the query images in each sequence are localized within 1 m and 5 degrees. The 90% accuracy of position error of our method is within 0.5 m and the attitude error is within 3 degrees, both of which are nearly half the pose error of the reference method. Specifically, all the cumulative position errors of the proposed method are better than those of the reference method. The cumulative attitude errors of the proposed method are better or comparable with those of the reference method. These good performances of the proposed indoor visual-positioning method also indicate the validity of the proposed database construction strategy.

Table 5. The cumulative pose errors of the reference method and the proposed method in six sequences. Each term contains a mean error and a median error. The unit of position error is meter and the unit of attitude error is degree. The numbers in bold indicate that these terms are better than those of the other method.

Six Sequences of TUM RGB-D Dataset	90% Accuracy of the Reference Method	90% Accuracy of the Proposed Method
freiburg1_plant	0.45 m 1.95°	**0.04 m 1.09°**
freiburg1_room	0.71 m 4.04°	**0.03 m 1.29°**
freiburg2_360_hemisphere	0.38 m **1.08°**	**0.21 m** 1.94°
freiburg2_flowerbouquet	0.26 m **2.54°**	**0.06 m** 2.76°
freiburg2_pioneer_slam3	0.66 m **1.54°**	**0.10 m** 2.75°
freiburg3_long_office_household	0.41 m 2.05°	**0.02 m 0.65°**

All the experiments were conducted by a laptop with an Intel Core i7-6820HQ CPU @ 2.7 GHz and 16 GB RAM. In the experiment, it takes about 0.4 s on average to implement the RGB-D database image-retrieval process in selecting the most similar database image with the input query image. The pose estimation process of one query image takes about 0.2 s on average. Considering that the RGB-D indoor positioning database is built offline, the time it costs is not taken into account. Therefore, our indoor-positioning program will take about 0.6 s to complete the two processes of image retrieval and pose estimation. Specifically, if we use a mobile platform to capture query image at a resolution of 640×480 and upload it into the laptop using 4G network, the process takes about 0.3 s. As for returning the location result from the laptop to the mobile platform, it takes about 0.1 s. Therefore, the whole procedure, which starts with capturing a query image by the mobile platform and finally obtains the position result from the laptop, takes about 1 s. In other words, the indoor-positioning frequency of the proposed method is about 1 Hz. This also shows that the proposed method has the ability of real-time indoor positioning while satisfying the need for high accuracy.

4. Conclusions

In this study, a novel indoor-positioning method with automated RGB-D image database construction was presented. The proposed method has two main innovations. First, the indoor-positioning database constructed by our method can reduce the workload of manually selecting database images and is more objective. The database is automatically constructed according to the preset rules, which reduces the redundancy of the database and improves the efficiency of the image-retrieval process. Second, by combining automatic database construction module, the CNN-based image retrieval module, and strict geometric relations based pose estimation module, we obtain a highly accurate indoor-positioning system.

In the experiment, the proposed indoor positioning method was evaluated with six typical indoor sequences of TUM RGB-D dataset. We presented the quantitative evaluation results of our method compared with a state-of-the-art indoor visual-positioning method. All the experimental results show that the proposed method obtains high-accuracy position and attitude results in common indoor scenes. The accuracy of the proposed method attained is generally higher than that of the reference method.

In the next version of our indoor-positioning method, we plan to combine the semantic information of the sequence to reduce the search space of visual positioning in large indoor scenes.

Author Contributions: R.W., K.D. and W.W. conceived the idea, designed the method and wrote the paper; R.W. and R.C. developed the software; R.W., W.W. and X.F. performed the experiments; W.W. and K.D. analyzed the data.

Funding: This research was funded by National Key Research and Development Program of China (No. 2016YFB0502102 and No. 2018YFB1305004).

Acknowledgments: We thank Computer Vision Group from Technical University of Munich for making the RGB-D SLAM dataset publically available.

Conflicts of Interest: The authors declare no conflict of interest.

Abbreviations

The following abbreviations are used in this manuscript:

RGB-D	red, green, blue and depth
GNSS	Global Navigation Satellite System
SIFT	scale-invariant feature transform
PCA	principal component analysis
SAD	sum of absolute difference
DoFs	degrees of freedom
SLAM	simultaneous localization and mapping
CNN	convolutional neural networks
TUM	Technical University of Munich
RMSE	root mean square errors
NetVLAD	vector of locally aggregated descriptors
RANSAC	random sample consensus
EPnP	efficient perspective-n-point method
CDF	cumulative distribution function

References

1. Jiang, B.; Yao, X. Location-based services and GIS in perspective. *Comput. Environ. Urban Syst.* **2006**, *30*, 712–725. [CrossRef]
2. Weiser, M. The computer for the 21st century. *IEEE Pervasive Comput.* **1999**, *3*, 3–11. [CrossRef]
3. Davidson, P.; Piché, R. A survey of selected indoor positioning methods for smartphones. *IEEE Commun. Surv. Tutor.* **2017**, *19*, 1347–1370. [CrossRef]
4. Atia, M.M.; Noureldin, A.; Korenberg, M.J. Dynamic Online-Calibrated Radio Maps for Indoor Positioning in Wireless Local Area Networks. *IEEE Trans. Mob. Comput.* **2013**, *12*, 1774–1787. [CrossRef]
5. Du, Y.; Yang, D.; Xiu, C. A Novel Method for Constructing a WIFI Positioning System with Efficient Manpower. *Sensors* **2015**, *15*, 8358–8381. [CrossRef]
6. Moghtadaiee, V.; Dempster, A.G.; Lim, S. Indoor localization using FM radio signals: A fingerprinting approach. In Proceedings of the 2011 International Conference on Indoor Positioning & Indoor Navigation (IPIN), Guimarães, Portugal, 21–23 September 2011; pp. 1–7.
7. Bahl, P.; Padmanabhan, V.N. In RADAR: An in-building RF-based user location and tracking system. *IEEE Infocom* **2000**, *2*, 775–784.
8. Youssef, M.; Agrawala, A. The Horus WLAN location determination system. In Proceedings of the 3rd International Conference on Mobile Systems, Application, and Services (MobiSys 2005), Seattle, WA, USA, 6–8 June 2005; pp. 205–218.
9. Cantón Paterna, V.; Calveras Augé, A.; Paradells Aspas, J.; Pérez Bullones, M.A. A Bluetooth Low Energy Indoor Positioning System with Channel Diversity, Weighted Trilateration and Kalman Filtering. *Sensors* **2017**, *17*, 2927. [CrossRef]
10. Chen, L.; Pei, L.; Kuusniemi, H.; Chen, Y.; Kröger, T.; Chen, R. Bayesian Fusion for Indoor Positioning Using Bluetooth Fingerprints. *Wirel. Pers. Commun.* **2013**, *70*, 1735–1745. [CrossRef]
11. Huang, X.; Guo, S.; Wu, Y.; Yang, Y. A fine-grained indoor fingerprinting localization based on magnetic field strength and channel state information. *Pervasive Mob. Comput.* **2017**, *41*, 150–165. [CrossRef]
12. Kim, H.-S.; Seo, W.; Baek, K.-R. Indoor Positioning System Using Magnetic Field Map Navigation and an Encoder System. *Sensors* **2017**, *17*, 651. [CrossRef]
13. Gronat, P.; Obozinski, G.; Sivic, J.; Pajdla, T. Learning and Calibrating Per-Location Classifiers for Visual Place Recognition. In Proceedings of the IEEE Computer Society Conference on Computer Vision and Pattern Recognition, Portland, OR, USA, 23–28 June 2013; pp. 907–914.
14. Vaca-Castano, G.; Zamir, A.R.; Shah, M. City scale geo-spatial trajectory estimation of a moving camera. In Proceedings of the IEEE Conference on Computer Vision and Pattern Recognition, Providence, RI, USA, 16–21 June 2012; pp. 1186–1193.

15. Arandjelović, R.; Zisserman, A. Three things everyone should know to improve pbject retrieval. In Proceedings of the IEEE Conference on Computer Vision and Pattern Recognition, Providence, RI, USA, 16–21 June 2012; pp. 2911–2918.
16. Zamir, A.R.; Shah, M. Accurate Image Localization Based on Google Maps Street View. In Proceedings of the 11th European Conference on Computer Vision, Heraklion, Crete, Greece, 5–11 September 2010; pp. 255–268.
17. Zamir, A.R.; Shah, M. Image Geo-Localization Based on MultipleNearest Neighbor Feature Matching UsingGeneralized Graphs. *IEEE Trans. Pattern Anal. Mach. Intell.* **2014**, *36*, 1546–1558. [CrossRef] [PubMed]
18. Babenko, A.; Slesarev, A.; Chigorin, A.; Lempitsky, V. Neural Codes for Image Retrieval. In Proceedings of the 13th European Conference on Computer Vision, Zurich, Switzerland, 6–12 September 2014; pp. 584–599.
19. Azzi, C.; Asmar, D.; Fakih, A.; Zelek, J. Filtering 3D keypoints using GIST for accurate image-based positioning. In Proceedings of the 27th British Machine Vision Conference, York, UK, 19–22 September 2016; pp. 1–12.
20. Lowe, D.G. Distinctive Image Features from Scale-Invariant Keypoints. *Int. J. Comput. Vis.* **2004**, *60*, 91–110. [CrossRef]
21. Hervé, J.; Douze, M.; Schmid, C. Product Quantization for Nearest Neighbor Search. *IEEE Trans. Pattern Anal. Mach. Intell.* **2010**, *33*, 117–128.
22. Nistér, D.; Stewénius, H. Scalable Recognition with a Vocabulary Tree. In Proceedings of the Computer Vision and Pattern Recognition, New York, NY, USA, 17–22 June 2006; pp. 2161–2168.
23. Kim, H.J.; Dunn, E.; Frahm, J.M. Predicting Good Features for Image Geo-Localization Using Per-Bundle VLAD. In Proceedings of the IEEE International Conference on Computer Vision (ICCV), Santiago, Chile, 13–16 December 2015; pp. 1170–1178.
24. Torii, A.; Arandjelovic, R.; Sivic, J.; Okutomi, M.; Pajdla, T. 24/7 place recognition by view synthesis. In Proceedings of the IEEE Conference on Computer Vision and Pattern Recognition (CVPR), Boston, MA, USA, 7–12 June 2015; pp. 1808–1817.
25. Milford, M.J.; Lowry, S.; Shirazi, S.; Pepperell, E.; Shen, C.; Lin, G.; Liu, F.; Cadena, C.; Reid, I. Sequence searching with deep-learnt depth for condition-and viewpoint-invariant route-based place recognition. In Proceedings of the IEEE Conference on Computer Vision and Pattern Recognition (CVPR), Boston, MA, USA, 7–12 June 2015; pp. 18–25.
26. Poglitsch, C.; Arth, C.; Schmalstieg, D.; Ventura, J. A Particle Filter Approach to Outdoor Localization Using Image-Based Rendering. In Proceedings of the IEEE International Symposium on Mixed & Augmented Reality, Fukuoka, Japan, 29 September–3 October 2015; pp. 132–135.
27. Razavian, A.S.; Azizpour, H.; Sullivan, J.; Carlsson, S. CNN features off-the-shelf: An astounding baseline for recognition. In Proceedings of the IEEE Conference on Computer Vision and Pattern RecognitionWorkshops, Columbus, OH, USA, 23–28 June 2014; pp. 512–519.
28. Yandex, A.B.; Lempitsky, V. Aggregating Local Deep Convolutional Features for Image Retrieval. In Proceedings of the IEEE International Conference on Computer Vision (ICCV), Santiago, Chile, 13–16 December 2015; pp. 1269–1277.
29. Cadena, C.; Carlone, L.; Carrillo, H.; Latif, Y.; Scaramuzza, D.; Neira, J.; Reid, I.; Leonard, J.J. Past, Present, and Future of Simultaneous Localization and Mapping: Toward the Robust-Perception Age. *IEEE Trans. Robot.* **2016**, *32*, 1309–1332. [CrossRef]
30. Younes, G.; Asmar, D.; Shammas, E.; Zelek, J. Keyframe-based monocular SLAM: Design, survey, and future directions. *Robot. Auton. Syst.* **2017**, *98*, 67–88. [CrossRef]
31. Campbell, D.; Petersson, L.; Kneip, L.; Li, H. Globally-optimal inlier set maximisation for simultaneous camera pose and feature correspondence. In Proceedings of the IEEE Conference on Computer Vision (ICCV), Venice, Italy, 22–29 October 2017; pp. 1–10.
32. Liu, L.; Li, H.; Dai, Y. Efficient Global 2D-3D Matching for Camera Localization in a Large-Scale 3D Map. In Proceedings of the IEEE Conference on Computer Vision (ICCV), Venice, Italy, 22–29 October 2017; pp. 2391–2400.
33. Yousif, K.; Taguchi, Y.; Ramalingam, S. MonoRGBD-SLAM: Simultaneous localization and mapping using both monocular and RGBD cameras. In Proceedings of the 2017 IEEE International Conference on Robotics and Automation (ICRA), Singapore, 29 May–3 June 2017; pp. 4495–4502.
34. Turan, M.; Almalioglu, Y.; Araujo, H.; Konukoglu, E.; Sitti, M. A non-rigid map fusion-based rgb-depth slam method for endoscopic capsule robots. *Int. J. Intell. Robot. Appl.* **2017**, *1*, 399. [CrossRef]

35. Sattler, T.; Torii, A.; Sivic, J.; Pollefeys, M.; Taira, H.; Okutomi, M.; Pajdla, T. Are Large-Scale 3D Models Really Necessary for Accurate Visual Localization? In Proceedings of the IEEE Conference on Computer Vision and Pattern Recognition (CVPR), Honolulu, HI, USA, 21–26 July 2017; pp. 6175–6184.
36. Degol, J.; Bretl, T.; Hoiem, D. ChromaTag: A Colored Marker and Fast Detection Algorithm. In Proceedings of the IEEE Conference on Computer Vision (ICCV), Venice, Italy, 22–29 October 2017; pp. 1481–1490.
37. Muñoz-Salinas, R.; Marin-Jimenez, M.J.; Yeguas-Bolivar, E.; Medina-Carnicer, R. Mapping and Localization from Planar Markers. *Pattern Recognit.* **2017**, *73*, 158–171. [CrossRef]
38. Schweighofer, G.; Pinz, A. Robust pose estimation from a planar target. *IEEE Trans. Pattern Anal. Mach. Intell.* **2007**, *28*, 2024–2030. [CrossRef]
39. Valentin, J.; Niebner, M.; Shotton, J.; Fitzgibbon, A.; Izadi, S.; Torr, P. Exploiting uncertainty in regression forests for accurate camera relocalization. In Proceedings of the IEEE Conference on Computer Vision and Pattern Recognition, Boston, MA, USA, 7–12 June 2015; pp. 4400–4408.
40. Tateno, K.; Tombari, F.; Laina, I.; Navab, N. CNN-SLAM: Real-Time Dense Monocular SLAM with Learned Depth Prediction. In Proceedings of the IEEE Conference on Computer Vision and Pattern Recognition (CVPR), Honolulu, HI, USA, 21–26 July 2017; pp. 6565–6574.
41. Shotton, J.; Glocker, B.; Zach, C.; Izadi, S.; Criminisi, A.; Fitzgibbon, A. Scene Coordinate Regression Forests for Camera Relocalization in RGB-D Images. In Proceedings of the IEEE Conference on Computer Vision and Pattern Recognition (CVPR), Portland, OR, USA, 23–28 June 2013; pp. 2930–2937.
42. Kendall, A.; Cipolla, R. Geometric loss functions for camera pose regression with deep learning. In Proceedings of the IEEE Conference on Computer Vision and Pattern Recognition, Honolulu, HI, USA, 21–26 July 2017; pp. 6555–6564.
43. Tekin, B.; Sinha, S.N.; Fua, P. Real-Time Seamless Single Shot 6D Object Pose Prediction. In Proceedings of the IEEE Conference on Computer Vision and Pattern Recognition (CVPR), Salt Lake City, UT, USA, 18–22 June 2018; pp. 292–301.
44. Piasco, N.; Sidibé, D.; Demonceaux, C.; Gouet-Brunet, V. A survey on Visual-Based Localization: On the benefit of heterogeneous data. *Pattern Recognit.* **2018**, *74*, 90–109. [CrossRef]
45. Ummenhofer, B.; Zhou, H.; Uhrig, J.; Mayer, N.; Ilg, E.; Dosovitskiy, A.; Brox, T. DeMoN: Depth and motion network for learning monocular stereo. In Proceedings of the IEEE Conference on Computer Vision and Pattern Recognition (CVPR), Honolulu, HI, USA, 21–26 July 2017; pp. 5622–5631.
46. Mahjourian, R.; Wicke, M.; Angelova, A. Unsupervised learning of depth and ego-motion from monocular video using 3d geometric constraints. In Proceedings of the IEEE Conference on Computer Vision and Pattern Recognition (CVPR), Salt Lake City, UT, USA, 18–22 June 2018; pp. 5667–5675.
47. Zhan, H.; Garg, R.; Weerasekera, C.S.; Li, K.; Agarwal, H.; Reid, I. Unsupervised learning of monocular depth estimation and visual odometry with deep feature reconstruction. In Proceedings of the IEEE Conference on Computer Vision and Pattern Recognition (CVPR), Salt Lake City, UT, USA, 18–22 June 2018; pp. 340–349.
48. Gao, X.; Zhang, T. Unsupervised learning to detect loops using deep neural networks for visual SLAM system. *Auton. Robot.* **2017**, *41*, 1–18. [CrossRef]
49. Wu, J.; Ma, L.; Hu, X. Delving deeper into convolutional neural networks for camera relocalization. In Proceedings of the 2017 IEEE International Conference on Robotics and Automation (ICRA), Singapore, 29 May–3 June 2017; pp. 5644–5651.
50. Chen, Y.; Chen, R.; Liu, M.; Xiao, A.; Wu, D.; Zhao, S. Indoor Visual Positioning Aided by CNN-Based Image Retrieval: Training-Free, 3D Modeling-Free. *Sensors* **2018**, *18*, 2692. [CrossRef] [PubMed]
51. Sturm, J.; Engelhard, N.; Endres, F.; Burgard, W.; Cremers, D. A benchmark for the evaluation of RGB-D SLAM systems. In Proceedings of the 2012 IEEE/RSJ International Conference on Intelligent Robots and Systems (IROS), Algarve, Portugal, 7–12 October 2012; pp. 573–580.
52. Arandjelovic, R.; Gronat, P.; Torii, A.; Pajdla, T.; Sivic, J. Netvlad: CNN architecture for weakly supervised place recognition. *IEEE Trans. Pattern Anal. Mach. Intell.* **2018**, *40*, 1437–1451. [CrossRef] [PubMed]
53. Richard, H.; Andrew, Z. *Multiple View Geometry in Computer Vision*, 2nd ed.; Cambridge University Press: Cambridge, UK, 2003; pp. 241–253.
54. Fischler, M.A.; Bolles, R.C. Random Sample Consensus: A Paradigm for Model Fitting with Applications to Image Analysis and Automated Cartography. *Commun. Acm* **1981**, *24*, 381–395. [CrossRef]

55. Lepetit, V.; Moreno-Noguer, F.; Fua, P. EPnP: An accurate O(n) solution to the PnP problem. *Int. J. Comput. Vis.* **2009**, *81*, 155–166. [CrossRef]
56. Kümmerle, R.; Grisetti, G.; Strasdat, H.; Konolige, K.; Burgard, W. G2o: A general framework for graph optimization. In Proceedings of the 2011 IEEE International Conference on Robotics and Automation (ICRA), Shanghai, China, 9–13 May 2011; pp. 3607–3613.

Article

Forward and Backward Visual Fusion Approach to Motion Estimation with High Robustness and Low Cost

Ke Wang [1,*,†], Xin Huang [1,†], JunLan Chen [2], Chuan Cao [1], Zhoubing Xiong [3] and Long Chen [4]

1 State Key Laboratory of Mechanical Transmission, School of Automobile Engineering, Chongqing University, Chongqing 400044, China; huangxinslam@cqu.edu.cn (X.H.); chuancao@cqu.edu.cn (C.C.)

2 School of Economics & Management, Chongqing Normal University, Chongqing 401331, China; junlanchen@cqnu.edu.cn

3 Intelligent Vehicle R&D Institute, Changan Auto Company, Chongqing 401120, China; xiongzb@changan.com.cn

4 State Key Laboratory of Vehicle NVH and Safety Technology, China Automotive Engineering Research Institute Company, Ltd., Chongqing 401122, China; chenlong@caeri.com.cn

* Correspondence: yeswangke@cqu.edu.cn

† These authors contributed equally to this work.

Received: 16 July 2019; Accepted:10 September 2019; Published: 13 September 2019

Abstract: We present a novel low-cost visual odometry method of estimating the ego-motion (self-motion) for ground vehicles by detecting the changes that motion induces on the images. Different from traditional localization methods that use differential global positioning system (GPS), precise inertial measurement unit (IMU) or 3D Lidar, the proposed method only leverage data from inexpensive visual sensors of forward and backward onboard cameras. Starting with the spatial-temporal synchronization, the scale factor of backward monocular visual odometry was estimated based on the MSE optimization method in a sliding window. Then, in trajectory estimation, an improved two-layers Kalman filter was proposed including orientation fusion and position fusion. Where, in the orientation fusion step, we utilized the trajectory error space represented by unit quaternion as the state of the filter. The resulting system enables high-accuracy, low-cost ego-pose estimation, along with providing robustness capability of handing camera module degradation by automatic reduce the confidence of failed sensor in the fusion pipeline. Therefore, it can operate in the presence of complex and highly dynamic motion such as enter-in-and-out tunnel entrance, texture-less, illumination change environments, bumpy road and even one of the cameras fails. The experiments carried out in this paper have proved that our algorithm can achieve the best performance on evaluation indexes of average in distance (AED), average in X direction (AEX), average in Y direction (AEY), and root mean square error (RMSE) compared to other state-of-the-art algorithms, which indicates that the output results of our approach is superior to other methods.

Keywords: motion estimation; trajectory fusion; mobile mapping; sensor fusion

1. Introduction

1.1. Motivations and Technical Challenges

This paper aims at developing a visual fusion approach for online ego-motion estimation with the data from onboard forward and backward cameras. Ego-motion represents and describes the self-motion of a moving object and the ego-motion problem can be stated as the recovery of observer rotation and direction of translation by at a given instant of time, as the observer moves through the environment, which is also called ego-motion estimation. The most ego-motion estimation methods comprise two steps, motion-field computation and motion field analysis [1]. In many real-world applications, the estimation of egomotion and localization is a pivot of major vision-based navigation system especially for autonomous ground vehicle and robotics [2–4], since it forms the basis of subsequent scene understanding and vehicle control [5]. In addition, ego-motion estimation in vehicles and robots is fundamental as it is usually the pre-requisite for higher-layer tasks, such as robot-based surveillance, autonomous navigation, path planning, for example, References [6,7]. A vision-based odometry system, compared to a traditional wheel-based or satellites-based localization system, has the advantages of an impervious character to inherent sensor inefficacies [8,9] (e.g., wheel encoder error because of uneven, slippery terrain or other adverse conditions) and can be used in a GPS-denied area [10,11] (e.g., underwater and tunnels in urban environments.) The proposed approach utilizes only visual perception cameras with lightweight, high robustness and low-cost characters.

Visual ego-motion estimation has been successfully proven to be able to estimate the movement of a road vehicle over a long distance under certain conditions [12], with a relative error ranging from 0.1 to 2% [13]. It incrementally estimates camera poses, yet small errors are inevitably accumulated and over time the estimated ego-motion slowly drifts away from their ground true [14]. With the purpose of improving the performance, appreciable progress has been made to enhance the system robustness, accuracy and efficiency. Many efforts, such as the direct method, semi-direct method and feature-based method, have been made for different visual ego-motion formulations [15]. Global map optimization techniques like loop closure, bundle adjustment (BA) and pose graph optimization were also proposed to make better the overall performance, which adjust the estimation results by taking into account the entire observation equations and eventual constraints [16].

However, until now, it was still hard to integrate into the motion estimation application of mobile robots and even less in that of an autonomous vehicle [17]. It suffers from many limitations [18–20]. First, most visual odometry methods encounter the inherent imperfection of motion drift and the inner fragility of sensor degradation. The data association model of visual ego-motion estimation is not completely stable and can fail under ubiquitous noise, texture-less scenery and the different level of sensor degradation. Second, the commonly used global drift optimization method of loop closure is often out of action, especially for large-scale outdoor use. Most of the time, there is no loop in many scenarios for a moving robot or vehicle in urban areas and the computation of large loops is time consuming, which make the robustness of system even more critical. Third, as a kind of passive measurement method, the motion estimation scheme usually requires enough keypoints [12], which are difficult to detect and track in some complicated circumstance, such as enter-in-and-out tunnel entrance scenery of sudden illumination changes, resulting in an enlargement of the outliers. Very often, other types of sensors, including differential GPS, precise IMU or 3D Lidar, are integrated to improve the system capability, along with a great increase of system cost.

1.2. Literature Review

Visual odometry is a particular case of structure-from-motion (SFM) [21], which aims to incrementally estimate the egomotion of an agent (e.g., vehicle, human and robot) using only vision stream and its origins can be dated back to works such as References [22,23]. The first visual odometry systems were successfully used by NASA in their Mars exploration program, endeavoring to provide all-terrain rovers with the capability of estimating its motion in the presence of wheel slippage terrains [24]. From then on, several methods, techniques and sensor models have been developed and employed to accomplish both vehicle and robotics' egomotion [25,26].

Some of the works were accomplished utilizing monocular vision technology with a single camera that they must measure the relative motion with 2-D bearing data [27]. Since the absolute scale is not clear, the monocular visual odometry has to calculate the relative scale and camera pose using either trifocal tensor method or the knowledge of 3-D structure [28]. Related works can generally be divided into three categories—appearance-based methods, feature-based methods and hybrid methods. Undeniably, some of them can deliver good results [29–31], but still they require high quality features to robustly perform and are more prone to errors [19]. Since our work is emphasizing the robust ego-motion of road vehicles in complex urban environments, the use of only a monocular vision system is not enough to meet the requirement and to alleviate the disturbances from a complex environment, including low textured scenes, non-stationary objects and others.

The idea of estimating egomotion from consecutive 3-D frames with a stereo vision method was successfully utilized in Reference [32]. A collaborative factor of these works is the utilization of feature-based methods because of the efficiency and robustness. As the works showing, both sparse features and dense features are used to establish corresponds between consecutive frames from input image sequence. Some of the most common choices are point [33], lines [34], contour [35], and hybrid [36], which can be tracked in both spatial and temporal domains. Point feature is the most commonly used of all, because of its simplicity. However, because of the complexity of dynamic road scenes, errors are unavoidable, appearing in every stage of detection, matching and tracking, especially in urban environments with texture-less, illumination change environments or bumpy road. It makes the robustness problem extremely critical.

Considering the problem of robustness, many researchers have established hypothesis-and-test and coarse-to-fine motion estimation mechanism based on the Random Sample Consensus (RANSAC) scheme, the bundle adjustment optimization approach (BA) and loop closure methods (LC). The 5-points based RANSAC algorithm for monocular visual odometry [37] and 3-points scheme for stereo method [38] has been proposed in the beginning. Yet, with the increase of complexity of environment the RANSAC iterations would grow exponentially, along with the increasing conflict of internal keypoints and the decreasing confidence of vision-based estimation results. Some other works use BA and LC methods to compensate for the drift of visual odometry once the loop is detected successful, but the absolute scale is very difficult to determine using mono based ego-motion estimation. Moreover, the paths of a moving agent do not always have loops in urban environment.

In this case, a combination of other sensors such as wheeled odometry [2], inertial sensing [18], and global position system [39] have been used. However, they have intrinsic limitations—wheeled odometry would be seriously affected in uneven, slippery terrain or other adverse conditions and prone to drift due to error accumulation. While the commonly used civil GPS can only update with a frequency of 1 Hz and the low-cost IMU is not accurate enough to fit the ego-motion estimation use. Some researchers integrated high defined maps (HDM) into the ego-motion estimation system, which can yield good results but have problems with the large-scale HDM collecting, high cost and local security laws [40]. In this work, we describe a novel visual odometry method only leveraging data from inexpensive visual sensors of forward and backward cameras, which enables high-accuracy, low-cost ego-pose estimation, along with providing robustness capability of handing camera module degradation. Moreover, different from the existing works, using only cameras instead of other sensors for computing egomotion allows a simple integration of egomotion data into other vision based algorithms, such as obstacle, pedestrian, and lane detection, without the need for calibration between sensors.

1.3. Main Contributions

We are interested in solving the ego-motion estimation with high robustness and low cost, along with solve the problem in real time and reliability. The issue is closely relevant to deal with the sensor degradation due to complex environment [41], such as illumination change, texture-less, and structure-less circumstance. To this end, in order to keep proposed approach cost competitiveness, we only leverage data from inexpensive visual sensors with different orientations. Our main contributions are summarized as follows:

(1) Utilizing the outstanding features of symmetry-adaption configuration of forward and backward cameras, we provided a new fusion mechanism of two-layers data processing to comprehensive utilize the nearby environment information. Therefore, it achieves high accuracy and low drift.

(2) The data processing pipeline was carefully designed to handle sensor degradation. Starting with the spatial-temporal synchronization, the scale factor of backward monocular visual odometry was estimated based on the MSE optimization method in sliding-window. With this help, both the forward and the backward camera own the capability of localization independently.

(3) Further, utilizing the fusion of two-layers Kalman Filter, the proposed pipeline can fully or partially bypass failure modules and make use of the rest of the pipeline to handle sensor degradation.

To the best of our knowledge, the proposed method is by far the cheapest method to enable such high robustness ego-motion estimation capable of handling sensor degradation under various lighting and structural conditions.

The remainder of this paper is organized as follows: Section 2 describes the detailed methodology including scale estimation and trajectory fusion. In Section 3, the experimental dataset, the performance evaluation results of the proposed method are presented. Sections 4 and 5 summarize this paper and discuss future research directions.

2. Materials and Methods

2.1. Assumptions and Coordinate Systems

Considering a sensor system including forward stereo camera and backward monocular camera, we assume that ego-motion estimation model is a rigid body motion model and the above cameras can be modeled by the pinhole camera model. Then, using the Zhang method [42], the intrinsic and extrinsic parameters can be easily calibrated in advance. With the calibration matrix, the relative pose transformation matrix between two camera systems can be obtained. Hence, we can use a single coordinate system for both the forward stereo camera and the backward monocular camera. For simplicity, in the 6DoF pose calculated by the proposed method is transformed to original global coordinate. The coordinate systems are defined in the following (see Figure 1 for illustration).

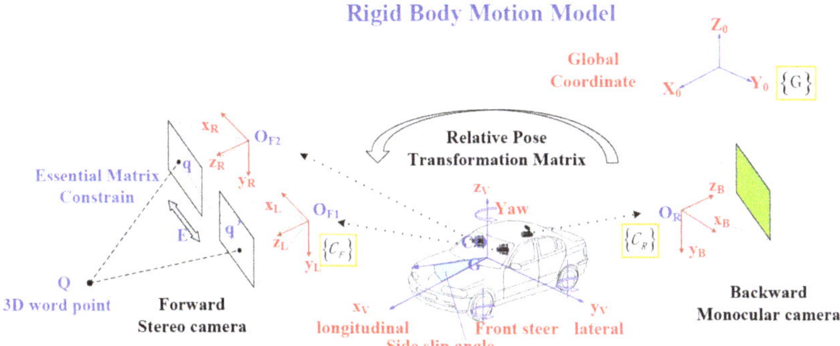

Figure 1. Coordinate system of rigid body motion model with forward stereo camera and backward monocular camera. Using the relative pose transformation matrix, we can transform the trajectories of backward monocular camera into the forward stereo camera Coordinate system. [Color figure can be viewed at www.mdpi.com].

There are three coordinates used in our system, the original global coordinate $\{G\}$, Forward camera coordinate system $\{C_F\}$ and Backward camera coordinate system $\{C_R\}$, which are defined as follows:

- We parallel X-O-Y plane of $\{G\}$ to the horizontal plane. The Z_0-axis points opposite to gravity. The X_0-axis points forward of the mobile platform, and the Y_0-axis is determined by the right-hand rule.
- Forward camera coordinate system $\{C_F\}$ is set originated at the left camera optical center of stereo camera system $\{O_{F1}\}$. The x-axis points to the left, the y-axis points upward, and the z-axis points forward coinciding with the camera principal axis.
- Backward camera coordinate system $\{C_R\}$ is originated at the camera optical center of monocular camera system $\{O_R\}$. The x-axis points to the left, the y-axis points upward, and the z-axis points forward coinciding with the camera principal axis.

2.2. Data Association of Forward-Facing and Backward-Facing Cameras

With the purpose of fitting and fusing the data of a backward-facing monocular camera and forward-facing stereo camera, the key point of data synchronization both in spatial and temporal space must be solved beforehand, which is shown Figure 2.

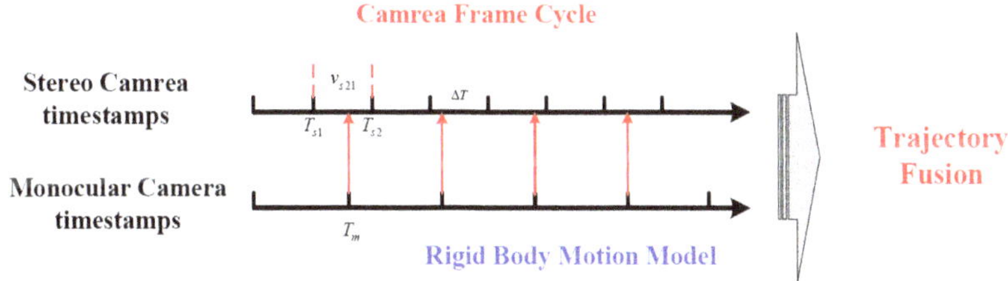

Figure 2. Rigid body motion model based trajectory data synchronization both in spatial and temporal space. [Color figure can be viewed at www.mdpi.com].

Within the stereo camera, left frame and right frame are triggered at the same time by the hardware flip-flop circuit. Then, we can assume that they were hardware synchronized and have the same timestamps. Hence, we only need to synchronize the data between monocular camera and stereo camera both in spatial and temporal space.

Using the camera intrinsic and extrinsic parameters, the data association in spatial space can be easily calculated and it will not change over time because of the rigid body motion assumption. Second, in temporal synchronization, we assume that the ego-motion has the same velocity between two consequent frames of stereo camera. So, after transforming the data from $\{C_R\}$ space into the $\{C_F\}$ space, there are two temporal synchronization steps that need to be dealt with—translation synchronization step and rotation synchronization.

In the translation synchronization step, linear interpolation method was given under the constant velocity assumption. In our method, we transform the data from monocular camera into stereo camera coordinate, as follows.

$$t_{\text{ms}} = \Delta T * v_{s21} + t_{s1} = (T_m - T_{s1})\frac{t_{s2} - t_{s1}}{T_{s2} - T_{s1}} + t_{s1} \tag{1}$$

Here, T_{s1} and T_{s2} is the consecutive timestamps of stereo camera frames, T_m is the timestamp of monocular camera frame pending to be associated. ΔT is the time difference calculated by $\Delta \mathrm{T} = \min(|T_m - T_{s1}|, |T_m - T_{s2}|)$ (in Equation (1), we take $T_m - T_{s1}$ for example). v_{s21} is the velocity value in this time gap. t_{s1} and t_{s2} are the motion states in the time T_{s1} and T_{s2} of stereo camera.

When it comes to the rotation synchronization step, however, the rotation matrix has a universal locking problem and cannot be directly used for interpolation. In our method, we use the spherical linear interpolation method with the rotation angles transformed to the unit quaternion representation space. Assuming $q_{s1} = (x_1, y_1, z_1, w_1)$ and $q_{s2} = (x_2, y_2, z_2, w_2)$ are the unit quaternion for the timestamps T_{s1} and T_{s2}, the rotation synchronization value q_{ms} can be calculated by the following equation.

$$q_{ms} = \frac{\sin[(1-e)\theta]q_{s1} + \sin(e\theta)q_{s2}}{\sin\theta} \tag{2}$$

In this equation, e is the interpolation coefficient and it can be calculated by $e = \frac{T_m - T_{s1}}{T_{s2} - T_{s1}}$. θ can be obtained from the inverse trigonometric function $\theta = \arccos(\frac{q_{s2} \cdot q_{s1}}{|q_{s2}||q_{s1}|})$. It has the advantage of convenient calculation and good performance.

2.3. Loosely Coupled Framework for Trajectory Fusion

The detail fusion processing of forward stereo camera and backward monocular camera is shown in Figure 3. It shows that our approach runs separate the stereo visual odometry and monocular visual odometry and fusion in the decision level. Therefore, it should be classified to loosely coupled fusion method.

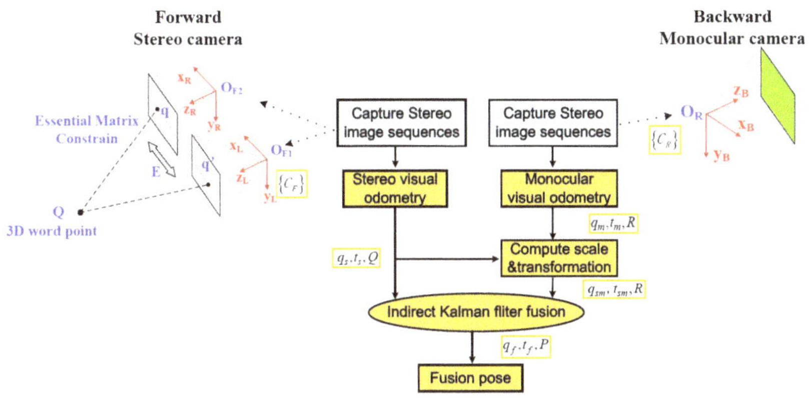

Figure 3. Framework of Loosely Coupled Forward and Backward Fusion Visual Odometry. [Color figure can be viewed at www.mdpi.com].

This framework choice would bring several good features. At first, loosely coupled fusion method support the stereo visual odometry and monocular visual odometry working as independent modules, which bring the fusion system with the capability of handing camera module degradation. Second, with the opposite orientation of forward and backward, the fusion pipeline can comprehensively utilize the nearby environment information and take advantage of the time difference from front forward stereo camera to rear backward monocular camera. Third, the system cost is much lower than other construction scheme, such as two sets of stereo vision system, lidar based odometry, or other kinds of precise localization sensor based systems.

2.4. Basic Visual Odometry Method

In our proposed loosely coupled fusion method, featured-based visual odometry was used as the fundamental motion estimation method to compute the translation and rotation of both forward facing stereo camera and backward facing monocular camera and it has a deep influence on the correctness and effectiveness of our method. It has three modules including feature extraction and matching, pose tracking, and local bundle adjustment.

Feature Extraction and Matching: In order to extract and match feature points rapidly, ORB points (Oriented Fast and Rotated BRIEF) were selected as image features, which have good multi-scale and oriented invariance to viewpoints. In the implementation, feature points in each layer of image pyramid space were extracted respectively and the maximum feature points in the unit grid were set to less than five points, making them uniform-distributed. Specifically, for the stereo camera, feature points in the left and right images were extracted simultaneously using two parallel threads and the matching process was employed to delete the isolate points under the epipolar geometry constraints. In the feature matching stage, the Hamming distance was applied to measure the distance of feature points and the best matching should not exceed 30%. Meanwhile, the Fast Library for Approximate Nearest Neighbors (FLANN) method was also used to speed up the feature matching process.

Pose Tracking: The motion model and keyframe based pose tracking method were used to estimate the successive pose of ego vehicle. Firstly, the uniform motion model was used to track feature points and get matching pairs in successive frames with the help of observed 3D map points. With the matching pairs, we can estimate the camera motion using PnP optimization method of minimizing the reprojection errors of all matching point pairs. However, if the matching point pairs are less than certain threshold, the keyframe based pose tracking method woude be activated. Using bag of words(BOW) mode, it can track and match the feature points between the current frame and the closest keyframe.

Local Bundle Adjustment: In this step, the 3D map points observed by closest keyframes were mapped to the current feature points and the poses and 3D map points were adjusted and optimized simultaneously to minimize the reprojection errors. In the application, in order to keep a real-time performance for the system, the maximum number of iterations should be set to a certain value.

2.5. Scale of Monocular Visual Odometry

To avoid the scale ambiguity in the monocular visual odometry and make it work independently, we use the sliding window based scale estimation method. In the sliding window, mean square error (MSE) was used to dynamically correct and update the scale coefficient under the rigid body motion assumption, as shown in Figure 4.

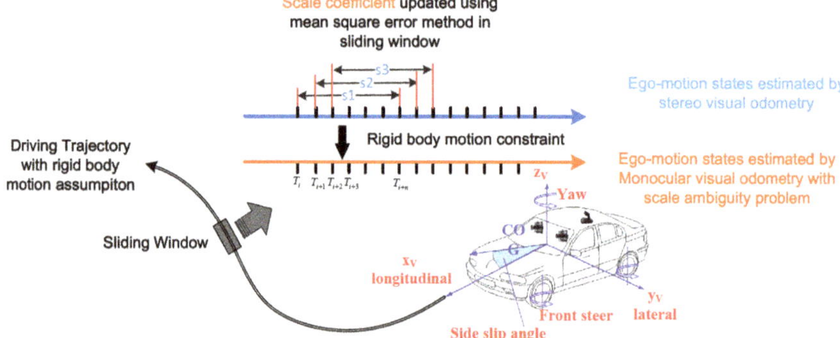

Figure 4. Sliding window based scale estimation of monocular visual odometry. [Color figure can be viewed at www.mdpi.com].

In the sliding window, we assume that there are n monocular visual odometry states of $t_{m,i}, t_{m,i+1}, t_{m,i+2} \cdots t_{m,i+n-1}$ and n stereo visual odometry states $t_{sm,i}, t_{sm,i+1}, t_{sm,i+2} \cdots t_{sm,i+n-1}$. Here, it is noteworthy that, the states of stereo visual odometry has been synchronized to monocular visual odometry states both in spatial and temporal space. Unbiased estimation of the mathematical expectations of those states can be calculated as follows:

$$\overline{t_m} = \frac{1}{n} \sum_{i}^{i+n-1} t_{m,i} \quad \text{and} \quad \overline{t_{sm}} = \frac{1}{n} \sum_{i}^{i+n-1} t_{sm,i} \tag{3}$$

The above states can be data centralized to $t'_{m,i}$ and $t'_{sm,i}$ by:

$$t'_{m,i} = t_{m,i} - \overline{t_m} \quad \text{and} \quad t'_{sm,i} = t_{sm,i} - \overline{t_{sm}} \tag{4}$$

Theoretically, under the rigid body assumption, the relative move distance of the forward stereo camera and backward monocular camera should be equal to each other within the sliding window. So, the problem can be transformed to align the n monocular visual odometry states to the corresponding n points in stereo visual odometry states space.

In practice, there must be some alignment bias and the objective of optimization function should be the minimization of alignment bias. Here, MSE was employed to solve the issue, as in the following equation.

$$\begin{aligned} \min \sum_{i}^{i+n-1} \|e_i\|^2 &= \min \sum_{i}^{i+n-1} \|t'_{sm,i} - (sRt'_{m,i} + r'_0)\|^2 \\ &= \min \sum_{i}^{i+n-1} \|t'_{sm,i} - sRt'_{m,i}\|^2 - 2r'_0 \sum_{i}^{i+n-1} [(t'_{sm,i} - sRt'_{m,i})] + n \|r'_0\|^2 \end{aligned} \tag{5}$$

In the equation, e_i is the alignment bias, s is the real scale coefficient of monocular visual odometry, R is the rotation matrix obtained from camera calibration, r'_0 is the data centralized result of translation value. In order to minimize the alignment bias e_i, we can set r'_0 equal to 0. Then, we get the following equation.

$$\begin{aligned} \min \sum_{i}^{i+n-1} \|e_i\|^2 &= \min \sum_{i}^{i+n-1} \|t'_{sm,i} - sRt'_{m,i}\|^2 \\ &= \min(\sum_{i}^{i+n-1} \|t'_{sm,i}\|^2 - 2s \sum_{i}^{i+n-1} (t'_{sm,i} \cdot sRt'_{m,i}) + s^2 \sum_{i}^{i+n-1} \|Rt'_{m,i}\|^2) \\ &= \min(\sum_{i}^{i+n-1} \|t'_{sm,i}\|^2 - 2s \sum_{i}^{i+n-1} (t'_{sm,i} \cdot sRt'_{m,i}) + s^2 \sum_{i}^{i+n-1} \|t'_{m,i}\|^2) \end{aligned} \tag{6}$$

Our goal is to solve the scale coefficient of s and we found that the above equation is a bivariate linear equation, with the following form.

$$\sum_{i}^{i+n-1} \|e_i\|^2 = A_{sm} - 2sB + s^2 A_m = (s\sqrt{A_m} - B/\sqrt{A_m})^2 + (A_{sm}A_m - B^2)/A_m \tag{7}$$

with

$$A_{sm} = \|t'_{sm,i}\|^2 \quad \text{and} \quad B = \sum_{i}^{i+n-1} (t'_{sm,i} \cdot sRt'_{m,i})$$

Then, through solving the bivariate linear problem, we can get the result of the scale coefficient of s as:

$$s = (\sum_{i}^{i+n-1} t'_{sm} \cdot Rt'_{m}) / \sum_{i}^{i+n-1} \|t'_{m,i}\|^2 \tag{8}$$

However, in Equation (8), the scale coefficient of s is not symmetric. It means that, there is no reciprocal relation between s (projecting data from monocular visual odometry to stereo visual odometry) and s' (projecting data from stereo visual odometry to monocular visual odometry). So, we rewrite the alignment bias e_i into the following form.

$$e_i = \frac{1}{\sqrt{s}} t'_{sm,i} - \sqrt{s} R t'_{m,i} \tag{9}$$

Then, the Equation (7) turns to:

$$\frac{1}{s} A_{sm} - 2B + sA_m = (\sqrt{s}A_m - \frac{1}{\sqrt{s}} A_{sm})^2 + 2(A_m A_{sm} - B) \tag{10}$$

Then, we can get the new result of scale coefficient of s' and take it as the scale of current monocular visual odometry sliding window end state $t_{m,i+n-1}$:

$$s' = A_{sm} / A_m = (\sum_{i}^{i+n-1} \|t'_{sm,i}\|^2 / \sum_{i}^{i+n-1} \|t'_{m,i}\|^2)^{1/2} \tag{11}$$

By comparing Formulas (8) and (11), we can know that the new Equation (11) can calculate the scale coefficient s' without solving the rotation matrix R. In practice, the scale estimation method is influenced seriously by vehicle motion state, especially stationary state or the changes of motion state. Therefore, we adjust the strategy of scale estimation and correct monocular scale step by step by considering the effect of vehicle motion state.

$$s^*_{m,i+n-1} = (s'_{m,i+n-1} - s^*_{m,i+n-2}) * \lambda * ve_{m,i+n-1} + s^*_{m,i+n-2} \tag{12}$$

In this equation, λ is the scale updating factor, and s^* is the updating scale coefficient of monocular visual odometry. The velocity $ve_{m,i+n-1}$ is computed in the following steps. At first, computing the initial velocity $ve'_{m,i+n-1}$ according to aligning stereo visual odometry position based on the uniform velocity model,

$$ve'_{m,i+n-1} = \frac{\sqrt{(x_{sm,i+n-1} - x_{sm,i+n-2})^2 + (y_{sm,i+n-1} - y_{sm,i+n-2})^2 + (z_{sm,i+n-1} - z_{sm,i+n-2})^2}}{T_{m,i+n-1} - T_{m,i+n-2}} \tag{13}$$

Then, we compare the initial velocity $ve'_{m,i+n-1}$ with the velocity threshold $ve*$ (0.04–0.06 m/s in our evaluation) and get the final velocity ve,

$$ve_{m,i+n-1} = \begin{cases} ve'_{m,i+n-1}, & ve'_{m,i+n-1} >= ve* \\ 0, & ve'_{m,i+n-1} < ve* \end{cases} \tag{14}$$

Finally, the real value trajectory of monocular visual odometry can be calculated using the updated scale coefficient s^*, as shown in the following equation.

$$t_{am,i+n-1} = R_e * s^*_{m,i+n-1} * R_{sm} * Rc * (t_{m,i+n-1} - t_{m,i}) + t_{am,i} \tag{15}$$

In this equation, R_e is the transformation matrix from forward stereo camere to the backward monocular camera. R_{sm} is the transformation matrix for the different initialization time points. R_c is the transformation matrix from the monocualr coordinate $\{C_R\}$ to the coordinate $z_B x_B y_B$ which is opposite to the global coordinate $\{G\}$ on X-O-Y plane.

2.6. Kalman Filter Based Trajectory Fusion

After giving a dynamically updated scale coefficient to the backward monocular visual odometry, the absolute ego-motion estimation can be correctly obtained. Then, in the sliding window, loosely coupled trajectory fusion was operated to get precise ego-motion estimation, using two-layers Kalman filter method.

2.6.1. Prediction Equation and Observation Equation

In the sliding window $[T_i, T_{i+n-1}]$, we first take the ego-motion state from forward stereo camera $\Delta x_{(i,i+n-1)s}$ as the state variable and establish state prediction equation as Formula (14). Then, we take the ego-motion state from backward monocular camera $\Delta x_{(i,i+n-1)m}$ as the observation and establish observation equation as Formula (15).

$$X_{(i,i+n-1)} = I\Delta x_{(i,i+n-1)_s} + w_{i,i+n-1)} \tag{16}$$

$$Z_{(i,i+n-1)} = I\Delta x_{(i,i+n-1)_m} + v_{(i,i+n-1)} \tag{17}$$

Here, $w_{(i,i+n-1)}$ is white Gaussian noise of state prediction equation, with covariance matrix of $Q_{(i,i+n-1)}$; $v_{(i,i+n-1)}$ is white Gaussian noise of observation equation, with covariance matrix of $R_{(i,i+n-1)}$. I is an identity matrix.

2.6.2. Calculation of Covariance Matrix

Accurately determining the covariance matrix Q and R would affect the accuracy of the trajectory fusion results. In the proposed method, the matching error of all image frames in the sliding window is considered during the calculation of visual odometry. We use minimum optimization error of each frame to measure the accuracy of the matching confidence and determine the covariance matrix.

Feature points based visual odometry is applied for both forward stereo camera and backward monocular camera and sparse 3-D map points are also reconstructed by different approaches. Here, we use back projection method for stereo camera and use triangulating method for monocular camera. By tracking the feature pairs between sequential images, the ego-motion estimation can be described as a 3D-to-2D problem. In the problem, the RANSAC method is employed to remove outliers. The 3D-to-2D feature pairs from both cameras are all considered in the same optimization function, shown in Equation (16), which tries to minimize the reprojection error of the images and will concern the rigid constraint of the system.

$$\{R, t\} = \arg\min_{R,t}(e_{sum}) = \arg\min_{R,t} \sum_{i\in\chi} \|p_i - \pi(RP_i + t)\|_\Sigma^2 \tag{18}$$

where, e_{sum} is total reprojection error, R and t are the rotation matrix and translation matrix between successive frames, χ is the number of inliners of the visual odometry, $i \in \chi$ represent the indexes of feature points in the frame that matching the 3D map points. p_i represent the matched feature points in the frame, P_i are the corresponding 3D points in the 3D sparse map space. π is the projection function from 3D to 2D. Σ is the total projection error.

From the optimization function Equation (16), we found that the more accuracy of the matching between 3D sparse map points and the corresponding 2D frame point, the better output of the pose obtained by solving the optimization equation. In practice, for the principle of comparability, we take into account of the number of frames in the sliding window and then we can get the mean reprojection error $\bar{e}_s$ as:

$$\bar{e}_{k\in\{s,m\}} = \frac{\eta}{n * \chi} \sum_{k\in n} \sum_{i\in\chi} \|p_i - \pi(RP_i + t)\|_\Sigma^2 \tag{19}$$

where η is the correction coefficient, n is the number of frames in the sliding window, $k = s$ when it represents stereo visual odometry and $k = m$ for monocular visual odometry. Then, we can use $\bar{e}_s$ as diagonal elements to construct the covariance matrix Q for process noise w and can use $\bar{e}_m$ as diagonal elements to construct the covariance matrix R for measurement noise v.

2.6.3. Two-Layers Kalman Filter Based Trajectory Fusion

After obtaining the parameters of prediction equation and observation equation, along with the covariance matrix Q and R, we can benefit from the data fusion pipeline. Here, we employed two fusion space for the motion propagation process of both the position estimation and orientation estimation.

In the first level of the fusion stage, the relative position $\Delta t = (\Delta tx, \Delta ty, \Delta tz)$ was estimated in the Euclidean linear space. In the sliding window, the Kalman filter was used to construct the prediction equation and observation equation. The detail was shown in Table 1. Where, $\Delta t_{(i,i+n-1)}$ is the estimation of changes of relative position within the sliding window, $t_{f,i}$ and $t_{f,i+n-1}$ are the estimation of ego-motion states. For the Kalman gain $K_{i,i+n-1}$, it is a diagonal matrix and all diagonal elements are same. Hence, we use $k * I$ to represent diagonal matrix K and the k is the coefficient of Kalman gain matrix.

Table 1. Trajectory obtained by our method in the second experiment.

Steps	Discription	Formula
1	calculate the current predicted value according to the prediction equation	$\widetilde{X}_{(i,i+n-1)} = I\Delta x_{(i,i+n-1)_s}$
2	update the covariance matrix of prediction equation	$\widetilde{P}_{(i,i+n-1)} = P_{(i-1,i+n-2)} + Q_{(i,i+n-1)}$
3	calculating kalman gain	$K_{(i,i+n-1)} = P_{(i,i+n-1)}(P_{(i,i+n-1)} + R_{(i,i+n-1)})^{-1}$
4	update the predicted value	$X_{(i,i+n-1)} = \widetilde{X}_{i,i+n-1} + K_{(i,i+n-1)}(Z_{(i,i+n-1)} - \widetilde{X}_{(i,i+n-1)})$
5	update the covariance of the prediction equation	$P_{(i,i+n-1)} = (I - K_{(i,i+n-1)})\widetilde{P}_{(i,i+n-1)}$

In the orientation estimation stage, we proposed an improved Kalman filter, which uses the orientation error space represented by unit quaternion as the state of the filter. In order to construct the motion propagation equation, the state vector of this filter need to be unified with four elements $\Delta q_{(i,i+n-1)} = (qx, qy, qz, qw)$, which can be described as follows:

$$\Delta q_{(i,i+n-1)} = q_{i+n-1} \otimes q_i^{-1} \tag{20}$$

where, $\otimes$ represents the multiplication of quaternion.Then, the estimation of orientation propagation states can be acquired by the following equation:

$$q_{f,i+n-1} = \Delta q_{(i,i+n-1)} \otimes q_{f,i} \tag{21}$$

Here, $\Delta q_{(i,i+n-1)}$ is the estimation of relative orientation changes within the sliding window, $t_{f,i}, t_{f,i+n-1}$ are the estimation of ego-motion states represented by the unit quaternion. In this form, the state propagation model and measurement model would be much simpler. Moreover, the processing of the data fusion occurred in the error space was represented by the error quaternion, which could be closer to linear space and, thus, more suitable to the Kalman filter.

With the proposed trajectory fusion method, the ego-motion states can be computed accurately. When one of the forward-facing stereo camera or backward-facing monocular camera fails, the Kalman filter can automatically update the Kalman gain and reduce the credibility of the impaired camera naturally. It means that the proposed pipeline can fully or partially bypass failure modules and make use of the rest of the pipeline handle sensor degradation.

3. Results

We evaluated the proposed method on the Oxford RobotCar Dataset and compared its performance to some state-of-the-art visual odometry systems. Some key performances, including monocular scale property, robustness capability, and accuracy performance were carried out and comprehensively evaluated.

3.1. Oxford RobotCar Dataset

Oxford RobotCar Dataset [43] was collected by the Mobile Robotics Group, University of Oxford, UK, and it focused on the long-term and large-scale real driving data for autonomous road vehicles, which contains over 1000 km driving data sequences. It was recorded from the car named Nissan LEAF equipping with a Point Grey Bumblebe XB3 trinocular stereo camera, three monocular cameras, three 2D Lidars, GPS and INS.

The reason we chose Oxford RobotCar Dateset as our testing dataset is that it was collected in all weather conditions, including heavy rain, night, direct sunlight and snow, which made it the ideal choice for our evaluation, especially for the robustness and accuracy test. The data in the Oxford RobotCar Dataset are shown in Figure 5.

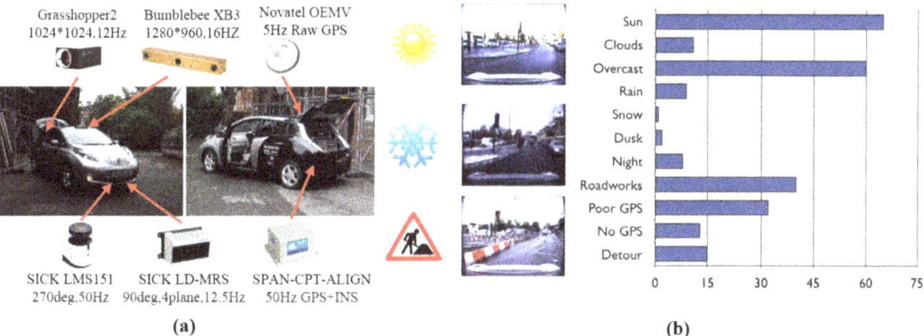

Figure 5. (**a**) shows the Oxford RobotCar platform with the equipped sensors, and (**b**) indicates the recording data sequences number in different conditions [43].

In our experiment setup, the BumbleeXb3 and the rear Grasshopper2 are selected as the forward and backward cameras in the proposed method. For the trinocular BumbleeXB3, the timestamp of three cameras are synchronized by inherent hardware and the left and right cameras are composed for wider baseline (24 cm) as the forward stereo camera. The relative positioning setup is shown in Figure 6 and the more accuracy calibration extrinsics and camera models are provided in the software development kit (SDK) on their website. In addition, the SDK also supplies us with the Matlab and Python functions for demosaicing and undistorting raw Bayer images.

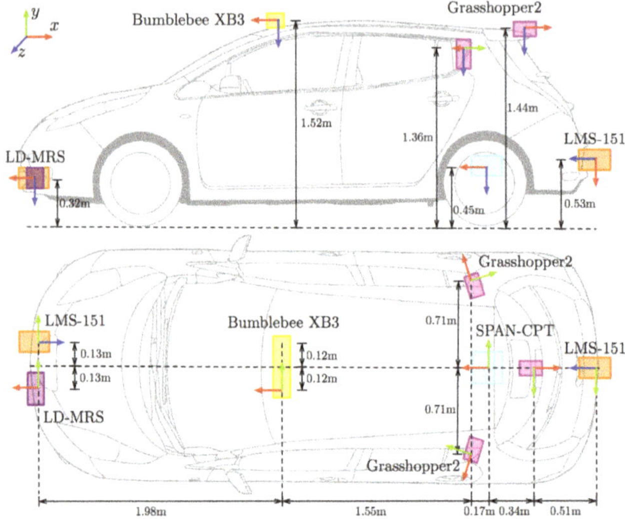

Figure 6. The RobotCar platform and sensor positioning setup.The global coordinates and sensor body coordinates are defined.All sensor extrinsics are provided as se(3) format in their software development kit (SDK) tools [43].

3.2. Evaluation of Scale Estimation Method

Monocular visual odometry always suffers from the scale ambiguity problem due to the unknown ego-motion translation length between frames. Hence, in this paper, real time scale estimation strategy for backward-facing monocular visual odometry was proposed. In the experiment, we show that the traditional scale estimation method always suffers from the vehicle motion state. For example, Figure 7a depicts that the scale coefficient may vary significantly when the ego-vehicle was waiting for a traffic light and kept still. After considering the effect of vehicle motion state in this paper, the scale coefficient can keep steady and smooth, as shown in Figure 7b.

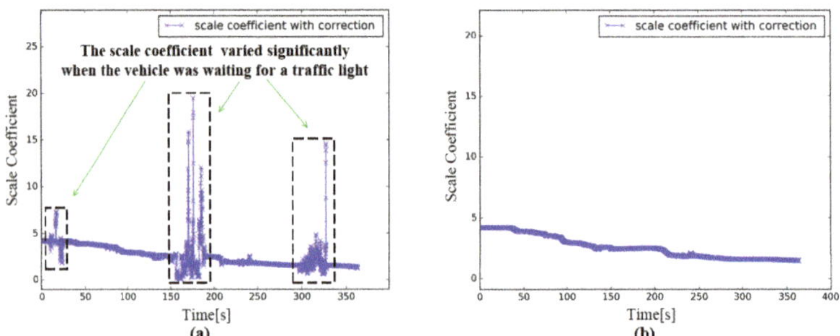

Figure 7. (**a**) shows the result of scale estimation method without considering the effect of vehicle motion state; (**b**) indicates the better result of our proposed method. [Color figure can be viewed at www.mdpi.com].

In order to test the validity of the proposed method, we compared our work with the traditional method proposed by Nist [28], which tends to set the scale of monocular visual odometry to a constant value. As a result, Figure 7 shows that the error caused by the ambiguous scale would continuously accumulated for Nist's method, which makes the system more prone to deviate from the ground truth. In this paper, the sliding window based scale estimation method was proposed to keep system work stable. In the sliding window, the MSE was used to dynamically correct and update the scale coefficient under the rigid body motion assumption. With this help, In Figure 8, we show that our method can significantly improve the performance of monocular visual odometry.

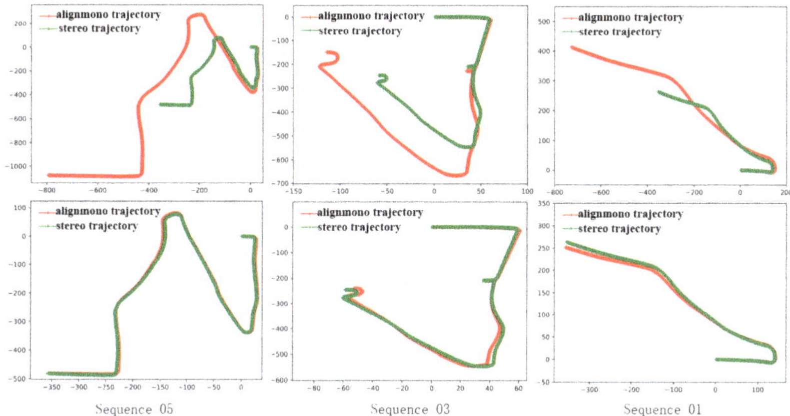

Figure 8. The top pictures show the results of Nist's method with constant scale. By contrast, the bottom pictures shows our method with a better validity and accuracy.

3.3. Robustness Evaluation

Utilizing the proposed fusion method, the system can fully or partially bypass failure modules and make use of the rest pipeline to handle sensor degradation. Figure 9 demonstrates the fusion procedure and Figure 9 depicts the corresponding fusion confidence. As it shows, with our proposed method, the coefficient of fusion confidence would be updated in real-time according to the matching error of all image frames in the sliding window. Even in the sensor performance degradation point, such as in point A in the following Figures, the system can automatically increase the confidence of the rest undamaged camera naturally to guarantee the safety of the system. In addition, the failed single stereoVO reinitializes successfully at point B.

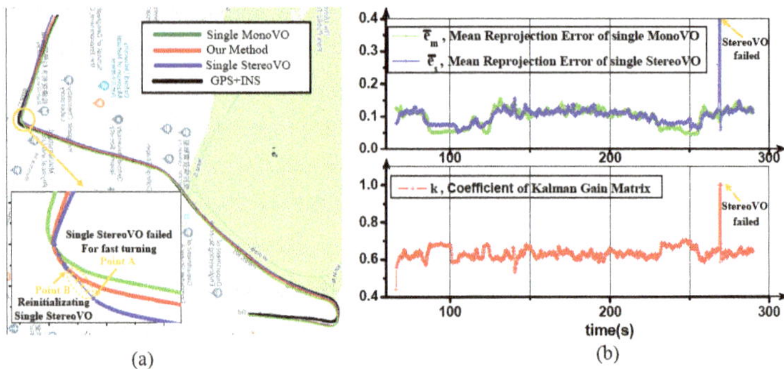

(a) (b)

Figure 9. (**a**) shows that four trajectories, including single MonoVO, single StereoVO, GPS+INS and our method. Single StereoVO fails at point A because of the fast change of scenes and reinitializes at the point B. (**b**) shows the mean reprojection error ($\overline{e}_{k\in\{s,m\}}$) of two visual odometry systems and the coefficient of Kalman gain matrix. The failing of Single StereoVO leads to the abrupt change of blue trajectory at the 289 s. For the coefficient k, it is equal to 1, which means that the system only utilizes the information only from single monocular visual odometry.

Further, In order to test the performance of robustness, we compared the proposed fusion method with Single Stereo-VO (single stereo visual odometry) and Single Mono-VO (monocular visual odometry) methods separately. Figure 10 shows that both the Single Stereo-VO and Single Mono-VO may fail under some complicated driving conditions. In sequence 02 of the Oxford RobotCar data set, the Single Stereo-VO failed due to the strong sunlight, which bring insufficient point features for the matching steps. The situation also happened in the sequence 16 for the Single Mono-VO system. In sequence 09 and sequence 12, the Single Stereo-VO lost effectiveness because of the big and fast corner turning. Where, during the quick turning, large motion between corresponding feature points of images would lead to the matching errors and made the system become invalid.

In our proposed system, the data processing pipeline was carefully designed to handle sensor degradation. Some comparative experiments in Figure 10 show that the proposed system has higher robustness than the single stereo and monocular visual odometry. It can still work even under the harsh conditions in which one of the cameras fails.

In order to test the robustness of the system in harsh environments, we also compared the proposed method with some state-of-the-art works including ORB-SLAM2 [33], DSO-Mono (direct sparse monocular odometry) and DSO-Stereo (direct sparse stereo odometry) [44] algorithms using our collected dataset. Our dataset contains 14 sequences of harsh driving environment, such as rain, dusk, night, snow, direct sunlight, texture-less, intense illumination, bumpy road, and fast turns. The performance of the above mentioned works in the dataset is shown in Table 2. During the experiment, some methods failed due to strong lighting and some failed in turning corners. In total, the DSO monocular method and DSO stereo method failed 3 times and 9 times respectively, and the ORB-SLAM2 failed 7 times. However, our proposed method always kept a good performance for all the testing sequences.

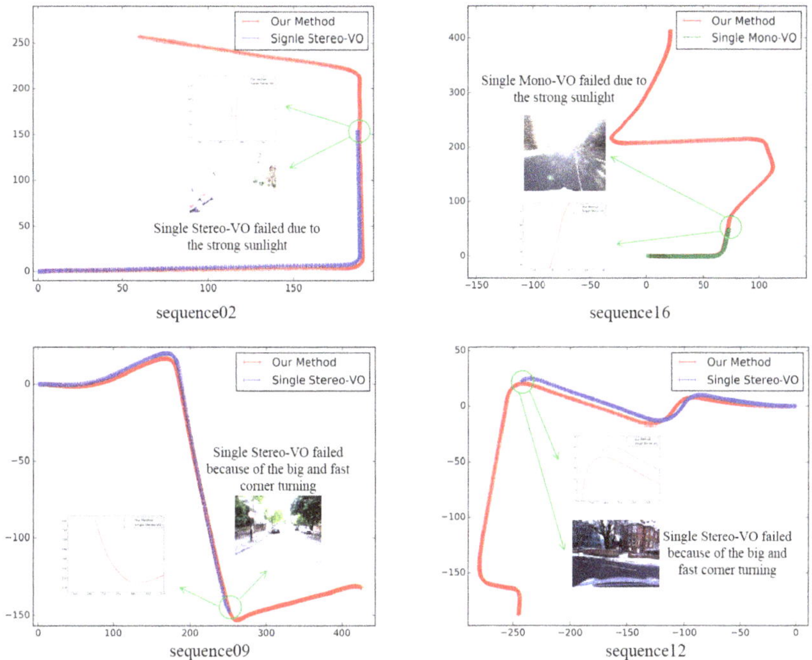

Figure 10. The top two pictures demonstrate that single stereo visual odometry and single monocular visual odometry sometimes might fail in strong sunlight. The bottom two pictures show they might fail at the corner of fast turning. However, our method always kept a good performance for all the testing sequences.

Table 2. Robustness Peformance Comparison of Different Methods in the Oxford Robotcar Dataset.

	Sequence Description	Duaring Time (s)	Frame Numeber	DSO Mono	DSO Stereo	ORBSLAM2 Stereo	OurMethod Fusion
Seq01	sun, traffic light	224	2485	T	F	F	T
Seq02	strong sunlight	206	2267	F	F	F	T
Seq03	ovrecast, sun	190	2096	T	T	T	T
Seq04	rain, overcast	109	1205	T	F	T	T
Seq05	overcast, traffic light	365	4027	T	F	F	T
Seq06	rain, overcast	151	1665	F	F	T	T
Seq07	dusk, rain	183	2020	T	F	T	T
Seq08	overcast, loop road	224	2474	T	T	F	T
Seq09	sun, clouds	90	2690	F	F	F	T
Seq10	night, dark	163	1795	T	F	T	T
Seq11	snow	119	1314	T	T	T	T
Seq12	snow, traffic light	252	2780	T	T	F	T
Seq13	illumination change	152	1672	T	T	T	T
Seq14	strong sunlight	188	2112	T	F	F	T
Total	–	–	–	11/14	5/14	7/14	14/14

3.4. Accuracy Evaluation

The accuracy of the proposed method was also evaluated. Firstly, the absolute translation root mean-square error (RMSE) [45] was employed as the quantitative metric to evaluate the performance of accuracy. Sequence 05 in Oxford RobotCar Dataset was used as the testing data, which was captured by forward-facing Bumblebee XB3 stereo camera and the backward-facing Grasshopper2 camera. The GPS + INS (NovAtel) data was used as the ground truth. In the testing dataset, the speed of ego-vehicle changed quickly and encountered a number of red lights. In order to evaluate the performance, some state-of-the-art monocular and stereo visual odometry methods were compared. In the experiment, the trajectories of monocular visual odometry were aligned to the ground truth using the similarity transformation method for the lacking of scale. The results are shown in Figure 11 and Table 3.

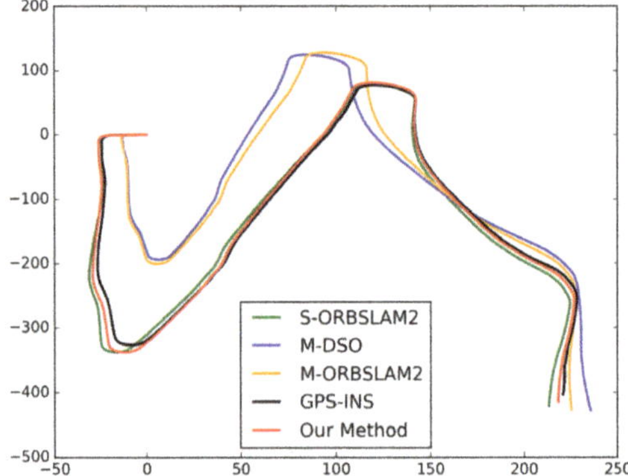

Figure 11. Trajectory obtained by four methods on sequence 05 in Oxofrd RobotCar Datasets.

Table 3. Comparison of RMSE of four methods.

Method	Setting	RMSE (m)	RMSE (%)
S-ORBSLAM2	Stereo	6.81	0.519
M-DSO	Monocular	44.22	3.372
M-ORBSLAM2	Monocular	41.90	3.195
Our fusion Method	Multicamera	5.49	0.419

From the results we can see that the accuracy of the monocular camera based methods of M-DSO and M-ORBSLAM2 obtained the RMSE of 3.372% and 3.195% respectively, which were poor compared to the stereo camera based methods because the monocular visual odometry always suffered from scale ambiguity problem. In contrast, our method output the best accuracy performance among the 4 state-of-the-art methods [33,44,46] with the RMSE of 0.419%, which is lower than the S-ORBSLAM2 method with the RMSE of 0.519%. This was because our method can take advantage of the outstanding features of symmetry-adaption configuration of forward and backward cameras and comprehensively utilizing the nearby environment information.

In the second experiment, we employed the average error (including AEX, AEY and AED) [47] as the quantitative metric to evaluated the performance of accuracy. Among three average error metrics, the assessment data of AED is more important than the others, since AEX and AEY depend on the choice of the coordinate system, while AED is invariant to it. The calculation equation is shown in the following:

$$AEX = \frac{\sum_{i=0}^{N} |X_i - X_i^{GT}|}{N} \quad \text{and} \quad AEY = \frac{\sum_{i=0}^{N} |Y_i - Y_i^{GT}|}{N} \tag{22}$$

$$AED = \frac{\sum_{i=0}^{N} \sqrt{(X_i - X_i^{GT})^2 + (Y_i - Y_i^{GT})^2}}{N} \tag{23}$$

Here, N is the number of frames, X_i^{GT} and Y_i^{GT} are the ground-truth values in X direction and Y direction obtained by the GPS + INS at the ith frame. Xi and Yi are the output results in X direction and Y direction obtained by the corresponding odometry method at the ith frame.

The experiment was carried out in sequence 18 of the Oxford RobotCar Dataset was on a drizzly day. We compared the proposed method with some state-of-the-art algorithms including S-ORBSLAM2 [33] and S-VINS (Stereo Visual-Inertial Systems) [46]. The visual results of the odometry estimations of different methods are shown in Figure 12, which shows the trajectories consisting of each (Xi, Yi) point. The ground-truth trajectories for each (X_i^{GT}, Y_i^{GT}) point were plotted as black line in Figure 12. The results of the corresponding average error (including AEX, AEY, and AED) are also given in Table 4. The results show that our method can achieve the best performance on AED, AEX and AEY among all the methods, which indicates that the output results of our method are the most stable compared with other methods [33,44,46]. In order to clearly demonstrate the results, we mapped the trajectories onto Google earth.

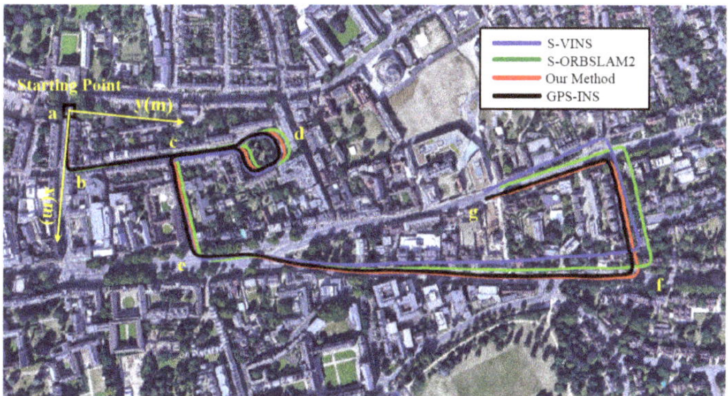

Figure 12. Trajectories of 3 methods comparing with the ground truth on the Google earth map. The results show that our proposed method can have smaller overall errors compared with other state-of-the-art methods [33,44,46].

Table 4. Average Trajectories Errors of Varied Method in the Oxford RobotCar Dataset.

Method	AEX (m)	AEY (m)	AED (m)
S-ORBSLAM2	5.84	12.41	14.30
S-VINS	4.55	12.19	13.89
Our Method	1.83	7.14	7.60

We also present the error curves of all the methods compared with the ground truths from GPS+INS. Figure 13 shows the distance error curve in each frame and the corresponding error for each method. The results show that our proposed method can have smaller overall errors and error boundary compared with other state-of-the-art methods [33,44,46]. So our method is capable of navigating in the real world for kilometers and performs better than other state-of-the-art algorithms.

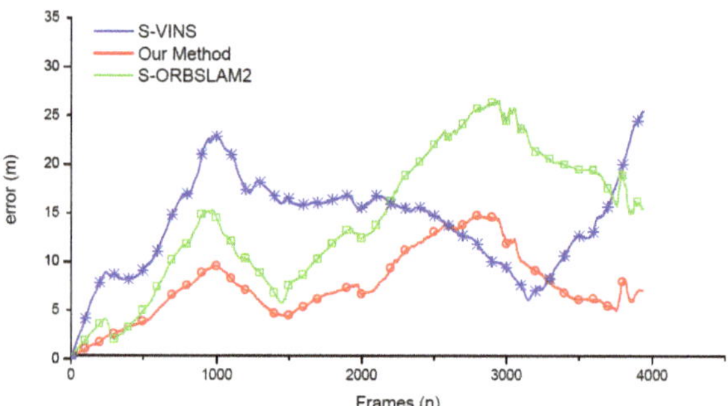

Figure 13. Error curves of varied methods comparing with the ground truth. The horizontal coordinate of the figure was the frame sequences of the camera. The vertical coordinate was the bias value between the ground truth and the estimation in the corresponding frame ($\sqrt{(X_i - X_i^{GT})^2 + (Y_i - Y_i^{GT})^2}$). The curve close to the zero line means the error was small.

3.5. Time Results

In this experiment, we tested all sequences 5 times in a computational platform with an Intel Core i5-7500 CPU (four cores @3.4 Hz) and 8 GB RAM to eliminate the randomness in the results. In addition, we set the extracting features as 2000, image pyramid layers as 8 for both stereo and monocular visual odometry systems. The consuming time results of the main modules of our proposed method are shown in Table 5. Besides, the monocular and stereo visual odometry systems ran at the same time on the distributed robot operating systems. So, the maximum processing frame rate would be decided by the system consuming more time. Therefore, the maximum processing frame rate would be determined by the system that consumes more time. From the table we can see that the total consumption time of our method is about 55 milliseconds with the capability of processing 20 frames per second.

Table 5. Consuming Time Results of Our Proposed Method in Oxford Robotcar Dataset.

Part	Module	Times (ms)
Monocular	Feature Extraction	20.42 ± 4.03
	Pose Tracking	1.54 ± 0.41
	Local Map Tracking	6.02 ± 1.27
	Keyframe Selecting	1.12 ± 0.94
	Total	29.10 ± 6.65
Stereo	Feature Extraction	24.19 ± 3.52
	Stereo Matching	15.73 ± 2.44
	Pose Tracking	2.01 ± 0.34
	Local Map Tracking	8.81 ± 3.12
	Keyframe Selecting	2.72 ± 2.68
	Total	53.46 ± 12.10
Fusion	Scale Computation	0.31 ± 0.12
	KF fusion	0.41 ± 0.17
	Total	0.72 ± 0.29

4. Discussion

Based on the obtained results, as well as proving its efficiency, robustness, and accuracy, it can be seen that our approach can achieve superior performance compared with other start-of -the-art methods. The method can fully utilize the information of driving conditions to promote the performance of our positioning system.

Regarding the scale estimation, the scale ambiguity problem of monocular visual odometry has been solved with the MSE optimization method. We show that, in the sliding window, MSE was used to dynamically correct and update the scale coefficient under the rigid body motion assumption. After considering the effect of vehicle motion state, the scale coefficient can keep steady and smooth, which makes the system not easily deviate from the ground truth.

Regarding the robustness, the proposed system can fully or partially bypass failure modules and make use of the rest pipeline to handle sensor degradation. It can still work even under the harsh conditions that one of the cameras fails. The proposed method was also tested in complex driving environments, such as rain, dusk, night, direct sunlight, texture-less, intense illumination, bumpy road, and fast turns. We show that our method can work well in the above mentioned complex and highly dynamic driving environments.

Regarding the accuracy, our method can utilize the remarkable characteristics of symmetry adaption configuration of forward and backward cameras. Meanwhile, a novel fusion mechanism of two-layers Kalman Fusion based data processing framework was employed to comprehensive utilize the nearby environment information. We compare the proposed method with some state-of-the-art works including M-ORBSLAM2, S-ORBSLAM2 [33], M-DSO, S-DSO [44] and S-VINS [46] algorithms. The results show that our method can achieve the best performance on evaluation indexes of AED, AEX, AEY and RMSE among all the methods, which indicates that the output results of our method are most accuracy compared with other methods.

5. Conclusions

We have presented a novel low-cost visual odometry method for estimating egomotion for ground vehicles in challenging environments. To improve the performance of system robustness and accuracy, the scale factor of backward monocular visual odometry was estimated based on the MSE optimization method in a sliding window. Then, in trajectory estimation, an improved two-layers Kalman filter was proposed including orientation fusion and position fusion. The experiments carried out in this paper have proved that our algorithm is superior to other state-of-the-art algorithms.

Different from traditional localization methods that use differential GPS, precise IMU or 3D Lidar, the proposed method only leverages data from inexpensive cameras. Meanwhile, our fusion system employed the outstanding features of symmetry-adaption configuration of forward and backward cameras and provided a new fusion mechanism of two-layers data processing framework to comprehensive utilize the nearby environment information. Therefore, the proposed pipeline can fully or partially bypass failure modules and make use of the rest pipeline to handle sensor degradation making the system more robustness and accuracy.

In future work, we will further optimize our proposed algorithm, reduce its computation complexity, and try to implement it in the compact embedded platform. We will try to integrate other lowcost sensors, such as conventional low-cost GPS, IMU to our system. Meanwhile, we will also explore the method to adjust accelerometer biases using the output of our system, since the velocity measured by our system should be equal to the velocity integrated from the bias-corrected.

Author Contributions: Conceptualization, K.W.; methodology, K.W.; software, X.H.; validation, X.H.; investigation, J.C., L.C.; resources, C.C.; data curation, Z.X.; writing—original draft preparation, K.W.; writing—review and editing, K.W.; funding acquisition, Z.X., L.C.

Funding: This research was funded by National Natural Science Foundation of China (Grant No. 51605054), Key Technical Innovation Projects of Chongqing Artificial Intelligent Technology (Grant No. cstc2017rgzn-zdyfX0039), The Science and Technology Research Program of Chongqing Education Commission of China (No. KJQN201800517 and No. KJQN201800107).

Acknowledgments: Authors thank Jianbo Lu, Weijian Han from Ford Motor Company, USA for useful discussions.

Conflicts of Interest: The authors declare no conflict of interest.

Abbreviations

The following abbreviations are used in this manuscript:

VO	Visual Odometry
GPS	Global Positioning System
IMU	Inertial Measurement Unit
MSE	Mean Square Error
NASA	National Aeronautics and Space Administration
SFM	Structure from Motion
RANSAC	Random Sample Consensus
BA	Bundle Adjustment optimization approach
LC	Loop Closure Method
RMSE	Root Mean Ssqare Error
AEX	Average in X Direction
AEY	Average in Y Direction
AED	Average in Distance

References

1. Gluckman, J.; Nayar, S.K. Ego-Motion and Omnidirectional Cameras. In Proceedings of the International Conference on Computer Vision, Bombay, India, 7 January 1998.
2. Gabriele, L.; Maria, S.A. Extended Kalman Filter-Based Methods for Pose Estimation Using Visual, Inertial and Magnetic Sensors: Comparative Analysis and Performance Evaluation. *Sensors* **2013**, *13*, 1919–1941.
3. Wang, K.; Xiong, Z.B. Visual Enhancement Method for Intelligent Vehicle's Safety Based on Brightness Guide Filtering Algorithm Thinking of The High Tribological and Attenuation Effects. *J. Balk. Tribol. Assoc.* **2016**, *22*, 2021–2031.

4. Chen, J.L.; Wang, K.; Bao, H.H.; Chen, T. A Design of Cooperative Overtaking Based on Complex Lane Detection and Collision Risk Estimation. *IEEE Access.* **2019**, 87951–87959. [CrossRef]
5. Wang, K.; Huang, Z.; Zhong, Z.H. Simultaneous Multi-vehicle Detection and Tracking Framework with Pavement Constraints Based on Machine Learning and Particle Filter Algorithm. *Chin. J. Mech. Eng.* **2014**, *27*, 1169–1177. [CrossRef]
6. Song, G.; Yin, K.; Zhou, Y.; Cheng, X. A Surveillance Robot with Hopping Capabilities for Home Security. *IEEE Trans. Consum. Electron.* **2010**, *55*, 2034–2039. [CrossRef]
7. Ciuonzo, D.; Buonanno, A.; D'Urso, M.; Palmieri, F.A.N. Distributed Classification of Multiple Moving Targets with Binary Wireless Sensor Networks. In Proceedings of the International Conference on Information Fusion, Chicago, IL, USA, 5–8 July 2011.
8. Kriechbaumer, T.; Blackburn, K.; Breckon, T.P.; Hamilton, O.; Rivas, C.M. Quantitative Evaluation of Stereo Visual Odometry for Autonomous Vessel Localisation in Inland Waterway Sensing Applications. *Sensors* **2015**, *15*, 31869–31887. [CrossRef] [PubMed]
9. Zhu, J.S.; Li, Q.; Cao, R.; Sun, K.; Liu, T.; Garibaldi, J.M.; Li, Q.Q.; Liu, B.Z.; Qiu, G.P. Indoor Topological Localization Using a Visual Landmark Sequence. *Remote Sens.* **2019**, *11*, 73. [CrossRef]
10. Perez-Grau, F.J.; Ragel, R.; Caballero, F.; Viguria, A.; Ollero, A. An architecture for robust UAV navigation in GPS-denied areas. *J. Field Robot.* **2018**, *35*, 121–145. [CrossRef]
11. Yang, G.C.; Chen, Z.J.; Li, Y.; Su, Z.D. Rapid Relocation Method for Mobile Robot Based on Improved ORB-SLAM2 Algorithm. *Remote Sens.* **2019**, *11*, 149. [CrossRef]
12. Li, Y.; Ruichek, Y. Occupancy Grid Mapping in Urban Environments from a Moving On-Board Stereo-Vision System. *Sensors* **2014**, *14*, 10454–10478. [CrossRef]
13. Scaramuzza, D.; Fraundorfer, F. Visual Odometry [Tutorial]. *Robot. Autom. Mag. IEEE* **2011**, *18*, 80–92. [CrossRef]
14. Chen, J.L.; Wang, K.; Xiong, Z.B. Collision probability prediction algorithm for cooperative overtaking based on TTC and conflict probability estimation method. *Int. J. Veh. Des.* **2018**, *77*, 195–210. [CrossRef]
15. Yang, N.; Wang, R.; Gao, X.; Cremers, D. Challenges in Monocular Visual Odometry: Photometric Calibration, Motion Bias and Rolling Shutter Effect. *IEEE Robot. Autom. Lett.* **2017**, 3, 2878–2885. [CrossRef]
16. Mou, X.Z.; Wang, H. Wide-Baseline Stereo-Based Obstacle Mapping for Unmanned Surface Vehicles. *Sensors* **2018**, *18*, 1085. [CrossRef] [PubMed]
17. Scaramuzza, D. 1-Point-RANSAC Structure from Motion for Vehicle-Mounted Cameras by Exploiting Non-holonomic Constraints. *Int. J. Comput. Vis.* **2011**, *95*, 74–85. [CrossRef]
18. Zhang, J.; Singh, S. Laser-visual-inertial odometry and mapping with high robustness and low drift. *J. Field Robot.* **2018**, *35*, 1242–1264. [CrossRef]
19. Siddiqui, R.; Khatibi, S. Robust visual odometry estimation of road vehicle from dominant surfaces for large-scale mapping. *IET Intell. Transp. Syst.* **2014**, *9*, 314–322. [CrossRef]
20. Ji, Z.; Singh, S. Visual-Lidar Odometry and Mapping: Low-Drift, Robust, and Fast. In Proceedings of the IEEE International Conference on Robotics and Automation, Seattle, WA, USA, 26–30 May 2015.
21. Demaeztu, L.; Elordi, U.; Nieto, M.; Barandiaran, J.; Otaegui, O. A temporally consistent grid-based visual odometry framework for multi-core architectures. *J. Real Time Image Process.* **2015**, *10*, 759–769. [CrossRef]
22. Longuet-Higgins, H.C. A computer algorithm for reconstructing a scene from two projections. *Nature* **1981**, *293*, 133–135. [CrossRef]
23. Harris, C.G.; Pike, J.M. 3D positional integration from image sequences. *Image Vis. Comput.* **1988**, *6*, 87–90. [CrossRef]
24. Maimone, M.W.; Cheng, Y.; Matthies, L. Two years of Visual Odometry on the Mars Exploration Rovers. *J. Field Robot.* **2010**, *24*, 169–186. [CrossRef]
25. Lategahn, H.; Stiller, C. Vision-Only Localization. *IEEE Trans. Intell. Transp. Syst.* **2014**, *15*, 1246–1257. [CrossRef]
26. Hasberg, C.; Hensel, S.; Stiller, C. Simultaneous Localization and Mapping for Path-Constrained Motion. *IEEE Trans. Intell. Transp. Syst.* **2012**, *13*, 541–552. [CrossRef]
27. Fraundorfer, F.; Scaramuzza, D. Visual Odometry: Part II: Matching, Robustness, Optimization, and Applications. *IEEE Robot. Autom. Mag.* **2012**, *19*, 78–90. [CrossRef]

28. Nistér, D.; Naroditsky, O.; Bergen, J.R. Visual odometry for ground vehicle applications. *J. Field Robot.* **2010**, *23*, 3–20. [CrossRef]
29. Scaramuzza, D.; Fraundorfer, F.; Siegwart, R. Real-Time Monocular Visual Odometry for on-Road Vehicles with 1-Point RANSAC. In Proceedings of the IEEE International Conference on Robotics and Automation, Kobe, Japan, 12–17 May 2009.
30. Forster, C.; Carlone, L.; Dellaert, F.; Scaramuzza, D. On-Manifold Preintegration for Real-Time Visual-Inertial Odometry. *IEEE Trans. Robot.* **2017**, *33*, 1–21. [CrossRef]
31. Pascoe, G.; Maddern, W.; Tanner, M.; Piniés, P.; Newman, P. Nid-Slam: Robust Monocular Slam Using Normalised Information Distance. In Proceedings of the IEEE Conference on Computer Vision and Pattern Recognition, Honolulu, HI, USA, 21–26 July 2017; pp. 1435–1444.
32. Nister, D.; Naroditsky, O.; Bergen, J. Visual Odometry. In Proceedings of the IEEE Computer Society Conference on Computer Vision and Pattern Recognition, Washington, DC, USA, 27 June–2 July 2004.
33. Mur-Artal, R.; Tardos, J.D. ORB-SLAM2: An Open-Source SLAM System for Monocular, Stereo, and RGB-D Cameras. *IEEE Trans. Robot.* **2017**, *33*, 1255–1262. [CrossRef]
34. Taylor, C.J.; Kriegman, D.J. Structure and motion from line segments in multiple images. *Pattern Anal. Mach. Intell. IEEE Trans.* **1995**, *17*, 1021–1032. [CrossRef]
35. Wong, K.Y.K.; Mendonça, P.R.S.; Cipolla, R. Structure and motion estimation from apparent contours under circular motion. *Image Vis. Comput.* **2002**, *20*, 441–448. [CrossRef]
36. Pradeep, V.; Lim, J. Egomotion Using Assorted Features. In Proceedings of the IEEE Computer Society Conference on Computer Vision and Pattern Recognition, San Francisco, CA, USA, 13–18 June 2010.
37. David, N. An efficient solution to the five-point relative pose problem. *IEEE Trans. Pattern Anal. Mach. Intell.* **2004**, *26*, 756–770.
38. Haralick, B.M.; Lee, C.N.; Ottenberg, K.; Nölle, M. Review and analysis of solutions of the three point perspective pose estimation problem. *Int. J. Comput. Vis.* **1994**, *13*, 331–356. [CrossRef]
39. Song, Y.; Nuske, S.; Scherer, S. A Multi-Sensor Fusion MAV State Estimation from Long-Range Stereo, IMU, GPS and Barometric Sensors. *Sensors* **2017**, *17*, 11. [CrossRef] [PubMed]
40. Khan, N.H.; Adnan, A. Ego-motion estimation concepts, algorithms and challenges: An overview. *Multimed. Tools Appl.* **2017**, *76*, 16581–16603. [CrossRef]
41. Liu, Y.; Chen, Z.; Zheng, W.J.; Wang, H.; Liu, J.G. Monocular Visual-Inertial SLAM: Continuous Preintegration and Reliable Initialization. *Sensors* **2017**, *17*, 2613. [CrossRef] [PubMed]
42. Zhang, Z. A flexible new technique for camera calibration. *IEEE Trans. Pattern Anal. Mach. Intel.* **2002**, *22*, 1330–1334. [CrossRef]
43. Maddern, W.; Pascoe, G.; Linegar, C.; Newman, P. 1 year, 1000 km: The Oxford RobotCar dataset. *Int. J. Robot. Res.* **2017**, *36*, 3–15. [CrossRef]
44. Engel, J.; Koltun, V.; Cremers, D. Direct Sparse Odometry. *IEEE Trans. Pattern Anal. Mach. Intell.* **2018**, *40*, 611–625. [CrossRef]
45. Sturm, J.; Engelhard, N.; Endres, F.; Burgard, W.; Cremers, D. A Benchmark for the Evaluation of RGB-D SLAM Systems. In Proceedings of the IEEE/RSJ International Conference on Intelligent Robots and Systems, Vilamoura, Portugal, 7–12 October 2012.
46. Qin, T.; Pan, J.; Cao, S.; Shen, S. A General Optimization-based Framework for Local Odometry Estimation with Multiple Sensors. *arXiv* **2019**, arXiv:1901.03638v1.
47. Yong, L.; Rong, X.; Yue, W.; Hong, H.; Xie, X.; Liu, X.; Zhang, G. Stereo Visual-Inertial Odometry with Multiple Kalman Filters Ensemble. *IEEE Trans. Ind. Electron.* **2016**, *63*, 6205–6216.

Article

Feasibility of Using Grammars to Infer Room Semantics

Xuke Hu [1], Hongchao Fan [2,*], Alexey Noskov [1], Alexander Zipf [1], Zhiyong Wang [3] and Jianga Shang [4,5]

1 GIScience Research Group, Institute of Geography, Heidelberg University, 69120 Heidelberg, Germany
2 Department of Civil and Environmental Engineering, Norwegian University of Science and Technology, 7491 Trondheim, Norway
3 Department of Human Geography and Spatial Planning, Faculty of Geosciences, Utrecht University, 3584 Utrecht, The Netherlands
4 School of Geography and Information Engineering, China University of Geosciences, Wuhan 430074, China
5 National Engineering Research Center for Geographic Information System, Wuhan 430074, China
* Correspondence: hongchao.fan@ntnu.no

Received: 13 May 2019; Accepted: 26 June 2019; Published: 28 June 2019

Abstract: Current indoor mapping approaches can detect accurate geometric information but are incapable of detecting the room type or dismiss this issue. This work investigates the feasibility of inferring the room type by using grammars based on geometric maps. Specifically, we take the research buildings at universities as examples and create a constrained attribute grammar to represent the spatial distribution characteristics of different room types as well as the topological relations among them. Based on the grammar, we propose a bottom-up approach to construct a parse forest and to infer the room type. During this process, Bayesian inference method is used to calculate the initial probability of belonging an enclosed room to a certain type given its geometric properties (e.g., area, length, and width) that are extracted from the geometric map. The approach was tested on 15 maps with 408 rooms. In 84% of cases, room types were defined correctly. It, to a certain degree, proves that grammars can benefit semantic enrichment (in particular, room type tagging).

Keywords: indoor mapping; room type tagging; semantic enrichment; grammar; Bayesian inference

1. Introduction

People spend most of their time indoors, such as offices, houses, and shopping malls [1]. New indoor mobile applications are being developed at a phenomenal rate, covering a wide range of indoor social scenarios, such as indoor navigation and location-enabled advertisement [2]. Semantically-rich indoor maps that contain the usage of rooms (e.g., office, restaurant, or book shop) are indispensable parts of indoor location-based services [3,4]. The floor plans modelled in computer-aided design (CAD), building information modeling (BIM)/industrial foundation classes (IFC), and GIS systems (e.g., ArcGIS and Google Maps) contain rich semantic information, including the type or function of rooms. However, only a small fraction of millions of indoor environments is mapped [5], let alone the type of rooms.

Currently, two mainstream indoor mapping methods include digitalization-based and measurement-based. The first provides digitized geometric maps comprising rooms, corridors, and doors extracted automatically from existing scanned maps [6–9]. Normally, it is incapable of extracting the room type information from the scanned map. According to the type of measurements, we can further divide the second group of approaches into three categories: LIDAR point cloud [10–13], image [14–16], and volunteers' trace [5,17]. LIDAR point cloud and image-based approaches can reconstruct accurate 3-D scenes that contain rich semantics, such as walls, windows, ceilings, doors, floors, and even the type of furniture in rooms (e.g., sofa, chairs, and desks) [11,18] but ignore the type of rooms. Utilizing volunteers' traces to reconstruct indoor maps has received much attention due to its low requirement on hardware and low computational complexity compared to LIDAR point cloud and image-based approaches. With the help of abundant traces, it can detect accurate geometric information (e.g., the dimension of rooms and corridors) and simple semantic information (e.g., stairs and doors). However, it is difficult to infer the type of rooms based on traces. Briefly, current indoor mapping approaches can detect accurate geometric information (e.g., dimension of rooms) and partial semantics (e.g., doors and corridors) but are incapable of detecting the room type (i.e., for digitalization-based and trace-based approaches) or ignore this issue (i.e., image and LIDAR point cloud-based approaches). To solve this problem, [19] proposed using a statistical relational learning approach to reason the type of rooms as well as buildings at schools and office buildings. [3,20] used check-in information to automatically identify the semantic labels of indoor venues in malls, i.e., business names (e.g., Starbucks) or categories (e.g., restaurant). However, it is problematic when check-in information is unavailable in indoor venues.

This work takes research buildings (e.g., laboratories and office buildings at universities) as examples, investigating the feasibility of using grammars to infer the type of rooms based on the geometric maps that can be obtained through the aforementioned reconstruction approaches. The geometric map we use contains the geometric information of rooms and simple semantic information (i.e., corridors and doors). We must reckon that it is impossible to manually construct a complete and reliable grammar that can represent the layout of the research buildings all over the world, which is not the aim of this work. Our goal is to prove that to a certain extent, grammars can benefit current indoor mapping approaches, at least the digitalization-based and traces-based methods, by providing the room type information. As for the creation of complete and reliable grammars, we plan to use grammatical inference techniques [21,22] to automatically learn a probabilistic grammar based on abundant training data in the future.

In this work, we use grammars to represent the topological and spatial distribution characteristics of different room types and use the Gaussian distribution to model the geometric characteristics of different room types. They are combined to infer the type of rooms. Grammar rules are mainly derived from guidebooks [23–26] about the design principles of research buildings. The idea is based on two assumptions: (1) Different room types follow certain spatial distribution characteristics and topological principles. For instance, offices are normally adjacent to external walls. Two offices are connected through doors. Multiple adjacent labs are clustered without being separated by other room types, such as toilets and copy rooms. (2) Different room types vary in geometric properties. For instance, a lecture room is normally much larger than a private office. We assume that the geometric properties (e.g., area, length, and width) of each room type follow the Gaussian distribution.

The input of the proposed approach is the geometric map of a single floor of a building without the room type information. The procedure of the proposed approach is as follows: We first obtain the frequency and the parameters of the multivariate Gaussian distribution of each room type from training rooms. Then, we improve the rules defined in the previous work [27] by removing a couple of useless rules and adding a couple of useful rules in semantic inference and changing the format of rules for the purpose of generating models to the format of rules for sematic inference. The next step is to partition these rules into multiple layers based on their dependency relationship. Then, we apply rules in primitive objects of each test floor-plan from the lowest layer to the highest layer to construct a parse forest. When applying rules at the lowest layer, Bayesian inference is used to calculate the initial probability of assigning an enclosed room with a certain type based on its geometric properties (e.g., area, length, and width) that are extracted from the geometric map. Low-ranking candidate types are removed to avoid the exponential growth of the parse trees. The constructed forest includes multiple parse trees with each corresponding to a complete semantic interpretation of the entire primitive rooms. The main contributions of this work include two parts:

(1) To the best of our knowledge, this is the first time inferring the types of rooms by using grammars given geometric maps.
(2) To a certain degree, we prove that grammars can benefit semantic enrichment.

The remainder of this paper is structured as follows: In Section 2, we present a relevant literature review. We introduce the semantic division of research buildings and the defined rules of the constraint attribute grammar in Section 3. In Section 4, we present the workflow and the details of each step of the proposed approach. In Section 5, we evaluate our approach using 15 floor-plans and discuss some issues in Section 6. We conclude the paper in Section 7.

2. Related Works

Models for indoor space. Currently, the mainstream geospatial standards that may cover indoor space and describe the spatial structure and semantics include CAD, BIM/IFC, city geography markup language (CityGML), and IndoorGML. CAD is normally used in the process of building construction, representing the geometric size and orientation of buildings' indoor entities. It uses the color and thickness of lines to distinguish different spatial entities. Apart from the notes related to indoor spaces, no further semantic information is represented in CAD. Compared to CAD, BIM is capable of restoring both geometric and rich semantic information of building components as well as their relationships [28]. It enables multi-schema representations of 3D geometry for indoor entities. The IFC is a major data exchange standard in BIM. It aims to facilitate the information exchange among stakeholders in AEC (architecture, construction, and engineering) industry [29]. Different from IFC, CityGML [30] is developed from a geospatial perspective. It defines the classes and relations for the most relevant topographic objects such as buildings, transportation, and vegetation in cities with respect to their geometrical, topological, semantic, and appearance properties [31]. CityGML has five levels of detail (LoDs), each for a different purpose. Particularly, the level of detail 4 (LoD 4) is defined to support the interior objects in buildings, such as doors, windows, ceiling, and walls. The only type for indoor space is room, which is surrounded by surfaces. However, they lack features related to indoor space model, navigation network, and semantics for indoor space, which are critical requirements of most applications of indoor spatial information [32]. In order to meet the requirements, IndoorGML is published by OGC (open geospatial consortium) as a standard data model and XML-based exchange format. It includes geometry, symbolic space, and network topology [33]. The basic goals of IndoorGML are to provide a common framework of semantic, topological, and geometric models for indoor spatial information, allowing for locating stationary or mobile features in indoor space and to provide spatial information services referring their positions in indoor space, instead of representing building architectural components.

Digitalization-based indoor modeling. The classical approach of parsing the scanned map or the image of floor-plans consists of two stages: Primitive detection and semantics recognition [6,34–37]. It starts from low-level image processing: Extracting the geometric primitives (i.e., segments and arcs) and vectorizing these primitives. Then, to identify the semantic classes of indoor spatial elements (e.g., walls, doors, windows, and furniture). In recent years, machine-learning techniques have been applied to detect the semantic classes (e.g., room, doors, and walls). For instance, de las Heras et al. [7,8,38] presented a segmentation-based approach that merges the vectorization and identification of indoor elements into one procedure. Specifically, it first tiles the image of floor plans into small patches. Then, specific feature descriptors are extracted to represent each patch in the feature space. Based on the extracted features, classifiers such as SVM can be trained and then used to predict the class of each patch. With the rapid development of deep learning in computer vision, deep neural networks have also been applied in parsing the image of floor plans. For instance, Dodge et al. [9] adopted the segmentation-based approach and fully convolutional network (FCN) to segment the pixels of walls. The approach achieves a high identification accuracy without adjusting parameters for different styles. Overall, the digitalization-based approach is useful considering the existence of substantial images of floor-plans. However, it is incapable of identifying the type of rooms if the image contains no text information that indicates the type of rooms.

Image-based indoor modeling. Image-based indoor modeling approaches can capture accurate geometric information by using smartphones. The advent of depth camera (RGB-D) further improves the accuracy and enables capturing rich semantics in indoor scenes. For instance, Sankar and Seitz [39] proposed an approach for modeling the indoor scenes including offices and houses by using cameras and inertial sensors equipped on smartphones. It allows users to create an accurate 2D and 3D model based on simple interactive photogrammetric modeling. Similarly, Pintore and Gobbetti [40] proposed generating metric scale floor plans based on individual room measurements by using commodity smartphones equipped with accelerometers, magnetometers, and cameras. Furukawa et al. [14] presented a Manhattan-world fusion technique for the purpose of generating floor plans for indoor scenes. It uses structure-from-motion, multiview stereo (MVS), and a stereo algorithm to generate an axis-aligned depth map, which is then merged with MVS points to generate a final 3D model. The experimental results of different indoor scenes are promising. Tsai et al. [41] proposed using motion cues to compute likelihoods of indoor structure hypotheses, based on simple geometric knowledge about points, lines, planes, and motion. Specifically, a Bayesian filtering algorithm is used to automatically discover 3-D lines from point features. Then, they are used to detect planar structures that forms the final model. Ikehata et al. [16] presented a novel 3D modeling framework that reconstructs an indoor scene from panorama RGB-D images and structure grammar that represents the semantic relation between different scene parts and the structure of the rooms. In the grammar, a scene geometry is represented as a graph, where nodes correspond to structural elements such as rooms, walls, and objects. However, these works focused on capturing mainly the geometric layout of rooms without semantic representation. To enrich the semantics of reconstructed indoor scenes, Zhang et al. [18] proposed an approach to estimate both the layout of rooms as well as the clutter (e.g., furniture) that compose the scene by using both appearance and depth features from RGB-D Sensors. Briefly, image-based approaches can accurately reconstruct the geometric model and even the objects in the scene, but they normally dismiss the estimation of room types.

Trace-based indoor modeling. Trace-based solutions assume that users' traces reflect accessible spaces, including unoccupied internal spaces, corridors, and halls. With enough traces, they can infer the shape of rooms, corridors, and halls. For instance, Alzantot and Youssef [4], Jiang et al. [42], and Gao et al. [5] used volunteers' motion traces and the location of landmarks derived from inertial sensor data or Wi-Fi to determine the accurate shape of rooms and corridors. However, the edge of a room sometimes is blocked by furniture or other obstacles. Users' traces could not cover these places, leading to inaccurate detection of room shapes. To resolve this problem, Chen et al. [43] proposed a CrowdMap system that combines crowdsourced sensory and images to track volunteers. Based on images and estimated motion traces, it can create an accurate floor plan. Recently, Gao et al. [44] proposed a Knitter system that can fast construct the indoor floor plan of a large building by a single random user's one-hour data collection efforts. The core part of the system is a map fusion framework. It combines the localization result from images, the traces from inertial sensors, and the recognition of landmarks by using a dynamic Bayesian network. Trace-based approaches can recognize partial semantic information, such as corridors, stairs, and elevators, but without the definition of room types.

LIDAR point cloud-based indoor modeling. These methods achieve a higher geometric accuracy of rooms than trace-based methods. For instance, Mura [45] et al. proposed reconstructing a clean architectural model for a complex indoor environment with a set of cluttered 3D input scans. It is able to reconstruct a room graph and accurate polyhedral representation of each room. Another work concerned mainly recovering semantically rich 3D models. For instance, Xiong et al. [10] proposed a method to automatically converting the raw 3D point data into a semantically rich information model. The points are derived from a laser scanner located at multiple locations throughout a building. It mainly models the structural components of an indoor environment, such as walls, floors, ceilings, windows, and doorways. Ambruş et al. [13] proposed an automatic approach to reconstructing a 2-D floor plan from raw point cloud data using 3D point information. They can achieve accurate and robust detection of building structural elements (e.g., wall and opening) by using energy minimization. One of the novelties of the approach is that it does not rely on viewpoint information and Manhattan frame assumption. Nikoohemat et al. [46] proposed using mobile laser scanners for data collection. It can detect openings (e.g., windows and doors) in cluttered indoor environments by using occlusion reasoning and the trajectories from the mobile laser scanners. The outcomes show that using structured learning methods for semantic classification is promising. Armeni et al. [11] proposed a new approach to semantic parsing of large-scale colored point clouds of an entire building using a hierarchical approach: Parsing point clouds into semantic spaces and then parsing those spaces into their structural (e.g., floor, walls, etc.) and building (e.g., furniture) elements. It can capture rich semantic information that includes not only walls, floors, and rooms, but also the furniture in the room, such as chairs, desks, and sofas. Qi et al. [12] proposed a multilayer perceptron (MLP) architecture named PointNet on point clouds for 3D classification and segmentation. It extracts a global feature vector from a 3D point and processes each point using the extracted feature vector and additional point level transformations. PointNet is a unified architecture that directly takes point clouds as input and outputs either class labels for the entire input or per point segment/part labels for each point of the input. Their method operates at the point level and, thus, inherently provides a fine-grained segmentation and highly accurate semantic scene understanding. Similar to the image-based approaches, they normally dismissed the estimation of room types.

Rule-based indoor modeling. This group of approaches uses the structural rules or knowledge of a certain building type to assist the reconstruction of maps. The rules can be gained through manual definitions [27,47–49] or machine learning techniques [19,50–52]. Yue et al. [48] proposed using a shape grammar that represents the style of Queen Anne House to reason the interior layout of residential houses with the help of a few observations, such as footprints and the location of windows. The work in [49] used split grammars to describe the spatial structures of rooms. The grammar rules of one floor can be learned automatically from reconstructed maps and then be used to derive the layout of the other floors. In this way, fewer sensor data are needed to reconstruct the indoor map of a building. However, the defined grammar consists of mainly splitting rules, producing geometric structure of rooms rather than rooms with semantic information. Similarly, Khoshelham and Díaz-Vilariño [53] used a shape grammar [54] to reconstruct indoor maps that contain walls, doors, and windows. The collected point clouds can be used to learn the parameters of rules. Dehbi et al. [55] proposed learning weighted attributed context-free grammar rules for 3D building reconstruction. They used support vector machines to generate a weighted context-free grammar and predict structured outputs such as parse trees. Then, based on a statistical relational learning method using Markov logic networks, the parameters and constraints for the grammar can be obtained. Rosser et al. [50] proposed learning the dimension, orientation, and occurrence of rooms from true floor plans of residential houses. Based on this, a Bayesian network is built to estimate room dimensions and orientations, which achieves a promising result. Luperto et al. [52] proposed a semantic mapping system that classifies rooms of indoor environments considering typology of buildings where a robot is operating. More precisely, they assume that a robot is moving in a building with a known typology, and the proposed system employs classifiers specific for that typology to semantically label rooms (small room, medium room, big room, corridor, hall.) identified from data acquired by laser range scanners. Furthermore, Luperto et al. [19] proposed using a statistical relational learning approach for global reasoning on the whole structure of buildings (e.g., office and school buildings). They assessed the potential of the proposed approach in three applications: Classification of rooms, classification of buildings, and validation of simulated worlds. Liu et al. [56] proposed a novel approach for automatically extracting semantic information (e.g., rooms and corridors) from more or less preprocessed sensor data. They propose to do this by means of a probabilistic generative model and MCMC-based reasoning techniques. The novelty of the approach is that they construct an abstracted semantic and top-down representation of the domain under consideration: A classical indoor environment consisting of several rooms, that are connected by doorways. Similarly, Liu et al. [57] proposed a generalizable knowledge framework for data abstraction. Based on the framework, the robot can reconstruct a semantic indoor model. Specifically, the framework is implemented by combining Markov logic networks and data-driven MCMC sampling. Based on MLNs, we formulate task-specific context knowledge as descriptive soft rules. Experiments on real world data and simulated data confirm the usefulness of the proposed framework.

3. Formal Representation of Layout Principles of Research Buildings

3.1. Definition of Research Buildings

Research buildings are the core buildings at universities, including laboratories [26] and office buildings. Specifically, laboratories refer to the academic laboratories of physical, biological, chemical, and medical institutes. They have a strict requirement on the configuration of labs [21]. According to [21], we categorize the enclosed rooms in research buildings into 11 types: Labs, lab support spaces, offices, seminar/lecture rooms, computer rooms, libraries, toilets, copy/print rooms, storage rooms, lounges, and kitchens. Labs refer to the standard labs that normally follow the module design principle [24,25] and are continuously occupied throughout working days. Thus, they are located in naturally lit perimeter areas. Lab support spaces consist of two parts. One is the specialist or non-standard laboratories, which do not adopt standard modules and are generally not continuously occupied. Therefore, they may be planned in internal areas. The other is ancillary spaces that support labs, such as equipment rooms, instrument rooms, cold rooms, glassware storage, and chemical storage [25].

This work focuses on the typical research building, which refers to those research buildings whose layouts are corridor based. Figure 1 shows the three types of layouts of typical research buildings based on the layout of corridors: Single-loaded, double-loaded, and triple-loaded [24], as shown in Figure 1. Most research buildings are the variations of them.

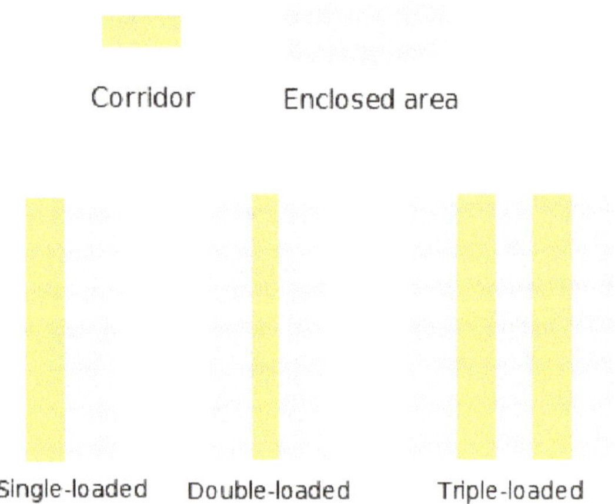

Figure 1. Three typical plans of research buildings.

3.2. Hierarchical Semantic Division of Research Buildings

We use a UML class diagram to represent the hierarchical semantic division of research buildings, as shown in Figure 2. Note that all the defined objects in the diagram are based on one floor of a building. We ignore the multi-level objects that cross multiple floors (e.g., atrium). A building consists of one or more building units that are adjacent or connected through overpasses. Each building unit has a core function, including lab-centered (e.g., laboratories), office-centered (e.g., office buildings), and academic-centered (e.g., lectures and libraries). Physically, a building unit contains freely accessible spaces (e.g., corridors and halls) and enclosed areas. Enclosed areas can be categorized into two types according to the physical location: The perimeter area on external walls and the central dark zone for lab support spaces and ancillary spaces [24]. A central dark zone does not mean the area is dark without light but refers to the area that is located in the center of a building and cannot readily receive natural light. A perimeter area is divided into a primary zone and optional ancillary spaces at the two ends of the primary zone. A primary zone has three variations: Lab zone, office zone, and academic zone. A primary zone is further divided into single room types: Labs, offices, lab support spaces, seminar rooms, and libraries.

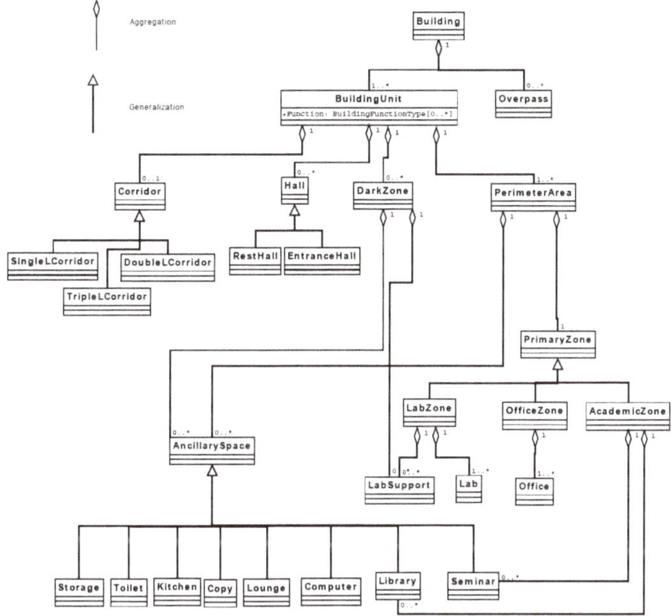

Figure 2. Semantic division of research buildings.

We take an example in Figure 3 to explain the semantic division of research buildings. Figure 3a shows a building consists of two adjacent building units. In Figure 3b, the left building unit is divided into three perimeter areas (1, 4, and 5), two central dark zones (2 and 3), one triple-loaded corridor, and one entrance hall. The right one is divided into one perimeter area and one single-loaded corridor. In Figure 3c, perimeter areas 1, 4, 5, and 6 are divided into a lab zone without ancillary spaces, an office zone without ancillary spaces, an office zone with ancillary spaces, and an academic zone without ancillary spaces, respectively. In Figure 3d, the lab zone is divided into single labs and lab support spaces. The two office zones are divided into multiple offices. The ancillary spaces are divided into a computer room and a seminar room. The academic zone is divided into two seminar rooms. The two central dark zones are divided into multiple lab supported spaces and toilets.

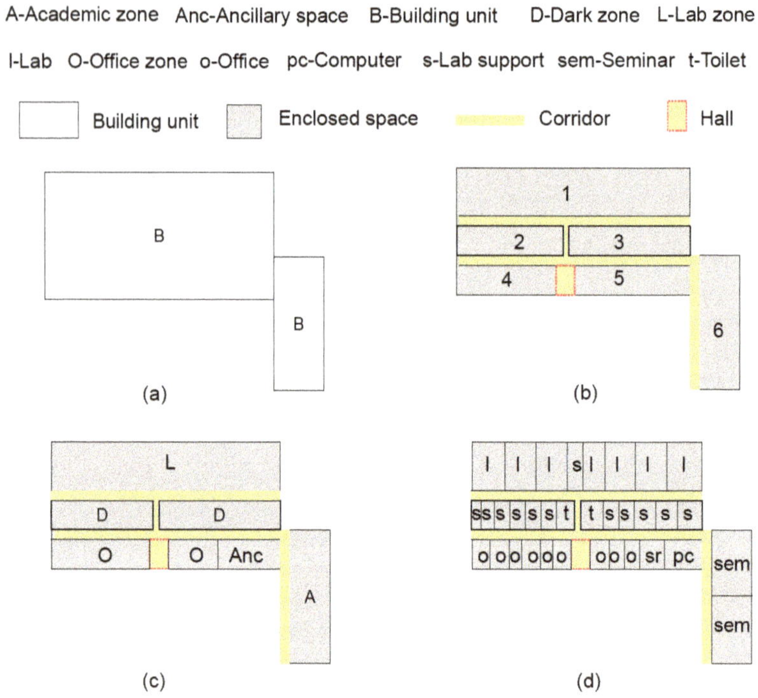

Figure 3. An example of semantic division of research buildings. (**a**) two building units in a floor. (**b**) corridors, halls, and enclosed zones in each building unit (**c**) type of each zone (**d**) single room types in each zone.

3.3. Constrained Attribute Grammar

Equation (1) formulates a typical rule of constrained attribute grammars [58–60]. p denotes the probability of applying the rule or generating the left-hand object with the right-hand objects. In this work, all the generated left-hand objects are assigned equal probability value of one except the generated objects that corresponds to the 11 room types (such as lab, office, and toilet) whose probability is estimated through the Bayesian inferring method given the geometric properties of primitive rooms. Z represents the parental or superior objects that can be generated by merging the right-hand objects denoted by X_k. x_k and z denote the instance of an object. Constraints define the preconditions that should be satisfied before applying this rule. The attribute part defines the operations that should be conducted on the attribute of the left-hand object.

$$p : Z\,z \rightarrow X_1\,x_1,\ X_2\,x_2 \ldots X_k\,x_k \langle \text{Constraints} \rangle \{Attribute\} \tag{1}$$

To simplify the description of rules, we define a collection operation *set*. It is used to represent multiple objects in the same type. For instance, $set(office, k)\ o$. defines a set of (k) *office* objects, denoted by o.

3.4. Predicates

A condition is a conjunction of predicates applied in rule variables x_i, possibly via attributes. Predicates primarily express geometric requirements but can generally represent any constraints on the underlying objects. In this work, we define several predicates by referring to guidebooks about the design principles of research buildings [23–26].

edgeAdj *(a,b)*: Object *a* is adjacent to object *b* via a shared edge without inclusive relationships between *a* and *b*.
inclusionAdj *(a,b,d)*: Object *a* includes object *b* and they are connected through an internal door *d*.
withExtDoor*(a)*: Object *a* has an external door connected to corridors.
onExtWall *(a)*: Object *a* is at the edge of external walls.
inCenter *(a)*: Most of the rooms in object *a* (zone) is not located at the external walls of buildings.
conByIntDoor($\{a_1, a_2 \ldots a_k\}$, $\{d_1, d_2 \ldots d_m\}$): Multiple objects $\{a_1, a_2 \ldots a_k\}$ are connected through internal doors $\{d_1, d_2 \ldots d_m\}$.
isTripleLoaded(*a*): Building *a* owns a triple-loaded circulation system.
isDoubleLoaded(*a*): Building *a* owns a double-loaded circulation system.
formFullArea($\{a_1, a_2 \ldots a_k\}$): Multiple objects form a complete area (e.g., a perimeter area or central dark zone), including all the primitive rooms and internal doors.

Figure 4 illustrates the defined predicates.

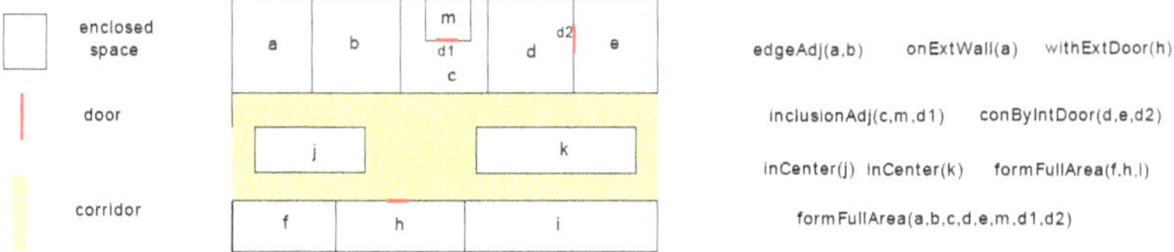

Figure 4. Predicates used in this work.

3.5. Defined Rules

We define 16 rules in total, which can be found in the Appendix A. Note that '|' denotes the OR operation. The objects in the rules correspond to the objects in Figure 2. Specifically, *Ancillary, Zone, Center, CZone, BUnit*, and *Building* objects in the rules correspond to the *AncillarySpace, PrimaryZone, DarkZones, PerimeterArea, BuildingUnit*, and the *Building* object in Figure 2, respectively. These rules are described as follows:

A1: A *room* object can be assigned with one of the eight types. When applying this rule, Bayesian inference methods are used to calculate the initial probability of belonging the room to corresponding type.

A2: A *Toilet* object is generated by merging one to three *room* objects when they satisfy the predicate *conByIntDoor* and only one of the *room* objects has an external door. Bayesian inference techniques are used to calculate the mean initial probability of each room to a toilet.

A3: Toilet, Copy, Storage, Kitchen, Lounge, Computer, Lecture, and Library objects are interpreted as *Ancillary* objects.

A4: A *Library* object is generated by merging a couple of *room* objects when they are connected by internal doors. Bayesian inference methods are used to calculate the mean probability of each room belonging to a library.

A5: A couple of *lecture* objects that are adjacent or connected by internal doors can be interpreted as an academic *Zone*.

A6: A *Library* object is interpreted as an academic *Zone*.

A7: A *Lab* object is generated by merging a single *room* r^l and an optional internal *room* r^w included by r^l when r^l. is on external walls. The Bayesian inference method is used to calculate the initial probability of r^l belonging to a lab.

A8: A *LGroup* object is generated by merging at least one *Lab* object and optional *Support* objects when they are connected by internal doors.

A9: A lab *Zone* is generated by merging multiple adjacent *LGroup* objects.

A10: A *room* object r^p with an optional internal *room* r^s contained by r^p can be explained as an *Office* object if r^p has an external door. The Bayesian inference method is used to calculate the initial probability of r^p belonging to an office.

A11: An *office* Zone can be generated by merging multiple *Office* objects if they are adjacent or connected through internal doors.

A12: A *Center* object can be generated by combining at most three *Ancillary* objects and optional adjacent or connected *Support* objects if the generated object satisfies the predicate *formFullArea*. If no *Support* objects exist, the type of the generated *Center* object is assigned *ancillary* otherwise *support*.

A13: A *CZone* object can be generated by combining at most three *Ancillary* objects and a *Zone* object if the generated object satisfies the predicate *formFullArea*.

A14: An office-centered or academic-centered building unit can be generated by merging at least one *CZone* object with the type of *office*, at most two *Center* objects with the type of *ancillary*, and at most two *CZone* objects with the type of *academic* if the generated object satisfies the predicate *formFullArea*.

A15: A lab-centered building unit can be generated by merging at least one *CZone* object with the type of *lab*, at least one *CZone* object with the type of *office*, and optional *CZone* objects with the type of *academic* if the generated object satisfies the predicate *formFullArea*. Note that if the building unit has a triple-loaded circulation system (with central dark areas), there exists at least one *Center* object with the type of *support*.

A16: A *Building* object can be generated by combining all the *BUnit* objects if they are adjacent.

4. Algorithm of Inferring Room Types

4.1. Workflow

The workflow of the proposed method is depicted in Figure 5. The input mainly consists of three parts. The first is the training data, including multiple rooms with their four properties (e.g., area, length, width, and room type). Based on the training data, we can extract the parameters of the Gaussian distribution for each room type. The second is the geometric map of the test scene. The third is the grammar rules, which are first partitioned into layers. Then, they are applied from the lowest layer to the highest layer in primitives to build a parse forest. The primitives are derived from the inputting geometric map, including enclosed rooms and internal doors. The reason that we do not infer corridors, halls, and stairs is that it is easy to identify them with point cloud and trace-based techniques [42,44]. When applying rules to assign rooms with certain types, the initial probability is calculated by using the Bayesian inferring method based on the geometric properties of the rooms (i.e., area, length, and width) that are extracted from the inputting geometric map. Finally, we can calculate the probability of belonging a room to a certain type based on the parse forest. The one with the highest probability is selected as the estimated type of the room.

4.2. Bayesian Inference

Different room types vary in geometric properties, such as length, width, and area. For instance, normally, the area of a seminar room is much larger than an office. We redefine the length and width of a rectangular room that is located at an external wall (denoted by w) as: The width of the room equals the edge that is parallel with w, while the length of the room corresponds to the other edge. For the rooms not located at external walls or located at multiple walls, the width and length follow their original definitions.

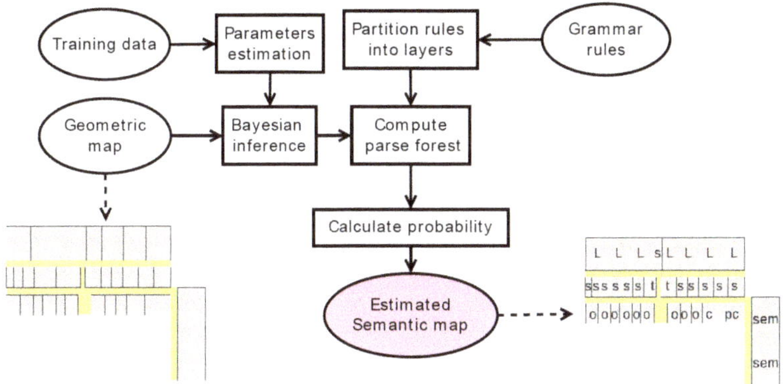

Figure 5. Workflow of proposed algorithm.

Given the geometric properties of a room, we can calculate the initial probability of the room belonging to a certain type by using the Bayesian probability theory, which will be invoked in the bottom-up approach when applying rules 1, 2, 4, 7, and 10. The estimated initial probability represents the probability of generating corresponding superior objects (room type) by applying rules 1, 2, 4, 7, and 10, which is attached to the generated objects. We use vector $x = (w, l, a)$ to denote the geometric properties of a room, where w, l, and a denote the width, length, and area, respectively. We use t to denote the type of rooms. Thus, the probability estimation equation can be written as:

$$p(t|x) = \frac{p(x|t)p(t)}{p(x)}. \tag{2}$$

In the equation, $P(t)$ represents the prior probability, which is approximated as the relative frequency of occurrence of each room type. $P(x)$ is obtained by integrating (or summing) $P(x|t)P(t)$ over all t, and plays the role of an ignorable normalizing constant. $P(x|t)$ refers to the likelihood function. It assesses the probability of the geometric properties of a room arising from the room types. To calculate the likelihood, we assume that the variables w, l, and a follow the normal distribution. The likelihood function is then written in the following notation:

$$p(x|t) = \frac{\exp(-\frac{1}{2}(x-u_t)^T \Sigma_t^{-1}(x-u_t))}{\sqrt{(2\pi)^k |\Sigma_t|}}. \tag{3}$$

In Equation (2), u_t is a 3-vector, denoting the mean value of the geometric properties of rooms in type t. Σ_t is the symmetric covariance matrix of the geometric properties of rooms in type t. Given a room with geometric properties $x = (w, l, a)$, we can first calculate the probabilities of the room belonging to one of the 11 types, denoted by $\hat{p}_i$, $1 < i < 11$. Then, the low-ranking candidate types are deleted, and the top T room types are kept. Their probabilities are then normalized. In this work, T is set to 5.

4.3. Compute Parse Forest

A parse tree corresponds to a semantic interpretation of one floor of a building. The proposed approach produces multiple interpretations that are represented by a forest. In this work, we use a bottom-up approach to construct the parse forest. Specifically, we continuously apply rules to merge the inferior objects into the superior objects of the rules if the inferior objects satisfy the preconditions of the rules. This process will terminate until no rules can be applied anymore. The inferior objects refer to the objects at the right-hand side of a rule and the superior objects refer to the objects at the left-hand side of a rule.

To improve the efficiency of searching proper rules during the merging procedure, we first partition the rules into multiple layers. The rules at the lower layers are applied ahead of the rules at the upper layers.

4.3.1. Partition Grammar Rules into Layers

Certain rules have more than one right-hand object, such as rule $\bar{r}$: $Z\,z \rightarrow X\,x,\ Y\,y,\ P\,p$. These rules can be applied only if all of their right-hand objects have been generated. That is, a rule denoted by $\acute{r}$ with X, Y, or P as the left-hand object should be applied ahead of rule $\bar{r}$. Then, we define that rule $\bar{r}$ dependents on rule $\acute{r}$. The dependency among the entire rules can be represented with a directed acyclic graph, in which a node denotes a rule and an edge with an arrow denotes the dependency. Based on the dependency graph, we can partition grammar rules into multiple layers. The rules at the lowest layer do not dependent on any rules. The process of partitioning rules into multiple layers is described as follows:

(1) Build dependency graph. Traversal each rule and draw a direct edge from current rule to the rules whose left-hand objects intersect the right-hand objects of this rule. If the right-hand objects of a rule include only primitive objects (e.g., rooms and doors), it is treated as a free rule.
(2) Delete free rules. Put the free rules at the lowest layer and then delete the free rules and all the edges connecting them from the graph.
(3) Handle new free rules. Identify new free rules and put them at the next layer. Similarly, delete the free rules and the corresponding edges. Repeat this step until no rules exist in the graph.

4.3.2. Apply Rules

After partitioning rules into layers, we then merge inferior objects into superior objects by applying rules from the lowest layer to the highest layer. During the procedure, if the generated superior objects correspond to a certain room type, the Bayesian inference method that is described in Section 4.2 is used to calculate the initial probability, which is assigned to the generated object. Otherwise, a probability value of one is assigned to the generated object. The process of computing a parse forest is as follows:

(1) Initialize an object list with the primitives and set the current layer as the first layer.
(2) Apply all the rules at the current layer to the objects in the list to generate superior objects.
(3) Fill the child list of the generated object with the inferior objects that form the generated objects.
(4) Assign a probability value to newly generated objects. When applying rules 1, 2, 4, 7, and 10, the probability is estimated through the Bayesian inference. Otherwise, we assign a probability of one to the generated objects.
(5) Add the newly generated objects to the object list.
(6) Move to the next layer and repeat steps (2)–(6).
(7) Create a *root* node and add all the *Building* objects to its child list.

We take a simplified floor-plan (shown in Figure 6) as an example to illustrate the procedure of creating the parse forest by using the proposed bottom-up method. The floor plan consists of three rooms, one internal and three external doors (denoted by solid red lines), and a corridor (with a yellow background). Based on this floor plan, hundreds of parse trees can be generated that form the parse forest. Here, we only choose three parse trees as examples. To clearly illustrate the bottom-up approach, we divide the procedure of constructing a forest into multiple sub procedures such that each procedure constructs a single tree. We denote the four primitives including three rooms and an internal door by $r1, r2, r3$, and d, respectively. In Figure 7a, the initial structure of a tree is the four primitives, playing the role of leaf nodes. In the second step, $r1, r2$, and $r3$ are interpreted as three offices denoted by $O1, O2$, and $O3$, respectively by applying rule **A10**. Next, $O1, O2, O3$, and d are merged into a *Zone* object, denoted by *Zone*1 by applying rule **A11**. Finally, a *Building* object can be created by applying rules **A13**, **A14**, and **A16** successively. Similarly, another two trees can be created as shown in Figure 7b,c, where *Anc*, *T*, and *Sem*, denote *Ancillary*, *Toilet*, and *Lecture* objects described in the rules, respectively. Note that, in the three trees, the nodes with the same name (e.g., the node with the name of *O*1 in the first and second trees) refer to the same node in the finally constructed forest. By merging the same nodes in these trees, we can obtain an initial forest (the left forest in Figure 8), where each node points to its children that form the node.

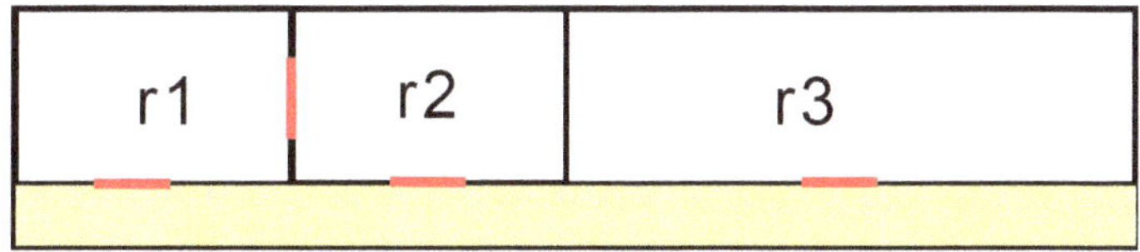

Figure 6. A simplified floor plan with three rooms.

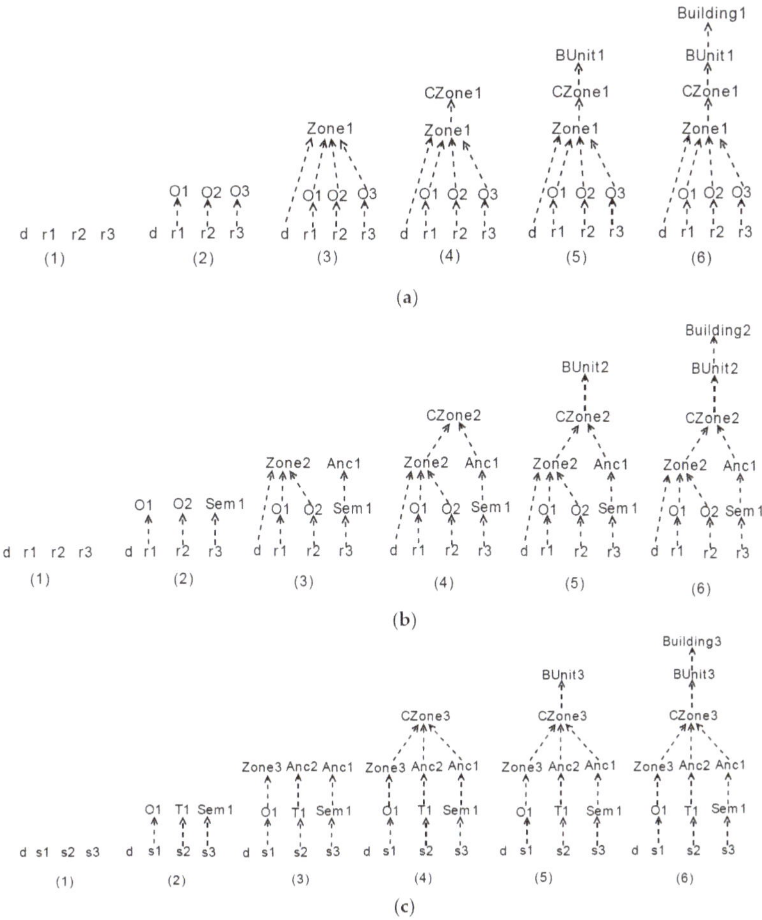

Figure 7. Procedure of creating parse forest by using bottom-up methods. (**a**) procedure of constructing first parse tree (**b**) procedure of constructing second parse tree (**c**) procedure of constructing third parse tree.

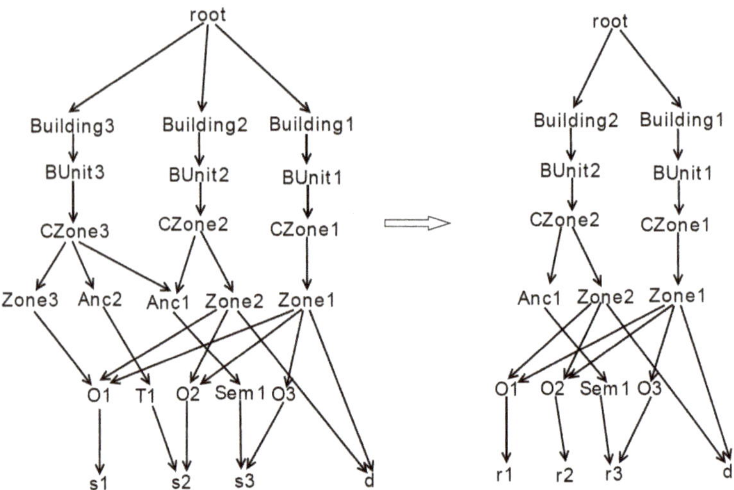

Figure 8. Pruning incomplete trees from parse forest.

The leaf nodes of a parse forest are primitives, including enclosed rooms and internal doors. The root node of the forest links to multiple *Building* nodes. Starting from a *Building* node, we can traversal its child list until the leaf nodes to find a parse tree. During the creation of the parse forest, immature or incomplete trees might be created if the semantic interpretation of these trees violate the defined rules. Thus, they are pruned from the forest. In this way, incorrect semantic interpretation (type) of the rooms can be removed. An incomplete tree refers to the tree whose root node is a *Building* node, but leaf nodes include only partial primitives. For example, the third tree in Figure 7 is an incomplete tree since its leaf nodes miss the internal door *d*. This can be explained by the fact that connecting an office and a toilet with internal doors rarely happens. Thus, this tree is pruned from the forest, as shown in Figure 8. The other two trees are valid parse trees.

The pseudocode of the algorithm that calculate the parse forest is described as follows:

```
Procedure F = ComputeParsingForest ((R_i)_{i≤i≤h},G);
Input:
(R_i)_{i≤i≤h} // partitioned rules. h denotes the number of layers.
G // all the primitives: rooms and internal doors
Output:
F // parse forest
    begin
O ← initializeObjects(G)  // initialize object list with G
for i = 1 to h do
for each rule r̄ ∈ R_i do
                    O ← applyRule(r̄, O)
end
end
                    F.child_list ← null
for each object ō ∈ O do
if the type of ō is a Building
                F.child_list ← F.child_list ∪ {ō}
end
end
end
```

Procedure *initializeObjects*(G) initials the object list with primitive objects (e.g., rooms and internal doors), which are treated as the leaf nodes of a parse forest. Procedure *applyRule*$(\overline{r}, O)$ searches the objects in O that satisfy the preconditions of rule $\overline{r}$. Then, they are merged to form superior objects that are at the left-hand side of rule $\overline{r}$. The probability of generating the superior object or applying the rule is estimated through the Bayesian inference method or is set to one.

4.4. Calculating Probability

Given a pruned forest with t parse trees, we can traversal each tree starting from a *Building* node until their leaf nodes (e.g., primitive rooms). For a primitive room r in the parse forest, the probability value attached to its parental object (a certain room type) in a tree is denoted by $\overline{p}_i, 1 \leq i \leq t$ and the probability value attached to its parental object (a certain room type) that matches its true type are denoted by $\widetilde{p}_k, 1 \leq k \leq m$, where m denotes the number of trees where room s is correctly assigned the type. The probability of room r belonging to its true type thus equals $\sum_{k=1}^{m} \widetilde{p}_k / \sum_{i=1}^{t} \overline{p}_i$. Assume that the true type of $r1, r2$, and $r3$ in Figure 6 are office (O), office (O), and lecture (Sem), respectively, and the forest in the right side of Figure 8 is the estimated forest. We denote the assigned probability value to nodes $O1, O2, O3$, and $Sem1$ by $\ddot{p}_1$, $\ddot{p}_2$, $\ddot{p}_3$, and $\ddot{p}_4$, respectively, which are estimated through the Bayesian inference method. Thus, the probability of room $r1, r2$, and $r3$ belonging to their true types equal $(\ddot{p}_1 + \ddot{p}_1)/(\ddot{p}_1 + \ddot{p}_1)$, $(\ddot{p}_2 + \ddot{p}_2)/(\ddot{p}_2 + \ddot{p}_2)$, and $\ddot{p}_4/(\ddot{p}_3 + \ddot{p}_4)$, respectively. Similarly, we can calculate the probability of belonging $r1, r2$, and $r3$ to the other room types (apart from the true type). Finally, for a room, the candidate type with the highest probability is selected as the estimated type of the room.

5. Experiments

5.1. Training Data

We collect 2304 rooms from our campuses. A 2304-by-4 matrix D is used to describe the training data. Each row of the matrix corresponds to a room, representing its four properties: Room type, area, length, and width. From the matrix, we can extract the relative frequency of occurrence of each room type, as shown in Figure 9. Further, for each room type, we can calculate the covariance matrix (3-by-3) and mean vector of the area, width, and length.

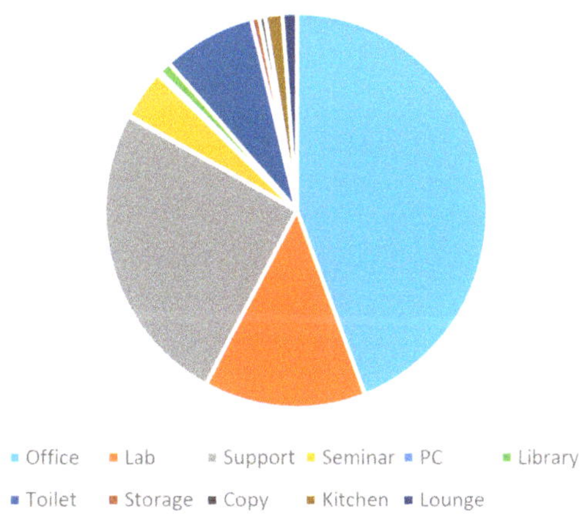

Figure 9. Proportion of different room types in training rooms.

5.2. Testbeds

We choose 15 buildings distributed in two campuses of Heidelberg University as the test bed, as shown in Figure 10. The footprints of these buildings include external passages, foyers, and external vertical passages; therefore, some are non-rectilinear polygons. We manually extract 15 rectilinear floors from these buildings by deleting the external parts, as shown in Figure 11. Table 1 shows the number of lab-centered, office-centered, and academic-centered building units in each floor plan.

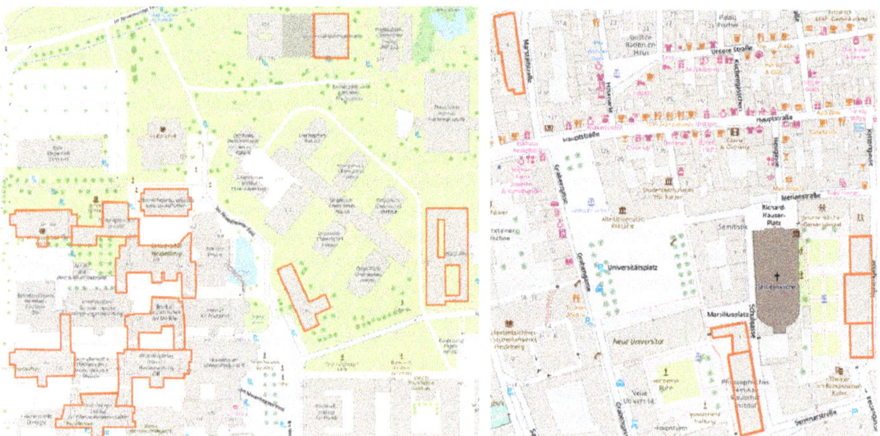

Figure 10. Distribution of test buildings.

Table 1. The number of different types of building units in each floor plan.

Floor Plan	Lab-Centered	Office-Centered	Academic-Centered
(a)	1	0	1
(b)	1	1	0
(c)	0	0	2
(d)	1	0	0
(e)	1	0	0
(f)	1	0	0
(g)	1	0	0
(h)	0	1	0
(i)	0	1	2
(j)	0	1	1
(k)	0	1	0
(l)	0	1	1
(m)	1	0	0
(n)	0	1	0
(o)	0	2	0

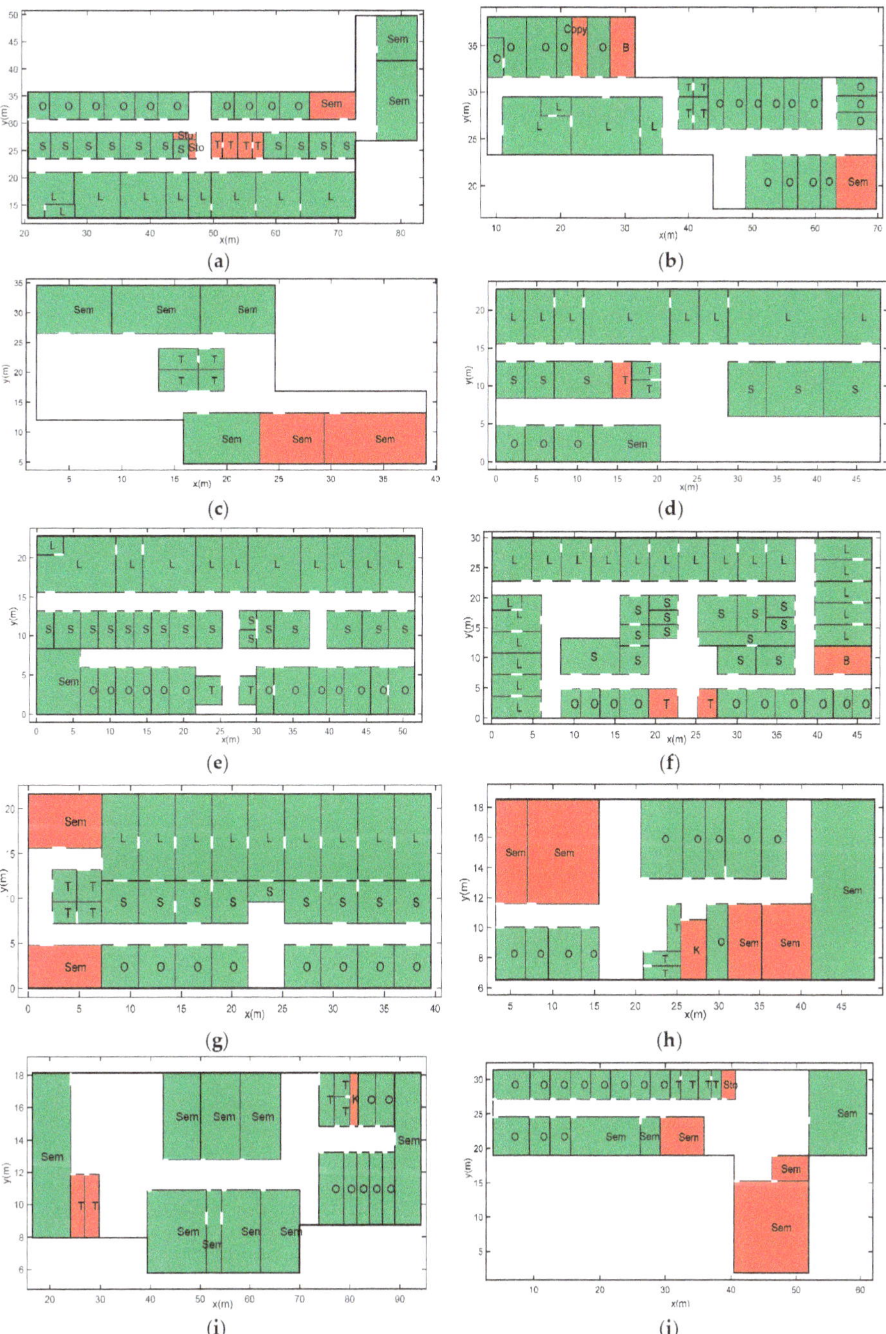

Figure 11. *Cont.*

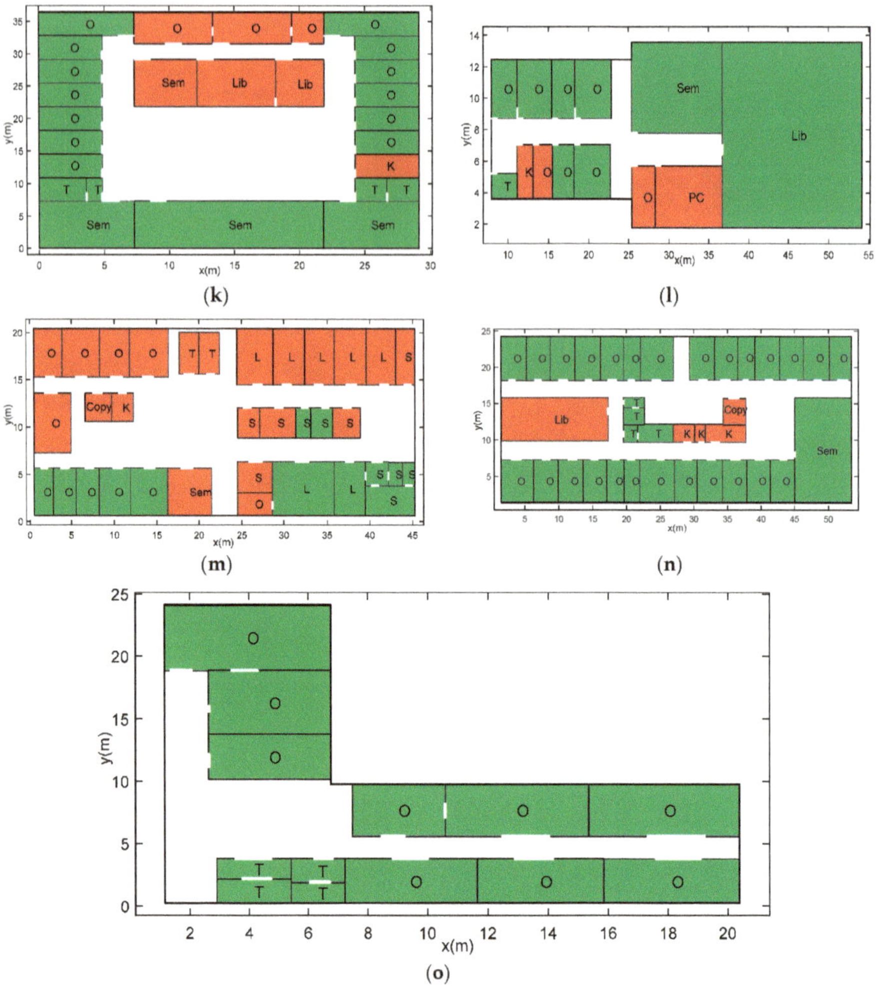

Figure 11. Floor plans (**a**–**o**) used for test.

We extract the geometric map from a scanned floor-plan by manually tagging the footprint of the building, the shape of rooms and corridors, and the location of both internal and external doors. All the lines are represented in pixels coordinates. In this procedure, we ignore the furniture in rooms. Then, based on the given area of a room that is tagged on the scanned map, we can convert the pixel coordinate of lines to a local geographic coordinate. Finally, the geometric size of rooms, corridors, and doors, as well as the topology relationship can be obtained. In this work, we assume that these spatial entities are already known since it is easy to detect them by current indoor mapping solutions [42,44,46]. Single rooms and internal doors are treated as the primitives. Internal doors refer to the doors connecting two rooms while external doors refer to the doors connecting a room and corridors. Doors are denoted by blank segments at the edge of rooms. Moreover, we delete a couple of spaces from the testbed since they are insignificant or easily detected by measurement-based approaches, such as electricity room and staircases. There are 408 rooms in total with each having a label, representing its type. Labels O, L, S, Sem, Lib, T, PC, Sto, C, K, and B denote offices, labs, lab support rooms, seminar rooms, libraries, toilets, computer rooms, storage rooms, kitchens, and lounges (break rooms), respectively. Note that the created grammars cover the room unit that consists of multiple sub-spaces (rooms) connected by internal doors. In this work, each sub-space is assigned to a certain type. In the test floor-plans, there are many room units, such as the one that consists of multiple labs connected by internal doors in floor-plan (a), the one that consists of multiple support spaces connected by internal doors in floor-plan (e), the one that consists of multiple labs and lab support spaces connected by internal doors in floor-plan (g), and the one that consists of multiple seminar rooms in floor-plan (i).

5.3. Experimental Results

As far as we know, currently, no works in indoor mapping (such as image- or Lidar-based approaches) have explicitly detected the room type in research buildings. Therefore, we only demonstrate the room tagging result of our proposed approach without comparing the results with other approaches. For a test floor plan, the identification accuracy denotes the proportion of the rooms whose type are corrected predicted among all the rooms in the floor plan. The average identification accuracy in 15 test floor plans reaches 84% by using our proposed method, as shown in Table 2. We use a green background and a red background to denote the room whose type is correctly and incorrectly identified, respectively, as shown in Figure 11.

Table 2. Identification accuracy of each floor plan.

Floor Plan	Identification Accuracy	Number of Rooms	Time Consumption(s)
Floor plan (a)	0.82	39	8.05
Floor plan (b)	0.90	29	3.93
Floor plan (c)	0.80	10	2.40
Floor plan (d)	0.95	21	3.10
Floor plan (e)	1.00	43	27.02
Floor plan (f)	0.94	48	7.18
Floor plan (g)	0.97	32	459.00
Floor plan (h)	0.74	19	3.68
Floor plan (i)	0.86	22	4.14
Floor plan (j)	0.82	22	4.45
Floor plan (k)	0.74	27	2.62
Floor plan (l)	0.69	13	5.66
Floor plan (m)	0.38	34	13.12
Floor plan (n)	0.86	36	2.33
Floor plan (o)	1.00	13	2.09
Overall	0.84	408	548

The high accuracy is achieved by fusing two kinds of characteristics of room types. The first is the spatial distribution characteristics and topological relationship among different room types, which are represented by grammar rules. We use only the grammars (geometry probability is set to 1) to calculate the probability of assigning rooms to a certain type, achieving an accuracy at around 0.3. The second is the distinguishable frequency and geometries properties (e.g., area, width, and length) of different room types. We use only the Bayesian inference method to estimate the room types of 408 testing rooms based on their geometric properties, achieving an accuracy of 0.38. Meanwhile, we use the random forest algorithm to train a model based on the geometric properties of 2300 rooms. Then, we calculate the probability of assigning each room in the test set (408 rooms) to a certain type given its geometric property, achieving an accuracy of 0.45, which is higher than that of Bayesian inference method. Furthermore, we replace the Bayesian inference method with the random forest method in the proposed solution. The result shows that there is no obvious improvement in the final accuracy.

We must reckon that cases that violate our defined rules still exist. For instance, in floor plan (b), a copy room is located between two office rooms, which is regarded as unreasonable according to our rules. Moreover, during the creation of parse forest, our method first deletes low-ranking candidate types for a room based on the initial probability calculated by the Bayesian inference method. This can greatly speed up the creation of the parse forest but can also rule out the right room type. For instance, in floor plan (k), a kitchen is incorrectly recognized because the estimated initial probability shows this room could not be a kitchen. Floor plan (m) gets a low identification accuracy. This is mainly because labs and offices have similar geometry properties and spatial distribution and topological characteristics. It is difficult to distinguish them.

The time used for calculating the parse forest and predicting the type of rooms based on the parse forest in each floor plan can be seen from the third column of Table 2. For most of the floor plans, it take about 10 s to build the parse forest and predict the type of room since we have ruled out the low-ranking type for a room based on the geometric probability estimated through Bayesian inference at the beginning of construction of the forest. This avoids an exponential growth of the number of parse trees. However, for floor-plan (g), it takes nearly 8 min. This is because one of the zones contains 23 rooms in total, with 18 connected, which enormously increases the number of possible combinations of room types.

The confusion matrix is shown in Figure 12, where the class labels 1 to 11 denote office, lab, lab support space, seminar, computer room, library, toilet, lounge, storage room, kitchen, and copy, respectively. The accuracy of identifying labs, offices, and lab support spaces is much higher than other types because (1) they are much more common than other types and (2) the defined rules are mainly derived from the guidebooks that focus on exploring the characteristics of these three kinds of rooms and the relationships among them. Moreover, internal doors play a vital role in identifying the type of rooms since only relevant types would be connected through inner doors, such as two offices, a lab and a lab support space, and multiple functional spaces in a toilet. For the ancillary spaces (i.e., lounge, storage room, kitchen, and copy room), the frequency of their occurrence is low, and their dimensional and topological characteristics are inapparent. Thus, the accuracy of identifying these ancillary spaces is much lower than that of other room types.

Figure 13 shows the constructed parse tree with the highest probability on floor-plan (d). *r*, *d*, *S*, *O*, *T*, *Sem*, *Anc*, *L*, and *LG* denote the objects of *room*, *door*, *Support*, *Office*, *Toilet*, *Lecture*, *Ancillary*, *Lab*, and *LGroup* in the rules, respectively. The text in the parentheses denotes the specific type of the object. The final parse forest consists of multiple parse trees, from which we can calculate the probability of assigning each room to a certain type. With the parse forest, we can not only infer the type of rooms, but also the type of zones and building units, as well as understand the whole scene since each parse tree represents a full semantic interpretation of the building. For instance, if we choose the parse tree with the highest probability as the estimated semantic interpretation of the scene, we can describe the scene as follows: Floor plan (d) has one lab-centered building unit, which consists of four enclosed areas or zones with one area mainly for offices, one area mainly for labs, and two areas mainly for lab support spaces located at the center of the building unit. We also infer the type (lab-centered, office-centered, and academic-centered) of building units based on the parse tree with the highest probability in other test floor-plans. Finally, 21 among 23 building units are correctly recognized.

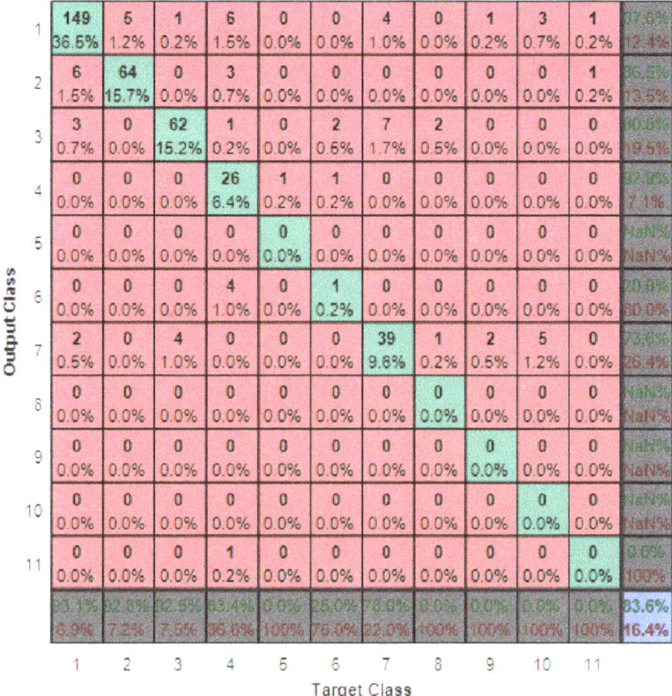

Figure 12. Confusion matrix of classification result.

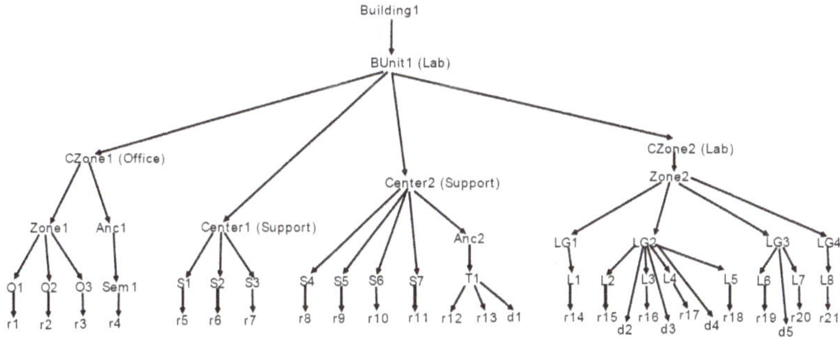

Figure 13. Parse tree with highest probability for floor-plan (d).

6. Discussions

Grammar learning: In this work, grammar rules are defined manually based on guidebooks about research buildings and our prior knowledge. This would produce two problems. One is that manual definition is a time-consuming task and requires a high level of expert knowledge. The other is that the deficiency of some significant rules and the constraints represented in the defined rules both lead to the reduction of the applicability and the accuracy of the proposed method since there exist always the cases that violate our defined rules and constraints. To overcome the two drawbacks, we plan to use grammatical inference techniques [21,22] to automatically learn a probabilistic grammar based on abundant training data in the future. Assigning each rule with a probability can better approximate the ground true since the frequency of occurrence of different rules in the real world varies.

For instance, multiple CZone objects can be merged into a lab-centered building or an academic-centered building. In this work, we assume that the probability of producing a lab-centered building and an academic-centered building is equal. However, the former appears much more frequently than the latter in the real world. Thus, the former should have earned a higher probability. We may argue that learning a reliable grammar for a certain building type is meaningful considering its great advantage in representation, which can benefit many application domains, such as reconstruction, semantic inference, computer-aided building design, and understanding a map by computers.

Deep learning: We may argue that the current advanced technology of deep learning can work for semantic labeling (specifically, room type) as in [61] if abundant images of each type of rooms are collected. However, these deep learning models are restricted in their capacity to reason, for example, to explain why the room should be an office or to further understand the map. Conversely, although grammar-based methods require users' intervention to create rules, they have the advantages of interpreting and representing. Therefore, they have a wide range of applications in GIS and building sectors. First, the grammars we create can not only be used to infer the semantics of rooms but also explain why a room is an office instead of a toilet. Second, the grammars can be used to formally represent a map and help computers to read or understand the map. Last but not least, grammars can benefit computer-aided building design [62].

7. Conclusions

This work investigates the feasibility of using grammars to infer the room type based on geometric maps. We take research buildings as example and create a set of grammar rules to represent the layout of research buildings. Then, we choose 15 floorplans and test the proposed approach. Results show it achieves an accuracy of 84% for 408 rooms. Although the grammar rules we create cannot cover all the research buildings in the world, we still believe the finding of this work is meaningful. It, to a certain extent, proves that grammar can benefit indoor mapping approaches in semantic enrichment. Furthermore, based on the constructed parse trees, we can not only infer the semantics of rooms, but also the type of zones and building units, as well as describe the whole scene.

Several tasks are scheduled for future works. First, we plan to mine useful knowledge from a university's website to enhance the identification of room types, such as the number of offices, the number of people in an office, and the number of conference rooms. This is because the information about researchers' offices and academic reports are accessible to everyone through a university's website. Based on the information, we can further prune parse forests to improve the identification accuracy. Second, a fully automatic solution will be proposed to learn the grammar rules from training data, based on which we can automatically build a more accurate and semantically richer map in a faster way with the help of fewer sensor measurements than the conventional measurement-based reconstruction approaches.

Author Contributions: Conceptualization, H.F.; data curation, X.H.; formal analysis, A.N. and Z.W.; funding acquisition, J.S.; methodology, X.H.; project administration, H.F.; resources, A.Z.; supervision, H.F. and A.Z.; validation, A.N.; writing—original draft, X.H.; writing—review and editing, A.N., Z.W., and J.S.

Funding: This study is supported by the National Key Research and Development Program of China (No.2016YFB0502200), the National Natural Science Foundation of China (Grant No. 41271440), and the China Scholarship Council.

Acknowledgments: We acknowledge financial support by Deutsche Forschungsgemeinschaft within the funding programme Open Access Publishing, by the Baden-Württemberg Ministry of Science, Research and the Arts and by Ruprecht-Karls-Universität Heidelberg.

Conflicts of Interest: The authors declare no conflict of interest.

Appendix A

$$Copy\ c|Storage\ s|Kitchen\ k|Lounge\ l|Lecture\ lec|Computer\ c|Support\ sup \rightarrow room\ r \tag{A1}$$

$$Toilet\ t \rightarrow set(room, k^r)r,\ set(door, k^d)\ d\left\langle 1 \leq k^r \leq 3; k^d \geq 0; conByIntDoor(r_i, r_{i+1}, d)_{0<i<k^r}; \sum_{i=1}^{k^r} withExtDoor(r_i) == 1 \right\rangle \tag{A2}$$

$$Ancillary\ a \rightarrow Toilet\ t|Copy\ c|Storage\ s|Kitchen\ k|Lounge\ l|\ Computer\ c|Lecture\ lec|Library\ lib \tag{A3}$$

$$Library\ l \rightarrow set(room, k^r)\ r, set(door, k^d)\ d\langle k^r \geq 1; k^d \geq 0; conByIntDoor(r_i, r_{i+1}, d)_{0<i<k^r}\rangle \tag{A4}$$

$$Zone^{aca}\ z \rightarrow set(Lecture, k^l)\ l\ set(door, k^d)\ d\left\langle \begin{matrix} k^l \geq 1; k^d \geq 0; withExtDoor(l_i)_{1\leq i\leq k^l}; \\ edgeAdj(l_i, l_{i+1})_{0<i<k^l}|conByIntDoor(l_i, l_{i+1}, d)_{0<i<k^l} \end{matrix} \right\rangle \{z.type = academic\} \tag{A5}$$

$$Zone^{aca}\ z \rightarrow Library\ lib\{z.type = academic\} \tag{A6}$$

$$Lab\ l \rightarrow room\ r^l, (room\ r^w, door\ d)|\varphi\ \langle inclusionAdj(r^l, r^w, d); onExtWall(r^l)\rangle \tag{A7}$$

$$LGroup\ g \rightarrow set(Lab, k^l)\ l,\ set(Support, k^s)\ s, set(door, k^d)\ d\ \langle k^l \geq 1;\ k^s >= 0; conByIntDoor(l, s, d)\rangle \tag{A8}$$

$$Zone^{lab}\ z \rightarrow set(LGroup, k^l)\ l\langle k^l \geq 1; edgeAdj(l_i, l_{i+1})_{0<i<k^l}\rangle\{z.type = lab\} \tag{A9}$$

$$Office\ o \rightarrow room\ r^p, (room\ r^s, door\ d)\ |\varphi\ \langle inclusionAdj(r^p, r^s, d); onExtWall(r^p); withExtDoor(r^p)\rangle \tag{A10}$$

$$Zone^{office}\ z \rightarrow set(Office, k^o)\ o, set(door\ , k^d)\ d\ \langle \mathrm{edgeAdj}(o_i, o_{i+1})_{0<i<k^o}|\ conByIntDoor(o_i, o_{i+1}, d)_{0<i<k^o}\rangle\{z.type = office\} \tag{A11}$$

$$\begin{aligned} &Center\ c \rightarrow set(Ancillary, k^a)\ a, set(Support, k^s)\ s \\ &\left\langle \begin{matrix} (0 \leq k^a \leq 3; k^s \geq 1)|(k^s == 0; k^a \geq 1)\ ; (edgeAdj(s_i, s_{i+1})\ |\ conByIntDoor(s_i, s_{i+1}, d))_{0<i<k^s}\ ; \\ formFullArea(a, s)\ ; inCenter(c); \end{matrix} \right\rangle \\ &\{c.type = support|ancillary\} \end{aligned} \tag{A12}$$

$$CZone\ c \rightarrow set(Ancillary, k)\ a, Zone\ z\langle 0 \leq k \leq 3; \mathrm{formFullArea}(a,z)\rangle\ \{c.type = z.type\} \tag{A13}$$

$$\begin{aligned} &BUnit\ b \rightarrow set(CZone^{office}, k^{o})\ z^{o}\ , set(Center^{ancillary}, k^{c})\ c, set(CZone^{aca}, k^{a})\ z^{a} \\ &\left\langle \begin{matrix} k^{o} \geq 0; 0 \leq k^{c} \leq 2; 0 \leq k^{a} \leq 2; k^{a} + k^{o} >= 1; ({z^{o}}_{i})_{1\leq i\leq k^{o}}.type == office; \\ (c_{i})_{1\leq i\leq k^{c}}.type == ancillary; ({z^{a}}_{i})_{1\leq i\leq k^{a}}.type == academic \end{matrix} \right\rangle \{b.type = office|academic\} \end{aligned} \tag{A14}$$

$$\begin{aligned} &BUnit^{lab}\ b \rightarrow set(CZone^{lab}, k^{l})\ z^{l}, set(CZone^{office}, k^{o})\ z^{o}, set(CZone^{academic}, k^{a})\ z^{a}, set(Center^{\mathrm{sup}}, k^{s})\ c^{s}, set(Center^{anc}, k^{anc})\ c^{a} \\ &\left\langle \begin{matrix} k^{l} \geq 1;\ k^{o} \geq 1;\ k^{a} \geq 0; k^{anc} \geq 0; ({z^{l}}_{i})_{1\leq i\leq k^{l}}.type == lab; ({z^{o}}_{i})_{1\leq i\leq k^{o}}.type == office; ({z^{a}}_{i})_{1\leq i\leq k^{a}}.type == academic; \\ ({c^{s}}_{i})_{1\leq i\leq k^{c}}.type == support; ({c^{a}}_{i})_{1\leq i\leq k^{anc}}.type == ancillary; \mathrm{isDoubleLoaded(b)}\,|\,(\mathrm{isTripleLoaded(b)};\ k^{s}\ \geq 1) \end{matrix} \right\rangle \{b.type = lab\} \end{aligned} \tag{A15}$$

$$Building\ f \rightarrow set(BUnit,\ k^{b})\ b \left\langle k^{b} \geq 1; edgeAdj(b_{i}, b_{i+1})_{0<i<k^{b}} \right\rangle \tag{A16}$$

References

1. Zhang, D.; Xia, F.; Yang, Z.; Yao, L.; Zhao, W. Localization technologies for indoor human tracking. In Proceedings of the 2010 5th International Conference on Future Information Technology, Busan, South Korea, 21–23 May 2010; pp. 1–6.
2. Yassin, A.; Nasser, Y.; Awad, M.; Al-Dubai, A.; Liu, R.; Yuen, C.; Raulefs, R.; Aboutanios, E. Recent advances in indoor localization: A survey on theoretical approaches and applications. *IEEE Commun. Surv. Tutor.* **2016**, *19*, 1327–1346. [CrossRef]
3. Elhamshary, M.; Youssef, M. SemSense: Automatic construction of semantic indoor floorplans. In Proceedings of the 2015 International Conference on Indoor Positioning and Indoor Navigation (IPIN), Banff, AB, Canada, 13–16 October 2015; pp. 1–11.
4. Youssef, M. Towards truly ubiquitous indoor localization on a worldwide scale. In Proceedings of the 23rd SIGSPATIAL International Conference on Advances in Geographic Information Systems, Seattle, WA, USA, 3–6 November 2015; ACM: New York, NY, USA, 2015.
5. Gao, R.; Zhao, M.; Ye, T.; Ye, F.; Wang, Y.; Bian, K.; Wang, T.; Li, X. Jigsaw: Indoor floor plan reconstruction via mobile crowdsensing. In Proceedings of the 20th Annual International Conference on Mobile Computing and Networking, Maui, HI, USA, 7–11 September 2014; ACM: New York, NY, USA, 2014; pp. 249–260.
6. Dosch, P.; Tombre, K.; Ah-Soon, C.; Masini, G. A complete system for the analysis of architectural drawings. *Int. J. Doc. Anal. Recognit.* **2000**, *3*, 102–116. [CrossRef]
7. De las Heras, L.P.; Ahmed, S.; Liwicki, M.; Valveny, E.; S'anchez, G. Statistical segmentation and structural recognition for floor plan interpretation. *Int. J. Doc. Anal. Recognit.* **2014**, *17*, 221–237. [CrossRef]
8. De las Heras, L.P.; Terrades, O.R.; Robles, S.; S'anchez, G. CVC-FP and SGT: A new database for structural floor plan analysis and its groundtruthing tool. *Int. J. Doc. Anal. Recognit.* **2015**, *18*, 15–30. [CrossRef]
9. Dodge, S.; Xu, J.; Stenger, B. Parsing floor plan images. In Proceedings of the IEEE IAPR International Conference on Machine Vision Application, Nagoya, Japan, 8–12 May 2017.
10. Xiong, X.; Adan, A.; Akinci, B.; Huber, D. Automatic creation of semantically rich 3D building models from laser scanner data. *Autom. Constr.* **2013**, *31*, 325–337. [CrossRef]
11. Armeni, I.; Sener, O.; Zamir, A.R.; Jiang, H.; Brilakis, I.; Fischer, M.; Savarese, S. 3D semantic parsing of large-scale indoor spaces. In Proceedings of the IEEE Conference on Computer Vision and Pattern Recognition, Las Vegas, NV, USA, 27–30 June 2016; pp. 1534–1543.
12. Qi, C.R.; Su, H.; Mo, K.; Guibas, L.J. Pointnet: Deep learning on point sets for 3d classification and segmentation. In Proceedings of the IEEE Conference on Computer Vision and Pattern Recognition, Honolulu, HI, USA, 21–26 July 2017; pp. 652–660.
13. Ambruş, R.; Claici, S.; Wendt, A. Automatic room segmentation from unstructured 3-d data of indoor environments. *IEEE Robot. Autom. Lett.* **2017**, *2*, 749–756. [CrossRef]
14. Furukawa, Y.; Curless, B.; Seitz, S.M.; Szeliski, R. Reconstructing building interiors from images. In Proceedings of the 2009 IEEE 12th International Conference on Computer Vision, Kyoto, Japan, 29 September–2 October 2009; pp. 80–87.
15. Henry, P.; Krainin, M.; Herbst, E.; Ren, X.; Fox, D. RGB-D mapping: Using Kinect-style depth cameras for dense 3D modeling of indoor environments. *Int. J. Robot. Res.* **2012**, *31*, 647–663. [CrossRef]
16. Ikehata, S.; Yang, H.; Furukawa, Y. Structured indoor modeling. In Proceedings of the IEEE International Conference on Computer Vision, Santiago, Chile, 7–13 December 2015; pp. 1323–1331.
17. Alzantot, M.; Youssef, M. Crowdinside: Automatic construction of indoor floorplans. In Proceedings of the 20th International Conference on Advances in Geographic Information Systems, Redondo Beach, CA, USA, 6–9 November 2012; ACM: New York, NY, USA, 2012; pp. 99–108.
18. Zhang, J.; Kan, C.; Schwing, A.G.; Urtasun, R. Estimating the 3d layout of indoor scenes and its clutter from depth sensors. In Proceedings of the IEEE International Conference on Computer Vision, Sydney, NSW, Australia, 1–8 December 2013; pp. 1273–1280.
19. Luperto, M.; Riva, A.; Amigoni, F. Semantic classification by reasoning on the whole structure of buildings using statistical relational learning techniques. In Proceedings of the 2017 IEEE International Conference on Robotics and Automation (ICRA), Singapore, 29 May–3 June 2017; pp. 2562–2568.
20. Elhamshary, M.; Basalmah, A.; Youssef, M. A fine-grained indoor location-based social network. *IEEE Trans. Mob. Comput.* **2017**, *16*, 1203–1217. [CrossRef]

21. De la Higuera, C. *Grammatical Inference: Learning Automata and Grammars*; Cambridge University Press: Cambridge, UK, 2010.
22. D'Ulizia, A.; Ferri, F.; Grifoni, P. A survey of grammatical inference methods for natural language learning. *Artif. Intell. Rev.* **2011**, *36*, 1–27. [CrossRef]
23. Charlotte, K. *New Laboratories: Historical and Critical Perspectives on Contemporary Developments*; Walter de Gruyter GmbH: Berlin, Germany, 2016.
24. Braun, H.; Grömling, D. *Research and Technology Buildings: A Design Manual*; Walter de Gruyter: Berlin, Germany, 2005.
25. Hain, W. *Laboratories: A Briefing and Design Guide*; Taylor & Francis: London, UK, 2003.
26. Watch, D.D. *Building Type Basics for Research Laboratories*; John Wiley & Sons: New York, NY, USA, 2002.
27. Hu, X.; Fan, H.; Zipf, A.; Shang, J.; Gu, F. A conceptual framework for indoor mapping by using grammars. In *ISPRS Annals of the Photogrammetry, Remote Sensing and Spatial Information Sciences*; ISPRS Geospatial Week: Wuhan, China, 2017; Volume IV-2/W4, pp. 335–342.
28. Azhar, S. Building information modeling (BIM): Trends, benefits, risks, and challenges for the AEC industry. *Leadersh. Manag. Eng.* **2011**, *11*, 241–252. [CrossRef]
29. Santos, R.; Costa, A.A.; Grilo, A. Bibliometric analysis and review of Building Information Modelling literature published between 2005 and 2015. *Autom. Constr.* **2017**, *80*, 118–136. [CrossRef]
30. Kolbe, T.H. Representing and exchanging 3D city models with CityGML. In *3D Geo-Information Sciences*; Springer: Berlin/Heidelberg, Germany, 2017; pp. 15–31.
31. Li, K.J.; Kim, T.H.; Ryu, H.G.; Kang, H.K. Comparison of cityGML and indoorGML—A use-case study on indoor spatial information construction at real sites. *Spat. Inf. Res.* **2015**, *23*, 91–101.
32. Kim, J.S.; Yoo, S.J.; Li, K.J. Integrating IndoorGML and CityGML for indoor space. In *International Symposium on Web and Wireless Geographical Information Systems*; Springer: Berlin/Heidelberg, Germany, 2014; pp. 184–196.
33. Kang, H.K.; Li, K.J. A standard indoor spatial data model—OGC IndoorGML and implementation approaches. *ISPRS Int. J. Geo-Inf.* **2017**, *6*, 116. [CrossRef]
34. Macé, S.; Locteau, H.; Valveny, E.; Tabbone, S. A system to detect rooms in architectural floor plan images. In Proceedings of the IEEE IAPR International Workshop on Document Analysis Systems, Boston, MA, USA, 9–11 June 2010.
35. Ahmed, S.; Liwicki, M.; Weber, M.; Dengel, A. Improved automatic analysis of architectural floor plans. In Proceedings of the IEEE International Conference on Document Analysis and Recognition, Beijing, China, 18–21 September 2011.
36. Ahmed, S.; Liwicki, M.; Weber, M.; Dengel, A. Automatic room detection and room labeling from architectural floor plans. In Proceedings of the IEEE IAPR International Workshop on Document Analysis Systems, Gold Cost, QLD, Australia, 27–29 March 2012.
37. Gimenez, L.; Robert, S.; Suard, F.; Zreik, K. Automatic reconstruction of 3D building models from scanned 2D floor plans. *Autom. Constr.* **2016**, *63*, 48–56. [CrossRef]
38. De las Heras, L.P.; Mas, J.; S'anchez, G.; Valveny, E. Notation-invariant patchbased wall detector in architectural floor plans. In Proceedings of the International Workshop on Graphics Recognition, Seoul, Korea, 15–16 September 2011.
39. Sankar, A.; Seitz, S. Capturing indoor scenes with smartphones. In Proceedings of the 25th Annual ACM Symposium on User Interface Software and Technology, Cambridge, MA, USA, 7–10 October 2012; ACM: New York, NY, USA, 2012; pp. 403–412.
40. Pintore, G.; Gobbetti, E. Effective mobile mapping of multi-room indoor structures. *Vis. Comput.* **2014**, *30*, 707–716. [CrossRef]
41. Tsai, G.; Xu, C.; Liu, J.; Kuipers, B. Real-Time Indoor Scene Understanding Using Bayesian Filtering with Motion Cues. In Proceedings of the 2011 International Conference on Computer Vision (ICCV), Barcelona, Spain, 6–13 November 2011; pp. 121–128.
42. Jiang, Y.; Xiang, Y.; Pan, X.; Li, K.; Lv, Q.; Dick, R.P.; Shang, L.; Hannigan, M. Hallway based automatic indoor floorplan construction using room fingerprints. In Proceedings of the 2013 ACM International Joint Conference on Pervasive and Ubiquitous Computing, Zurich, Switzerland, 8–12 September 2013; ACM: New York, NY, USA, 2013; pp. 315–324.

43. Chen, S.; Li, M.; Ren, K.; Qiao, C. Crowd map: Accurate reconstruction of indoor floor plans from crowdsourced sensor-rich videos. In Proceedings of the IEEE 35th International Conference on Distributed Computing Systems (ICDCS), Columbus, OH, USA, 29 June–2 July 2015; pp. 1–10.
44. Gao, R.; Zhou, B.; Ye, F.; Wang, Y. Knitter: Fast, resilient single-user indoor floor plan construction. In Proceedings of the INFOCOM 2017—IEEE Conference on Computer Communications, Atlanta, GA, USA, 1–4 May 2017; pp. 1–9.
45. Mura, C.; Mattausch, O.; Villanueva, A.J.; Gobbetti, E.; Pajarola, R. Automatic room detection and reconstruction in cluttered indoor environments with complex room layouts. *Comput. Graph.* **2014**, *44*, 20–32. [CrossRef]
46. Nikoohemat, S.; Peter, M.; Elberink, S.O.; Vosselman, G. Exploiting Indoor Mobile Laser Scanner Trajectories for Semantic Interpretation of Point Clouds. *ISPRS Ann. Photogramm. Remote Sens. Spat. Inf. Sci.* **2017**, *IV-2/W4*, 355–362. [CrossRef]
47. Becker, S.; Peter, M.; Fritsch, D. Grammar-supported 3d Indoor Reconstruction from Point Clouds for" as-built" BIM. *ISPRS Ann. Photogramm. Remote Sens. Spat. Inf. Sci.* **2015**, *2*, 17. [CrossRef]
48. Yue, K.; Krishnamurti, R.; Grobler, F. Estimating the interior layout of buildings using a shape grammar to capture building style. *J. Comput. Civ. Eng.* **2011**, *26*, 113–130. [CrossRef]
49. Philipp, D.; Baier, P.; Dibak, C.; Dürr, F.; Rothermel, K.; Becker, S.; Peter, M.; Fritsch, D. Mapgenie: Grammar-enhanced indoor map construction from crowd-sourced data. In Proceedings of the 2014 IEEE International Conference on Pervasive Computing and Communications, Budapest, Hungary, 24–28 March 2014; pp. 139–147.
50. Rosser, J.F. Data-driven estimation of building interior plans. *Int. J. Geogr. Inf. Sci.* **2017**, *31*, 1652–1674. [CrossRef]
51. Luperto, M.; Amigoni, F. Exploiting structural properties of buildings towards general semantic mapping systems. In *Intelligent Autonomous Systems 13*; Springer: Cham, Switzerland, 2016; pp. 375–387.
52. Luperto, M.; Li, A.Q.; Amigoni, F. A system for building semantic maps of indoor environments exploiting the concept of building typology. In *Robot Soccer World Cup*; Springer: Berlin/Heidelberg, Germany, 2013; pp. 504–515.
53. Khoshelham, K.; Díaz-Vilariño, L. 3D modelling of interior spaces: Learning the language of indoor architecture. *Int. Arch. Photogramm. Remote Sens. Spat. Inf. Sci.* **2014**, *40*, 321. [CrossRef]
54. Mitchell, W.J. *The Logic of Architecture: Design, Computation, and Cognition*; MIT Press: Cambridge, MA, US, 1990.
55. Dehbi, Y.; Hadiji, F.; Gröger, G.; Kersting, K.; Plümer, L. Statistical relational learning of grammar rules for 3D building reconstruction. *Trans. GIS* **2017**, *21*, 134–150. [CrossRef]
56. Liu, Z.; von Wichert, G. Extracting semantic indoor maps from occupancy grids. *Robot. Auton. Syst.* **2014**, *62*, 663–674. [CrossRef]
57. Liu, Z.; von Wichert, G. A generalizable knowledge framework for semantic indoor mapping based on Markov logic networks and data driven MCMC. *Future Gener. Comput. Syst.* **2014**, *36*, 42–56. [CrossRef]
58. Deransart, P.; Jourdan, M.; Lorho, B. *Attribute Grammars: Definitions, Systems and Bibliography*; Springer Science & Business Media: Heidelberg, Germany, 1988; Volume 323.
59. Deransart, P.; Jourdan, M. Attribute grammars and their applications. In *Lecture Notes in Computer Science*; Springer: Heidelberg, Germany, 1990; Volume 461.
60. Boulch, A.; Houllier, S.; Marlet, R.; Tournaire, O. Semantizing complex 3D scenes using constrained attribute grammars. *Comput. Graph. Forum* **2013**, *32*, 33–42. [CrossRef]
61. Russakovsky, O.; Deng, J.; Su, H.; Krause, J.; Satheesh, S.; Ma, S.; Huang, Z.; Karpathy, A.; Khosla, A.; Bernstein, M.; et al. Imagenet large scale visual recognition challenge. *Int. J. Comput. Vis.* **2015**, *115*, 211–252. [CrossRef]
62. Müller, P.; Wonka, P.; Haegler, S.; Ulmer, A.; Van Gool, L. Procedural modeling of buildings. *ACM Trans. Graph.* **2006**, *25*, 614–623. [CrossRef]

Article

An Accurate Visual-Inertial Integrated Geo-Tagging Method for Crowdsourcing-Based Indoor Localization

Tao Liu [1], Xing Zhang [2,*], Qingquan Li [2], Zhixiang Fang [3] and Nadeem Tahir [4]

1 College of Resources and Environment, Henan University of Economics and Law, Zhengzhou 450002, China
2 Shenzhen Key Laboratory of Spatial Information Smart Sensing and Services & Key Laboratory for Geo-Environment Monitoring of Coastal Zone of the National Administration of Surveying, Mapping and GeoInformation, the School of Architecture and Urban Planning, Shenzhen University, Shenzhen 518060, China
3 State Key Laboratory of Information Engineering in Surveying, Mapping, and Remote Sensing, Wuhan University, Wuhan 430072, China
4 College of Mechanical and Electrical Engineering, Henan Agricultural University, Zhengzhou 450002, China
* Correspondence: xzhang@szu.edu.cn

Received: 21 June 2019; Accepted: 13 August 2019; Published: 16 August 2019

Abstract: One of the unavoidable bottlenecks in the public application of passive signal (e.g., received signal strength, magnetic) fingerprinting-based indoor localization technologies is the extensive human effort that is required to construct and update database for indoor positioning. In this paper, we propose an accurate visual-inertial integrated geo-tagging method that can be used to collect fingerprints and construct the radio map by exploiting the crowdsourced trajectory of smartphone users. By integrating multisource information from the smartphone sensors (e.g., camera, accelerometer, and gyroscope), this system can accurately reconstruct the geometry of trajectories. An algorithm is proposed to estimate the spatial location of trajectories in the reference coordinate system and construct the radio map and geo-tagged image database for indoor positioning. With the help of several initial reference points, this algorithm can be implemented in an unknown indoor environment without any prior knowledge of the floorplan or the initial location of crowdsourced trajectories. The experimental results show that the average calibration error of the fingerprints is 0.67 m. A weighted k-nearest neighbor method (without any optimization) and the image matching method are used to evaluate the performance of constructed multisource database. The average localization error of received signal strength (RSS) based indoor positioning and image based positioning are 3.2 m and 1.2 m, respectively, showing that the quality of the constructed indoor radio map is at the same level as those that were constructed by site surveying. Compared with the traditional site survey based positioning cost, this system can greatly reduce the human labor cost, with the least external information.

Keywords: indoor localization; crowdsourcing trajectory; fingerprinting; smartphone

1. Introduction

Nowadays, indoor localization has become a common issue for various location-based services and applications. A number of technologies have been proposed for indoor localization, which are based on different principles, such as Wi-Fi [1], geomagnetic [2], ultra wide band (UWB) [3], ultrasound [4] and so on. Among these localization technologies, ultrasound and UWB can be used to estimate the distance between the source and terminals, which can provide accurate localization results. However, such technologies require an extra deployment of localization devices, which restricts their large-scale applications. Many studies have focused on developing localization scheme that do not rely on extra devices or only use the existing infrastructures, such as Wi-Fi fingerprinting [5–8], geomagnetic [9], or visual positioning [10–12]. For example, crowd participants walking in the indoor environment can collect the fingerprints and construct radio map or magnetic map by uploading their inertial data, Wi-Fi received signal strength (RSS) data or magnetic readings, which can be directly used for indoor localization. For image matching-based visual positioning, pictures that are taken by a user can be matched against the geo-tagged images stored in a database. The location of a geo-tagged image can be used as the localization results when the two images (the query image and the geo-tagged image) are successfully matched with each other.

The collection of data is an essential bottleneck for developing localization solutions with free of additional devices. For example, Wi-Fi fingerprinting-based localization requires a radio map of the whole environment. Visual-based positioning relies on the image database or other semantic information to represent locations. Geomagnetic-based localization depends on a similar map of magnetic field strength of the environment. The collection and updating of the required data are quite labor-intensive and time-consuming which hinders these localization solutions from large-scale application. To solve this issue, many studies propose the use of crowdsourcing-based approaches to reduce the labor and time cost that needed for data collection [13–22]. For example, it has been exploited that smartphones collect RSS and inertial data to construct radio map with unconscious cooperation among volunteers [19–21]. These methods can construct the fingerprint database effectively with less time consuming. However, most of these methods still require user intervention [16] or prior knowledge e.g., initial radio map [21,23], access point's (AP) location [22], initial location of volunteers [21], or indoor floorplans [13,16–18]. In practice, it is usually difficult to collect all of the required data, such as the method proposed in [23] uses image matching to improve the localization accuracy of Wi-Fi fingerprinting. However, it requires an initial Wi-Fi radio map as an initial input, which is suitable for the updating Wi-Fi database. GROPING [24] is a self-contained indoor navigation system that relies on geomagnetic fingerprinting. It exploits crowdsensing walking trajectories to construct floor maps and semantic navigation map while using user contributed sensor data and semantic labels. However, although visual positioning can achieve good positioning accuracy and it does not rely on extra infrastructure, the study of crowdsourcing-based visual positioning is much fewer than that of Wi-Fi and magnetic positioning. The existing studies [11,25,26] mainly concentrate on developing new algorithm or model to improve the accuracy of visual positioning. Less attention has been devoted to developing an efficient and reliable indoor image collection and geo-tagging method. The lack of a large mount of indoor geo-tagged image databases is an essential bottleneck in the application of visual positioning. If a crowdsourcing-based image collection and geo-tagging method can be proposed, the difficulty for deploying visual positioning systems or services may be significantly reduced.

As a wide-spread mobile device, the smartphone is suitable for collecting crowdsourcing data, including wireless signals, inertial data, magnetic field, image, and so on. While people are walking in different indoor environments, their smartphones can collect the required data continuously at a certain sampling rate. The collected signals are associated with the corresponding sampling points with timestamps. However, the spatial location of sampling points cannot be directly received from smartphone built-in sensors in indoor spaces. The geo-tagging of trajectory sampling points is a key issue for crowdsourcing-based localization approaches. The intervention of users has been considered in some studies to facilitate the geo-tagging of trajectory. For example, Redpin [27] and OIL [28] prompt the users to recognize their current location for trajectory tracking based on a prior build-in displayed map. Without initial indoor map, Elekspot [29] and FreeLoc [16] take the advantage of the semantic labels that are associated with a trajectory, such as room or corridor. Consequently, the localization result of these methods is at a semantic level. Instead of user intervention, other studies realize trajectory geo-tagging by using smartphone sensors and the map matching method. For example, RCILS [13], LiFS [17], Zee [18], and WILL [30] used the pedestrian dead-reckoning (PDR) method to recover the trajectory of smartphone users. The recovered trajectory can be spatially matched to an indoor floor plan by using an activity recognition mechanism. However, this map matching mechanism highly depends on an assumption that all activities of a user occur at the special locations in an indoor space (e.g., intersections or corners). This assumption is vulnerable to the randomness of human activity. For example, if one person makes a free turn not at a special location, this method may match this activity to an incorrect location. Much effort has been made to improve the performance of trajectory estimation [31–34]. For example, the trajectory alignment and calibration method proposed in [32] can align a crowdsourcing trajectory into a coordinate system by using a foot-mounted inertial sensor and Wi-Fi RSS measurements. CrowdMap [33] jointly leverage crowdsourced sensor data and video data to track the movements of users. It takes the latest known GPS position as an initial location and the user can modify it if it is incorrect. The video frames can be used to improve the localization accuracy. Pan et al. [34] provides a collaborative filtering with graph-based semi-supervised learning method to estimate the location of user, as well as the location of APs. A training phase is needed to calibrate a probabilistic location estimation system. In summary, although progress has been made in pedestrian tracking and trajectory estimation, but the requirement of extra devices, user intervention, or prior knowledge limits the practical use of these approaches. It remains a question as to how to improve the accuracy and robustness of smartphone-based geo-tagging with least device or prior knowledge requirements.

Another problem is the diversity of smartphone devices. The collected wireless signal probably will be different for two smartphone devices even at a same spatial location due to the difference of built-in sensors in smartphone. It will obviously affect the geo-tagging accuracy of the Wi-Fi or geomagnetic clustering based crowdsourcing methods. However, it is difficult to meet the condition, where all crowdsourcing users have the same type of smartphone devices. Some studies have tried to solve this problem by calibrating RSS fingerprints that were collected by different devices. For example, in [29], a calibration matrix has been constructed for all supported devices. The elements from the matrix represent the linear regression relation between different devices. In [35], the diversity of devices and its effect on localization have been analyzed. A kernel function has been proposed to model the distribution of RSS. It applies local variations as a compensation for linear transformation between two devices. However, the accurately calibration of Wi-Fi fingerprints remains a question due to the inherent instability of the real world. Recently, channel state information (CSI) based indoor localization [36,37] have attracted much attention. When compared with RSS-based solution, CSI maintains more stability and sensitivity, which can provide detailed and fine-grain subcarrier information. For example, in [36], a CSI based indoor localization technique is developed by employing both the intrasubcarrier statistics features and the inter-subcarrier network features. Their results showed that it could achieve 96% classification accuracy. In [37], the proposed DeepFi system uses CSI (collected from three antennas) and the deep learning method to train the fingerprinting database. A probabilistic method is used in the online localization phase, which can achieve a mean error of about 1.8m. However, the CSI based indoor localization methods require a Wi-Fi network interface cards (NIC) to receive CSI signals, which is not a built-in part for the current smartphones. In summary, although significant effort has been devoted to solve the problem of device diversity, it remains an important issue for the practical application of crowdsourcing-based indoor and it has a negligible effect on the localization performance.

The present study proposes involving visual information in the geo-tagging of sampling points for crowdsourcing based localization scheme. The video frames that were collected by smartphone camera contain abundant visual information regarding different spatial location with an advantage of minimum difference in video frame in spite of diversity in smartphone devices. It has been observed that under different condition of smartphone camera (e.g., resolution ratio, size, etc.), the visual features that were extracted from these images can accurately reflect the spatial information of the sampling points. Consequently, this study attempts to develop a visual-based geo-tagging method for crowdsourcing-based indoor localization. This method can be used to reconstruct geometrical trajectories (associated with the collected signals) of multiple crowdsourcing users with different smartphone devices. Some prior knowledge, such as an existing database (e.g., Wi-Fi radio map), floorplan of the environment, and the initial locations of all crowdsourcing users are assumed to be unknown to improve the practicability of this method. A reference coordinate system (RCS) is defined for the geometry reconstruction of crowdsourced trajectories, which can be easily deployed in indoor spaces. The only requirement before reconstructing trajectories and indoor localization is several initial reference points (IRPs), which can be used as the origin of the RCS. An algorithm is also proposed to geo-tag the spatial location of sampling points from all the trajectories. A Wi-Fi radio map and an image dataset were constructed based on the crowdsourced trajectories after geo-tagging to evaluate the proposed method. The RSS-based localization accuracy and the image matching based localization accuracy demonstrate the effectiveness of this method. The proposed method can be used to accurately geo-tag the sampling points that were extracted from crowdsourced trajectories and generate different maps or datasets, such as radio map or geo-tagged image database, for indoor positioning. Based on the positioning results, various indoor location-based services can be provided to the public users, such as indoor navigation, intelligent parking, shopping mall or museum tourism, management of mobile objects, and so on. Besides, it can also be employed to collect and generate other indoor maps, such as noise map, illumination map, or other maps with the help of the corresponding sensors (e.g., PM 2.5). These maps can be used for indoor management, architectural design, or other analysis of indoor environment.

The remainder of this paper is organized, as follows. Section 2 describes the theoretical framework of the visual-inertial integrated geo-tagging method and the principle of the visual-based method for trajectory geometry recovery. Section 3 describes the spatial estimation of crowdsourced trajectories and the construction of radio map. Section 4 presents and discusses the experimental results. Finally, Section 5 summarizes the main conclusions and future work.

2. Visual Based Trajectory Geometry Recovery

Figure 1 illustrates the framework of this method. Smartphones are used to collect various sensor data, including video frames, inertial sensors data, Wi-Fi RSS, etc. In this study, some volunteers are required to collect the experiment data with smartphone. During data collection, the volunteers hold a smartphone in hand and walk normally in an indoor area. Firstly, a trajectory reconstruction method was designed to geometrically recover the trajectories from the corresponding crowdsourcing data. Heading angle estimation, step detection and step length estimation are the necessary steps for recovery of trajectory geometry. Both the video frames and the inertial data were employed in order to improve the accuracy of trajectory geometry recovery. Image matching and structure from motion (SFM) methods were used to calculate heading direction of trajectory geometry. This method can recover relative location of trajectory sampling points without prior knowledge, such as floorplan or initial location of the trajectory. After trajectory geometry recovering, a trajectory calibration algorithm was proposed to spatially estimate the location of trajectory in the RCS. By using the initial reference point (IRP), the trajectory geometry can be calibrated into RCS. Importantly, sampling points from the calibrated trajectory can be used as supplementary reference points for the following crowdsourced trajectories. Finally, the sensor data that are associated with the sampling points (from the calibrated trajectories) can be used for the mapping of different database, such as Wi-Fi radio map or image database with spatial labels.

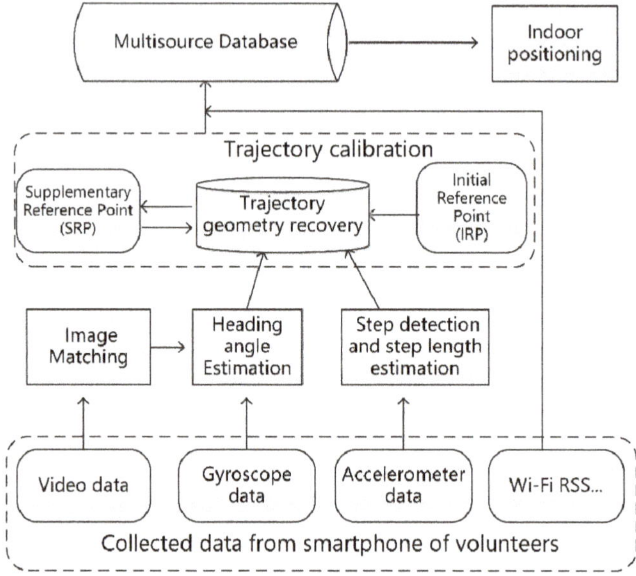

Figure 1. The framework of the proposed method.

During trajectory geometry recovery, the accurate estimation of heading angle of each sampling point from a trajectory is an important issue. The estimations by employing angular velocity (from gyroscope) are usually not accurate due to the drift error of smartphone sensors and the accuracy degradation over time [38]. The method that is proposed in this study uses an SFM-based method for the estimation of heading angle by using video frames. A sliding-window filter-based algorithm is proposed to improve further the performance of heading angle estimation. Finally, the geometry of a trajectory can be recovered by integrating the heading angle of each sampling point and the distance between every pair of adjacent sampling point.

The proposed approach aims to estimate, at every sampling instant t, the pose of sampling point $s_t = (x_t, y_t, \theta_t)$ with regard to the initial pose s_0, where (x_t, y_t) represents the position of s_t relative to the initial position (x_0, y_0), and θ_t is the orientation of s_t relative to the initial heading angle θ_0. Note that the initial pose s_0 is unknown for each trajectory. Section 3 describes the spatial estimation of a trajectory in a coordinate system.

2.1. Heading Angle Estimation

The main idea of this method is that the heading angle of a sampling point can be represented by the heading angle of the image taken at this sampling point. Therefore, the heading angle change of a sampling point sequence can be estimated by calculating the heading angle change of the corresponding image sequence that is extracted from video frames. Similar to [8,11], we use the SFM-based method to estimate the heading angle change in sampling points from a trajectory. An image matching method was implemented on two adjacent images from image sequence. For a pair of images, the homogeneous matching points are used for calculating the fundamental matrix F:

$$\left[u'_i, v'_i, 1\right] \cdot F \cdot \begin{bmatrix} u_i \\ v_i \\ 1 \end{bmatrix} = 0 \tag{1}$$

where $m_i(u_i, v_i, 1)^T$, $m_i'\left(u_i', v_i', 1\right)$ are the homogeneous matching points from the image matching result $\left\{m_i, m_i' \middle| i = 1, 2, \ldots n\right\}$, F is a 3×3 order matrix. It is possible to linearly calculate the matrix F if there are enough matched points [39]. After obtaining the fundamental matrix, the essential matrix E can be calculated as:

$$\mathrm{E} = K^T F K \tag{2}$$

where K is the intrinsic matrix of a smartphone camera, which can be obtained based on the MATLAB Camera Calibrator (MATLAB 8.x on Windows) [40]. The rotation matrix R can be calculated by utilizing singular value decomposition (SVD) of essential matrix E. Importantly, the heading angle of a sampling point can be expressed by a rotation matrix:

$$R = \begin{bmatrix} cos\Delta\theta & 0 & sin\Delta\theta \\ sin\Delta\vartheta sin\Delta\theta & cos\Delta\vartheta & -sin\Delta\vartheta cos\Delta\theta \\ -cos\Delta\vartheta sin\Delta\theta & sin\Delta\vartheta & cos\Delta\vartheta cos\Delta\theta \end{bmatrix} \tag{3}$$

where $\Delta\theta$ is the heading angle change of sampling point P_t relative to the last sampling point P_{t-1}. The schematic diagram of this SFM-based heading angle estimation method is shown in Figure 2. If the initial sampling point heading angle is θ_0, the heading angle of sampling point P_t can be calculated as:

$$\theta_t = \theta_0 + \sum_{i=1}^{t-1} \Delta\theta_i \tag{4}$$

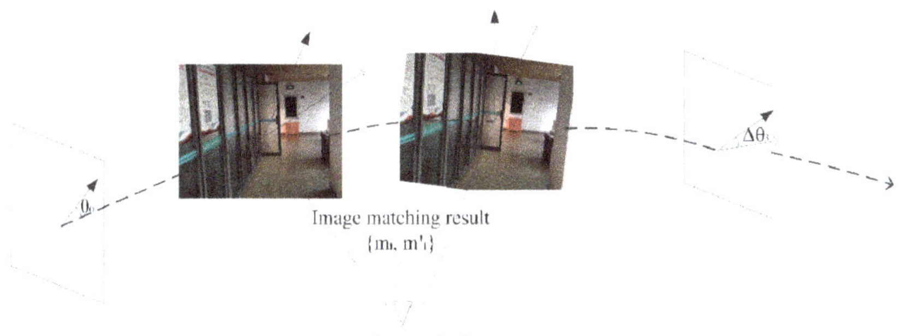

Figure 2. The schematic diagram of structure from motion (SFM)-based heading angle estimation.

2.2. *Trajectory geometry recovery*

In this study, the aim of trajectory geometry recovery is to estimate the relative location of each sampling point from a trajectory. The relative location of a sampling point can be calculated, as follows:

$$\begin{cases} x_t = x_{t-1} + L \cdot sin(\theta_{t-1} + \Delta\theta) \\ y_t = y_{t-1} + L \cdot cos(\theta_{t-1} + \Delta\theta) \end{cases} \tag{5}$$

where $(x_t,\ y_t)$ is the location of sampling point P_t, θ_{t-1} is the heading angle of sampling point P_{t-1}, and $\Delta\theta$ is the heading angle change of P_t that is relative to P_{t-1}. L is the distance between P_t and P_{t-1}. The SFM-based heading angle estimation method can be used to calculate the $\Delta\theta$ of a sampling point, where accuracy of this method depends on the performance of image matching. The accuracy will be affected if the quality of video frames is poor. To solve this issue, the collected inertial data is also used to estimate the heading angle, which is independent from the visual estimation results.

Usually, smartphone gyroscope-based heading angle estimation can be calculated as integral of the angular velocity (rad/s) with respect to time. The frequency of smartphone gyroscope is more than

100 HZ, which is higher than video frame 30 fps. This method is only suitable for estimating heading angles in a short-term condition due to the drift error in smartphone gyroscope. With the increase in the integration time, the error of the heading angle rapidly and continuously grows. In order to avoid this problem, we have employed different strategies for different route segments. As shown in Figure 3, a pedestrian trajectory consists of two types of segments: turning segments (TSs) and non-turning segments (NTSs). A TS segment refers to a turning period with a relatively long turning time; an NTS segment refers to a straight (or approximately straight) walking period that may contain several slight turning actions with very short turning times. The strategy of our method is described as follows:

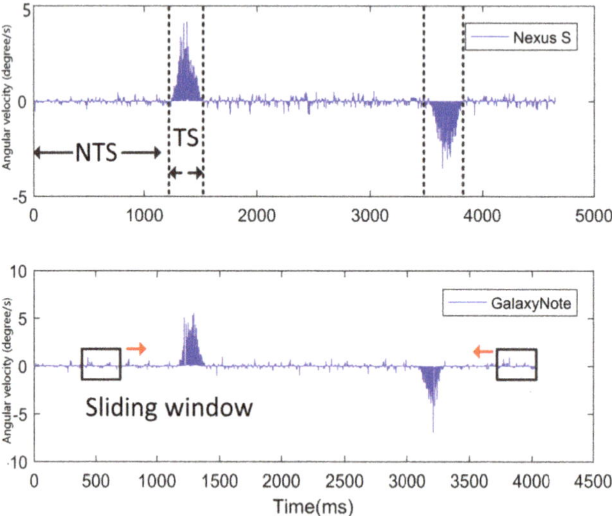

Figure 3. Sliding-window filter-based turning detection.

(1) for each TS sampling point (i.e., sampling points from a TS segment): the SFM-based method is used to estimate the heading angle if there is no image-matching failure.

(2) for TS sampling points with image-matching failure: the gyroscope-based method is used for heading angle estimation.

(3) for NTS sampling point: the lowest from two outputs (SFM-based and gyroscope-based) is used as the final output.

Based on the integration of the two sources, the robustness of the heading angle estimation method can be improved, especially when there is failure of image matching.

One important issue with this method is to accurately detect each turning moment (i.e., the joining point between each pair of TS segment and NTS segment) of a trajectory. When considering the high sampling rates (100 Hz) of gyroscope, this method uses gyroscope readings to detect turning moments. As shown in Figure 3, the angular velocity from an NTS segment fluctuates slightly around zero. However, for a TS segment, the angular velocity is always higher (or lower) than zero and the absolute value is much higher as compared to the NTS segment. According this regularity, a sliding-window filter is designed to detect the starting and ending moments of each turning action of a trajectory, detailed description can be found as follows:

Algorithm 1 Sliding-window filter-based turning moment detection

```
Input: gyroscope angles
Input: sliding window
Output: Turn[,]; //a two-dimensional vector which records the starting and ending moment of each TS
segment of a trajectory
definition: size_win; //the size of the sliding window
count(); // The function to count the number of positive values or negative values
Pair(); // The function to find the starting and ending moment of each TS segment
Turn_S=[]; // The vector to record the candidate moments of start turning
Turn_E=[];// The vector to record the candidate moments of end turning
Np=0;// the number of angular velocity readings which are higher than 0
Ne=0;// the number of angular velocity readings which are below 0
for i=1:length(gyroscope angle)
  sliding window=gyroscope angle[i,(size_win+i)];
  Np=count(sliding window);
  Ne=count(sliding window);
  if Np==size_win || Ne=size_win
    Turn_S.add(i);
  end if
end for
for i=length(gyroscope angle):-1:1
  sliding window=gyroscope angle[(i-size_win),i];
    Np=count(sliding window);
    Ne=count(sliding window);
  if Np==size_win || Ne=size_win
    Turn_E.add(i);
  end if
end for
Turn=Pair(Turn_S, Turn_E);
```

The input of Algorithm 1 includes gyroscope-based heading angle estimations and a sliding window. The gyroscope-based angles are calculated by integrating the angular velocity readings with respect to the timespan between two sampling points. Similar to [41], the value of the sliding window was set to 50 in the experiment. The output of the Algorithm 1 is a two-dimensional vector *Turn*, which records the starting and ending moment for each TS segment. The main idea of Algorithm 1 is to monitor the fluctuation of gyroscope angles within the sliding window. If all of the gyroscope angles within the sliding window are in the same interval (>0˚°or <0˚°), the first moment of the sliding window can be treated as a candidate for the starting or ending moment of a TS segment, which are stored in vector *Turn_S* and *Turn_E*, respectively. For a TS segment, the first candidate in *Turn_S* is treated as its starting moment. Similarly, the first candidate in *Turn_E* is treated as its ending moment.

Based on the turning moment detection algorithm, the visual and inertial estimations can be integrated to calculate the heading angle for each sampling point from a trajectory. Walking steps can be detected by a peak and valley detection algorithm [38], which is an important step in restricting the distance between the adjacent sampling points. The length of each step can be estimated based on a Weinberg model [42]:

$$\text{step\_length}_i = K^4 \sqrt{A_{max} - A_{min}} \tag{6}$$

where step_length_i is the length of the i-th step of a trajectory (i.e., step_i), A_{max} and A_{min} are the maximum and minimum values of the Z-axis acceleration during one step period. K is the ratio of the real distance and the estimated distance. After the estimation of steps and step length, the location of the trajectory sampling points can be calculated. When considering the high sampling rate of gyroscope, we assumed the sampling points are equally spaced in a step. The distance between two sampling points from a trajectory can be calculated as:

$$\text{distance}_{t-1,t} = \frac{1}{n} step\_length_i P_t,\ P_{t-1} \in S_i^P \quad (7)$$

where $\text{distance}_{t-1,t}$ is the distance between two adjacent sampling points P_t and P_{t-1}, which are in the i-th step of a trajectory. The S_i^P is a set of sampling points within step_i, n is the number of sampling points in S_i^P.

3. Trajectory Calibration and Geo-Tagging

In this section, a method is proposed to estimate spatial a trajectory in a reference coordinate system. Each sampling point from a trajectory can be geo-tagged while using an iterative algorithm. The geotagged sampling points, which are associated with the corresponding RSS and image data, can be used to construct multisource datasets for indoor localization.

3.1. Indoor Reference Coordinate System

In most cases, indoor location-based services and applications mainly focus on the location of targets in the local coordinate system (e.g., inside a building), but not the location in a world coordinate system (e.g., WGS84). In this study, firstly, we define a trajectory geometry coordinate system (GCS), which represents the location of a trajectory relative to its initial location. It uses the initial location of a trajectory as its origin and the X and Y axes directly along the east and north, respectively. Subsequently, a two-dimensional reference coordinate system (RCS) is defined to determine the location of a point in the whole indoor space. It uses the location of an initial reference point (IRP) as its origin. Note that the IRP can be arbitrarily selected from an indoor space. For local applications (e.g., navigation services in a building), there is no need to measure its location in a world coordinate system. For global applications, the location of the whole indoor space can be estimated based on the global location of an IRP. The purpose of using IRP is to reduce the application difficulty for the geo-tagging method: it is difficult for many participants to measure the global location of the collected sampling points in a crowdsourcing condition. By using the IRP as an origin, the location of all the collected sampling points (from different participants) can be estimated in the RCS using the proposed algorithm. The location of an IRP needs to be measured once only. For the geo-tagging algorithm, several images are collected at the IRP (in different directions) and they are used as reference images. The algorithm can also be implemented in a multiple IRP condition. The influence of the number of IRP will be evaluated in Section 4. Prior knowledge, such as a floorplan or the initial location of the crowdsourced trajectories, is not required for the geo-tagging algorithm.

The main idea of this algorithm is to determine whether a trajectory crosses a reference point by matching image keypoints between the reference images (images of reference points) and the sampling images (images from a trajectory). If there is a matching success, the location of the sampling points (from the trajectory) can then be estimated in the RCS by using the bundle adjustment (BA) algorithm [43]. More importantly, the spatially estimated sampling points can also be used as reference points, called supplementary reference points (SRPs), to estimate the location of the sampling points from the following trajectories. The sampling images of the SRPs can be directly used as their reference images. The coverage of reference points, including IRPs and SRPs, continuously increases with the increase in crowdsourced trajectories, which makes the algorithm more efficient and robust.

3.2. Geo-Tagging Sampling Points in Reference Coordinate System

To geo-tag sampling points from a trajectory, keypoints from each sample image are matched against those from the reference images (i.e., reference images of IRPs and SRPs) by using the image matching technique that is detailed in Section 2. If the number of successfully matched keypoints is higher than a threshold r, the reference point is used to estimate the location of each sampling point (from the trajectory) by using a BA method. The main idea of BA is to calculate the three-dimensional (3D) location of keypoints and to refine the relative location between images by minimizing the projection error of the keypoints and the tracked keypoints on the images. The result of BA is the optimal 3D location of keypoints and the relative pose among the cameras. The spatial relation between a reference point and a sampling point can be represented by the cameras' relative pose that is calculated by the use of BA. If a sampling point is successfully matched with reference point, the location of all the sampling points from the trajectory can be estimated in the RCS. The sampling points from the estimated trajectory will be used as SRPs for the following trajectories.

As shown in Figure 4a, *Tr1* is a trajectory, and its geometry has been reconstructed by using the method that is proposed in Section 2. An image-matching method is used to find the best matching result (with the highest number of matched keypoints) between the sampling point image P_i and the initial reference point image P_s. Note that its adjacent sampling point P_{i+1} will also be selected as a candidate that may across the IRP P_s if the matching result between P_{i+1} and P_s is higher as compared to P_{i-1} and P_s. Otherwise, P_{i-1} is selected. In Figure 4b, the optimal 3D point cloud and the relative location among P_i, P_{i+1}, and P_s are calculated based on the BA method. After the BA process, the 3D coordinates of the two sampling points are: (x_i, y_i, z_i), $(x_{i+1}, y_{i+1}, z_{i+1})$, respectively. The 3D coordinates should be transformed to the coordinates in the RCS due to the lacking of scale information for the BA method. The scale parameter between the two coordinate systems can be calculated, as follows:

$$\sigma = \frac{\mathrm{D}(\mathrm{i}, \mathrm{i}+1)}{\sqrt{(x_i - x_{i+1})^2 + (y_i - y_{i+1})^2 + (z_i - z_{i+1})^2}} \tag{8}$$

where σ is the scale parameter of BA, $\mathrm{D}(\mathrm{i}, \mathrm{i}+1)$ is the distance between P_i and P_{i+1}. Therefore, the coordinates of a sampling point P_i in the RCS can be calculated, as follows:

$$\begin{bmatrix} X_i \\ Y_i \end{bmatrix} = \sigma \begin{bmatrix} cos\beta & sin\beta \\ -sin\beta & cos\beta \end{bmatrix} \begin{bmatrix} x'_i \\ y'_i \end{bmatrix} \tag{9}$$

where (X_i, Y_i) is the coordinates of P_i in the RCS, β is the heading angle of the matched reference image of IRP P_s, and (x'_i, y'_i) is the two-dimensional (2D) projection of the 3D coordinates (x_i, y_i, z_i) in the RCS. In equation (9), the coordinates of P_{i+1} in the RCS are represented as (X_{i+1}, Y_{i+1}). The coordinate of P_i and P_{i+1} in Section 2 can be calculated as (gx_i, gy_i) and (gx_{i+1}, gy_{i+1}). The transformation parameters, including the rotation angle ϑ and the shiftings $(\mathrm{t}_x, \mathrm{t}_y)$, can be calculated according to the coordinates of P_i and P_{i+1} in the GCS and the RCS [44]. Based on the location estimation results of P_i, the location of each sampling point from the trajectory can be estimated in the RCS, as follows:

$$\begin{bmatrix} X_i \\ Y_i \end{bmatrix} = \begin{bmatrix} \mathrm{t}_x \\ \mathrm{t}_y \end{bmatrix} + \begin{bmatrix} cos\vartheta & sin\vartheta \\ -sin\vartheta & cos\vartheta \end{bmatrix} \begin{bmatrix} gx_i \\ gy_i \end{bmatrix} \tag{10}$$

where (X_i, Y_i) is the coordinates of the sampling point in the RCS, (gx_i, gy_i) is the coordinates of the sampling point in the GCS. Figure 4c shows an example of a recovered trajectory in the RCS.

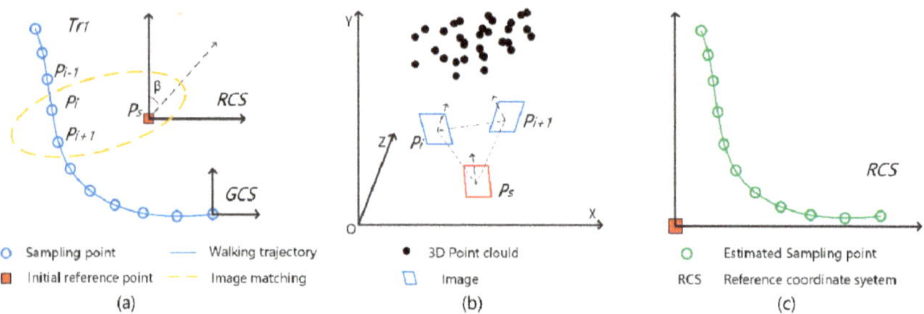

Figure 4. Trajectory estimation in the reference coordinate system (RCS). (**a**) Image matching between an initial reference point (IRP) and the sampling points. (**b**) Calculating the relative pose of Ps by using bundle adjustment (BA) method. (**c**) The estimated trajectory in the RCS.

Once a trajectory has been estimated successfully in the RCS, its sampling points can be used as SRPs. This type of reference points is used to estimate the trajectories that do not cross IRP. As shown in Figure 5, a trajectory may cross multiple SRPs from different trajectories. To increase the robustness of this method, the location of a trajectory is calculated as the average of the estimation results by using each SRP. For example, utilizing supplementary reference point S_1, the coordinates of the sampling points can be calculated as $\{(X_1^1, Y_1^1), (X_2^1, Y_2^1) \ldots (X_k^1, Y_k^1)\}$, where k is the number of sampling points in this trajectory. For example, using SRP S_j, the coordinates of each sampling point from a trajectory can calculated as: $\{(X_1^j, Y_1^j), (X_2^j, Y_2^j) \ldots (X_k^j, Y_k^j)\}$, where k is the number of sampling points from this trajectory. If there are m SRPs, the coordinates of a sampling point can be estimated in the RCS, as follows:

$$\begin{cases} X_i = \frac{1}{m}\left(X_i^1 + X_i^2 + \cdots + X_i^m\right) \\ Y_i = \frac{1}{m}\left(Y_i^1 + Y_i^2 + \cdots + Y_i^m\right) \end{cases} i \in [1,k] \tag{11}$$

where (X_i, Y_i) is the coordinates of the i-th sampling point, m is the number of SRPs that cross the trajectory.

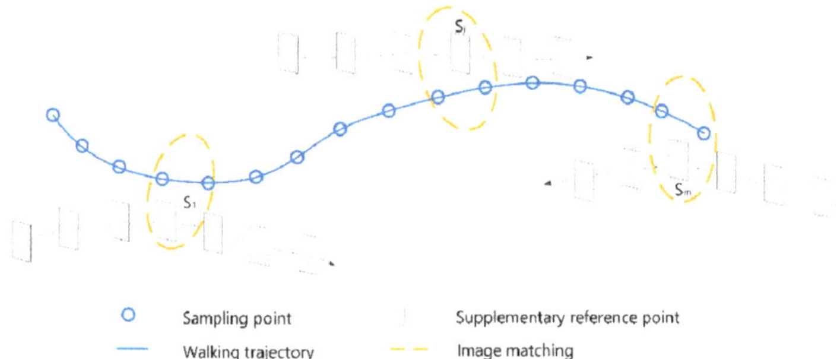

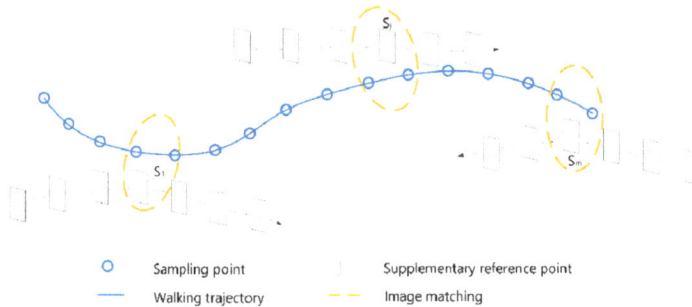

Figure 5. Trajectory estimation in the RCS by using supplementary reference points (SRPs).

The algorithm for trajectory estimation in the RCS is described in Algorithm 2. The inputs of this algorithm include N trajectories (geometry recovered), and at least one IRP where estimated trajectories are the outputs in RCS. The number of available SRPs continuously increases as the iteration of the algorithm. The algorithm ends when all trajectories are estimated.

Algorithm 2 Trajectory estimation in the RCS

```
input: N trajectories with recovered geometry
input: IRP[] //initial reference point
output: Estimated trajectories in the RCS
definition: Multi-IM() is the multi-constrained image-matching function, which returns the number of matched keypoints.
BA() is a bundle adjustment function which returns the location of two adjacent sampling points relative to an IRP
SRP=[]; // supplementary reference point
Label_trajectory=[]; //label a trajectory if it has been estimated in the RCS
While true
    for ss=1:length(IRP)
     for i=1 to N
     if i does not exist in Label_trajectory
        NUM=the number of sampling points of trajectory{i}
        candidate=[];
        for k=1 to NUM
          n=Multi-IM(point{k}, IRP{ss}); // returns the number of matched keypoints
          if n>r //the number of matched keypoints is higher than threshold r
            candidate.add(point{k});
          end if
          if SRP.size>0
            n=Multi-IM(point{k}, SRP);
            if n>r //the number of matched keypoints is higher than threshold r
              candidate.add(point{k});
            end if
          end
        end for
        flag=0; //label whether the two sampling points have been estimated in the RCS
        for j=1: candidate
          if points k and (k+1) exist in candidate[]
            dist= || point{k}, point{k+1} ||; //calculate the distance between point{k} and point{k+1}
            BA(point{k}, point{k+1}, dist, IRP); //calculate the relative location of point{k} and point{k+1} by using the bundle adjustment function
            flag=1;
          end if
          if flag==1
            estimate the trajectory{i} in the RCS;
            SRP.add (sampling points of trajectory{i});
            Label_trajectory.add(i);
            break;
          end if
        end for
        candidate.clear();
    end if
  end for
end for
  if Label_trajectory.number==N
    break;
  end
end while
```

3.3. Generating Multi-Source Datasets for Indoor Positioning

After geo-tagging, the sampling points from crowdsourced trajectories can be used to generate datasets, including Wi-Fi radio map and geo-tagged image datasets. Table 1 shows the attributes of sampling points. The collected images, which are associated with their spatial location and direction attributes, can be directly used to generate geo-tagged image datasets. In order to reduce the time cost required for image matching-based indoor localization, a spatial index is constructed for the geo-tagged images.

Table 1. The attributes of the sampling points.

Sampling Point ID	Trajectory ID	Coordinates	RSS	Image	Direction
P1	Tr_1	(X_1, Y_1)	$\{(rss_1, ap_1), (rss_2, ap_2) \ldots\}$	I1	Azimuth1
P2	Tr_2	(X_2, Y_2)	$\{(rss_1, ap_1), (rss_2, ap_2) \ldots\}$	I2	Azimuth2
P3	Tr_3	(X_3, Y_3)	$\{(rss_1, ap_1), (rss_2, ap_2) \ldots\}$	I3	Azimuth2

In order to construct Wi-Fi radio map, an indoor space can be partitioned into many regular grids. The center of a grid is treated as the location of a Wi-Fi fingerprint. As this method is designed for crowdsourcing-based data collection and localization systems, it is assumed that the spatial distribution of fingerprints in an indoor space is not equal. Some grids may be passed through by multiple crowdsourced trajectories. Additionally, some grids may not be covered by any trajectory. In the first condition, the sampling points from all related trajectories can be integrated to generate fingerprints, we defined as *integrated fingerprint*. If there are m sampling points within the spatial extent of a grid, the RSS of AP i for this grid can be calculated, as follows:

$$\mathrm{rss}_i = \frac{1}{m}\sum\nolimits_{k\in G} rss_i^k \tag{12}$$

where rss_i is the RSS value of AP i, G is the set of AP in the grid, rss_i^k is the rss_i of the k-th sampling point. If a grid does not contain any sampling point, we defined as *interpolated fingerprint*, the RSS of this grid can be interpolated by using its four-neighborhoods:

$$\mathrm{rss}_i = \frac{\sum_j w\left(d_j\right)\cdot RSS\{\}}{\sum_j w\left(d_j\right)} \tag{13}$$

where rss_i is the interpolated RSS value of AP i, $RSS\{\}$ is the four-neighborhoods grid of *interpolated fingerprint*, j is the index of a grid, d is the distance between *interpolated fingerprint* and grid j, $w(x)$ is the weight function which inverse the distance. The purpose of interpolation is to generate a radio map for indoor localization when space has not been completely covered by crowdsourced sampling points. The interpolated RSS of a grid will be replaced by an actual measured RSS when it is covered by the following trajectories.

4. Evaluation

In this section, several experiments are designed to evaluate the performance of the proposed method for indoor geo-tagging and positioning. As shown in Figure 6, a typical indoor environment with 106×61 m was selected as the experimental area with long corridors and wide areas. Two android based smartphones (SUMSUNG and HUAWEI) were used for collecting data, including inertial data, video frames and Wi-Fi RSS. The sampling rates of inertial data was 100 HZ, and the sampling frequency of video frame was 30 frames per second (FPS). During data collection, five volunteers (three males and two females) were invited to collect the experimental data. There are 10 different trajectories and each volunteer performed two trajectories. The participants held a smartphone vertically in front of them and kept the camera forward facing. Three IRPs were selected from the experimental area, as shown in Figure 6.

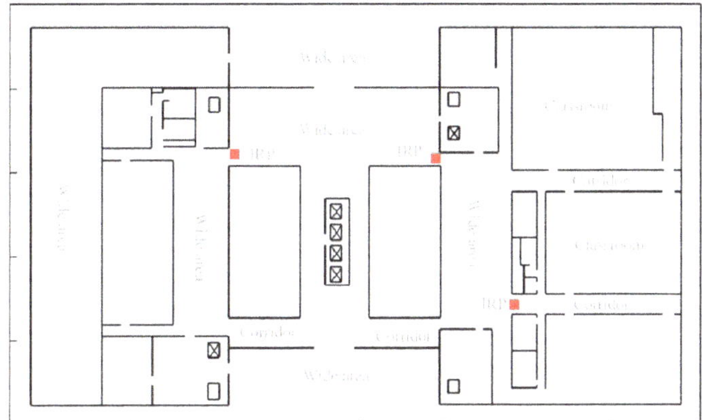

Figure 6. Experimental area with multiple corridors and wide areas.

After data collection, the geometry recovery and geo-tagging of the trajectories were performed offline while using a laptop (4-core i7 CPU and 8G RAM). Based on the result of trajectory calibration and geo-tagging, a Wi-Fi radio map and a geo-tagged image dataset were constructed. An RSS based localization and an image based localization experiments were performed to evaluate the quality of the geo-tagged datasets.

4.1. Evaluation of Trajectory Estimation

In this experiment, ten different trajectories were collected in the study area. As shown in Figure 7a, some markers with known location were set along each trajectory to collect the ground-truth data. The visual results of trajectories recovering are shown in Figure 7b. The average geo-tagging error of sampling points from ten trajectories is around 0.6m, the standard deviation of location error is about 0.4 m. The computation time of recovering 10 trajectories geometry is about 5.7 min.

Figure 7. Ten routes to verify this trajectory recovery method. (**a**) is the ground-truth data, (**b**) is the recovered trajectory.

The results were compared with gyroscope-based method and the SFM-based method in order to further evaluate the performance of the geometry recovery method. The gyroscope based method only uses the gyroscope data from smartphone to calculate the heading angle and restore walking trajectory. The SFM-based method employs a SFM process to estimate the heading angle using video frames.

The shape discrepancy metric (SDM) [45] was used to verify the accuracy of these methods, which is defined as the Euclidean distance between a sampling point and its corresponding ground-truth point. Figure 8 shows the cumulative distribution function (CDF) of the SDM for 10 trajectories while using the three methods. The SDM error in gyroscope-based method is much higher as compared to other methods. For the SFM-based method, the maximum SDM error is about 3 m; the 80-percentile SDM error is around 2 m; and, the mean SDM error is about 1.1 m. This indicates that visual information can help to improve the estimation performance of the trajectory recovery. Moreover, the SDM error can be further reduced by integrating both visual and inertial information: the maximum SDM error is about 1.5 m; the 80-percentile SDM error is around 1 m; and, the mean SDM error is about 0.6 m. Figure 9 shows the increasing speed of trajectory recovery error using different methods. Three routes (#1, #2, and #3) are taken as examples. As can be seen from the figure, the increasing speed of the SFM-based method and the proposed method is obviously slower than the gyroscope-based method. These results demonstrate that the fusion of the visual and inertial information helps to overcome the shortcomings in a single-source based method, e.g., drift error from the gyroscope or matching failure of images. Furthermore, the experimental trajectories covered wide spaces in the study area. The results demonstrate that this approach can perform well in a wide indoor space, which may be difficult for the PDR-based methods.

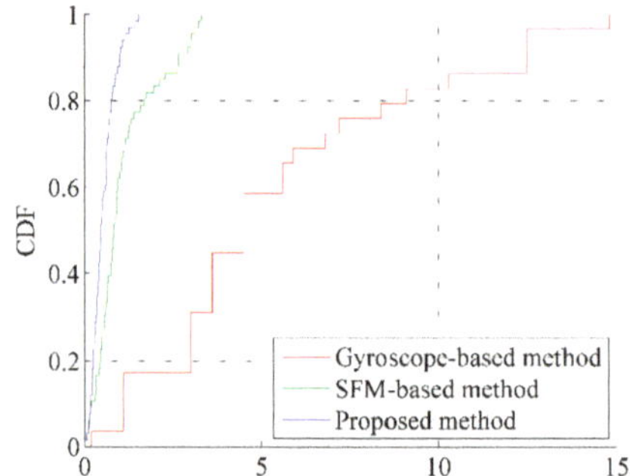

Figure 8. The CDF error of 10 trajectories.

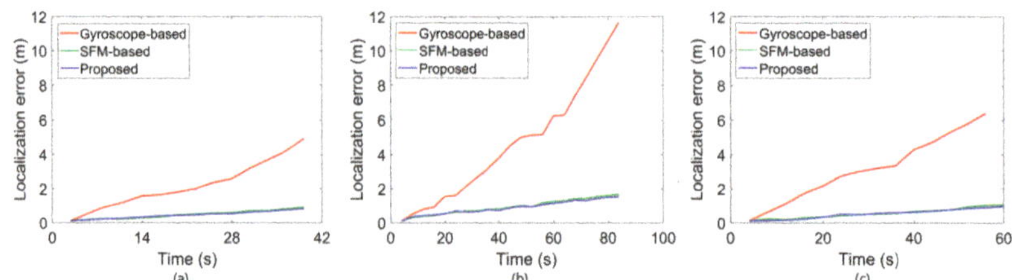

Figure 9. Geometry recovery error drift with time. (**a**) Route#1, (**b**) Route#2, (**c**) Route#3.

Using the algorithm proposed in Section 3, the recovered crowdsourcing trajectories can be geo-tagged in the RCS. Note that the initial locations of these trajectories were unknown for the algorithm. As shown in Figure 10b, three IRPs (points A, B, and C) were set in the study area. Each IRP was associated with 12 reference images (intervals of 30°). Firstly, the algorithm uses one IRP (A) to estimate these trajectories and evaluate the performance of geo-tagging. After that, the two others IRPs are also used to test the influence of the number of IRPs on trajectory estimation and geo-tagging.

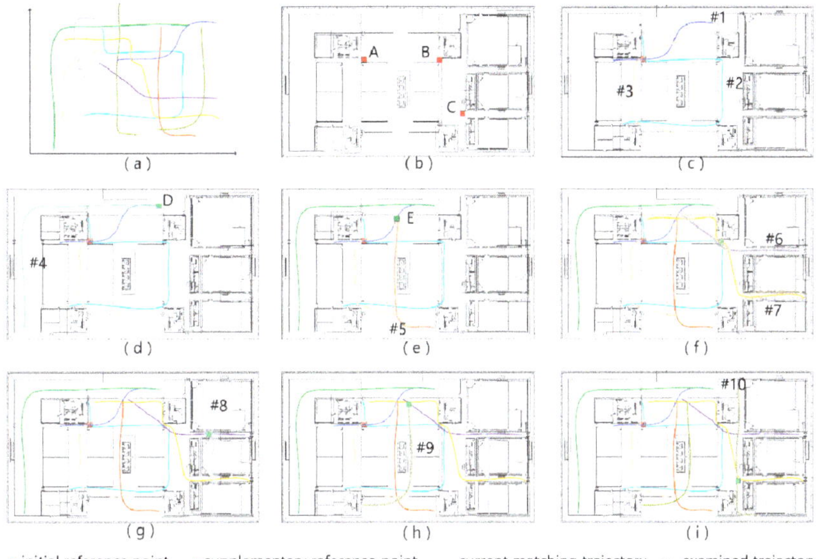

Figure 10. Estimation of 10 trajectories in the RCS. (**a**) The geometrically recovered trajectories. (**b**) Configuration of the three IRPs. (**c**)–(i) The estimation process of the trajectories.

The spatial estimation results of all trajectories are shown in Table 2. The results were evaluated by the maximum, minimum, average estimation error of the ground-truth points, and the computation time of each trajectory. The average error of all the trajectories is 1.03 m. The computation time for estimating these trajectories is 9.3 min.

Table 2. The estimation results of 10 trajectories based on one IRP.

	IRP Trajectory			SRP Trajectory						
Trajectory	#1	#2	#3	#4	#5	#6	#7	#8	#9	#10
max error (m)	1.35	1.42	1.28	1.45	2.1	2.39	2.52	2.85	3.05	2.98
min error (m)	0.2	0.36	0.3	0.32	0.67	0.77	0.68	0.65	0.72	0.58
avg error (m)	0.61	0.77	0.65	0.85	1.09	0.98	1.12	1.55	1.46	1.28
Length (m)	41.7	102.8	56.9	101.1	59.0	55.2	85.6	65.4	57.8	57.6
Time (s)	21	52	29	153	180	980	1505	924	435	1260

As shown in Figure 10, the trajectories that cross an IRP (e.g., trajectories #1, #2, and #3) are termed IRP trajectories. Similarly, the trajectories that cross SRPs are termed SRP trajectories (e.g., trajectories #4–#10). The maximum, minimum, and average error of the IRP trajectories (1.45 m for #3, 0.36 m for #2, 0.85 m for #3) are clearly smaller than those of the SRP trajectories (3.05 m for #8, 0.77 m for #5, 1.55 m for #7), respectively. Nevertheless, the average error of all the SRP trajectories is below 1.56 m, which suggested that the proposed algorithm could achieve reasonable estimation results under the condition of only one reference point. By using SRPs, the spatial location of the SRP trajectories can also be estimated in the RCS.

Figure 10c–i shows the estimation process of all the trajectories. IRP A was firstly used for the estimation of trajectories. As shown in Figure 10c, only three trajectories (#1, #2, and #3) crossed IRP A. By using the method described in Section 3.2, these trajectories were first estimated in the RCS and the sampling points from the trajectories can be used as SRPs. By verifying relation between examined trajectories (nos. #1, #2, #3) and unexamined trajectories (nos. #4–#10), it was found that trajectories #4 and #5 are intersected with trajectory #1 at SRPs D and E, respectively (Figure 10d–e). The sampling points from the newly examined trajectories can also be used as SRPs, which continuously increases the coverage of the reference points in the study area. Note that although trajectories #6 and #7 also intersected with trajectory #1, the intersection relationship were not detected by the algorithm. This may be due to the orientation of the sample images of trajectory #6 (or #7), as it was not consistent with trajectory #1. As shown in Figure 10f–i, after several iterations, the remaining trajectories were all calibrated in the RCS.

To test the influence of the number of IRPs on the trajectory estimation, two other IRPs (B and C) have been added to the environment, as shown in Figure 10b. Same trajectories were estimated by the algorithm by using three IRPs. The average error of all trajectories reduced from 1.03 m (one IRP) to 0.67 m (three IRPs). As shown in Figure 11, after addition of two IRPs, the average error of trajectories #2, #6, #7, and #10 reduced from 0.77 m, 0.98 m, 1.12 m, and 1.28 m to 0.69 m, 0.73 m, 0.61 m, and 0.7 m, respectively. Figure 12 shows the increasing speed of trajectory estimation error. Four routes (#2, #6, #7, and #10) are taken as examples. As can be seen from the figure, once a trajectory passes through a reference point, the location error obviously reduces. The results show that the increase in number of IRPs helps to further improve the performance of the trajectory estimation. However, more IRPs also require more workload for the collection of reference points, including their location and reference images. Accordingly, this may increase the difficulty of crowdsourcing-based indoor positioning systems. Therefore, it is practical to set IRPs at the places where most people walk past, such as indoor intersections and entrances/exits, to reduce the number of required IRPs for large indoor environments.

4.2. *Performance of Constructed Databases for Indoor Positioning*

The sampling points from calibrated trajectory can be used to construct multisource databases for indoor positioning. In this section, two experiments are conducted to evaluate the quality of the constructed datasets, including RSS-based positioning test and image matching based positioning test.

4.2.1. RSS-Based Indoor Positioning

Fifty calibrated trajectories (ten trajectories are shown in Figure 10, where each trajectory is repeated five times) were used to construct an indoor RSS database. The study area was partitioned into a 2.4 m × 2.4 m mesh grid. By using the method described in Section 3.3, the fingerprints were generated based on the integration of sampling points. The results of the fingerprint generation are shown in Figure 13. There were 89 integrated fingerprints (based on the integration of sampling points) and 55 interpolated fingerprints. Figure 14 shows the RSS distribution of two APs.

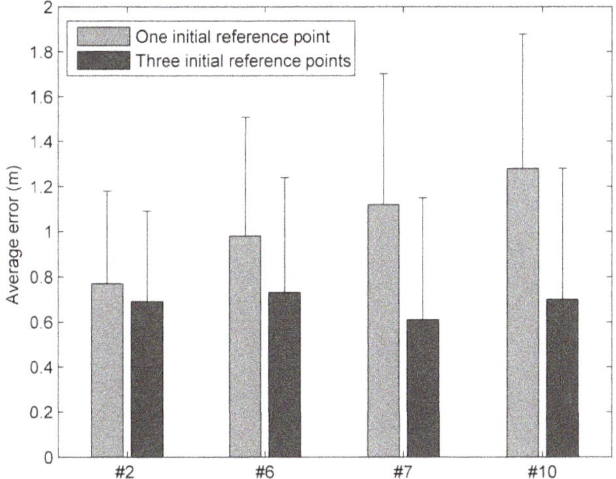

Figure 11. The estimation results under two conditions.

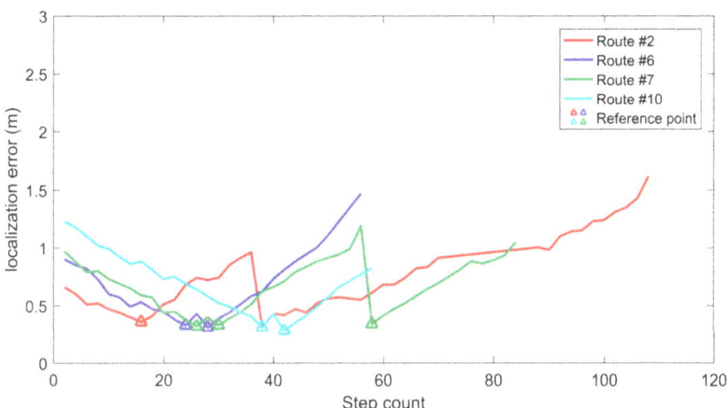

Figure 12. Trajectory estimation error drift with time.

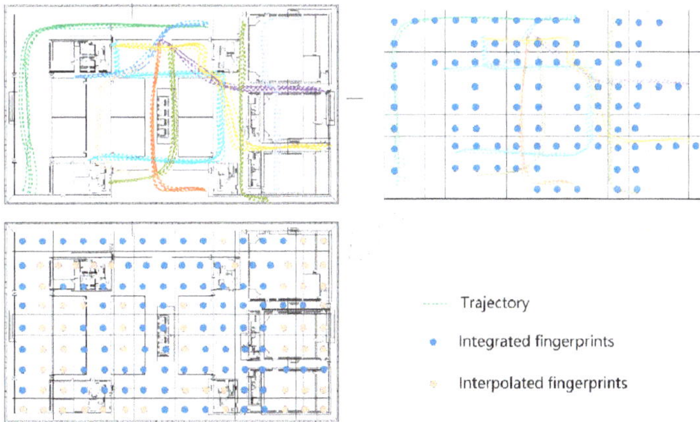

Figure 13. The generation of fingerprints based on the integration of sampling points.

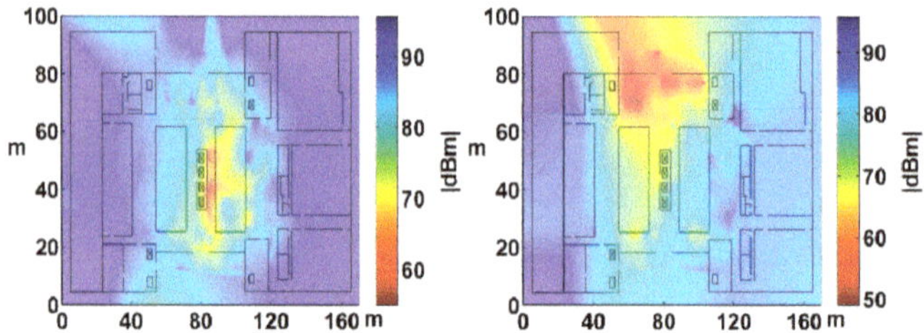

Figure 14. RSS distribution of two APs.

To evaluate the quality of the constructed RSS database, a positioning test was conducted based on a weighted k-nearest neighbor method at the 66 test points (the centers of 66 grids). Each grid was covered by more than five trajectories. The reason to select these grids was to verify the improvement in the quality of constructed radio map with increase in crowdsourced trajectories. For comparison, a site survey process was also conducted based on the same mesh grids. The positioning error was calculated, as follows:

$$\mathrm{Err}_i = \sqrt{\left(x_i^r - x_i^e\right)^2 + \left(y_i^r - y_i^e\right)^2} \tag{14}$$

where Err_i is the location error of point i, $\left(x_i^r, y_i^r\right)$ is the actual physical location of point i, and $\left(x_i^e, y_i^e\right)$ is the estimated physical location of point i.

Figure 15 shows the performance of the two methods (the site survey based method and the proposed method). R0 represents the localization error of the site survey based method. R1 to R5 represent the localization error of the proposed method. Here, R1 refers to the constructed radio map by using sampling points from one trajectory. Similarly, R2, R3, R4, and R5 refer to the radio maps constructed by using two, three, four, and five trajectories, respectively. As it can be seen from Table 3, the localization error of the site survey based method ranges from 0 to 4.9 m and the average error is 2.6 m. The average error of R1 is 4.3 m, which is higher as compared to R0. However, as the increase of the trajectories (from R1 to R5), the average error gradually decreases and reaches 3.2 m as the sampling points are extracted from five trajectories. It indicates that the quality of the constructed database is comparable to site survey based database provided with sufficient crowdsourced data. Once there are enough crowdsourced trajectories, the quality of the constructed radio map will become stable and it may not improve as the further increasing of crowdsourcing data. The proposed system can considerably reduce the human labor that is needed for database construction. Moreover, it performs well in wide indoor spaces, which increases the potential for applying this system to large indoor environments, such as shopping malls, underground parking garages, or supermarkets.

Table 3. Average location error of different method.

Database	R0	R1	R2	R3	R4	R5
Max error	4.9	8.4	7.8	7.2	7.2	7.2
Average error	2.6	4.3	3.7	3.4	3.2	3.2
Error std	1.3	2.2	1.9	1.8	1.8	1.8

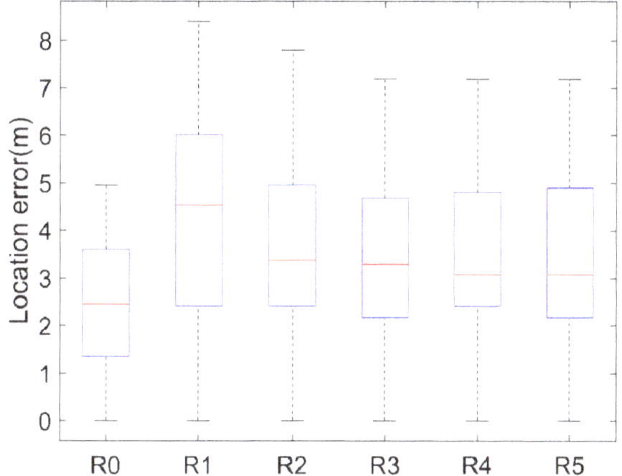

Figure 15. Localization results of the two methods, R0 is the localization error of the site survey-based method, R1 to R5 is the localization error using the crowdsourced radio maps.

4.2.2. Image Matching Based Indoor Positioning

By using the proposed method, each collected sampling point contains single image associated with the corresponding location and direction information. The geo-tagged images can be used to construct image datasets for indoor positioning while using image matching. Most of the image matching based positioning methods use similarity as the metric for location estimation. In the experiment, the number of matched keypoints were calculated to find similarity between a query image and the images from the constructed dataset. The location of image (from the dataset) with the highest number of matched keypoints has been used as the positioning result. In the experiment, 100 different images with known coordinates are used as query images. A SURF [46] based image matching method is used to calculate the similarity among query images and reference images.

Table 4 shows the results. The average location error of image matching based method is 1.2 m, the accurate matching rate is 94%. As compared with other image matching based method [10], the proposed method achieves a relative higher accuracy and matching rate. This can be due to high spatial sampling rate along trajectory which helps to construct image datasets with high spatial resolution. The current methods also help to improve the performance of image matching and reduce the location error of image matching based positioning. As compared with [47], this method does not need laser backpack to construct a 3D model. The time and equipment cost of this method are relatively low, which is important to crowdsourcing-based data collection and indoor localization.

Table 4. Image matching based location error of three method.

Method	Matching Rate	Mean Error	Maximal Error
Proposed	94%	1.2 m	3 m
Reference [10]	80%	Room-level	Quarter-room-level
Reference [47]	94%	1 m	2 m

5. Conclusions

The collection and updating of the indoor positioning database are an unavoidable bottleneck for indoor localization. The traditional site survey is quite labor-intensive and time-consuming, which limits the indoor localization for its commercial and industrial application. In this paper, an efficient geo-tagging method is proposed for crowdsourcing-based indoor positioning. This method can recover the geometry of trajectories and spatially estimate the location of sampling points in the RCS.

Multi-source datasets can be geo-tagged and constructed by using this method in different types of indoor spaces, such as corridors, rooms, or wide spaces. It further minimizes the dependence on prior knowledge, such as floorplans or initial locations of crowdsourced trajectories, which makes the proposed method applicable. The experimental results demonstrated that the integration of visual and inertial information can improve the performance of trajectory recovery and geo-tagging significantly. The average location error of the RSS based positioning and image based positioning are 3.2 m and 1.2 m, respectively.

The proposed method can considerably reduce the workload needed for indoor positioning dataset constructing and updating. We believe that it could serve as a tool for crowdsourcing-based indoor positioning systems and facilitate the participation of the public in the collection of multi-source datasets. In future work, the energy and time cost for crowdsourcing-based data collection and geo-tagging will be studied, which is important for the practical use of the localization system.

Author Contributions: Conceptualization, T.L. and X.Z.; Methodology, T.L. and X.Z; Software, Q.L. and Z.F.; Validation, T.L. and X.Z.; Formal analysis, T.L. and X.Z.; Investigation, T.L. and X.Z.; Resources, Q.L. and Z.F.; Data curation, T.L.; Writing—original draft preparation, T.L.; Writing—review and editing, T.L., X.Z. and N.T.; funding acquisition, T.L., X.Z., Q.L. and Z.F.

Funding: This research was funded by National Science Foundation of China (grants 41801376, 41301511, 41771473), National Key Research Development Program of China (2016YFB0502203), Natural Science Foundation of Guangdong Province (2018A030313289), Shenzhen Scientific Research and Development Funding Program (JCYJ20170818144544900, JCYJ20180305125033478), Open Research Fund of state key laboratory of information engineering in surveying, mapping and remote sensing, Wuhan University (18S03). Key Research Projects of Henan Higher Education Institutions (19A420004). Open Research Fund Program of Shenzhen Key Laboratory of Spatial Smart Sensing and Services (Shenzhen University).

Conflicts of Interest: The authors declare no conflict of interest.

References

1. Wang, A.Y.; Wang, L. Research on indoor localization algorithm based on WIFI signal fingerprinting and INS. In Proceedings of the International Conference on Intelligent Transportation, Big Data & Smart City (ICITBS), Xiamen, China, 25–26 January 2018; pp. 206–209.
2. Lee, N.; Ahn, S.; Han, D. AMID: Accurate magnetic indoor localization using deep learning. *Sensors* **2018**, *18*, 1598. [CrossRef] [PubMed]
3. Chen, P.; Kuang, Y.; Chen, X. A UWB/Improved PDR integration algorithm applied to dynamic indoor positioning for pedestrians. *Sensors* **2017**, *17*, 2065. [CrossRef] [PubMed]
4. Díaz, E.; Pérez, M.C.; Gualda, D.; Villadangos, J.M.; Ureña, J.; García, J.J. Ultrasonic indoor positioning for smart environments: A mobile application. In Proceedings of the IEEE 4th Experiment@ International Conference, Faro, Algarve, Portugal, 6–8 June 2017; pp. 280–285.
5. Bahl, P.; Padmanabhan, V.N. RADAR: An in-building RF-based user location and tracking system. In Proceedings of the IEEE INFOCOM 2000. Conference on Computer Communications. Nineteenth Annual Joint Conference of the IEEE Computer and Communications Societie, Tel Aviv, Israel, 26–30 March 2000; pp. 775–784.
6. Youssef, M.; Ashok, A. The Horus WLAN location determination system. In Proceedings of the 3rd International Conference on Mobile Systems, Applications, and Services, Seattle, WA, USA, 6–8 June 2005; pp. 205–218.
7. He, S.; Chan, S.H.G. Wi-Fi fingerprint-based indoor positioning: Recent advances and comparisons. *IEEE Commun. Surv. Tutor.* **2017**, *18*, 466–490. [CrossRef]
8. Tao, L.; Xing, Z.; Qingquan, L.; Zhixiang, F. A visual-based approach for indoor radio map construction using smartphones. *Sensors* **2017**, *17*, 1790.
9. IndoorAtlas. Available online: https://www.indooratlas.com/ (accessed on 21 June 2019).
10. Ravi, N.; Shankar, P.; Frankel, A.; Elgammal, A.; Iftode, L. Indoor localization using camera phones. In Proceedings of the IEEE Workshop on Mobile Computing Systems & Applications, Orcas Island, WA, USA, 1 August 2006, Orcas Island, WA, USA, 1 August 2006; pp. 1–7.

11. Chen, Y.; Chen, R.; Liu, M.; Xiao, A.; Wu, D.; Zhao, S. Indoor visual positioning aided by CNN-based image retrieval: Training-free, 3D modeling-free. *Sensors* **2018**, *18*, 2692. [CrossRef] [PubMed]
12. Ruotsalainen, L.; Kuusniemi, H.; Bhuiyan, M.Z.H.; Chen, L.; Chen, R. A two-dimensional pedestrian navigation solution aided with a visual gyroscope and a visual odometer. *GPS Solut.* **2013**, *17*, 575–586. [CrossRef]
13. Zhou, B.; Li, Q.; Mao, Q.; Tu, W. A robust crowdsourcing-based indoor localization system. *Sensors* **2017**, *17*, 864. [CrossRef]
14. Zhuang, Y.; Syed, Z.; Georgy, J.; El-Sheimy, N. Autonomous smartphone-based WiFi positioning system by using access points localization and crowdsourcing. *Pervasive Mob. Comput.* **2015**, *18*, 118–136. [CrossRef]
15. Jung, S.H.; Han, D. Automated construction and maintenance of Wi-Fi radio maps for crowdsourcing-based indoor positioning systems. *IEEE Access* **2018**, *6*, 1764–1777. [CrossRef]
16. Yang, S.; Dessai, P.; Verma, M.; Gerla, M. FreeLoc: Calibration-free crowdsourced indoor localization. In Proceedings of the IEEE INFOCOM 2013, Turin, Italy, 14–19 April 2013; pp. 2481–2489.
17. Wu, C.; Yang, Z.; Liu, Y. Smartphones based crowdsourcing for indoor localization. *IEEE Trans. Mob. Comput.* **2015**, *14*, 444–457. [CrossRef]
18. Rai, A.; Chintalapudi, K.K.; Padmanabhan, V.N.; Sen, R. Zee: Zero-effort crowdsourcing for indoor localization. In Proceedings of the 18th Annual International Conference on Mobile Computing and Networking, Istanbul, Turkey, 22–26 August 2012; pp. 293–304.
19. Yang, D.; Xue, G.; Fang, X.; Tang, J. Incentive mechanisms for crowdsensing: Crowdsourcing with smartphones. *IEEE/ACM Trans. Netw.* **2015**, *99*, 1–13. [CrossRef]
20. Zhao, W.; Han, S.; Hu, R.Q.; Meng, W.; Jia, Z. Crowdsourcing and multi-source fusion based fingerprint sensing in smartphone localization. *IEEE Sens. J.* **2018**, *18*, 3236–3247. [CrossRef]
21. Lim, J.S.; Jang, W.H.; Yoon, G.W.; Han, D.S. Radio map update automation for WiFi positioning systems. *IEEE Commun. Lett.* **2013**, *17*, 693–696. [CrossRef]
22. Zhuang, Y.; Syed, Z.; Li, Y.; El-Sheimy, N. Evaluation of two WiFi positioning systems based on autonomous crowd sourcing on handheld devices for indoor navigation. *IEEE Trans. Mob. Comput.* **2015**, *15*, 1982–1995. [CrossRef]
23. Chen, W.; Wang, W.; Li, Q.; Chang, Q.; Hou, H. A crowd-sourcing indoor localization algorithm via optical camera on a smartphone assisted by Wi-Fi fingerprint RSSI. *Sensors* **2016**, *16*, 410.
24. Zhang, C.; Subbu, K.P.; Luo, J.; Wu, J. GROPING: Geomagnetism and crowdsensing powered indoor navigation. *IEEE Trans. Mob. Comput.* **2014**, *14*, 387–400. [CrossRef]
25. Wu, T.; Liu, J.; Li, Z.; Liu, K.; Xu, B. Accurate smartphone indoor visual positioning based on a high-precision 3D photorealistic map. *Sensors* **2018**, *18*, 1974. [CrossRef]
26. Gao, R.; Tian, Y.; Ye, F.; Luo, G.; Bian, K.; Wang, Y.; Li, X. Sextant: Towards ubiquitous indoor localization service by photo-taking of the environment. *IEEE Trans. Mob. Comput.* **2015**, *15*, 460–474. [CrossRef]
27. Bollinger, P. Redpin–adaptive, zero-configuration indoor localization through user collaboration. In Proceedings of the First ACM International Workshop on Mobile Entity Localization and Tracking in GPS-Less Environments, San Francisco, CA, USA, 14–19 September 2008; pp. 55–60.
28. Park, J.G.; Charrow, B.; Curtis, D.; Battat, J.; Minkov, E.; Hicks, J.; Ledlie, J. Growing an organic indoor location system. In Proceedings of the 8th International Conference on Mobile Systems, Applications, and Services, San Francisco, CA, USA, 15–18 June 2010; pp. 271–284.
29. Lee, M.; Jung, S.H.; Lee, S.; Han, D. Elekspot: A platform for urban place recognition via crowdsourcing. In Proceedings of the 2012 IEEE/IPSJ 12th International Symposium on Applications and the Internet, Izmir, Turkey, 16–20 July 2012; pp. 190–195.
30. Wu, C.; Yang, Z.; Liu, Y.; Xi, W. WILL: Wireless indoor localization without site survey. *IEEE Trans. Parallel Distrib. Syst.* **2012**, *24*, 839–848.
31. Liu, T.; Zhang, X.; Li, Q.; Fang, Z.X. Modeling of structure landmark for indoor pedestrian localization. *IEEE Access* **2019**, *7*, 15654–15668. [CrossRef]
32. Gu, Y.; Zhou, C.; Wieser, A.; Zhou, Z. WiFi based trajectory alignment, calibration and crowdsourced site survey using smart phones and foot-mounted IMUs. In Proceedings of the International Conference on Indoor Positioning and Indoor Navigation (IPIN), Sapporo, Japan, 18–21 September 2017; pp. 1–6.

33. Chen, S.; Li, M.; Ren, K.; Qiao, C. Crowd map: Accurate reconstruction of indoor floor plans from crowdsourced sensor-rich videos. In Proceedings of the IEEE 35th International Conference on Distributed Computing Systems, Columbus, OH, USA, 29 June–2 July 2015; pp. 1–10.
34. Pan, J.J.; Pan, S.J.; Yin, J.; Ni, L.M.; Yang, Q. Tracking mobile users in wireless networks via semi-supervised colocalization. *IEEE Trans. Pattern Anal. Mach. Intell.* **2011**, *34*, 587–600. [CrossRef] [PubMed]
35. Park, J.G.; Curtis, D.; Teller, S.; Ledlie, J. Implications of device diversity for organic localization. In Proceedings of the IEEE INFOCOM 2011, Shanghai, China, 10–15 April 2011; pp. 3182–3190.
36. Wu, Z.; Jiang, L.; Jiang, Z.; Chen, B.; Liu, K.; Xuan, Q.; Xiang, Y. Accurate indoor localization based on CSI and visibility graph. *Sensors* **2018**, *18*, 2549. [CrossRef] [PubMed]
37. Wang, X.; Gao, L.; Mao, S.; Pandey, S. CSI-based fingerprinting for indoor localization: A deep learning approach. *IEEE Trans. Veh. Technol.* **2016**, *66*, 763–776. [CrossRef]
38. Kang, W.; Han, Y. SmartPDR: Smartphone-based pedestrian dead reckoning for indoor localization. *IEEE Sens. J.* **2015**, *15*, 2906–2916. [CrossRef]
39. Luong, Q.T.; Faugeras, O.D. The fundamental matrix: Theory, algorithms, and stability analysis. *Int. J. Compt. Vis.* **1996**, *17*, 43–75. [CrossRef]
40. Bouguet, J.Y. Camera Calibration Toolbox for Matlab. Available online: http://www.vision.caltech.edu/bouguetj/calib_doc/ (accessed on 1 June 2017).
41. Mladenov, M.; Mock, M. A step counter service for Java-enabled devices using a built-in accelerometer. In Proceedings of the 1st International Workshop on Context-Aware Middleware and Services: Affiliated with the 4th International Conference on Communication System Software and Middleware (COMSWARE 2009), Dublin, Ireland, 16 June 2009; pp. 1–5.
42. Jahn, J.; Batzer, U.; Seitz, J.; Patino-Studencka, L.; Boronat, J.G. Comparison and evaluation of acceleration based step length estimators for handheld devices. In Proceedings of the International Conference on Indoor Positioning and Indoor Navigation, Zurich, Switzerland, 15–17 September 2010; pp. 1–6.
43. Torr, P.H.; Zisserman, A. *Vision Algorithms: Theory and Practice*; Springer: Berlin, Germany, 1999.
44. Akyilmaz, O. Total least squares solution of coordinate transformation. *Surv. Rev.* **2007**, *39*, 68–80. [CrossRef]
45. Shen, G.; Chen, Z.; Zhang, P.; Moscibroda, T.; Zhang, Y. Walkie-Markie: Indoor pathway mapping made easy. In Proceedings of the 10th Symposium on Networked Systems Design and Implementation, Lombard, IL, USA, 2–5 April 2013; pp. 85–98.
46. Bay, H.; Tuytelaars, T.; Van Gool, L. Surf: Speeded up robust features. In Proceedings of the European Conference on Computer Vision, Graz, Austria, 7–13 May 2006; pp. 404–417.
47. Liang, J.Z.; Corso, N.; Turner, E.; Zakhor, A. Image based localization in indoor environments. In Proceedings of the 2013 Fourth International Conference on Computing for Geospatial Research and Application, Washington, DC, USA, 22–24 July 2013; pp. 70–75.

Article

DRE-SLAM: Dynamic RGB-D Encoder SLAM for a Differential-Drive Robot

Dongsheng Yang, Shusheng Bi, Wei Wang, Chang Yuan, Wei Wang, Xianyu Qi and Yueri Cai *

Robotics Institute, Beihang University, Beijing 100191, China; ydsf16@buaa.edu.cn (D.Y.); ssbi@buaa.edu.cn (S.B.); wangweilab@buaa.edu.cn (W.W.); yuanchang@buaa.edu.cn (C.Y.); wangwbh@buaa.edu.cn (W.W.); qixianyu@buaa.edu.cn (X.Q.)

* Correspondence: caiyueri@buaa.edu.cn; Tel.: +86-10-82314554

Received: 16 January 2019; Accepted: 9 February 2019; Published: 13 February 2019

Abstract: The state-of-the-art visual simultaneous localization and mapping (V-SLAM) systems have high accuracy localization capabilities and impressive mapping effects. However, most of these systems assume that the operating environment is static, thereby limiting their application in the real dynamic world. In this paper, by fusing the information of an RGB-D camera and two encoders that are mounted on a differential-drive robot, we aim to estimate the motion of the robot and construct a static background OctoMap in both dynamic and static environments. A tightly coupled feature-based method is proposed to fuse the two types of information based on the optimization. Dynamic pixels occupied by dynamic objects are detected and culled to cope with dynamic environments. The ability to identify the dynamic pixels on both predefined and undefined dynamic objects is available, which is attributed to the combination of the CPU-based object detection method and a multiview constraint-based approach. We first construct local sub-OctoMaps by using the keyframes and then fuse the sub-OctoMaps into a full OctoMap. This submap-based approach gives the OctoMap the ability to deform, and significantly reduces the map updating time and memory costs. We evaluated the proposed system in various dynamic and static scenes. The results show that our system possesses competitive pose accuracy and high robustness, as well as the ability to construct a clean static OctoMap in dynamic scenes.

Keywords: visual simultaneous localization and mapping; dynamic environment; RGB-D camera; encoder; OctoMap

1. Introduction

Visual simultaneous localization and mapping (V-SLAM) provides localization and perception capabilities for indoor mobile robots. The state-of-the-art V-SLAM algorithms enable high-precision pose estimations and provide impressive maps [1–3]. Most of the V-SLAM methods perform well in static sceneries, while they tend to fail in dynamic scenes that are full of moving objects. Typical dynamic scenes include homes, offices, and factories, where people, animals, machines, or vehicles are in motion. Moreover, mobile robot navigation requires a suitable dense map representation. Although there have been many breakthrough methods for dense mapping [2,3], most of them cannot cope with dynamic scenes. Most dense mapping methods assume a static world. They build dynamic objects into the map, which is not suitable for robot navigation. This paper aims to simultaneously estimate the robot pose and construct a static background dense map for a differential-drive robot working in real dynamic indoor scenes.

Pure V-SLAM is fragile in challenging scenarios, such as texture-less scenes or where fast motion is involved [4]. Especially in dynamic environments, there may not be enough static pixels for the estimation. Applying the inertial measurement unit (IMU) can effectively increase the robustness of the V-SLAM system [5]. As a substitution for the IMU, wheel-encoders possess unique advantages for differential-drive robots moving on a two-dimensional (2D) plane. Since the data-processing is simple, the encoders do not require initialization, the encoder measurements do not drift with temperature, and the encoder integration does not diverge with time. Whereas the encoder integration diverges with distance, the encoder is only suitable for 2D motion.

Localization and dense mapping for dynamic environments require pixelwise dynamic data detection. Semantic segmentation [6] and instance segmentation [7–11] allow predefined dynamic objects to be detected at the pixel level. However, these methods heavily rely on the graphics processing unit (GPU), which is costly for most indoor mobile robots. Furthermore, the states of many objects, in motion or still, are difficult to predefine. For example, a chair may be pulled by someone, or it may be stationary. In these cases, semantic segmentation and instance segmentation may not work well.

Point cloud [12], surfel [2], and volumetric [13,14] representations are three commonly used dense environment representations [4]. The volumetric representation can be used directly for path planning [8], unlike the other two representations. OctoMap [14] is a popular volumetric-based representation for robot navigation. It allows compact memory expression and provides the following three types of information: occupied, free, and unknown. However, OctoMap suffers the following shortcoming: it is not deformable, which means the entire map should be reconstructed through reprocessing all of the original raw data after the loop closure. The process is time-consuming and memory intensive.

To address these problems, we present a SLAM system that can operate both in dynamic and static environments. The proposed system leverages information of an RGB-D camera and two wheel-encoders, which are common configurations for differential-drive robots. The outputs are 2D robot poses and an OctoMap of the static background. We name the system DRE-SLAM (Dynamic RGB-D Encoder SLAM). DRE-SLAM is built based on the feature-based sparse V-SLAM method. It tightly fuses the information of the RGB-D camera and the wheel-encoders by using optimization, which results in a series of keyframes (KFs) with high-precision poses. The dynamic pixels in the keyframes are detected and deleted to eliminate interference from moving objects. Based on the dynamic-free keyframes, we build a static background OctoMap. The main contributions are summarized as follows:

- A robust and accurate framework based on optimization to fuse the information of an RGB-D camera and two wheel-encoders for dynamic environments.
- A dynamic pixel detection and culling method that does not need GPUs and can detect dynamic pixels both on predefined and undefined moving objects.
- A submap-based OctoMap construction method, which can speed up the rebuilding process and decrease memory consumption.

Various evaluations were performed to test the performance of the DRE-SLAM both in dynamic and static environments. The results show that our system achieves competitive accuracy and robustness both in dynamic and static environments. The OctoMap is unaffected by moving objects. The open-source implementation, dataset, and video are available at: https://github.com/ydsf16/dre_slam.

The remainder of this paper is organized as follows. Section 2 briefly reviews the sensor fusion methods, the key technologies, and the popular SLAM methods for static and dynamic environments. Section 3 presents preliminaries, assumptions, and definition of the problem to be solved. Section 4 describes the proposed DRE-SLAM system in detail. Section 5 compares the proposed method with state-of-the-art systems in terms of accuracy, map quality, and robustness. Additionally, the performances of the main modules are tested. Finally, we conclude our work in Section 6.

2. Related Work

2.1. Sensor Fusion

A single sensor has unique advantages and limitations. Combining multiple sensors can effectively improve the performance of a SLAM system [15]. The commonly used sensor configuration is visual plus IMU, which results in visual inertial SLAM (VI-SLAM). The VI-SLAM methods, such as VINS-Mono [5], are robust in dynamic environments. Recently, Tristan, et al. [16] proposed a dense RGB-D-Inertial-SLAM system. However, it is limited to static environments. Compared to VI-SLAM, studies on visual encoder SLAM (VE-SLAM) methods are still relatively rare. The encoders contain special advantageous compared with the IMU for robots moving on a 2D plane. The representative advantage is that the integration of the encoder measurements diverge with distance, whereas the integration of the IMU measurements diverge over time. The VE-SLAM method can deal with indoor untextured scenes, such as white walls, over a long period. RTAB-MAP [15] is a versatile system. It can robustly combine RGB-D images with wheel odometry. However, it cannot handle dynamic environments. Moreover, the local robot pose estimation only uses the wheel odometry without coupling the visual information. In this paper, we use both encoder and RGB-D information to estimate the robot pose of every frame.

2.2. Dynamic Pixels Detection

Identifying the dynamic image pixels is the key to coping with dynamic environments. In recent years, several methods have been explored. The first method relies on deep learning techniques, such as semantic segmentation [6] and instance segmentation [7–11]. This method can detect predefined dynamic objects at the pixel level, but it cannot do anything for undefined moving objects or objects with uncertain motion properties. Moreover, the substantial demand on GPUs limits its application on consumer indoor robots. The second method takes advantage of the constraints introduced by multiview geometry [6,9,17]. It assumes that static pixels satisfy the model of multiview geometry, while dynamic pixels do not. However, this method generally uses a threshold to determine the dynamic and static properties of the pixels, which easily causes over recognition or less recognition. For example, a person who remains still for a long time may be incorrectly considered to be static and added to the map. The third method uses background [18] or foreground [19] detection algorithms. Most of its implementations rely on GPUs and cannot run in real time. The fourth method is the scene flow [8,20,21] or optical flow [22] method, which is based on the fact that dynamic objects obey different motion patterns. This method suffers from similar drawbacks to the multiview constraint-based method. Since a single method has its limitations, recent studies have attempted to combine the deep learning method with multiview constraint-based method [6,9] or with scene flow method [8]. Our approach falls into this category. However, we do not use semantic segmentation and instance segmentation. We use the lightweight object detection method, which can operate on a CPU, eliminating the reliance on GPUs. Furthermore, we propose a cluster-based method to speed up the multiview constraint-based detection process.

2.3. Simultaneous Localization and Dense Mapping

The emergence of the RGB-D sensors makes indoor 3D dense mapping affordable. The first breakthrough work is KinectFusion [13]. It tracks the camera motion by using a coarse-to-fine iterative closest point (ICP) [23] method and fuses the depth images to a truncated signed distance function (TSDF) [24] map. The TSDF map divides the 3D space into voxels. Each voxel is encoded as the distance to the closest surface. However, the massive memory consumption and no loop closure limit the application scenarios to small scales. These shortcomings are then solved by the subsequent works [25,26]. Unlike the traditional SLAM methods, ElasticFusion [2] does not optimize the pose-graph in loop closure but optimizes a deformable graph. This method performs tracking using the frame to model approach. It implements local and global loop closures using the model to model approach. Its map is represented by the surfel model. However, the surfel representation contains only two types of information, i.e., free and occupied. The unknown information, which is required by exploration tasks, is not available. Furthermore, most of these methods are implemented on the GPU to ensure real-time operation. In contrast, OctoMap can run on CPUs and provides all three types of information: occupied, free, and unknown [14]. Similar to TSDF, OctoMap divides the space into voxels. The difference is that each voxel of OctoMap is given an occupied probability and a Bayesian filter is used to update the occupied probability with multiple measurements. The update policy can filter out a small amount of dynamic data. However, OctoMap is not deformable. After finishing the pose adjustment in the loop closure, the full map needs to be rebuilt with the original raw data, which consumes a great amount of storage space and computing time. In this paper, we use a submap representation. The raw data are first integrated into many small submaps. After the pose adjustment, we only need to reassemble these submaps into a full map. The method can improve the map update efficiency and decrease the memory consumption. RTAB-MAP [15] uses a similar submap-based representation to construct OctoMap. However, its submap only contains one frame. It does not analyze the impact of the submap size and voxel size on the reconstruction time and memory consumption. RGBDSLAMv2 [27] first uses a sparse feature-based SLAM to obtain high-precision camera poses and then constructs the OctoMap based on these poses. However, this method cannot update dense maps efficiently after the pose graph is changed. Similar to RGBDSLAMv2, our method also first implements sparse feature-based SLAM and then constructs an OctoMap using the keyframes. The difference is that our approach introduces encoder information, enables efficient OctoMap update, and handles dynamic environments.

2.4. Simultaneous Localization and Dense Mapping for Dynamic Scenes

The simultaneous localization and dense mapping systems for dynamic environments can be divided into three categories.

The first type of system detects and removes dynamic objects and constructs a dense map of the static background. DynaSLAM [9] extends the capabilities of ORB-SLAM2 [1] to build static dense point cloud maps in dynamic environments. This system uses the instance segmentation method Mask-RCNN [28], multiview constraints, or both of them to detect moving pixels in each RGB-D frame. However, the system is not real-time. Furthermore, the dense point cloud map is too rough for robot navigation. DS-SLAM [6] is another algorithm derived from ORB-SLAM2. This system uses the semantic segmentation method SegNet [29] to detect pixels in which people are located from the RGB image; then, only the features that fall on the people are checked by moving consistency to determine if they are dynamic. An OctoMap-based semantic map is finally constructed using static pixels in the depth images. This system only considers predefined dynamic objects, such as person. StaticFusion [17] jointly estimates the camera motion as well as a probabilistic static/dynamic segmentation of the RGB-D image pairs. The segmentation is then used to fuse a surfel-based static background model. To achieve real-time performance, the authors reduced the image size to 320×240 pixels. Moreover, the people that are stationary for a long time may be detected as static and fused to the map. Most of the above systems require powerful GPU support. Our system falls into this category. We can run on the CPU and detect dynamic pixels on both predefined and undefined moving objects.

The second type of system not only constructs a dense static background but also tracks the moving objects [7,8,10]. This system aims to support path planning and operational tasks for the robot in a dynamic environment. Unlike them, our goal is to provide a static map as a basic reference for mobile robot navigation.

The third type of system constructs a nonrigid deformable environment [30,31]. However, the operating environment is limited to a small area. Moreover, the nonrigid deformable environment representation may not be applied to mobile robots, as mobile robots own a more extensive operating space and more environmental changes.

3. Preliminaries

Figure 1 presents a differential-drive robot equipped with an RGB-D camera. Each wheel is mounted with an encoder. We use the following three types of coordinate frames: the world frame $\{w\}$, the camera frame $\{c\}$, and the robot frame $\{r\}$. The world frame $\{w\}$ is coincident with the robot frame $\{r\}$ at the start time. The extrinsic parameters and the kinematics parameters of the robot are precalibrated by [32]. The extrinsic parameters, which are expressed by a 4×4 transformation matrix ${}^{r}_{c}\mathbf{T} \in SE(3)$, are the transformation from the camera frame $\{c\}$ to the robot frame $\{r\}$. The kinematic parameters are the left and right wheel factors k_l, k_r and wheel space b. The wheel factors transform the encoder displacement in unit tick to the wheel displacement in unit m. The robot pose is the transformation from the robot frame $\{r\}$ to the world frame $\{w\}$, which is expressed by a 4×4 transformation matrix ${}^{w}_{r}\mathbf{T} \in SE(3)$. In fact, the robot pose only contains three degrees-of-freedom (DoF), as the robot only moves in a 2D plane. Thus, the robot pose can be expressed by a 3×1 vector: ${}^{w}_{r}\zeta = [x, y, \theta]^T \in \mathbb{R}^3$, $[x, y]^T \in \mathbb{R}^2$ is the translation, and $\theta \in (-\pi, \pi]$ is the yaw angle. Then, ${}^{w}_{r}\mathbf{T}$ takes the following form:

$$
{}^{w}_{r}\mathbf{T} = \begin{bmatrix} \cos(\theta) & -\sin(\theta) & 0 & x \\ \sin(\theta) & \cos(\theta) & 0 & y \\ 0 & 0 & 1 & 0 \\ 0 & 0 & 0 & 1 \end{bmatrix}, \tag{1}
$$

${}^{w}_{r}\mathbf{T}$ and ${}^{w}_{r}\zeta$ are equivalent. Given ${}^{w}_{r}\mathbf{T}$, we can obtain ${}^{w}_{r}\zeta$ by the function $F(\cdot)$:

$$
{}^{w}_{r}\zeta = F\left({}^{w}_{r}\mathbf{T}\right) = \begin{bmatrix} {}^{w}_{r}\mathbf{T}(1,4) \\ {}^{w}_{r}\mathbf{T}(2,4) \\ \arctan 2\left({}^{w}_{r}\mathbf{T}(2,1), {}^{w}_{r}\mathbf{T}(1,1)\right) \end{bmatrix}, \tag{2}
$$

where ${}^{w}_{r}\mathbf{T}(i,j)$ is the element in the matrix ${}^{w}_{r}\mathbf{T}$ indexed by (i,j).

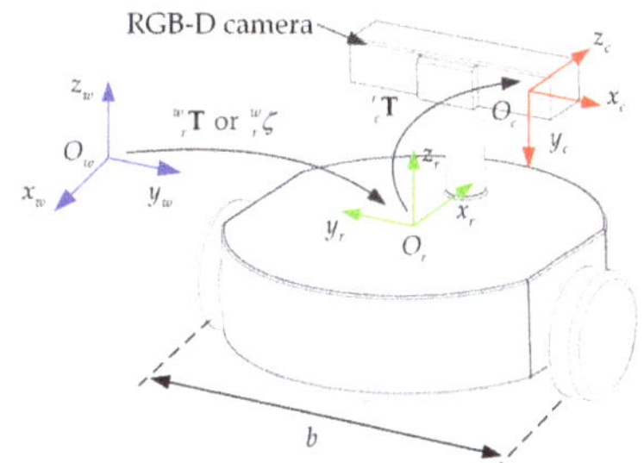

Figure 1. Differential-drive robot equipped with an RGB-D camera.

The RGB-D camera outputs pairs of color image C and depth image D: $\{C, D\}$. The depth image is preregistered on the color image. We use the pinhole camera model $\pi(\cdot)$, which projects a 3D point ${}^{c}\mathbf{p} \in \mathbb{R}^3$ in the camera frame $\{c\}$ to a 2D point $\mathbf{u} \in \mathbb{R}^2$ on the image:

$$
\mathbf{u} = \pi\left({}^{c}\mathbf{p}\right). \tag{3}
$$

The inverse camera model projects a 2D pixel to a 3D point with depth z:

$$
{}^{c}\mathbf{p} = \pi^{-1}(\mathbf{u}, z). \tag{4}
$$

The aim of this paper is to simultaneously estimate the robot pose ${}^{w}_{r}\mathbf{T}$ or ${}^{w}_{r}\zeta$ and create a static OctoMap in both dynamic and static environments. The map is expressed in the world frame $\{w\}$.

4. Dynamic RGB-D Encoder SLAM (DRE-SLAM)

4.1. System Overview

Figure 2 provides an overview of the proposed DRE-SLAM system. This system leverages two types of inputs: the RGB-D image pairs and the encoder measurements. The outputs are a series of robot poses and an OctoMap. The system is based on the sparse feature-based V-SLAM. Coupled with the addition of encoder information and the culling of dynamic pixels, we develop a sparse RGB-D Encoder SLAM for dynamic environments. We then build an OctoMap based on the keyframes. The pipeline is divided into four modules: RGB-D Encoder tracking, dynamic pixels culling, sparse mapping, and OctoMap construction.

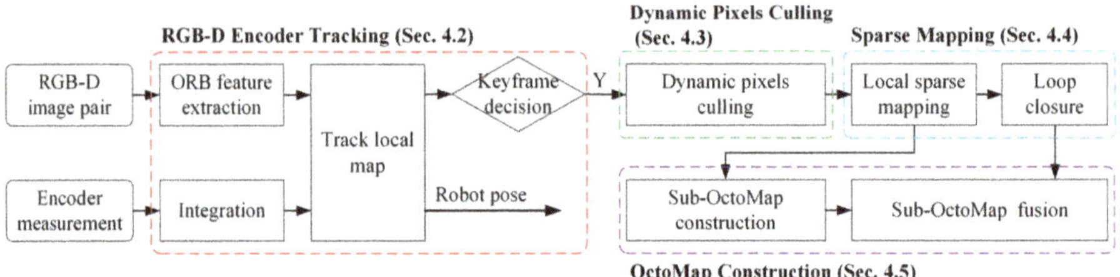

Figure 2. Pipeline of the proposed DRE-SLAM.

RGB-D Encoder Tracking. When an RGB-D pair $\{C, D\}$ comes in, it will be constructed as the current frame and then will be extracted a sparse set of ORB features. As the output frequency of the encoder is generally much higher than that of the camera, we integrate the encoder measurements between the last keyframe and the current frame to obtain a reliable initial robot pose of the current frame to reduce the calculation complexity. The 3D local map points are projected to the current frame to match with the 2D ORB features. The robot pose of the current frame is calculated by jointly minimizing projection errors and encoder errors. Then, whether the current frame is become a new keyframe or not is decided.

Dynamic Pixels Culling. Only the keyframes are used to construct the OctoMap and the sparse map in our system. The key to building a static map in a dynamic environment is identifying and removing the dynamic pixels. This module first detects the predefined dynamic objects in the color image C using an object detection method and then uses multiview constraints to identify dynamic pixels in the depth image D. The dynamic pixels detected by the two approaches are all culled.

Sparse Mapping. Given a new keyframe, this module performs a sliding window local bundle adjustment (BA), which minimizes projection errors and encoder errors. In this way, the robot pose of the new keyframe is refined. The refined new keyframe is sent to the OctoMap construction module. Then, we construct new map points. After this, we detect the loops between the new keyframe and the history keyframes. If a loop is detected, a pose graph optimization will be implemented.

OctoMap Construction. This module uses keyframes to construct sub-OctoMaps first and then assembles these sub-OctoMaps into a full OctoMap. Once the loop closure has completed, we only need to reassemble the sub-OctoMaps.

Our approach runs on six separative threads to obtain real-time operation. The threads are RGB-D encoder tracking, dynamic pixels culling, local sparse mapping, loop closure, sub-OctoMap construction, and sub-OctoMap fusion.

4.2. RGB-D Encoder Tracking

4.2.1. ORB Feature Extraction

We extract 1600 ORB features from the image with resolution of 960 × 540 pixels produced by a Kinect 2.0 camera. Note that when there are uniform surfaces or fewer textures in the environment, we may not be able to extract enough features in the image. Fortunately, we introduce the encoder information, which enables the system to work properly in these cases. The 2D positions of these features are organized in a KD-Tree to speed up the search process. Homogeneous feature distribution is much more critical for SLAM systems in dynamic scenes because it can reduce the number of feature points on dynamic objects, especially when the dynamic objects contain rich textures. We use the same feature detection policy as the ORB-SLAM2 [1] to obtain a homogeneous feature distribution. The policy divides the image into cells and detects the features in each cell.

4.2.2. Encoder Integration

This section integrates the encoder measurements between the last keyframe and the current frame to obtain the change of the robot pose.

Let the robot pose at time t be $\zeta_t = [x_t, y_t, \theta_t]^T$, the robot pose at time $t+1$ is given by the odometry model [33]:

$$\begin{bmatrix} x_{t+1} \\ y_{t+1} \\ \theta_{t+1} \end{bmatrix} = \begin{bmatrix} x_t \\ y_t \\ \theta_t \end{bmatrix} + \begin{bmatrix} \Delta s \cdot \cos(\theta_t + 0.5\Delta\theta) \\ \Delta s \cdot \sin(\theta_t + 0.5\Delta\theta) \\ \Delta\theta \end{bmatrix}, \quad \begin{cases} \Delta s = \dfrac{\Delta s_r + \Delta s_l}{2} \\ \Delta\theta = \dfrac{\Delta s_r - \Delta s_l}{b} \end{cases}, \tag{5}$$

where Δs is the translation distance, $\Delta\theta$ is the rotation angle, and $\Delta s_{l/r}$ is the left/right wheel displacement in unit m. $\Delta s_{l/r}$ is given by

$$\begin{cases} \Delta s_r = k_r \cdot \Delta e_r + \delta_r,\ \delta_r \sim \mathcal{N}\left(0, \|K \cdot k_r \cdot \Delta e_r\|^2\right) \\ \Delta s_l = k_l \cdot \Delta e_l + \delta_l,\ \delta_l \sim \mathcal{N}\left(0, \|K \cdot k_l \cdot \Delta e_l\|^2\right) \end{cases}, \tag{6}$$

where $\Delta e_{l/r}$ is the left/right encoder displacement in unit tick. $\Delta s_{l/r}$ is affected by a zero mean Gaussian noise $\delta_{l/r}$, whose standard deviation is proportional to the displacement of the left/right wheel, and the proportionality factor is K. The sources of the noises mainly include the deformation of the wheel, transmission error, encoder error, slight slippage, and so on [33]. The occasional large slippages are not considered in this paper. If a large slippage occurs, the noise model will not be satisfied, which results in large estimation errors. We will solve the slipping problem by the visual information or by adding new sensors in our future work.

Assuming that the robot pose obeys a Gaussian distribution, and the robot pose at time t is $\zeta_t \sim \mathcal{N}(\bar{\zeta}_t, \mathbf{\Sigma}_t)$. Then, the covariance of the robot pose at time $t+1$ is given by

$$\mathbf{\Sigma}_{t+1} = \mathbf{G}^{\zeta}\mathbf{\Sigma}_t\left(\mathbf{G}^{\zeta}\right)^T + \mathbf{G}^{s}\mathbf{\Sigma}^{s}(\mathbf{G}^{s})^T, \tag{7}$$

where $\mathbf{G}^{\zeta}$ is the Jacobian of Equation (5) with respect to the robot pose ζ_t, $\mathbf{G}^s$ is the Jacobian of Equation (5) with respect to the left and right wheel displacements $[\Delta s_l, \Delta s_r]^T$, $\mathbf{\Sigma}^s$ is the covariance of $[\Delta s_l, \Delta s_r]^T$:

$$\mathbf{\Sigma}^s = \begin{bmatrix} \|K \cdot k_l \cdot \Delta e_l\|^2 & 0 \\ 0 & \|K \cdot k_r \cdot \Delta e_r\|^2 \end{bmatrix}. \tag{8}$$

Equations (5) and (7) are iteratively used for each encoder measurement between the last keyframe and the current frame. Then, the transformation from the current robot coordinate frame $\{r_c\}$ to the robot coordinate frame $\{r_k\}$ of the last keyframe is obtained: ${}^{r_k}_{r_c}\mathbf{T}^e$ or ${}^{r_k}_{r_c}\zeta^e \sim \mathcal{N}\left({}^{r_k}_{r_c}\bar{\zeta}^e, {}^{r_k}_{r_c}\mathbf{\Sigma}^e\right)$. We also call it encoder observation in this paper. The superscript symbol e indicates that it is obtained by the encoder measurements.

We only perform encoder integration over a short distance between the current frame with the last keyframe, and between two consecutive keyframes. It ensures the drift of the encoder integration remains within a small range. Furthermore, this drift will be reduced by joint optimization with visual information and by the loop closure.

4.2.3. Local Map Tracking

This section estimates the current robot pose by minimizing the encoder and the projection errors.

The initial robot pose ${}^{w}_{r_c}\mathbf{T}'$ of the current frame is propagated from the robot pose ${}^{w}_{r_k}\mathbf{T}$ of the last keyframe by the encoder integration result:

$$ {}^{w}_{r_c}\mathbf{T}' = {}^{w}_{r_k}\mathbf{T}\,{}^{r_k}_{r_c}\mathbf{T}^e. \tag{9} $$

Then, we track the local map. The local map refers to the keyframes (local keyframes) that are close to the current frame in distance and view angle, as well as the map points (local map points) viewed by the local keyframes. The distance threshold d_{lm} is set to 4 m, and the angle threshold θ_{lm} is set to 1 rad in our implementation. The local map points are projected onto the current image to match with the ORB features. Note that the dynamic and static properties of the ORB features of the current frame are unknown, so there may be some erroneous matches. To solve this problem, we propose two policies: (1) limiting the matching search area and (2) using the depth image data to remove erroneous matches. We detail them as follows.

Limiting the matching search area. Thanks to the use of encoder integration, we can search for matching in a small area around the projection position, which improves the matching quality. As shown in Figure 3, a local map point ${}^{w}\mathbf{p}$ is projected onto the current frame, which results in a 2D position $\mathbf{u}$:

$$ \mathbf{u} = \pi\left(\left({}^{w}_{r_k}\mathbf{T}\,{}^{r_k}_{r_c}\mathbf{T}^e\,{}^{r}_{c}\mathbf{T}\right)^{-1}{}^{w}\mathbf{p}\right). \tag{10} $$

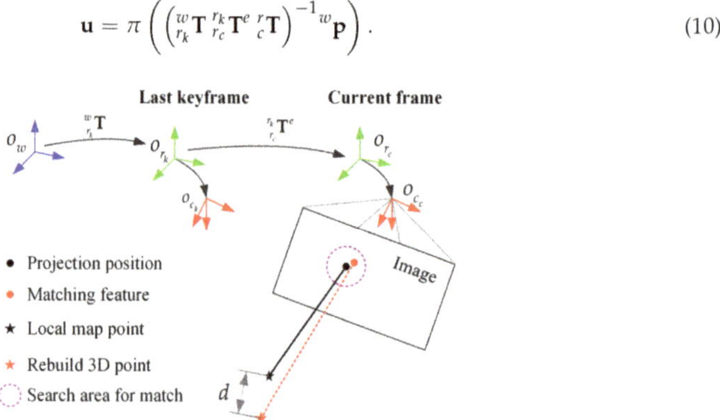

Figure 3. Principle of matching of the local map points with the ORB features.

We assume that (1) the local map points are affected by Gaussian noise: ${}^{w}\mathbf{p} \sim \mathcal{N}\left({}^{w}\bar{\mathbf{p}}, \mathbf{\Sigma}^p\right)$, (2) the pose of the last keyframe is accurate, and (3) the projection position $\mathbf{u}$ obeys the Gaussian distribution $\mathbf{u} \sim \mathcal{N}\left(\bar{\mathbf{u}}, \mathbf{\Sigma}^u\right)$. Then, the covariance of $\mathbf{u}$ is obtained by

$$ \mathbf{\Sigma}^u = \mathbf{G}^e\,{}^{r_k}_{r_c}\mathbf{\Sigma}^e\left(\mathbf{G}^e\right)^T + \mathbf{G}^p\,\mathbf{\Sigma}^p\left(\mathbf{G}^p\right)^T, \tag{11} $$

where $\mathbf{G}^e$ is the Jacobian of Equation (10) with respect to the encoder observation ${}^{r_k}_{r_c}\mathbf{T}^e$ or ${}^{r_k}_{r_c}\zeta^e$, $\mathbf{G}^p$ is the Jacobian of Equation (10) with respect to the local map point position ${}^{w}\mathbf{p}$. Then, we can search for matching within the 3σ boundary around the projection position $\mathbf{u}$. However, the 3σ boundary is typically an ellipse, which is difficult to search in. To speed up the matching, we use the radius search method of KD-Tree by using the envelope circle of the 3σ ellipse as the matching search area. Thus, a matching ORB feature is obtained by searching the best match in this area.

Using the depth image data to remove erroneous matches. As shown in Figure 3, if an effective depth value z' can be found on the position $\mathbf{u}'$ of the matching feature in the depth image, we can rebuild a 3D map point:

$$ {}^{w}\mathbf{p}' = {}^{w}_{r_k}\mathbf{T}\, {}^{r_k}_{r_c}\mathbf{T}^e\, {}^{r}_{c}\mathbf{T}\, \pi^{-1}\left(\mathbf{u}', z'\right). \tag{12} $$

If the matching feature and the map point are both static and the match is correct, the difference between ${}^{w}\mathbf{p}$ and ${}^{w}\mathbf{p}'$ is generally small. However, if there are dynamic feature or map point in the match or the match is incorrect, the difference tends to be large. We use the distance d between ${}^{w}\mathbf{p}$ and ${}^{w}\mathbf{p}'$ to measure the difference. Then, we can use a threshold d_{th} to determine the correct match and the erroneous match. The distance d is influenced by the error of the robot pose, the map point, and the depth image for the correct matches. The assumption of using the depth image to filter out erroneous matches is that the accuracy of the RGB-D sensor must be high enough. If the depth error is too large, the distance d of a correct match will be large, and the correct match may be mistakenly considered to be an erroneous match. The Kinect 2.0 camera used in this paper has sufficient accuracy to remove the erroneous matches. We set the threshold d_{th} to grow linearly with the depth value z', as the depth error of RGB-D camera increases with distance.

$$ d_{th} = d_b + k_d \cdot z'. \tag{13} $$

where d_b is the base threshold and k_d is the scale factor. We set $d_b = 0.2$ m and $k_d = 0.025$ in this paper.

Given the encoder observation between the last keyframe and the current frame (see Section 4.2.2), as well as the 3D–2D matches, the pose of the current frame is refined by minimizing the encoder error and the projection errors:

$$ {}^{w}_{r_c}\mathbf{T}^* = \underset{{}^{w}_{r_c}\mathbf{T}}{\arg\min} \left(E_{enc} + \sum_{i \in \mathcal{R}} E_{i,proj} \right), \tag{14} $$

where $\mathcal{R}$ contains all 3D–2D matches. The encode error is defined by

$$ E_{enc} = \rho\left((\mathbf{e}^e)^T \left({}^{r_k}_{r_c}\boldsymbol{\Sigma}^e\right)^{-1} \mathbf{e}^e \right), \ \mathbf{e}^e = {}^{r_k}_{r_c}\zeta^e - F\left(\left({}^{w}_{r_k}\mathbf{T}\right)^{-1} {}^{w}_{r_c}\mathbf{T} \right), \tag{15} $$

where $\rho(\cdot)$ is the Huber robust cost function. $F(\cdot)$ converts a 4×4 homogenous transformation matrix to a 3×1 vector. The projection error for each 3D–2D match is defined by

$$ E_{i,proj} = \rho\left((\mathbf{e}^v_i)^T (\boldsymbol{\Sigma}^v_i)^{-1} \mathbf{e}^v_i \right), \ \mathbf{e}^v_i = \mathbf{u}'_i - \pi\left(\left({}^{w}_{r_c}\mathbf{T}\, {}^{r}_{c}\mathbf{T}\right)^{-1} {}^{w}\mathbf{p}_i \right), \tag{16} $$

where $\rho(\cdot)$ is the Huber robust cost function, and $\boldsymbol{\Sigma}^v_i$ is the feature covariance, which is associated with the ORB feature scale.

4.2.4. Keyframe Decision

The current frame will be constructed as a new keyframe if the following conditions are all satisfied.

1. N_{pf} frames have passed from the last keyframe. This condition guarantees enough time between two adjacent keyframes allow for the dynamic pixels culling, sparse mapping, and OctoMap construction. We set N_{pf} to be the output frequency of the RGB-D camera to ensure at least 1 s of

time is between the two adjacent keyframes. If the robot moves too fast, the distance between two consecutive keyframes may be too large, which results in too much encoder integration error. The two erroneous match detection policies in Section 4.2.3 may be invalid. The system accuracy is reduced. Then, there will be no enough co-view part for the mapping. Thus, we recommend operating the robot at a slow speed, e.g., lower than 0.4 m/s.

2. The distance or angle between the current frame and the last keyframe is larger than threshold d_{kf} or θ_{kf}, respectively. On the one hand, this ensures that the encoder integration does not diverge too much, thus ensuring the functional of erroneous match detection and the accuracy of robot pose tracking. On the other hand, it ensures enough viewing overlaps between keyframes for OctoMap construction. In our implementation, we set these two parameters as $d_{kf} = 0.3$ m and $\theta_{kf} = 0.5$ rad.

4.3. Dynamic Pixels Culling

Figure 4 shows the dynamic pixels culling pipeline. The object detection and multiview geometry constraints are combined to detect dynamic pixels in the new keyframe.

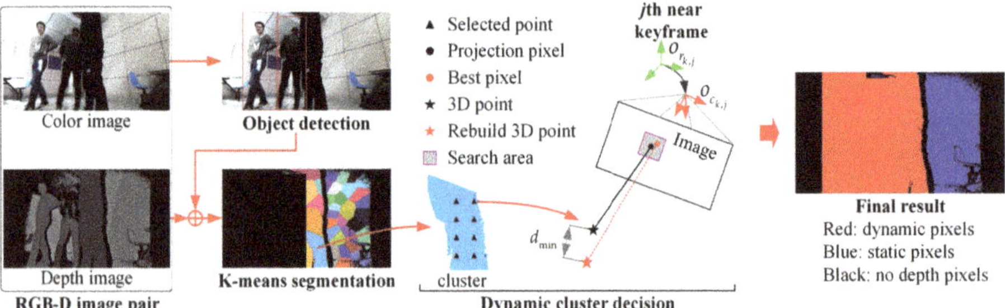

Figure 4. Pipeline of dynamic pixels culling.

4.3.1. Objects Detection

We detect predefined dynamic objects from the color image. In an indoor environment, dynamic objects may be person, cat, dog, etc. We detect these predefined dynamic objects by the object detection method YOLOv3 [34] that is implanted in OpenCV. The publicly available model trained on the COCO dataset [35] is directly used. The YOLOv3 can process 6–7 images per second on our Intel i7-8700 CPU, without the need of a GPU. However, there may be some miss-detections, especially when only a part of a dynamic object is viewed, as shown in Figure 4. Furthermore, some objects that are not included in the predefined dynamic objects may also be moving, such as books, chairs, and desks. Thus, a method based on a multiview constraint is proposed to detect dynamic pixels using the depth image in the following.

4.3.2. Multiview Constraint-Based Dynamic Pixels Culling

As shown in Figure 4, the pixels in bounding boxes of dynamic objects are culled first. Then, the remanding pixels are projected to 3D to create a point cloud. We use the K-means method to segment the point cloud into several clusters. The clusters are assumed to be rigid bodies, which means that the pixels in the same cluster have the same motion property. This assumption is reasonable because we focus on removing dynamic pixels and building static maps but not tracking dynamic objects. The number of clusters, k, is determined according to the size s_p of the point cloud: $k = s_p/n_{pt}$. n_{pt} is the average point number of a cluster to be adjusted. Reducing n_{pt} guarantees a better approximation, but it also increases the computational burden. We set n_{pt} to be 6000 to balance the computational consumption and precision.

We only need to detect which clusters are dynamic ones. To speed up the dynamic cluster detection process, we only select a small number of pixels (e.g., 100) uniformly in a cluster and then use the multiview constraint to judge their dynamic and static attributes. If the number of dynamic pixels is large, the cluster is determined to be dynamic. Otherwise, the cluster is determined to be static.

We detect whether a pixel is dynamic by projecting it to the last N_{pc} keyframes (near keyframes) for comparison. We set $N_{pc} = 5$ in this paper. The pixels $\mathbf{u}$ are back projected to be a 3D point in the world frame using its depth z in the depth image and the robot pose ${}^{w}_{r_{kn}}\mathbf{T}$ of the new keyframe:

$$ {}^{w}\mathbf{p} = {}^{w}_{r_{kn}}\mathbf{T}\,{}^{r}_{c}\mathbf{T}\,\pi^{-1}\left(\mathbf{u}, z\right). \tag{17} $$

The 3D point ${}^{w}\mathbf{p}$ can be projected onto the image of the jth near keyframe:

$$ \mathbf{u}' = \pi\left(\left({}^{w}_{r_{k,j}}\mathbf{T}\,{}^{r}_{c}\mathbf{T}\right)^{-1}{}^{w}\mathbf{p}\right), \tag{18} $$

where ${}^{w}_{r_{k,j}}\mathbf{T}$ is the robot pose of the jth near keyframe. If the pixel $\mathbf{u}'$ of the depth image of the jth keyframe has a valid depth z', a rebuilt 3D point ${}^{w}\mathbf{p}'$ can be obtained as

$$ {}^{w}\mathbf{p}' = {}^{w}_{r_{k,j}}\mathbf{T}\,{}^{r}_{c}\mathbf{T}\,\pi^{-1}\left(\mathbf{u}', z'\right). \tag{19} $$

Note that the dynamic pixels of the depth image of the jth keyframe have been culled before. If a pixel is static, the distance d between ${}^{w}\mathbf{p}'$ and ${}^{w}\mathbf{p}$ should be very small. Otherwise, this pixel is a dynamic pixel. Considering that both the depth image and the poses of the keyframes have errors, $\mathbf{u}'$ may not be the pixel corresponding to $\mathbf{u}$, so we search for a pixel within a square around $\mathbf{u}'$ so that d takes the minimum value d_{min}. The square size s_z is affected by the keyframe accuracy and depth image accuracy. We empirically set the square size s_z to 9 pixels. If d_{min} is larger than a threshold d_{mth}, the pixel $\mathbf{u}'$ is judged as a dynamic pixel in this projection step, otherwise it is a static pixel. The threshold d_{mth} is determined by the keyframe accuracy and depth image accuracy. Same as Equation (10), we also set the threshold d_{mth} to grow linearly with the depth value z'. The other situations are that no valid depth values can be found in the square search area or $\mathbf{u}'$ is out of the range of the image. In this situation, the pixel $\mathbf{u}$ is judged as invalid. We can judge whether a pixel $\mathbf{u}$ is dynamic, static, or invalid by projecting it to one keyframe now. Given that the result of one keyframe is not reliable enough and may produce invalid results, we apply the above process to all near keyframes. Finally, the property of a pixel is determined by voting. Assuming the number of static results is N_s, the number of dynamic results is N_d, and the number of invalid results is N_i, the final decision of the pixel $\mathbf{u}$ is given by

$$ \begin{cases} static,\ if(N_s \geq N_d,\ N_s > 0) \\ dynamic,\ if(N_d > N_s,\ N_d > 0) \\ invalid,\ if(N_s = N_d = 0) \end{cases}. \tag{20} $$

Similarly, we also use the voting approach to determine the properties of a cluster. Among the selected pixels in one cluster, assuming that the number of static ones are N'_s, the number of dynamic ones are N'_d, and the number of invalid ones are N'_i, the final property of the cluster is

$$ \begin{cases} static,\ if(N'_s \geq N'_d) \\ dynamic,\ if(N'_d > N'_s) \end{cases}. \tag{21} $$

The final static pixels are obtained by culling the pixels in the dynamic clusters and in the bounding boxes of the potential dynamic objects, as shown in Figure 4. It should be noted that some static pixels in the bounding box are also removed as dynamic pixels. This is acceptable because we focus on eliminating dynamic interference and dynamic data in the OctoMap. The remaining static pixels can satisfy the needs of the OctoMap construction.

4.4. Sparse Mapping

4.4.1. Sparse Mapping

A sliding window local bundle adjustment is applied as shown in Figure 5. The states participating in the optimization are the robot poses of the last N_{opt} keyframes, the last $(N_{opt}+1)$th keyframe, as well as the map points not only observed by the last N_{opt} keyframes but also observed by at least two keyframes. The last $(N_{opt}+1)$th keyframe and the map points created before the N_{opt}th keyframe are fixed during the optimization. The cost function is defined by

$$\arg\min\left(\sum_{l=1}^{N_{opt}} E_{enc}\left(l,l+1\right)+\sum_{(i,l)\in\chi} E_{proj}\left(i,l\right)\right). \quad (22)$$

$E_{enc}\left(l,l+1\right)$ defines the encoder error between the lth and the $(l+1)$th keyframe. It is the same as Equation (15). $E_{proj}(i,l)$ is the projection error term of the ith map point to the last lth keyframe. It is the same as Equation (16). χ contains all projection pairs. We set the sliding window size N_{opt} to 8 KFs to balance the computing time and the accuracy.

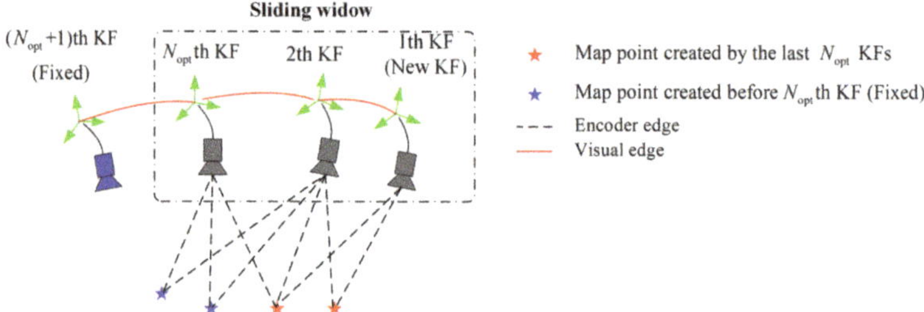

Figure 5. An illustration of the local bundle adjustment.

After local optimization, the robot pose of the new static keyframe is refined. It is then sent to the OctoMap construction module. After that, new map points are created. As dynamic pixels have been discarded, the map points created are all static map points.

4.4.2. Loop Closure

The loop detection is carried out for every new keyframe using the bag-of-words approach DBow2 [36]. Once a loop is detected, the pose graph optimization will be performed. Since the map points are all static, the loop closure avoids interferences from moving objects. All the abovementioned optimizations are performed by the Ceres Solver [37].

4.5. OctoMap Construction

OctoMap divides the 3D space into voxels and stores these voxels through the octree structure, which effectively reduces the memory cost. Each voxel is given an occupancy probability. This probability is updated by integrating multiple observations $z_{1:t}$ through a binary Bayesian filter [14]:

$$P\left(o|z_{1:t}\right)=\left[1+\frac{1-P\left(o|z_t\right)}{P\left(o|z_t\right)}\frac{1-P\left(o|z_{1:t-1}\right)}{P\left(o|z_{1:t-1}\right)}\frac{P\left(o\right)}{1-P\left(o\right)}\right], \quad (23)$$

where $P(o)$ is the prior, which is often set to 0.5. By using log-odds notation, Equation (23) can be rewritten as

$$L\left(o|z_{1:t}\right)=L\left(o|z_{1:t-1}\right)+L\left(o|z_t\right), L\left(o\right)=\log\left(\frac{P\left(o\right)}{1-P\left(o\right)}\right). \quad (24)$$

The measurement integration into a voxel becomes a simple log-odds addition. Typically, a clamping update policy that defines the upper and lower bound on the occupancy probability is applied:

$$L(o|z_{1:t}) = \max(\min(L(o|z_{1:t}),\, l_{max}),\, l_{min}), \tag{25}$$

where l_{max} and l_{min} are the upper and lower log-odds bounds. The clamping update policy ensures the confidence in the map remains bounded. As a consequence, the model can adapt to environment changes quickly [14].

OctoMap is suitable for robot navigation. However, it is not deformable. When a loop is detected and the poses of the keyframes are adjusted, the full OctoMap should be rebuild by retracing all the raw data. To tackle this problem, as shown in Figure 6, this section first creates a series of sub-OctoMaps, then fuses them to form a full OctoMap. The sub-OctoMap integrates raw measurements to a small OctoMap, which saves memory space compared with storing raw data. We just need to refuse the sub-OctoMaps after the loop closure, which speeds up the rebuilding process.

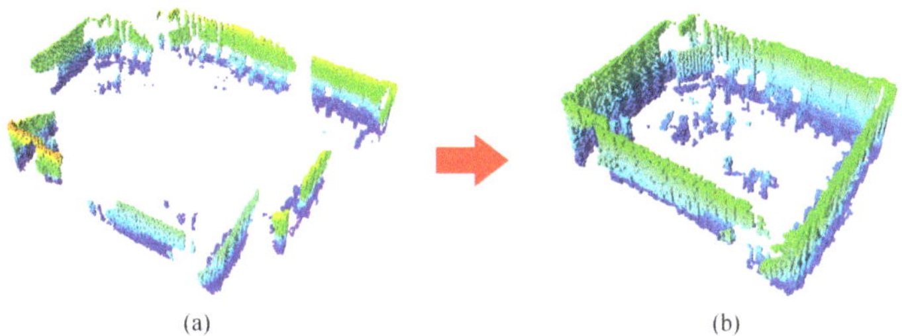

Figure 6. An illustration of the OctoMap construction. (**a**) sub-OctoMaps, (**b**) full OctoMap.

4.5.1. Sub-OctoMap Construction

Generally, within the consecutive N_{sub} keyframes, the robot pose error is bounded within a small range. Therefore, we construct the N_{sub} keyframes as a local sub-OctoMap S_i. We call N_{sub} the submap size. The sub-OctoMap is built relative to a local coordinate frame that is set to be the robot coordinate frame of the first keyframe in this sub-OctoMap. We denote its pose relative to the world coordinate frame as ${}^{w}_{S_i}\mathbf{T}$. The measurements of the depth images in these keyframes are integrated into the sub-OctoMap by performing a ray castings operation from the RGB-D camera to the end points.

4.5.2. Sub-OctoMap Fusion

This section fuses all sub-OctoMaps $\left\{\left\{S_1, {}^{w}_{S_1}\mathbf{T}\right\}, \left\{S_2, {}^{w}_{S_2}\mathbf{T}\right\}, \cdots \left\{S_n, {}^{w}_{S_n}\mathbf{T}\right\}\right\}$ to form a full OctoMap. The full OctoMap is constructed in reference to the world frame $\{w\}$. The process is to transform all sub-OctoMap voxels into the full OctoMap and fuse the occupancy probabilities. For one submap $\left\{S_i, {}^{w}_{S_i}\mathbf{T}\right\}$, the center position of one voxel V_{S_i} in its local frame is ${}^{S_i}\mathbf{p}$, and we transform it to the world frame:

$${}^{w}\mathbf{p} = {}^{w}_{S_i}\mathbf{T}\, {}^{S_i}\mathbf{p}. \tag{26}$$

Assuming ${}^{w}\mathbf{p}$ falls into the voxel V_{F_j} of the full OctoMap, the log-odds of V_{S_i} is L_{S_i}, and the log-odds L_{F_j} of voxel V_{F_j} is then updated by simple addition:

$$L_{F_j} = L_{F_j} + L_{S_i}. \tag{27}$$

The clamping update policy is also used. In this way, we transform the voxels of all sub-OctoMaps into the world coordinate frame to fuse the sub-OctoMaps. Once the pose graph has been adjusted, we just need to refuse these sub-OctoMaps.

This section has two parameters to be turned. One is the voxel size of the OctoMap. The smaller the voxel is, the larger the memory and the amount of computation required. The other is the size of the submap; a submap that is too large will cause distortion of the full map. The impact of these two parameters on the time and memory costs will be explored in the experiment (see Section 5.4).

It should be noted that this submap-based method reduces the precision of the full OctoMap, because the precision of the point cloud is compressed to the voxel size when building the sub-OctoMaps. Furthermore, we use the center point of a voxel as its position when fusing the sub-OctoMaps.

4.6. System Initialization

The system is initialized by constructing the first RGB-D image pair as the first keyframe. The dynamic pixels are culling only by objects detection, as no other keyframes are available. After the dynamic pixels have been culled, we create a number of map points.

5. Experiments

This section evaluates the proposed DRE-SLAM system by self-collected data in various dynamic and static environments. The data were collected by the RedBot using Rosbag tools [38], as shown in Figure 7. Each wheel of RedBot has been mounted an encoder that outputs measurements in a frequency of 100 Hz. A Kinect 2.0 camera is fixed at the front of the robot. The camera outputs registered RGB-D image pairs at a rate of 20 Hz with a resolution of 960×540 pixels by using the iai-kinect2 driver [39]. We implemented the proposed DRE-SLAM in C++ and performed all experiments on a computer with Intel Core i7-8700 CPU, 16 GB RAM. The operating system is Ubuntu 16.04.

Figure 7. RedBot, Kinect 2.0 camera, and ground truth acquisition device.

5.1. Comparative Tests

This section compares the proposed DRE-SLAM with three other state-of-the-art open source SLAM systems in terms of pose accuracy, map quality, and robustness. The chosen systems are: RTAB-MAP [15], StaticFusion [17], and Co-Fusion [7]. The StaticFusion and Co-Fusion were performed on a GTX1070 GPU. Since some scenes are not close to the camera in our experiments, we used depth pixels within a distance 8.0 m of the depth map for all of the four methods. The other special configurations are as follows:

DRE-SLAM. In the OctoMap construction module, the voxel size of the OctoMap was set to 0.05 m, the submap size was set to five keyframes.

RTAB-MAP. This system takes RGB-D images and the wheel odometry as inputs. We integrated the wheel-encoder measurements to form the wheel odometry for the RTAB-MAP. Its OctoMap voxel size was also set to 0.05 m. All other parameters were set to original.

StaticFusion. This system only uses RGB-D images and outputs a static surfel-based background map. The original implementation is for images of 640 × 480 pixels that are downsampled to 320 × 240 pixels in the program. We modified the interface to accept images of 960 × 540 pixels, and we also downsampled the images to 480 × 270 pixels in the program. The other parameters were not changed.

Co-Fusion. This system constructs a surfel-based background map and tracks moving objects only using RGB-D image pairs. As we only consider the background construction, we deactivated its tracking part.

As shown in Figure 8, we collected data in three different environments: office (O), laboratory (L), and corridor (C). The office environment is clean and tidy. The lab environment is messy and rich in texture and structure information. The corridor environment is full of white walls and less in texture and structure information. We collected three types of data, static (ST), low dynamic (LD), and high dynamic (HD), in each environment. In the static data, there are no moving objects. In the low dynamic data, only one person is moving. In the high dynamic data, three persons are moving. The persons move with tablets, chairs or other undefined objects. The purpose is to verify the recognition capability of the dynamic pixels on undefined objects. Three sequences were collected for each type of data. A total of 27 sequences were collected for the comparative tests. Each data sequence is named using the environment type and data type. For example, O-ST1 is the first sequence collected in the static (ST) office (O) environment. The four approaches were tested in all these 27 sequences.

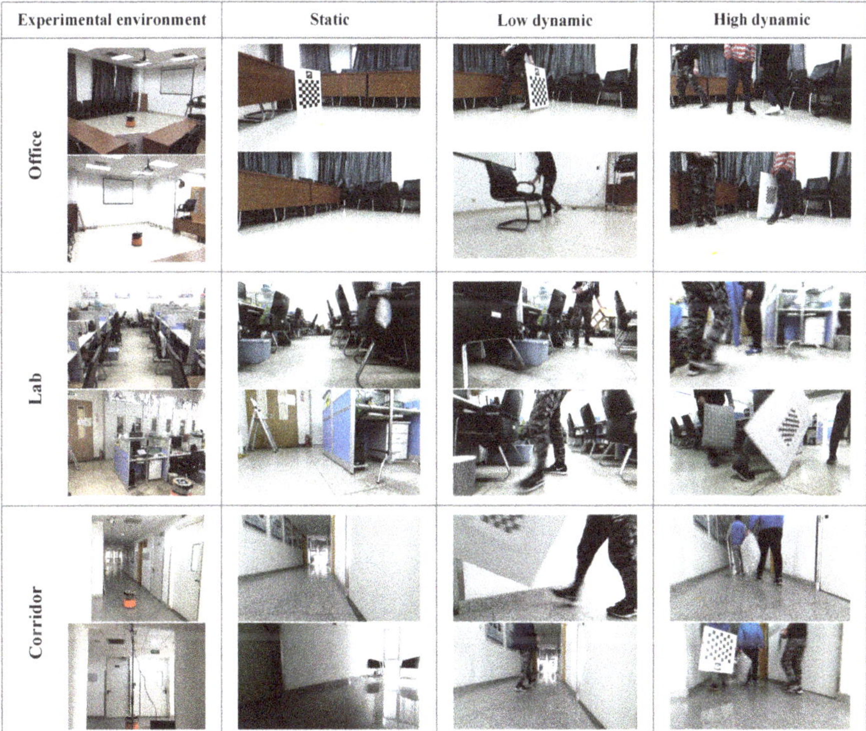

Figure 8. Experimental environments and RGB images in the static, low dynamic, and high dynamic data sequences.

5.1.1. Pose Accuracy

As shown in Figure 7, we mounted an ArUco [40] marker on the top of the RedBot and used a fixed downview camera to capture the marker to produce the pose ground truth by using a plane measurement method [41]. The absolute pose error (APE) metric is used to evaluate the pose accuracy with the evo tool [42]. The results are shown in Table 1. The root-mean-square-error (RMSE) comparison is shown in Figure 9. It can be seen that DRE-SLAM obtains the highest accuracy in most sequences against the other three methods. Moreover, the pose errors are not much different, whether in static environments or in dynamic environments. This shows that DRE-SLAM is rarely affected by dynamic objects in terms of pose accuracy. The pose errors of RTAB-MAP in most sequences are also relatively small. However, the pose errors increase in dynamic sequences in the office environment, especially in high dynamic sequences. This is because RTAB-MAP does not consider dynamic interference in the loop and proximity detection, whereas DRE-SLAM eliminates dynamic pixels and is, therefore, unaffected by dynamic objects. Furthermore, RTAB-MAP only uses the wheel odometry for the current robot pose estimation, whereas DRE-SLAM tightly couples the information of the RGB-D camera with the wheel-encoders. StaticFusion possesses large pose errors in most sequences. Co-Fusion performs small errors most static environments, whereas the errors increase dramatically in dynamic environments. The reasons for the large errors in StaticFusion and Co-Fusion may be that (1) the scene is too far away or (2) there is no encoder data. This also shows that using the encoder improves the system accuracy, especially in dynamic environments. The use of the encoder information can provide good prior and strong constraints. When the static pixel is missing, the encoder can also obtain a reliable pose estimation. We also provide the trajectories of DRE-SLAM with the ground truths in Figure 10. The trajectories of DRE-SLAM are smooth and fit well with the ground truth.

Table 1. The root-mean-square-error (RMSE) of the translation (m) and the rotation (rad) of the four methods in all the sequences. The numbers in bold represent the minimum results.

Sequences	DRE-SLAM		RTAB-MAP		StaticFusion		Co-Fusion	
	Trans.	Rot.	Trans.	Rot.	Trans.	Rot.	Trans.	Rot.
O-ST1	**0.0171**	**0.0252**	0.0574	0.0934	0.6896	0.2304	0.2136	0.1091
O-ST2	**0.0233**	**0.0436**	0.0429	0.0633	2.2755	2.8240	0.2114	0.1266
O-ST3	**0.0218**	**0.0335**	0.0964	0.1406	3.4684	1.8022	0.3499	0.2061
L-ST1	**0.0274**	**0.0461**	0.0422	0.0672	0.1575	0.2352	0.1361	0.0950
L-ST2	**0.0116**	0.0403	0.0216	**0.0364**	0.2429	0.2784	0.1726	0.1017
L-ST3	**0.0212**	**0.0315**	0.0447	0.0648	0.1951	0.1846	0.1641	0.1427
C-ST1	**0.0162**	**0.0209**	0.0339	0.0433	0.8425	0.6428	0.2174	0.1427
C-ST2	**0.0088**	**0.0192**	0.1342	0.0991	0.4140	0.2310	0.1889	0.1287
C-ST3	**0.0129**	**0.0246**	0.0228	0.0401	0.3572	0.2095	0.1980	0.1852
O-LD1	**0.0108**	**0.0185**	0.0655	0.0915	1.0082	1.1706	0.3272	0.2191
O-LD2	**0.0162**	**0.0309**	0.0683	0.0797	0.7561	0.8079	0.7212	0.4193
O-LD3	**0.0175**	**0.0236**	0.0915	0.0818	2.0426	2.6095	1.2238	0.3720
L-LD1	**0.0133**	**0.0270**	0.0321	0.0561	0.1912	0.2503	0.3934	0.1662
L-LD2	**0.0110**	**0.0262**	0.0325	0.0396	0.2037	0.1894	0.5229	0.6387
L-LD3	**0.0206**	**0.0358**	0.0227	0.0451	0.2105	0.1836	0.3281	0.2751
C-LD1	0.0231	0.0317	**0.0087**	**0.0252**	1.4949	1.2988	0.4433	0.3749
C-LD2	**0.0184**	0.0274	0.0374	**0.0211**	2.4187	1.4065	0.6043	0.2630
C-LD3	0.0164	0.0549	**0.0072**	**0.0246**	3.2005	2.5659	0.2238	0.3615
O-HD1	**0.0141**	**0.0229**	0.1627	0.1739	4.7680	2.4754	0.7468	0.3635
O-HD2	**0.0284**	**0.0331**	0.1396	0.1329	2.2350	2.6707	0.7736	1.0450
O-HD3	**0.0127**	**0.0256**	0.0896	0.0846	0.8233	1.3533	0.4835	0.3002
L-HD1	**0.0136**	**0.0317**	0.0277	0.0332	2.6896	1.2771	1.2393	1.1438
L-HD2	0.0262	0.0380	**0.0211**	**0.0258**	1.4187	2.0547	0.8072	1.0425
L-HD3	**0.0190**	**0.0314**	0.0340	0.0471	2.1998	2.1153	0.7263	0.6762
C-HD1	**0.0095**	**0.0292**	0.0127	0.0357	0.5868	0.2994	0.8212	0.4649
C-HD2	0.0225	0.0519	**0.0111**	**0.0195**	1.8197	2.8022	1.3695	0.4685
C-HD3	**0.0222**	0.0575	0.0257	**0.0416**	1.3683	2.3261	0.5095	0.2080

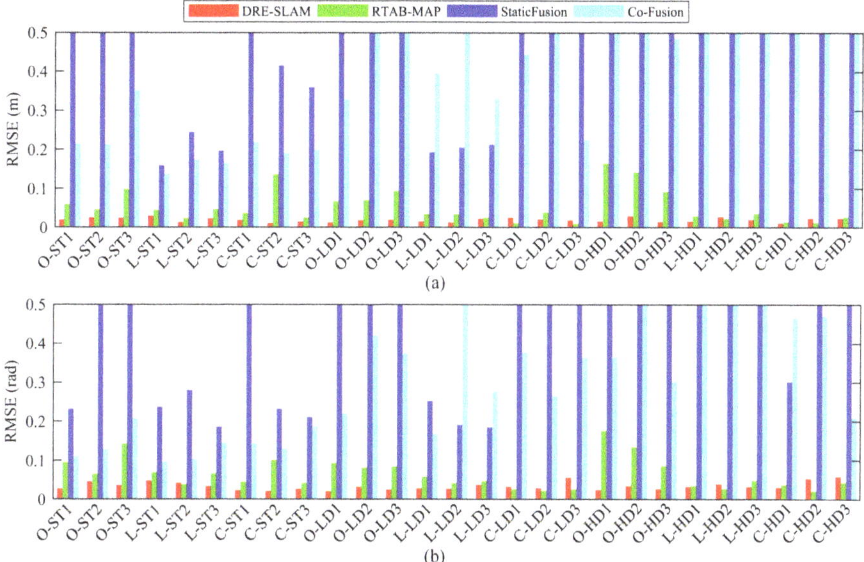

Figure 9. The root-mean-square-error (RMSE) of (**a**) the translation (m) and (**b**) the rotation (rad) comparison of the four methods in all the sequences.

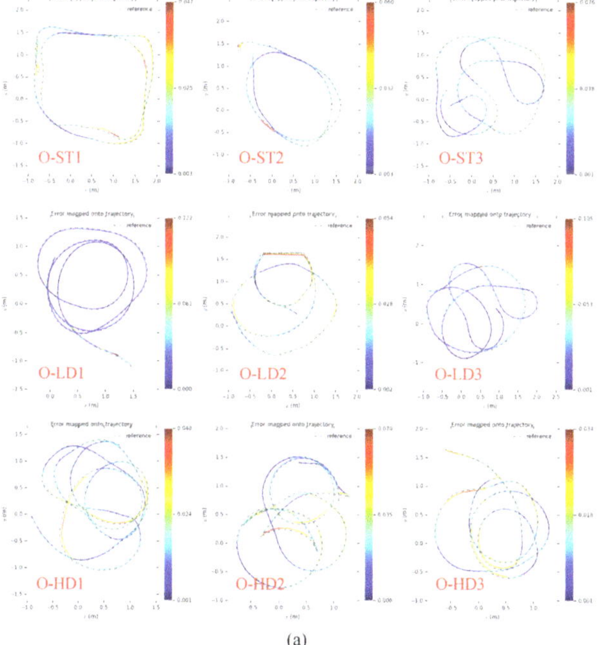

(a)

Figure 10. *Cont.*

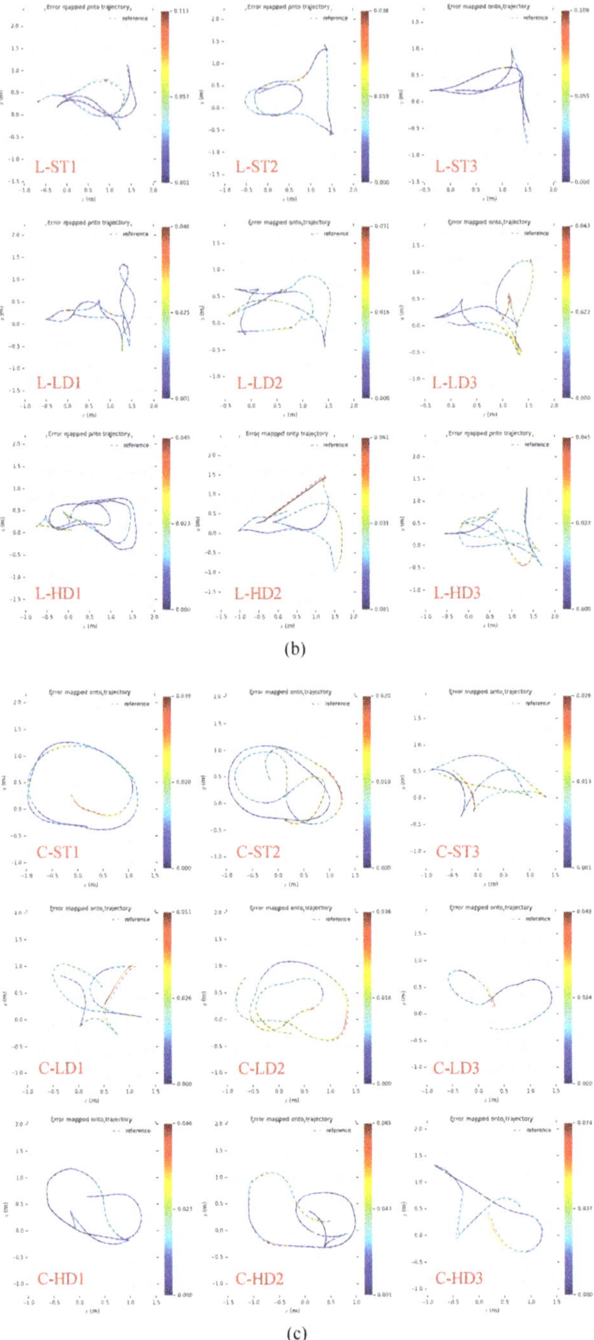

Figure 10. Comparison of the trajectories of the proposed method with the ground truth in the (**a**) office, (**b**) lab, and (**c**) corridor environments. The absolute pose errors are encoded by colors.

5.1.2. Map Quality

Figure 11 provides all maps created by the four methods with all the data sequences. DRE-SLAM constructed clean background maps in most sequences. There are no obvious distortions in these maps. This proves that DRE-SLAM can reliably construct static maps in both dynamic and static environments. The RTAB-MAP constructs nice maps in most static sequences. However, many dynamic objects are introduced in its maps in the low and high dynamic sequences. Some maps even contain obvious distortions, e.g., maps of O-HD1and C-ST2. StaticFusion obtains nice maps in the static and low dynamic lab sequences. This may benefit from the rich structure of the lab environment. However, in other sequences, its maps suffer a lot of distortions. We find that StaticFusion treats many static pixels as dynamic pixels. Co-Fusion can build correct maps in most static sequences, but the maps are obviously confusing in the low and high dynamic sequences. We can also see that DRE-SLAM and RTAB-MAP, which fuse encoder information, produce maps with fewer distortions. Whereas the pure RGB-D methods, StaticFusion and Co-Fusion, create maps with more distortions. This phenomenon indicates that using the encoder is helpful to improve the map quality. The quality of the map is mainly determined by the pose accuracy of the keyframes and the dynamic pixels' culling effect. The sounder map quality of our system in Figure 11 further shows that our method produces high pose accuracy and a dynamic pixel culling effect.

5.1.3. Robustness

DRE-SLAM achieves accurate pose estimation and impressive static maps in all sequences, even when there are dynamic objects in the environments. This shows the robustness of the system. RTAB-MAP also works reliably on most sequences because its local pose estimation only relies on wheel odometry. Only the loop closure and proximity detection use visual information that is confused by dynamic objects. However, the pure RGB-D methods, StaticFusion and Co-Fusion, are fragile. They cannot work reliably even in a static environment. This is because it is difficult to achieve reliable estimation if the number of identified static pixels is too small, which leads to the whole system collapse. However, if the encoder information is added, the encoder can work reliably over a short distance even if the visual information is completely lost. In short, combining of the RGB-D camera and encoders creates a highly robust system.

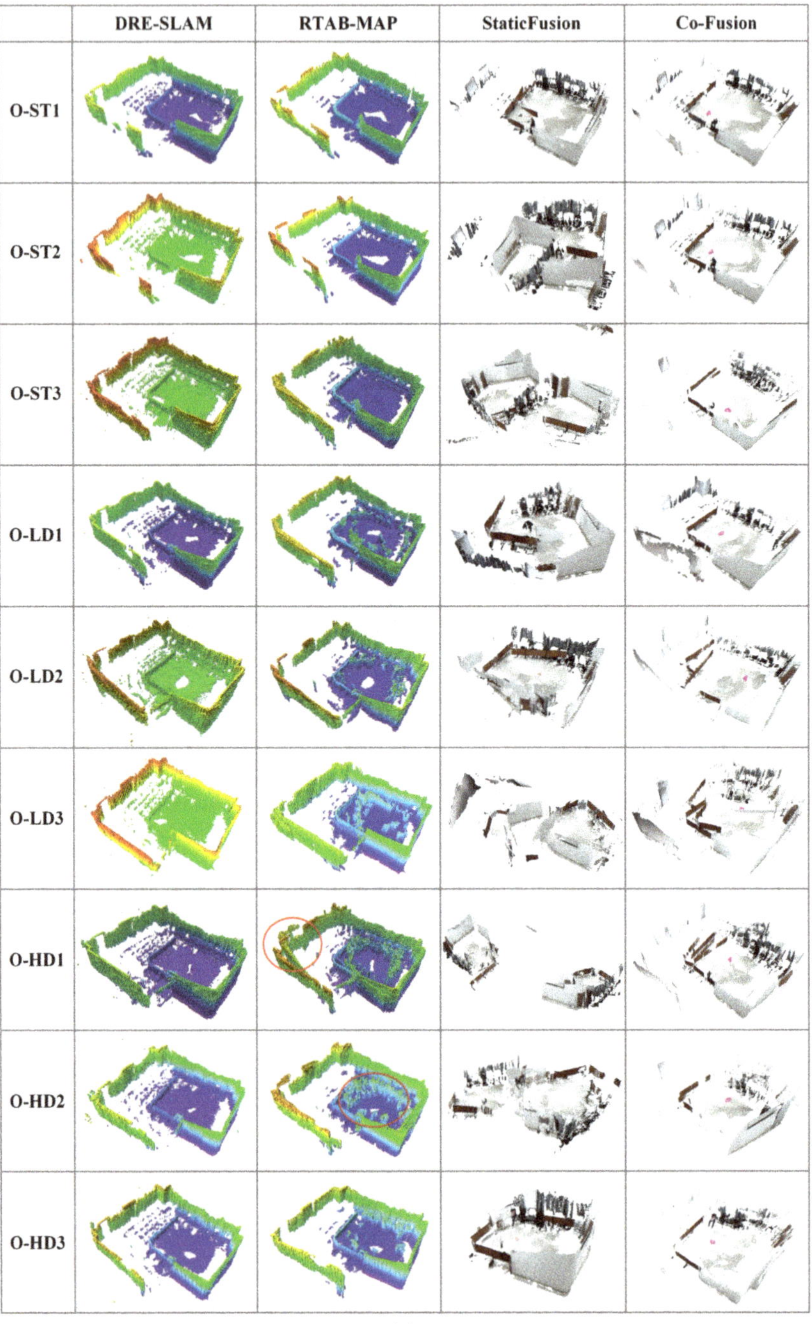

(a)

Figure 11. *Cont.*

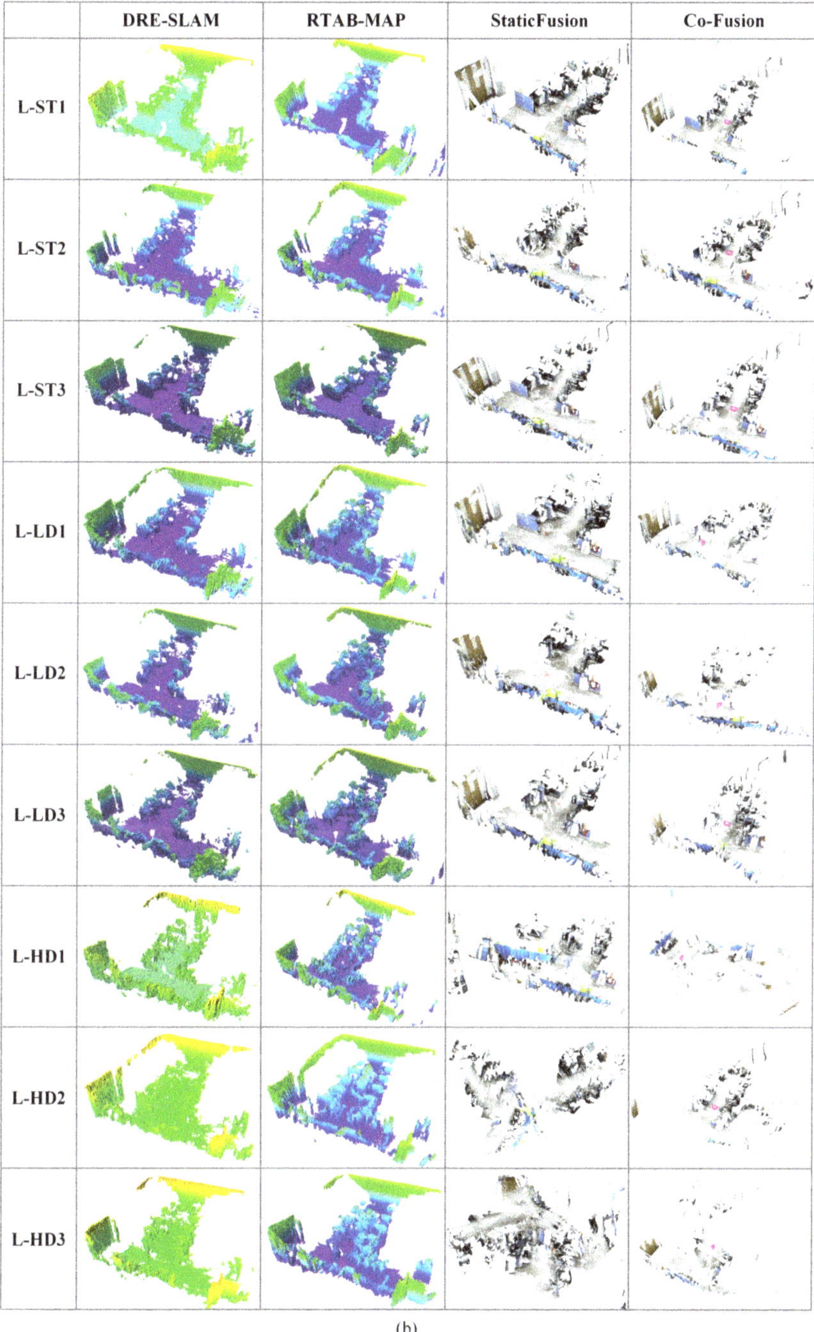

(b)

Figure 11. *Cont.*

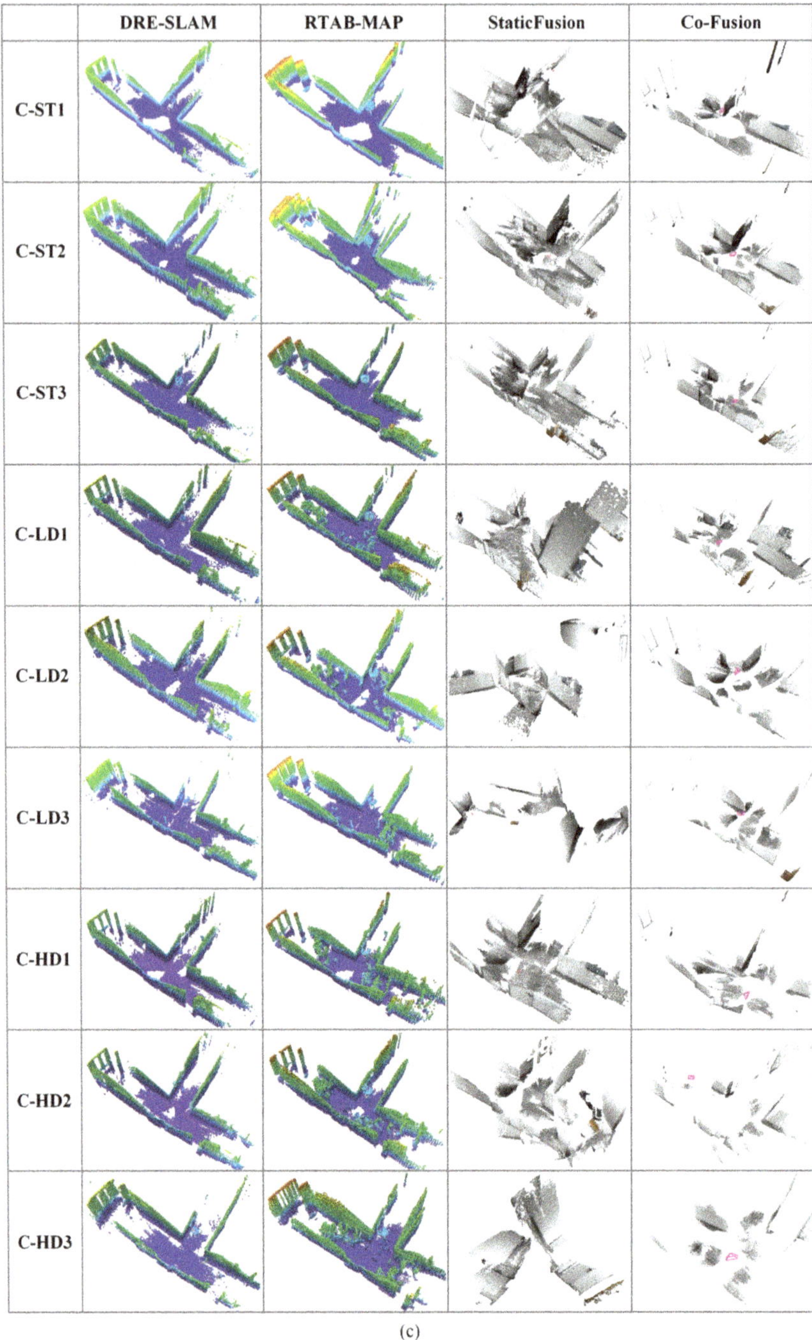

(c)

Figure 11. Maps obtained by the four methods in the (**a**) office, (**b**) lab, and (**c**) corridor environments. The red circles show the deformation and the dynamic data introduced in the OctoMaps created by RTAB-MAP.

5.2. Performance of Dynamic Pixels Culling

Figure 12 gives the results of the dynamic pixels culling module. In addition to predefined dynamic objects, such as persons, undefined moving objects, such as tablets and chairs, are also in the environments. YOLOv3 can recognize most predefined objects, but partially visible objects may not be identified (see the second column in Figure 12). Thanks to the multiview constraint-based strategy, we can handle the leak-recognized and undefined moving objects (see the first column in Figure 12). However, our approach also produces erroneous detections. For example, in the third column of Figure 12, the pixels of a part of the moving tablet are identified as static pixels. Because these pixels and some pixels on the ground are classified into one cluster by the K-means algorithm, and the ground pixels are the majority, our method determines this cluster as static. This problem can be solved by increasing the number of clusters, but it also increases the computation consumption. Moreover, our method cannot detect slow moving undefined dynamic objects due to the use of a fixed threshold. This small amount of mistaken recognitions do not pose a severe problem for the OctoMap construction, because OctoMap uses a filter. Only if a voxel has been observed as an obstacle multiple times will the occupancy probability of the voxel be high enough to label the voxel occupied.

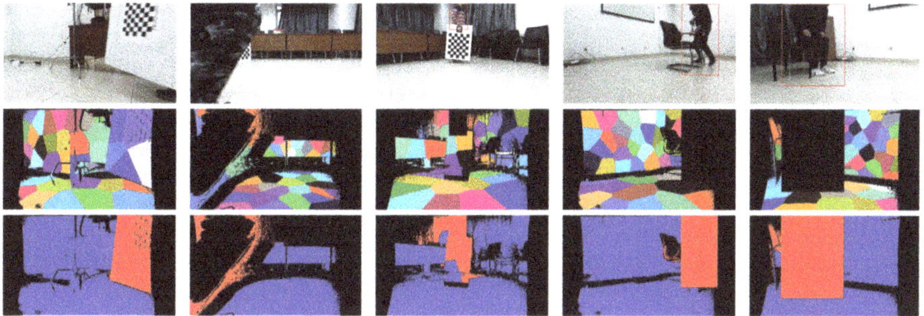

Figure 12. Typical results of the dynamic pixels culling module. The first row includes the predefined dynamic object detection results. The second row includes the results of K-means segmentation using the depth images. The third row shows the final dynamic pixel detection results. The red pixels are dynamic, the blue pixels are static, and the black pixels are pixels with invalided depth values.

5.3. Extension Tests

In this section, we further test the proposed system in an 8 m × 6 m office, a 15 m × 9 m hall, and a 37 m × 27 m corridor. In these sequences, several persons are moving at different speeds and directions. Many of them have only a part of their body in the view of the camera. Sometimes, the moving persons take up almost the entire field of view of the camera. There are some doors being opened and closed. These sequences are very challenging. Both the StaticFusion and Co-Fusion failed. Although RTAB-MAP can work, its trajectories and maps contain many twists. There are many dynamic objects be built in its maps. Conversely, DRE-SLAM successfully performs pose estimation, map construction, and loop closure, as shown in Figure 13. The robot trajectories are smooth and the maps hold no apparent distortions. Most dynamic objects are filtered out in the maps.

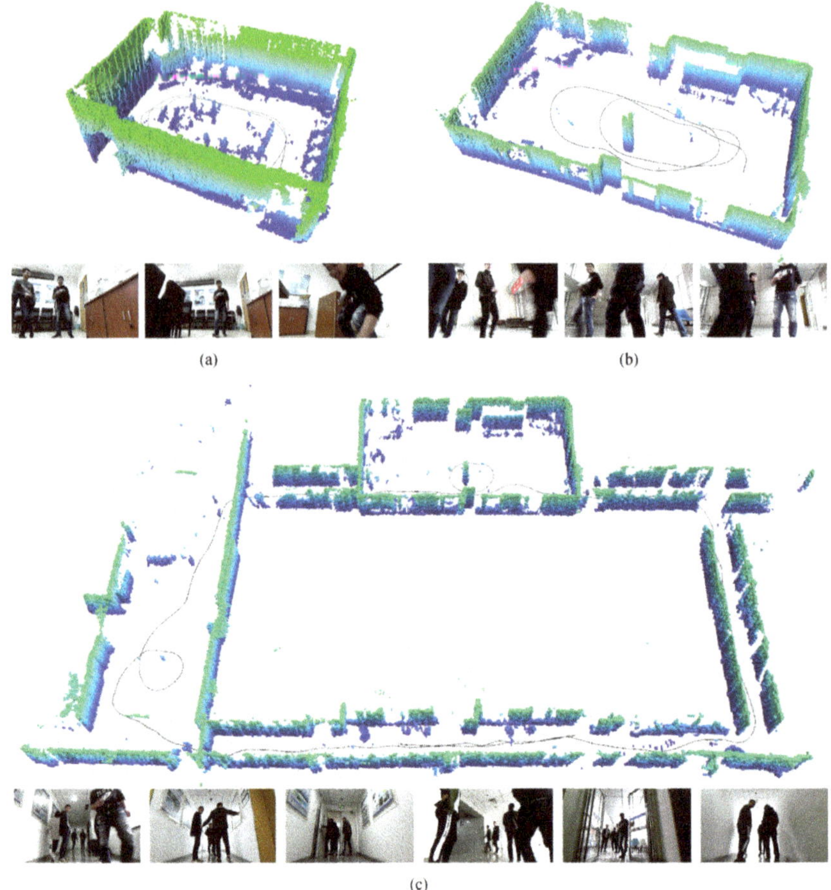

Figure 13. Results of the extension tests. (**a**) 8 m × 6 m office (Voxel size: 0.05 m), (**b**) 15 m × 9 m hall (Voxel size: 0.05 m), and (**c**) 37 m × 27 corridor (Voxel size: 0.1 m). The black lines are the robot trajectories of keyframes.

5.4. *Performance of OctoMap Construction*

The voxel size and the submap size are two key parameters in our submap-based mapping algorithm. In this section, we experimentally explore the effects of these two parameters on the time cost, memory cost, and the quality of the map. We chose the longest sequence, i.e., the corridor, mentioned in Section 5.3, for this experiment. The DRE-SLAM has a loop of 408 keyframes in this sequence. We also compare our submap-based approach with the traditional retrace approach that rebuilds the entire map using all the raw data. We set four types of submap sizes, i.e., 5, 10, 15, and 20, and four types of voxel size, i.e., 0.05 m, 0.10 m, 0.15 m, and 0.20 m. We run DRE-SLAM a total of 16 times with the combinations of these different parameter configurations.

The time costs of the map update, when the loop has been closed, are shown in Table 2 and Figure 14a. Compared with the retrace method, our approach dramatically reduces the time consumption. The improvement rate is over 80% in all the configurations. The improvement is more obvious when raising the voxel size or increasing the submap size. As seen from Figure 14a, the parameter that mainly influences the time cost is the voxel size. The submap size num has less effect. In other words, the time consumption will not increase too much when we choose a smaller submap size.

Table 2. Time costs of the retrace approach and our method with various voxel sizes and submap size (ms).

Voxel Size (m)		Submap Size (KFs)			
		5	10	15	20
0.05	Retrace	78,458.12	79,088.49	79,063.69	78,568.92
	Our approach	13,963.45	11,558.43	10,189.10	10,308.59
	Improvement	82.20%	85.39%	87.11%	86.88%
0.10	Retrace	12,754.71	12,607.22	12,712.09	12,730.30
	Our approach	2198.32	1682.83	1470.14	1416.18
	Improvement	82.76%	86.65%	88.44%	88.88%
0.15	Retrace	5476.40	5453.38	5549.65	5472.54
	Our approach	768.58	583.72	489.83	429.21
	Improvement	85.97%	89.30%	91.17%	92.16%
0.20	Retrace	3576.74	3694.08	3647.32	3728.31
	Our approach	324.03	238.94	246.99	196.73
	Improvement	90.94%	93.53%	93.23%	94.72%

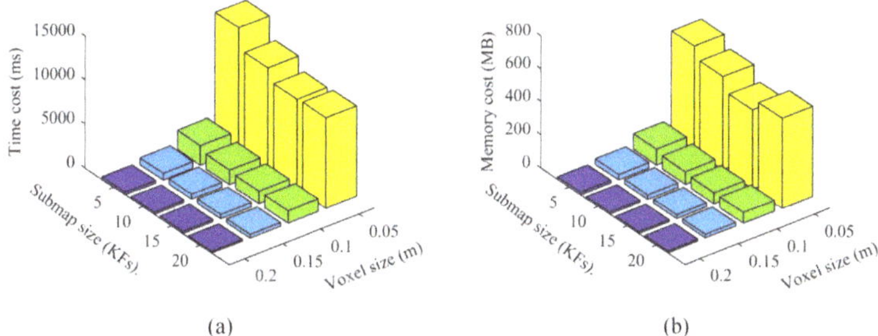

Figure 14. (**a**) time cost and (**b**) memory cost of our approach with various voxel sizes and submap sizes.

Table 3 and Figure 14b show the memory costs. In our approach, all the sub-OctoMaps and a full OctoMap are maintained in the memory. In the retrace approach, all the point clouds in all keyframes and a full OctoMap are stored in the memory. The full OctoMaps of the two methods are the same, which are not considered. We only calculate the memory needed to save the sub-OctoMaps for our approach and the point clouds for the retrace method. We assume one point occupies three float type of memory to count the memory of the point clouds. The function, *memoryUsage()*, provided by the OctoMap library is used to calculate the memory consumption of all the sub-OctoMaps. For configurations with a voxel size of 0.05 m, our method does not significantly improve the memory consumption. However, for configurations with resolutions of 0.10 m, 0.15 m, and 0.20 m, our algorithm dramatically reduces the memory consumption. The improvement rate is over 85%. This improvement is getting better with increasing the voxel size or increasing the submap size. As seen in Figure 14b, the voxel size plays a major role in the memory consumption, whereas the submap size has less impact. In summary, our approach effectively reduces the time cost and memory consumption. Furthermore, we can choose a smaller submap size without a significant increase in time and memory consumption.

Table 3. Memory costs of the retrace approach and our method with various voxel sizes and submap sizes (MB).

Voxel Size (m)		Submap Size (KFs)			
		5	10	15	20
0.05	Retrace	768.57	772.14	769.51	781.48
	Our approach	630.22	562.91	479.58	546.55
	Improvement	18.00%	27.10%	37.68%	30.06%
0.10	Retrace	770.69	768.68	773.55	773.81
	Our approach	95.24	77.92	67.56	66.47
	Improvement	87.64%	89.86%	91.27%	91.41%
0.15	Retrace	776.46	771.22	771.98	782.25
	Our approach	33.73	27.44	23.88	22.34
	Improvement	95.66%	96.44%	96.91%	97.14%
0.20	Retrace	770.64	770.67	772.14	772.39
	Our approach	16.52	13.04	11.55	11.01
	Improvement	97.86%	98.31%	98.50%	98.57%

Figure 15 compares the map of retrace with the maps of our method in different submap sizes. The voxel size of the maps is 0.05 m. It can be seen that as the submap size increases, the distortion of the map tends to become larger. This is because the error of the sub-OctoMap increases with the submap size. There are no visible differences between our map with the retrace map when the submap size is 5. Combined with the above findings, we recommend setting the submap size to 5 and the voxel size to 0.1 m to obtain a balanced mapping performance.

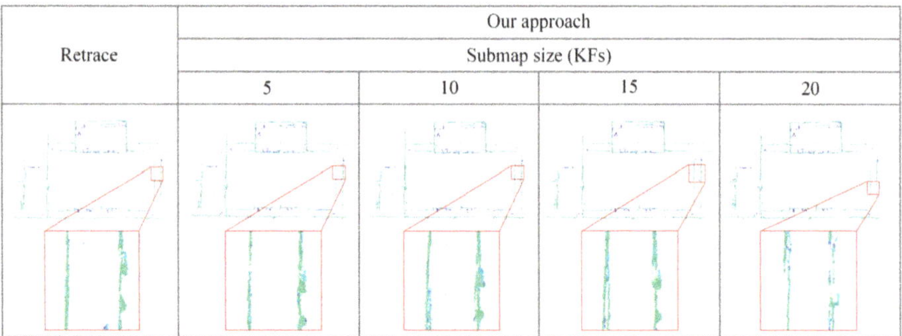

Retrace
Our approach
Submap size (KFs)
5
10
15
20

Figure 15. Top view of a map built by the retrace approach and top views of maps built by our approach with different submap sizes. The voxel size is 0.05 m.

5.5. Timing Results

Table 4 presents the time costs of the DRE-SLAM. These time costs are calculated from the results of all the extension test sequences in Section 5.3. Our method costs approximately 34 ms per frame in the tracking part, in which the ORB feature extraction costs approximately 17 ms. The dynamic pixel culling takes approximately 480 ms per frame, in which the time costs of the object detection and the K-means segmentation are the majority. Due to the existence of dynamic objects, the number of map points participating in the local BA may vary greatly, so the time consumption of the Local BA has a large standard deviation. The loop closure detection time consumption is related to the number of keyframes in the loop. We provide the time result of the longest loop in the corridor sequence. The time costs in Table 4 are calculated in the dynamic scenes, which are smaller than the time costs in static scenes because less static map points and features can be used in the dynamic scenes. The tracking time is approximately 45 ms in static environments.

Table 4. Timing results of each thread in milliseconds (mean ± STD).

Thread	Operation	Time Cost (ms)
RGB-D Encoder Tracking	Encoder integration	0.01 ± 0.00
	Feature extraction (1600 per frame)	17.28 ± 6.91
	3D–2D match	8.24 ± 4.48
	Pose estimation	5.86 ± 3.64
	Total	33.24 ± 12.12
Dynamic Pixels Culling	Object detection	162.17 ± 8.88
	K-means	286.71 ± 48.61
	Multiview constraints	28.10 ± 14.58
	Total	480.08 ± 54.27
Local Sparse Mapping	New map points creation	0.23 ± 0.06
	Local BA (Window size: 8 KFs)	138.54 ± 73.80
	Total	138.94 ± 73.83
Loop Closure (Loop size: 408 KFs)	Loop check	80.87
	Pose graph optimization	143.28
	Total	224.19
Sub-OctoMap Construction (voxel: 0.1 m)	Total	52.04 ± 27.84
Sub-OctoMap Fusion (408 KFs; voxel: 0.1 m)	Total	2198.32

6. Conclusions

In this paper, we propose the dynamic RGB-D Encoder SLAM system, i.e., DRE-SLAM. The system works robustly and with high accuracy both in dynamic and static environments. DRE-SLAM takes measurements of an RGD-camera and two wheel-encoders as inputs and produces robot poses and a static background OctoMap. The system runs purely on a CPU, with no need for GPUs. The robustness and accuracy achieved are attributed to the tightly coupled approach with the encoders and the RGB-D camera. By combining the YOLOv3 object detection method with the multiview constraint-based method, we identify dynamic pixels on not only predefined dynamic objects but also undefined dynamic objects. The capability of working in dynamic environments is gained by culling the pixels occupied by these dynamic objects. The submap-based OctoMap construction approach significantly lowers the rebuilding time and memory consumption. We further find that the number of keyframes used to construct sub-OctoMaps has little effect on time and memory costs. Thus, we can reduce the number of keyframes in a submap to decrease the map distortion without adding too much computation and memory burdens. The proposed system is especially suitable for mobile robots that are operating in indoor dynamic environments.

In future work, we will extend our system in two directions. First, we will deal with the problem of wheel slipping to improve the robustness and accuracy. Second, we will explore the function of tracking dynamic objects.

Author Contributions: D.Y., S.B., W.W. (third author), and Y.C. designed the algorithm. D.Y. developed the open source code and wrote the draft. D.Y., C.Y., W.W. (fifth author), and X.Q. performed the experiments and analyzed the data. All the author revised the manuscript.

Funding: This research was funded by Scientific and Technological Project of Hunan Province on Strategic Emerging Industry under grant number 2016GK4007, Beijing Natural Science Foundation under grant number 3182019, and National Natural Science Foundation of China under grant number 91748101.

Conflicts of Interest: The authors declare no conflict of interest.

References

1. Mur-Artal, R.; Tardós, J.D. Orb-slam2: An open-source slam system for monocular, stereo, and rgb-d cameras. *IEEE Trans. Robot.* **2017**, *33*, 1255–1262. [CrossRef]
2. Whelan, T.; Salas-Moreno, R.F.; Glocker, B.; Davison, A.J.; Leutenegger, S. Elasticfusion: Real-time dense slam and light source estimation. *Int. J. Robot. Res.* **2016**, *35*, 1697–1716. [CrossRef]
3. Fu, X.; Zhu, F.; Wu, Q.; Sun, Y.; Lu, R.; Yang, R. Real-Time Large-Scale Dense Mapping with Surfels. *Sensors* **2018**, *18*, 1493. [CrossRef] [PubMed]
4. Cadena, C.; Carlone, L.; Carrillo, H.; Latif, Y.; Scaramuzza, D.; Neira, J.; Reid, I.; Leonard, J.J. Past, present, and future of simultaneous localization and mapping: Toward the robust-perception age. *IEEE Trans. Robot.* **2016**, *32*, 1309–1332. [CrossRef]
5. Qin, T.; Li, P.; Shen, S. Vins-mono: A robust and versatile monocular visual-inertial state estimator. *IEEE Trans. Robot.* **2018**, *34*, 1004–1020. [CrossRef]
6. Yu, C.; Liu, Z.; Liu, X.-J.; Xie, F.; Yang, Y.; Wei, Q.; Fei, Q. Ds-slam: A semantic visual slam towards dynamic environments. In Proceedings of the IEEE/RSJ International Conference on Intelligent Robots and Systems (IROS), Madrid, Spain, 1–5 October 2018; pp. 1168–1174.
7. Rünz, M.; Agapito, L. Co-fusion: Real-time segmentation, tracking and fusion of multiple objects. In Proceedings of the IEEE International Conference on Robotics and Automation(ICRA), Singapore, 29 May–3 June 2017; pp. 4471–4478.
8. Bârsan, I.A.; Liu, P.; Pollefeys, M.; Geiger, A. Robust dense mapping for large-scale dynamic environments. In Proceedings of the IEEE International Conference on Robotics and Automation (ICRA), Brisbane, Australia, 21–25 May 2018; pp. 7510–7517.
9. Bescos, B.; Fácil, J.M.; Civera, J.; Neira, J. Dynaslam: Tracking, mapping and inpainting in dynamic scenes. In Proceedings of the IEEE/RSJ International Conference on Intelligent Robots and Systems (IROS), Madrid, Spain, 1–5 October 2018.
10. Rünz, M.; Agapito, L. Maskfusion: Real-time recognition, tracking and reconstruction of multiple moving objects. In Proceedings of the IEEE International Symposium on Mixed and Augmented Reality (ISMAR), Munich, Germany, 16–20 October 2018.
11. Zhou, G.; Bescos, B.; Dymczyk, M.; Pfeiffer, M.; Neira, J.; Siegwart, R. Dynamic objects segmentation for visual localization in urban environments. *arXiv* **2018**, arXiv:1807.02996.
12. Pizzoli, M.; Forster, C.; Scaramuzza, D. Remode: Probabilistic, monocular dense reconstruction in real time. In Proceedings of the IEEE/RSJ International Conference on Intelligent Robots and Systems, Chicago, IL, USA, 14–18 September 2014; pp. 2609–2616.
13. Newcombe, R.A.; Izadi, S.; Hilliges, O.; Molyneaux, D.; Kim, D.; Davison, A.J.; Kohi, P.; Shotton, J.; Hodges, S.; Fitzgibbon, A. Kinectfusion: Real-time dense surface mapping and tracking. In Proceedings of the IEEE International Symposium on Mixed and Augmented Reality (ISMAR), Basel, Switzerland, 26–29 October 2011; pp. 127–136.
14. Hornung, A.; Wurm, K.M.; Bennewitz, M.; Stachniss, C.; Burgard, W. Octomap: An efficient probabilistic 3d mapping framework based on octrees. *Auton. Robots* **2013**, *34*, 189–206. [CrossRef]
15. Labbé, M.; Michaud, F. Rtab-map as an open-source lidar and visual simultaneous localization and mapping library for large-scale and long-term online operation. *J. Field Robot.* **2018**, *36*, 416–446. [CrossRef]
16. Laidlow, T.; Bloesch, M.; Li, W.; Leutenegger, S. Dense rgb-d-inertial slam with map deformations. In Proceedings of the IEEE/RSJ International Conference on Intelligent Robots and Systems (IROS), Vancouver, BC, Canada, 24–28 September 2017; pp. 6741–6748.
17. Scona, R.; Jaimez, M.; Petillot, Y.R.; Fallon, M.; Cremers, D. Staticfusion: Background reconstruction for dense rgb-d slam in dynamic environments. In Proceedings of the IEEE International Conference on Robotics and Automation (ICRA), Brisbane, Australia, 21–25 May 2018; pp. 1–9.
18. Kim, D.-H.; Kim, J.-H. Effective background model-based rgb-d dense visual odometry in a dynamic environment. *IEEE Trans. Robot.* **2016**, *32*, 1565–1573. [CrossRef]
19. Sun, Y.; Liu, M.; Meng, M.Q.-H. Improving rgb-d slam in dynamic environments: A motion removal approach. *Robot. Auton. Syst.* **2017**, *89*, 110–122. [CrossRef]
20. Xiao, Z.; Dai, B.; Wu, T.; Xiao, L.; Chen, T. Dense scene flow based coarse-to-fine rigid moving object detection for autonomous vehicle. *IEEE Access* **2017**, *5*, 23492–23501. [CrossRef]

21. Alcantarilla, P.F.; Yebes, J.J.; Almazán, J.; Bergasa, L.M. On combining visual slam and dense scene flow to increase the robustness of localization and mapping in dynamic environments. In Proceedings of the IEEE International Conference on Robotics and Automation (ICRA), Saint Paul, MN, USA, 14–18 May 2012; pp. 1290–1297.
22. Wang, Y.; Huang, S. Towards dense moving object segmentation based robust dense rgb-d slam in dynamic scenarios. In Proceedings of the International Conference on Control Automation Robotics & Vision (ICARCV), Singapore, 10–12 December 2014; pp. 1841–1846.
23. Besl, P.; Mckay, N. A method for registration of 3D shapes. *IEEE Trans. Pattern Anal. Mach. Intell.* **1992**, *14*, 239–256. [CrossRef]
24. Curless, B.; Levoy, M. A volumetric method for building complex models from range images. In Proceedings of the 23rd Annual Conference on Computer Graphics and Interactive Techniques, New Orleans, LA, USA, 4–9 August 1996; pp. 303–312.
25. Nießner, M.; Zollhöfer, M.; Izadi, S.; Stamminger, M. Real-time 3d reconstruction at scale using voxel hashing. *ACM Trans. Graph.* **2013**, *32*, 169. [CrossRef]
26. Whelan, T.; Kaess, M.; Johannsson, H.; Fallon, M.; Leonard, J.J.; McDonald, J. Real-time large-scale dense rgb-d slam with volumetric fusion. *Int. J. Robot. Res.* **2015**, *34*, 598–626. [CrossRef]
27. Endres, F.; Hess, J.; Sturm, J.; Cremers, D.; Burgard, W. 3D mapping with an rgb-d camera. *IEEE Trans. Robot.* **2014**, *30*, 177–187. [CrossRef]
28. He, K.; Gkioxari, G.; Dollar, P.; Girshick, R. Mask r-cnn. In Proceedings of the IEEE International Conference on Computer Vision (ICCV), Venice, Italy, 22–29 October 2017; pp. 2980–2988.
29. Badrinarayanan, V.; Kendall, A.; Cipolla, R. Segnet: A deep convolutional encoder-decoder architecture for image segmentation. *IEEE Trans. Pattern Anal. Mach. Intell.* **2017**, *39*, 2481–2495. [CrossRef] [PubMed]
30. Newcombe, R.A.; Fox, D.; Seitz, S.M. Dynamicfusion: Reconstruction and tracking of non-rigid scenes in real-time. In Proceedings of the IEEE Conference on Computer Vision and Pattern Recognition (CVPR), Boston, MA, USA, 7–12 June 2015; pp. 343–352.
31. Dou, M.; Khamis, S.; Degtyarev, Y.; Davidson, P.; Fanello, S.R.; Kowdle, A.; Escolano, S.O.; Rhemann, C.; Kim, D.; Taylor, J. Fusion4d: Real-time performance capture of challenging scenes. *ACM Trans. Graph.* **2016**, *35*, 114. [CrossRef]
32. Bi, S.; Yang, D.; Cai, Y. Automatic Calibration of Odometry and Robot Extrinsic Parameters Using Multi-Composite-Targets for a Differential-Drive Robot with a Camera. *Sensors* **2018**, *18*, 3097. [CrossRef] [PubMed]
33. Siegwart, R.; Nourbakhsh, I.R. *Introduction to Autonomous Mobile Robots*, 2nd ed.; MIT Press: Cambridge, MA, USA, 2004; pp. 270–275.
34. Redmon, J.; Farhadi, A. Yolov3: An incremental improvement. *arXiv* **2018**, arXiv:1804.02767.
35. Lin, T.-Y.; Maire, M.; Belongie, S.; Hays, J.; Perona, P.; Ramanan, D.; Dollár, P.; Zitnick, C.L. Microsoft coco: Common objects in context. In Proceedings of the European Conference on Computer Vision (ECCV), Zurich, Switzerland, 6–12 September 2014; pp. 740–755.
36. Gálvez-López, D.; Tardos, J.D. Bags of binary words for fast place recognition in image sequences. *IEEE Trans. Robot.* **2012**, *28*, 1188–1197. [CrossRef]
37. Ceres Solver. Available online: http://ceres-solver.org (accessed on 15 January 2019).
38. Rosbag. Available online: http://wiki.ros.org/rosbag (accessed on 15 January 2019).
39. iai_kinect2. Available online: https://github.com/code-iai/iai_kinect2/ (accessed on 15 January 2019).
40. Garrido-Jurado, S.; Muñoz-Salinas, R.; Madrid-Cuevas, F.J.; Marín-Jiménez, M.J. Automatic generation and detection of highly reliable fiducial markers under occlusion. *Pattern Recognit.* **2014**, *47*, 2280–2292. [CrossRef]
41. Hartley, R.; Zisserman, A. *Multiple View Geometry in Computer Vision*, 2nd ed.; Cambridge University Press: Cambridge, UK, 2003; pp. 87–131.
42. evo. Available online: https://michaelgrupp.github.io/evo/ (accessed on 15 January 2019).

Article

Rapid Relocation Method for Mobile Robot Based on Improved ORB-SLAM2 Algorithm

Guanci Yang *, Zhanjie Chen *, Yang Li and Zhidong Su

Key Laboratory of Advanced Manufacturing Technology of Ministry of Education, Guizhou University, Guiyang 550025, China; liyanggzu@163.com (Y.L.); suzhidong2016@163.com (Z.S.)

* Correspondence: gcyang@gzu.edu.cn (G.Y.); chenzhanjie0320@163.com (Z.C.); Tel.: +86-851-84737007 (G.Y.)

Received: 24 November 2018; Accepted: 10 January 2019; Published: 14 January 2019

Abstract: In order to realize fast real-time positioning after a mobile robot starts, this paper proposes an improved ORB-SLAM2 algorithm. Firstly, we proposed a binary vocabulary storage method and vocabulary training algorithm based on an improved Oriented FAST and Rotated BRIEF (ORB) operator to reduce the vocabulary size and improve the loading speed of the vocabulary and tracking accuracy. Secondly, we proposed an offline map construction algorithm based on the map element and keyframe database; then, we designed a fast reposition method of the mobile robot based on the offline map. Finally, we presented an offline visualization method for map elements and mapping trajectories. In order to check the performance of the algorithm in this paper, we built a mobile robot platform based on the EAI-B1 mobile chassis, and we implemented the rapid relocation method of the mobile robot based on improved ORB SLAM2 algorithm by using C++ programming language. The experimental results showed that the improved ORB SLAM2 system outperforms the original system regarding start-up speed, tracking and positioning accuracy, and human–computer interaction. The improved system was able to build and load offline maps, as well as perform rapid relocation and global positioning tracking. In addition, our experiment also shows that the improved system is robust against a dynamic environment.

Keywords: ORB-SLAM2; binary vocabulary; small-scale vocabulary; rapid relocation

---

1. Introduction

With the rapid development of robotics, the localization and navigation of mobile robots have attracted the attention of many scholars [1,2], and it has become a hot-spot in the field of robotics research. Currently, the localization and navigation of robots mainly rely on SLAM (Simultaneous Localization and Mapping) [3], which can conduct real-time localization and environmental reconstruction in the unknown environment.

The current SLAM technology mainly includes Lidar-based SLAM [4] and Vision-based SLAM [5]. Laser SLAM is a relatively mature method of SLAM, which has been successfully applied to a variety of commercial products. However, the map information constructed by Lidar-based SLAM is too simple to enable the robot to obtain more abundant environmental information for other intelligent tasks. Visual SLAM has been a rapidly developing SLAM solution in the past ten years, which can reconstruct 3D environment maps using a camera sensor. Besides, the image contains a wealth of object information which can help the robot complete a variety of intelligent tasks based on vision. ORB-SLAM [6], proposed in 2015, is the representative of visual SLAM. With the support of various optimization mechanisms, its real-time and positioning accuracy is at a high level, but there are still some problems, such as monocular scale uncertainty and tracking loss. ORB-SLAM2 [7], which was proposed in 2016 as an improved version of ORB-SLAM, added support for the stereo camera and RGB-D camera, as well as improved robustness against environmental changes and violent movement; however, when the target platform is a mobile robot, ORB-SLAM2 has the following disadvantages:

1. Slow start-up. The ORB-SLAM2 system needs to read a large-scale text format vocabulary for loop closure detection during start-up, but after testing, the process is very time-consuming and takes up most of the time spent in the start-up process.
2. The vocabulary has a large amount of invalid data when the robot's working environment is relatively fixed. ORB-SLAM2 provides a vocabulary based on a large dataset, which enables ORB-SLAM2 to maintain high accuracy in different environments. However, when the working environment of the robot is relatively fixed, it still takes much time to read a large amount of invalid data in the vocabulary.
3. The map cannot be saved and reused. ORB-SLAM2 cannot save and load maps, so the robot needs to "relearn" its work environment when it starts up every time. If the ORB-SLAM2 system could save the map and reuse it, it would save many computing resources and improve the efficiency of the robot.
4. Lack of offline visualization of map and trajectory. ORB-SLAM2 provides real-time visualization, but users usually do not pay attention to the real-time status of the mapping but view the map and track information after the system is finished. However, ORB-SLAM2 does not provide offline visualization of maps and mapping trajectories.

Because of the above disadvantages of the ORB-SLAM2 system, this paper proposed a rapid relocation method for the mobile robot based on an improved ORB SLAM2 algorithm.

The main contributions of this paper are summarized as follows:

- A binary-based vocabulary storage method is designed to convert the text-format vocabulary into a binary format without data loss. This method can improve the vocabulary loading speed of the system.
- A vocabulary training algorithm is proposed based on an improved ORB operator, which is used to train the small-scale vocabulary to improve localization accuracy and reduce vocabulary size.
- An offline map construction algorithm is proposed based on map elements and a keyframe database, and we also designed a fast relocation method based on the offline map.
- We designed an offline visualization method for the map and mapping trajectory of the ORB-SLAM2 system.

The rest of the paper is organized as follows. Section 2 is a review and analysis of related research. Section 3 introduces the main threads and core algorithms of ORB-SLAM2. Section 4 presents our improvements to the ORB-SLAM2 algorithm. In Section 5, we test and analyze the method of this paper based on the robot platform. Section 6 summarizes this paper and discusses future research directions.

2. Related Work

In order to make Visual SLAM better for mobile robots, Davison et al. proposed MonoSLAM [8] as the first high frame-rate, real-time Monocular SLAM solution. The algorithm creates a sparse but stable map of landmarks based on a probabilistic framework, and solved the problem of monocular feature initialization, among others. However, as this work was based on the EKF (Extended Kalman Filter) algorithm [9], it caused a problem with error drift; the landmarks map are also sparse, making it easy for the system to lose tracking. In order to solve the linearization error caused by the EKF, UKF (Unscented Kalman Filter) [10] and improved UKF [11] were proposed and applied to Visual SLAM to improve the linearization uncertainty. Moreover, Sim et al. [12] proposed a particle filter-based monocular SLAM algorithm to avoid linearization. However, although the above method improves accuracy, it also greatly increases the computational complexity. Klein and Murray et al. [13] proposed a keyframe-based monocular SLAM algorithm, called PTAM (parallel tracking and mapping). The algorithm introduced a keyframe extraction method, and for the first time, divided the tracking and mapping into two parallel threads. Besides, PTAM used nonlinear optimization instead of the EKF method to eliminate the linearization error problem, which improved the location accuracy of the system. However, this study did not pay attention to the global optimization problem of the PTAM, which caused the system to lose tracking easily in a large-scale environment. Engel et al. proposed the Large-Scale Direct Monocular SLAM (LSD-SLAM) [14] algorithm, which is the monocular SLAM algorithm based on the direct method. LSD-SLAM can construct a semi-dense global consistent map, which is a more complete representation of the environmental structure than a point-cloud map based on the feature method. In addition, LSD-SLAM presents a novel direct tracking method which can accurately detect scale drift, and the algorithm can run in real time on the Central Processing Unit (CPU). However, this work still did not solve the gray-scale invariant hypothesis of the direct method, which made the performance of LSD-SLAM decrease rapidly when the robot was working in an environment with frequent illumination changes.

Focusing on the localization and navigation of the robot in an indoor environment, Brand et al. [15] proposed an on-board SLAM algorithm for mobile robots operating in an unknown indoor environment, where the algorithm introduced local 2.5D maps which could be directly employed for fast obstacle avoidance and local path planning, and which also constituted a suitable input for a 2D grid map SLAM. Lee et al. [16] proposed a SLAM algorithm based on autonomous detection and the registration of objects. This algorithm helps the robot to identify the object without relying on any a priori information, and inserts the detected objects as landmarks into the grid map. The new map has a certain improvement in positioning and navigation accuracy compared to the traditional pure visual SLAM. However, both studies ignored the importance of the 3D map, and they also cannot reconstruct a complete environmental structure, which made them lose the most important advantage of visual SLAM. Considering how the sparse map cannot be applied to obstacle avoidance and navigation, Qiang et al. [17] proposed a dense 3D reconstruction method based on the ORB-SLAM algorithm. The method enables the ORB-SLAM system to construct an octomap [18] based on the octrees and probabilistic occupancy estimation, and the improved ORB-SLAM can complete the map reconstruction in real time using a Kinect 2.0 camera in the real world. However, this work cannot express the working environment intuitively because of the use of octomap, so we could not directly check the mapping effect, which reduced the interactivity of the robot.

Researchers have made a lot of effort to try and optimize the performance of ORB-SLAM2. Wang et al. [19] proposed a monocular SLAM algorithm based on the fusion of IMU (Inertial Measurement Unit) information and ORB-SLAM2 pose information. The algorithm can determine the scale information and make-up for the pose information when the pure monocular visual SLAM tracking is lost. Caldato et al. [20] proposed a tightly sensor fusion method to improve the tracking results of visual odometry when the robot works in featureless environments. This algorithm integrated image data and odometer data to improve graph constraints between frames and prevent tracking loss. However, although the above two studies improved the robustness of the ORB-SLAM2 system, they increased the occupation of robot computing resources. The improvements of the two papers were also mainly focused on dealing with unexpected situations of tracking, and did not improve the performance of the system under a normal tracking state. In order to solve the problem of how ORB-SLAM2 cannot distinguish 3DOF (degrees of freedom) image frames and 6DOF image frames, Zeng et al. [21] proposed a 6DOF keyframe selection method to avoid wrong triangulation to 3DOF keyframes. However, since the 3DOF keyframes are filtered out, the number of keyframes in the map or keyframe database is reduced, thereby reducing the accuracy of loop detection and relocation, which ultimately leads to degradation of the performance of the ORB-SLAM2 system. Senbo et al. [22] introduced a unified spherical camera model to extend the ORB-SLAM2 framework. The model enables the system to obtain a larger perceiving region by using fisheye cameras. In addition, it proposed a semi-dense feature-matching method, which can make use of high-gradient regions as semi-dense features to construct the semi-dense map, thereby providing richer environment information than a sparse feature map. However, the mapping error is relatively high when the robot is working in a large-scale environment, and although this work reconstructed a complete environment, it ignored the extended application and reuse of semi-dense map information.

3. ORB-SLAM2

ORB-SLAM2 [7], which was proposed by R Mur-Artal, added support for the stereo camera and RGB-D camera outside of the monocular camera. As shown in Figure 1, the system framework of ORB-SLAM2 mainly contains three parallel threads: Tracking, Local Mapping, and Loop Closing.

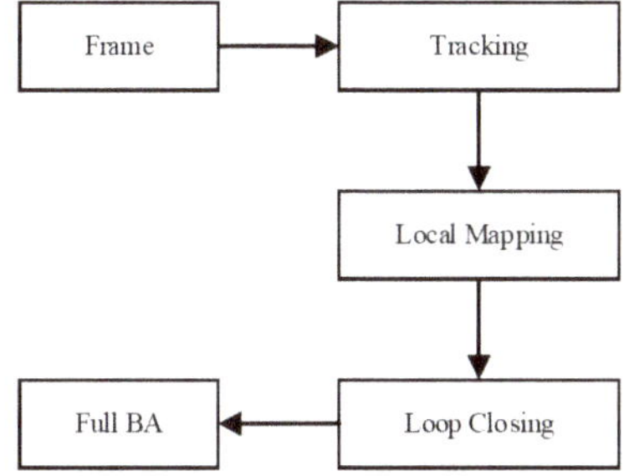

Figure 1. The system framework of ORB-SLAM2.

3.1. Tracking

The main task of the tracking thread is to estimate the camera pose for each frame of the input image based on the feature method. There are three tracking models in the tracking thread: the motion model, keyframe model, and relocation model. The tracking state of the system is affected by factors such as environmental changes, and the tracking thread selects the tracking model according to different tracking states.

ORB-SLAM2 estimates the camera pose by building a PnP (Perspective-n-Point) [23] model. There are several methods for the PnP problem, such as P3P [24], DLT (Direct Linear Transformation) [25], EPnP (Efficient PnP) [26], etc. EPnP is the pose estimation algorithm used by the relocation tracking model, and it is also one of the best accurate PnP problem-solving algorithms.

3.2. Local Mapping

The local mapping thread is mainly responsible for receiving and processing new keyframes, checking new map points, maintaining the accuracy of local maps, and controlling the quality and scale of keyframe sets. The workflow is as follows:

(1) Process new keyframes: First, calculate the BoW (Bag-of-Words) [27] vector of the current keyframe, update the map point observation values of the keyframe, and put these map points on the new map point list; then, update the Covisibility graph and Essential graph, and add the current keyframe to the map.

(2) Filter map points: Rule out redundant points by checking the list of new map points. The culling rules are as follows: (a) The map point is marked as a bad point; (b) the number of keyframes that can observe the map point is no more than 25% or three (the threshold is two when the sensor is a monocular camera).

(3) Restore new map points based on the current keyframe: First, select the keyframes connected to the current keyframe from the Covisibility graph, and then perform feature-matching on the current keyframe and the selected keyframes, and calculate the pose of the current keyframe by the epipolar geometry; after that, the feature-point depth is restored by the triangulation method. Finally, the re-projection error of the new map point is calculated according to the depth of the feature point obtained, and whether the map point is eliminated or not is determined according to the relationship between the error and the given threshold value.

(4) Local BA (Bundle Adjustment): When new keyframes are added to the Covisibility graph, iterate optimization is performed and the outer points are removed to optimize the pose of the locally connected keyframes.

(5) Filter local keyframe: The rule for ORB-SLAM2 to remove redundant keyframes is, if 90% of the map points observed by the keyframe can be observed by the other three or more keyframes simultaneously, delete this keyframe.

3.3. Loop Closing

Loop Closing contains loop detection and loop correction.

The task of loop detection is to screen and confirm the loop closure. First, calculate the BoW score of the current keyframe and the connected keyframe, and select the closed-loop candidate frame by the lowest threshold. The independent keyframes of low quality are eliminated by calculating the number of shared words and the cluster score, and the remaining candidate keyframes are continuously detected. After detecting the loop closure, the similarity transformation Sim3 is solved by the RANSAC (Random Sample Consensus) [28] framework, and then Sim3 is optimized by re-matching and g2o (General Graphic Optimization) [29] to correct the pose of the current keyframe.

Loop correction is responsible for eliminating global cumulative errors. First, adjust the pose of the keyframe connected to the current keyframe by the propagation method. Then, project the updated map points to the corresponding keyframes and fuse the matching map points. Finally, update the connection relationship of keyframes according to the adjusted map points. After the map fusion is completed, perform the pose graph optimization by the essential graph.

4. Improved ORB-SLAM2 Algorithms

In this section, we proposed an improved ORB-SLAM2 algorithm to improve the disadvantages of ORB-SLAM2 mentioned above. The specific improvements include the binary-based vocabulary storage method, vocabulary training algorithm for improved ORB operator, offline map construction and preservation method, and robot relocation method, based on the offline map.

4.1. Binary-Based Vocabulary Storage Method

ORB-SLAM2 provides a vocabulary for training through large-scale data, and the authors implemented a function to save the vocabulary as a text format in order to enable the system to load the vocabulary directly. However, the text file needs to process the data format and line breaks during the loading process, and the time consumption is very large when the vocabulary data size is large. At the same time, we noticed that the binary file is non-interpretable and can be read directly without complicated data conversion and line-break processing, so it has high reading efficiency. Based on the above facts, in order to improve the vocabulary loading speed of the ORB-SLAM2 system, a binary-based vocabulary storage method (Algorithm 1) is proposed.

In Step 1 of Algorithms 1, we used the class of *TemplatedVocabulary* provided by DBoW2 [30] to load the vocabulary. In Step 3, the value of K, L, W_{node}, and S_{sim} are inherited from the text vocabulary. In Step 4, if the Boolean value is true, that means the node is a leaf. In practical application, the system also needs to load the binary vocabulary file. As the loading is the inverse process of the above vocabulary saving method, we will not describe it in detail.

Algorithm 1: Binary-Based Vocabulary Storage Method

Input: A text format vocabulary
Output: A binary format vocabulary

1) Load the vocabulary data in text format and create a vector V_{node} that includes all the nodes, then read the number of nodes N_{node} and the size of nodes N_{size};
2) Create an empty binary file F_{b};
3) Write the N_{node}, N_{size}, number of branches K, vocabulary tree depth L, node weight type W_{node}, similarity score calculation method S_{sim} into F_{b};
4) For each node $V^i_{\text{node}} \in V_{\text{node}}$ (i from 1 to N_{node}):
    a) Write the parent node of V^i_{node} into F_{b};
    b) Write the feature descriptor of V^i_{node} into F_{b};
    c) Write the weight of V^i_{node} into F_{b};
    d) Write a Boolean value to indicate the node property to know whether it is a leaf node.
5) Save F_{b} as the binary format vocabulary.

4.2. Vocabulary Training Algorithm Based on Improved ORB Operator

When the working environment of the robot is relatively fixed, there are large amounts of data that cannot be used in the large-scale vocabulary provided by ORB-SLAM2. Therefore, in order to obtain a small-scale vocabulary for a relatively fixed environment, we proposed a vocabulary training algorithm based on the improved ORB operator, where the specific steps are shown in Algorithm 2.

Algorithm 2: Vocabulary training algorithm based on improved ORB operator

Input: Training dataset X of images for the vocabulary
Output: A binary vocabulary, V, of the given training dataset, X

1) Establish the dataset X using RGB images.
2) Initialize a null feature vector F.
3) For each image $X_i \in X$:
    a) Read image X_i, and then apply the improved ORB feature extractor to get the feature vector F_i of image X_i;
    b) $F = F \cup F_i$;
4) Use the K-means++ [31] algorithm to deal with feature vector F to obtain the vocabulary tree V_{tree};
5) Call the function *create*() provided by DBoW2 to create the vocabulary V by creating leaf nodes of the vocabulary tree V_{tree} and assigning the weight of each node. The weight type is set to be TF-IDF (Term Frequency-Inverse Document Frequency).
6) Output the vocabulary V.

In Step 1, the images are captured from a relatively fixed environment. In Step 3, the improved ORB feature extractor is a part of ORB-SLAM2, which employs the quadtree to extract features to improve the quality and distribution of the feature.

In Step 5, we followed the original rules of ORB-SLAM2 in terms of the vocabulary parameter setting, where the weight type was TF-IDF (Term Frequency-Inverse Document Frequency) [32]. *TF* represents the degree of differentiation of a feature in an image, and *IDF* indicates the degree of discrimination of a word in the vocabulary. Therefore, for the word w_i in an image, its weight η_i can be calculated as follows:

$$\eta_i = TF_i \times IDF_i = \frac{n_{w_i}}{n_w} \times \log \frac{nfeatures}{nfeatures_{w_i}} \tag{1}$$

where n_{w_i} is the number of times the word w_i appears in an image A, n_w represents the sum of the number of occurrences of all words in image A, $nfeatures$ indicates the number of all features in the vocabulary, and $nfeatures_{w_i}$ is the number of features included in w_i of the vocabulary. For the image A, after calculating the word weight, the BoW vector v_A containing N words can be expressed as:

$$v_A \triangleq \{(w_1, \eta_1), (w_2, \eta_2), \ldots, (w_N, \eta_N)\} \tag{2}$$

We set the similarity calculation method to the L1-norm. Therefore, for two images A and B, the similarity $s(v_A, v_B)$ is calculated as follows:

$$s(v_A, v_B) = 2\sum_{i=1}^{N} \left|v_{A_i}\right| + \left|v_{B_i}\right| - \left|v_{A_i} - v_{B_i}\right| \tag{3}$$

4.3. *Offline Map Construction Method*

In order to enable the system to load offline data and restore to the previous running state, we proposed a method for constructing offline maps (Algorithm 3), and the method flow is shown in Figure 2.

Algorithm 3: Offline Map Construction Method

Input: Map and keyframe database.
Output: Binary format offline map.

1) Create an empty binary file F_{mapb} to save the data of the offline map;
2) Read the data from the threads: 3D Map points M, keyframes K in the map, Covisibility Graph and Spanning tree of keyframes, the storage vector V_{KFDB} of keyframe database;
3) For each map point $M_i \in M$:
    a) Write the index of M_i to F_{mapb};
    b) Write the coordinates of M_i in the world coordinate system to F_{mapb};
4) For each keyframe $K_i \in K$:
    a) Write the pose of K_i to F_{mapb};
    b) Write all feature descriptors of K_i to F_{mapb};
    c) Write the index of all map points connected to K_i to F_{mapb};
    d) Write the Covisibility Graph and Spanning tree of K_i to F_{mapb};
5) Write the V_{KFDB} to F_{mapb};
6) Output F_{mapb} as a binary file.

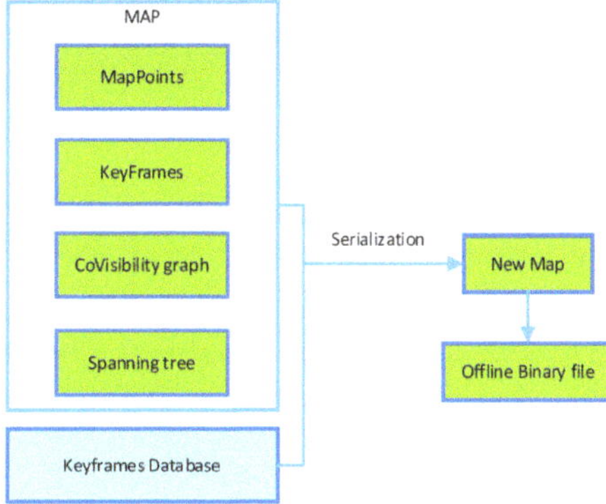

Figure 2. Construction flow of the offline map.

In Step 2, we first checked the data in the map and keyframe database, and ended the system when the data to be saved was empty. In Steps 3 to 5, the map contains M, K, Covisibility graph, and Spanning tree; V_{KFDB} is a collection of all keyframes corresponding to each word in the ORB dictionary. It is used for relocation and loop detection, providing data support for global positioning tracking after the system restarts. In Step 4, because the BoW vector of keyframes can be obtained by calculating the feature descriptor, we did not save them. In Step 6, the binary file is the obtained offline map. Considering the need to save the association of map such as the Covisibility Graph, we used the binary file to store the offline map.

In the implementation of Algorithm 3, we used a serialization method to build the offline map to ensure that the data could be recovered correctly by deserialization.

4.4. Robot Fast Relocation Method Based on Offline Map

In order to be able to use the offline map constructed by the previous method, this section presents a robotic fast relocation algorithm based on offline maps to fit the data of two adjacent runs of the system to avoid repeated mapping. Specific steps are shown in Algorithm 4.

Algorithm 4: Fast Relocation Method Based on Offline Map

---

1) Start the ORB-SLAM2 system, and load the offline map file F_{mapb};
2) If F_{mapb} is null, then:

   Call the original method of ORB-SLAM2 to perform the complete SLAM process;
   Return;
3) Load the vocabulary, and construct the storage vector V_{KFDB} using the index of words in the vocabulary and the keyframe database in F_{mapb};
4) Calculate the BoW vector of the keyframes to completely restore the data of the previous run;
5) Initialize and start up all threads;
6) Set the tracking state *mState*=LOST to touch off the relocation tracking model;
7) Call the relocation function *Relocalization*() to restore the position of the robot;
8) Return.

ORB-SLAM2 designed a relocation model to restore tracking for tracking lost conditions. In Step 5, we proposed a new triggering mechanism for the relocation tracking model, where the system enters the tracking lost state directly when the data in F_{mapb} is restored successfully.

4.5. Offline Visualization Method for Map and Mapping Trajectory

In order to enable users to view the mapping effect and the tracking trajectories of the robot while being offline, this section proposes an offline visualization method for the map and mapping trajectory; the specific steps are shown in Algorithms 5 and 6.

Algorithm 5: Offline visualization method of map elements

---

1) Load the parameters file F_{va}, which includes the camera intrinsic matrix;
2) Load map data *Map* of offline map file F_{mapb};
3) If *Map*= NULL, then Return;
4) Read observation parameters *ViewpointX*, *ViewpointY*, *ViewpointZ*, and camera parameter *ViewpointF* from F_{va};
5) Create a map visualization window *ViewerMap*;
6) Based on the *MapDrawer* class of ORB-SLAM2, and using *mpMap* and F_{va} as parameters:
    a) Call the member function *DrawMapPoints*() to draw map points and reference map points in *ViewerMap*;
    b) Call the member function *DrawKeyFrames*() to draw the poses and connection relationship of keyframes in *ViewerMap*;
7) Output the visual map file.

In Step 2, data such as map points and keyframes that need to be visualized belong to the map element, so the data of the keyframe database is not loaded; in Step 5, the visualization window is created by the open source library *Pangolin*; in Step 7, users can view the 3D map in the visualization window or export the JPG file from a fixed perspective.

Algorithm 6: Offline visualization method of mapping trajectory

---

1) Load keyframes file F_{KT};
2) Create a pose vector V_{P};
3) For each line data $\in F_{\text{KT}}$
    a) Create rotation quaternion q and a transport vector t;
    b) Construct keyframe pose P using q and t;
    c) Save P to V_{P};
4) Create a trajectory visualization window *ViewerTrajectory*;
5) For each pose $P_i \in V_{\text{P}}$
    a) Set drawing color and line format;
    b) Draw P_i to *ViewerTrajectory* in 3D point format;
6) Output trajectory file.

In Step 1, the keyframe file F_{KT} is automatically saved by the ORB-SLAM2 system at the end of the run. In Step 3, each row of data in F_{KT} is composed of a time stamp, displacement, and rotation, where displacement and rotation can be constructed into a pose.

5. Experiments and Results Analysis

5.1. Platform of the Used Robot

In order to test the performance of the proposed algorithm, we built a social robot platform, the MAT social robot [33], as shown in Figure 3. The platform consists of mobile chassis, host computer, sensors, and mechanical bracket, etc. The mobile chassis is the EAI DashGO B1 ((The manufacturer of the equipment is the EnjoyAI company in Shenzhen, China)), which is capable of remote mobile control and provides a 5 V–24 V independent power interface. The audio sensor is the ring microphone array board (The manufacturer of the equipment is the iFLYTEK company in Hefei, China). The vision sensor is a depth camera Kinect 2.0 with a maximum resolution of 1920*1080 and a transmission frame rate of 30 fps. The host computer of the service robot is a Next Unit of Computing (NUC) produced by Intel Corporation, and it has a Core i7-6770HQ processor that has 2.6–3.5 GHz frequency, and the graphics processor is the Intel IRIS Pro. Besides, for the host computer, we installed 16 GB of memory with 2133 Mhz frequency.

Figure 3. The used social robot platform MAT.

We implemented the proposed algorithms with C++ programming and integrated the implemented algorithms with the ORB-SLAM2 system to obtain an improved ORB-SLAM2 system. Finally, we installed the improved system in the Ubuntu16.04 system on the MAT social robot's host computer.

In particular, before the experiments, we performed an intrinsic matrix calibration of the depth camera according to the steps of the official document. Kinect2.0 and Intel NUC were powered by mobile chassis.

5.2. Experimental Design

In order to evaluate the performance of the improved ORB-SLAM2 system, we designed the following experiments:

(1) The impact of different formats' vocabulary on the system startup speed and scale. Considering that loading the vocabulary is the most time-consuming part of the system startup process when testing the system startup speed, we compared the time it took for the system to read the different formats of the vocabulary.

(2) Performance comparison between small-scale vocabulary based on the algorithm of this paper and original large-scale vocabulary of ORB-SLAM2. The training and test data used in the experiments were from the public RGB-D dataset provided by Technische Universität München (TUM) [34]: *fr1_room, fr1_xyz, fr1_360, fr2_rpy, fr2_desk*. The reason why we used this database was that every sequence of the TUM database contained the real trajectory file, so we could compare the test results with the real situation. We used the *fr1_room* sequence as an input to the vocabulary training algorithm, where the *fr1_room* contains 1362 RGB images taken by the Kinect 1.0 depth camera, the image sequence collection environment is a complete indoor office environment, and Figure 4 is part of the scene. Also, in order to visualize the comparison results, we used the root mean square error (RMSE) of the actual mapping trace and the real trace as a performance comparison metric.

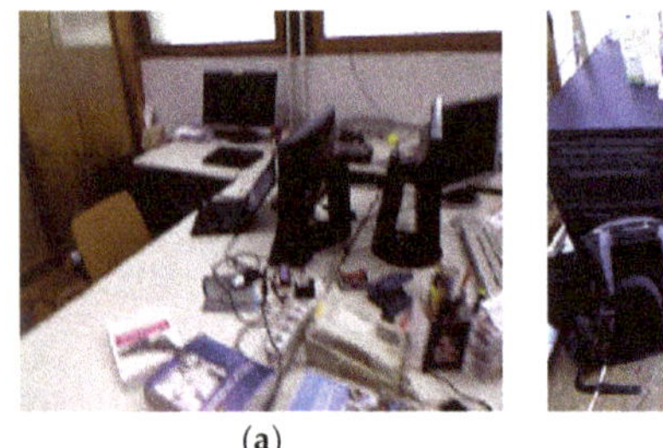
(a)

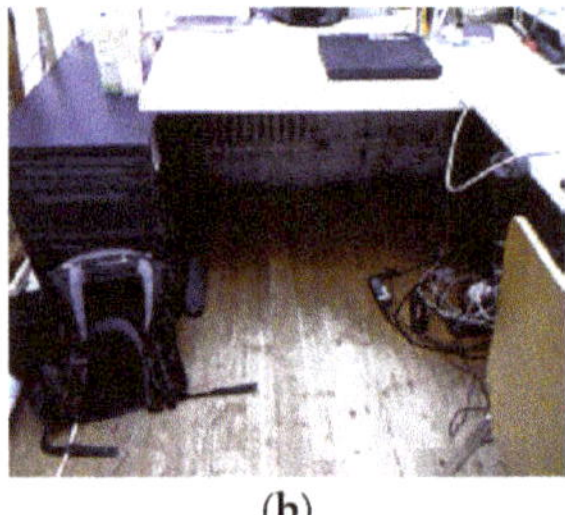
(b)

(c)

Figure 4. Part of the scene in *fr1_room*.

(3) Offline map construction and fast relocation experiments. In the real world, we ran the robot to test whether the system could build an offline map, and whether it could load the offline map to quickly relocate the robot after rebooting. In addition, based on the saved offline map and keyframes file, we will verify that the method in this paper can visualize the map elements and mapping trajectory offline.

In this experiment, the actual test environment we used is shown in Figure 5. The office is a typical indoor office environment of approximately 43 square meters. The interior includes desks, chairs, computers, printers, etc. The wall is white and contains large transparent glass windows, so there will be a change in the number of feature points extracted and changes in illumination in the tracking process of the system, which will have an impact on the accuracy of localization and mapping. However, in reality, the working environment of mobile robots is usually non-idealized, so our test environment is of practical significance.

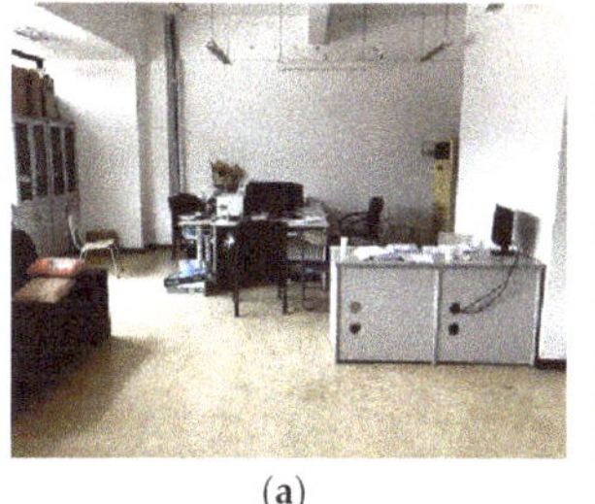
(a)

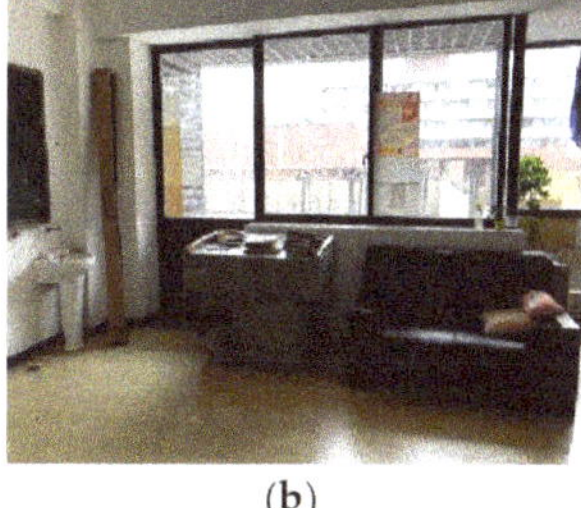
(b)

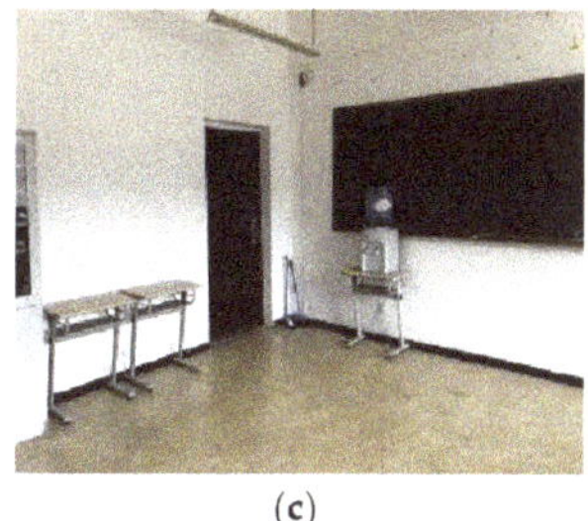
(c)

Figure 5. Part of the scene in the actual test environment. (**a**) An office area with tables, chairs, computers, and bookcases; (**b**) the wall with large transparent glass windows; (**c**) the door of the office, which is the starting point and end point of the robot's trajectory.

(4) Restore mapping experiment. We took into consideration that in actual applications, the working environment of the robot will be dynamic, e.g., chairs will often move. Therefore, in order to test whether the proposed method could adapt to such environmental changes, we designed a restore mapping experiment based on the expected results of Experiment 3. The experimental environment is shown in Figure 6. This is an empty room with a relatively regular ground marking. The reason for choosing this environment was that it was easy to observe changes in the map. We placed several objects of different shapes and colors on the blackboard to ensure that the system could detect enough feature points. In addition, as seen in Figure 6b, we set the motion area of the robot to be the light-colored track portion on the ground and kept the robot's viewing angle unchanged during the motion. These settings are for visual comparison of the actual environment and the map.

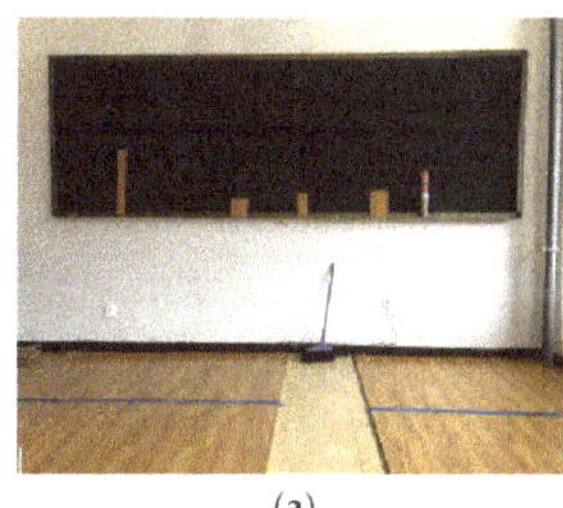
(a)

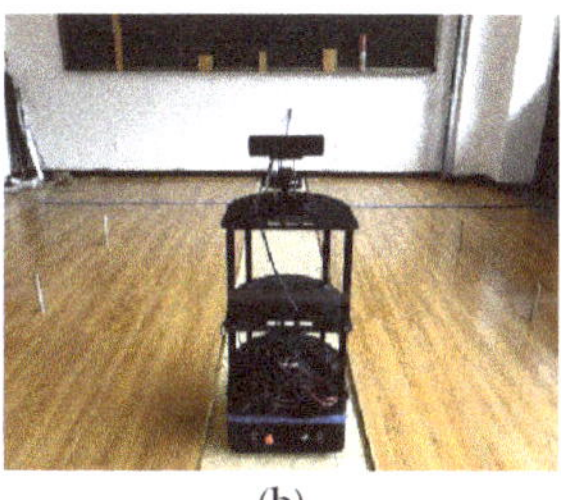
(b)

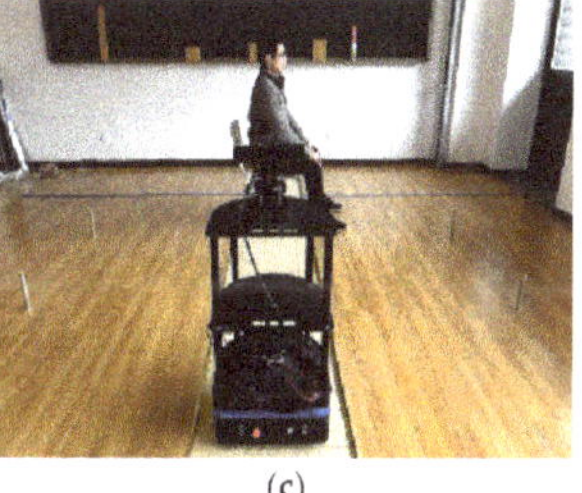
(c)

Figure 6. Scene of the restore mapping experiment. (**a**) The experimental environment we designed; (**b**) the state of robot and experimental environment before adding changes; (**c**) the state of robot and experimental environment after adding changes.

In the experiment, we first started the improved ORB-SLAM2 system and controlled the robot to move on the track to complete the construction. After completing the mapping task, we controlled the robot to stop moving to keep the map data stable, and closed the system to construct an offline map. Next, we restarted the system for rapid relocation of the robot, and added an "obstacle" (a person sitting in a chair) to the visible range of the robot to change the original environment, as shown in Figure 6c. Then, we started the mapping mode in the system's real-time visual interface, and controlled the robot to continue to move on the track, while observing the real-time changes of the map data.

5.3. Experimental Results Analysis

5.3.1. Optimization Experiment in Binary Format Vocabulary

We converted the original text format vocabulary T_{VOC} of ORB-SLAM2 to the binary format, B_{VOC} using the method of this paper, and the size comparison of the two vocabularies is shown in

Table 1. Then, we used two kinds of vocabulary to start the system five times each. Time cost statistics are shown in Table 2.

Table 1. The size comparison between text vocabulary and binary vocabulary.

Sizes of Different Format Vocabularies (MB)		Percentage of Size Decrease
T_{VOC}	B_{VOC}	
145.3	44.4	↓ 69.44%

Table 2. Time overhead comparison of loading two kinds of vocabularies.

No.	Time Overhead of Loading Vocabulary (s)	
	T_{VOC}	B_{VOC}
1	8.05213	0.432957
2	8.04686	0.429398
3	8.09128	0.428909
4	8.05895	0.431752
5	8.06749	0.430972
Average	8.06334	0.430798

The results in Table 1 show that when the vocabulary is converted to binary format, the volume of the vocabulary is reduced by 69.44%, which is a reduction from 145.3 MB to 44.4 MB, thus greatly saving the space occupied by the vocabulary and helping the system to be lightweight.

The results in Table 2 show that reading the binary format vocabulary takes an average of 0.43 s, while the text vocabulary has an average reading time of about 8.06 s, which is 18.7 times the time overhead of reading the binary vocabulary.

5.3.2. Small-Scale Vocabulary Performance Test Based on Improved Training Algorithm

We obtained the binary format small-scale vocabulary $Fr1_{VOC}$ using the method of this paper, and compared the performance between $Fr1_{VOC}$ and the original vocabulary B_{VOC} of ORB-SLAM2.

Firstly, we experimented with the training dataset *fr1_room* as a test dataset, making the vocabulary training environment exactly the same as the system working environment. Then, we took $Fr1_{VOC}$ and B_{VOC} as inputs, respectively, and started up the ORB-SLAM2 system to localization and mapping, and finally calculated the RMSE value of the actual mapping trajectory and the real trajectory. The comparison between the actual mapping trajectory and the real trajectory is shown in Figure 5, and Table 3 compares the size of the two vocabularies and the trajectory RMSE values when using two vocabularies, respectively.

Table 3. Comparison of trajectory error and size.

	B_{VOC}	$Fr1_{VOC}$	Change
RMSE	0.081027m	0.049486m	↓ 38.92%
Size	44.4 MB	11.9 MB	↓ 73.20%

According to Figure 7, it can be seen that when using the large-scale vocabulary B_{VOC} to track the *fr1_room* dataset, the distance between the actual running trajectory and the real trajectory in most of the road segments is larger than the distance when using the small-scale vocabulary $Fr1_{VOC}$. The specific data comparison in Table 3 verifies our point of view; the trajectory error when using B_{VOC} is 0.081, while it is 0.049 when using $Fr1_{VOC}$, which is a 38.92% reduction; at the same time, the vocabulary volume has been reduced from 44.4 MB to 11.9 MB, a decrease of 73.2%. The above data and analysis show that in an ideal environment (training the vocabulary using the image dataset of the working environment), the small-scale vocabulary can achieve a large performance improvement for the ORB-SLAM2 system.

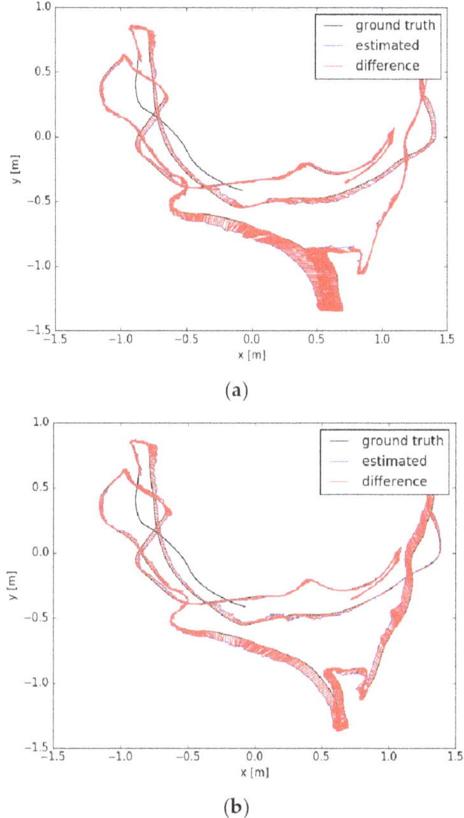

Figure 7. The comparison of the actual mapping trajectory and the real trajectory. (**a**) Trajectory deviation when using B_{VOC}; (**b**) trajectory deviation when using $Fr1_{\text{VOC}}$.

Considering the difference in practical applications, we could not collect working environment data for each robot and train the corresponding vocabulary, but different environments can be classified, and each type of environment has the same characteristics. In order to test the performance of the small-scale vocabulary in similar environments (not the same), image sequences *fr1_xyz*, *fr1_360*, *fr2_rpy*, and *fr2_desk* were used as test datasets, and then we started up the ORB-SLAM2 system to track and construct a map using $Fr1_{\text{VOC}}$ and B_{VOC}, respectively, after which we could finally calculate the RMSE values of trajectory errors, where the results are shown in Table 4.

Table 4. The performance tests of small-scale vocabulary in similar environments.

Datasets	RMSE(B_{VOC})	RMSE($Fr1_{\text{VOC}}$)	Change of RMSE
fr1_xyz	0.009544m	0.009916m	↑ 3.90%
fr1_360	0.275813m	0.259340m	↓ 5.97%
fr2_rpy	0.013706m	0.012929m	↓ 5.67%
fr2_desk	0.085194m	0.084459m	↓ 0.86%

The data in Table 4 show that when using the small-scale vocabulary $Fr1_{VOC}$, the trajectory errors value for the *fr1_360*, *fr2_rpy*, and *fr2_desk* datasets is slightly reduced by 0.86% to 5.97%, compared to using the B_{VOC} vocabulary. Only when the test dataset is *fr1_xyz*, the trajectory error is slightly increased by 3.9%. The above results show that small-scale vocabulary, based on specific environment training, can improve the running accuracy in most cases when applied to similar environments.

The above experiments and results analysis show that the small-scale vocabulary obtained by training the image of the specific environment based on the algorithm of this paper can improve the running accuracy of the system when working in a specific environment. Moreover, it also eliminates the problem of how the original large-scale vocabulary of ORB-SLAM2 includes a large amount of ineffective data.

5.3.3. The Experiments of Offline Map Construction and Robot Rapid Relocation

Firstly, we started the ORB-SLAM2 system which implemented the algorithm of this paper, and controlled the Mat Social robot to walk around the test environment to form a trajectory closed loop; then we shut down the system, and used this method to visualize the map and the keyframes trajectory. The results are shown in Figure 8.

Figure 8. Visualization of the offline map and mapping trajectory. (**a**) Point-cloud map; (**b**) mapping trajectory.

As shown in Figure 8a, the visualization method of this paper enabled the ORB-SLAM2 system to successfully load the offline map and restore 3D map points and keyframes. The visual map completely expresses the structure of the test environment, and the relative positions of the obstacles are correct. The robot trajectory consisting of keyframes in Figure 8a is consistent with the trajectory shown in Figure 8b. The above results show that the ORB-SLAM2 system, based on this method, can construct the offline map and visualize the map and mapping trajectory offline.

We experimented with the rapid relocation of the robot based on the saved offline map. Considering that large independent obstacles are easily distinguishable in the point-cloud map, in order to distinguish the relocation result of the robot, we first controlled the robot to leave the mapping trajectory and stay near a separate desk. Then we started the ORB-SLAM2 system through the data of a real-time visualization window, checked whether the system had loaded the offline map data, and confirmed the robot relocation effect. The result is shown in Figure 9.

According to the keyframe, map point, and feature-matching data in the red box below Figure 9a and the complete point-cloud map shown in Figure 9b, it can be explained that the offline map data has been fully loaded and restored correctly. Firstly, we infer the position of the robot in the point-cloud map based on the RGB in Figure 9a. The robot is facing the office area, and the area marked 1 on the right front is the stand-alone desk mentioned above, which we recorded as *desk 1*. Since the robot is close to *desk 1*, it needs to be bypassed when building the map, meaning the location of the robot should be close to the corner of the mapping trajectory. Figure 9b shows the map's real-time visualization interface of the ORB-SLAM2 system, where the green image block in the red elliptical circle represents the robot, and the location of the robot is near the corner of the construction trajectory. At the right front of the robot is the position of *desk 1* in the map. This verified our judgment of the location of the robot based on the information in Figure 9a. The above test results and analysis show that the method of this paper can make the robot achieve rapid relocation.

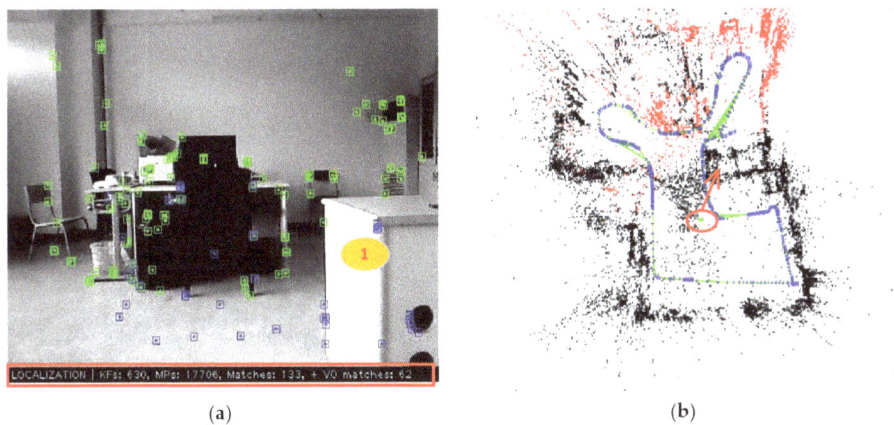

Figure 9. Real-time running status after a system restart. (**a**) Current RGB frame and data of system; (**b**) real-time status of map and tracking.

We tested whether the system could restore global tracking after the robot relocated correctly. Firstly, we controlled the robot to retreat a distance without changing the direction, and then observed the tracking state of the robot in the map. The results are shown in Figure 10.

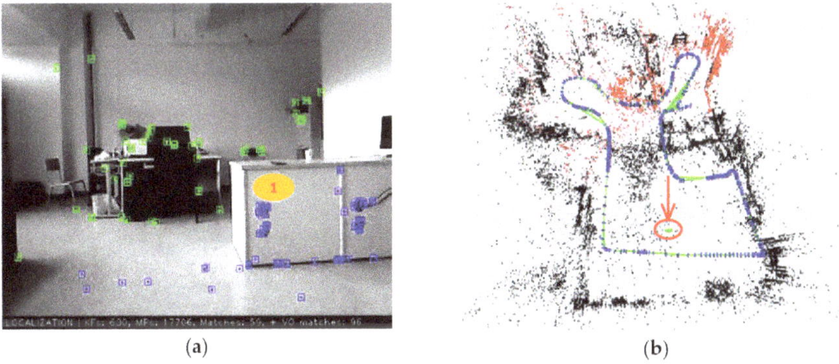

Figure 10. System global tracking status after relocation. (**a**) Current frame image after the complete robot moving; (**b**) real-time status of map and tracking.

Figure 10 shows the real-time visual interface of the ORB-SLAM2 system after the robot has moved a distance. Comparing Figure 10a with Figure 9a, the movement of the robot follows the experimental design: moving backward without changing the direction. Although we cannot accurately judge the moving distance of the robot based on the image, it can be determined that the robot location should be close to the mapping trajectory behind it. As shown by the red arrow in Figure 10b, the position of the robot has changed, moving from the original position to the rear and nearing the mapping trajectory below the map, and this is consistent with the actual movement of the robot.

The above experimental results and analysis show that the proposed method can make the robot relocate quickly and restore global location and tracking. This avoids repeated mapping, saving computational resources for the robot and improving work efficiency.

5.3.4. Restore-Mapping Experiment

Firstly, we controlled the robot to move onto the orbit and built the map according to the experimental design, and then reproduced the map scene with the offline visualization method proposed in this paper. The results are shown in Figure 2.

Figure 11 shows the map status after the mapping is completed. Among them, Figure 11a shows the real-time RGB image of the system, and the figure shows that the number of feature points is sufficient. At the bottom of Figure 11a, we marked the map element data with a red box, in which the number of keyframes was 116 and the number of map points was 1092. Figure 11b is the sparse map with the same angle-of-view as Figure 11a. Compared with the actual environment, the environmental structure of the map is correct and clear. Figure 11c is a map with an overlooking angle, and we can find that the map points are mainly concentrated in the part close to the wall. There is some blank space between the point-cloud concentration region and the position of the robot. This result is what we expected because it helps us to use this blank area for subsequent experiments, and we can directly observe changes in the map without interference from other point-cloud distributions.

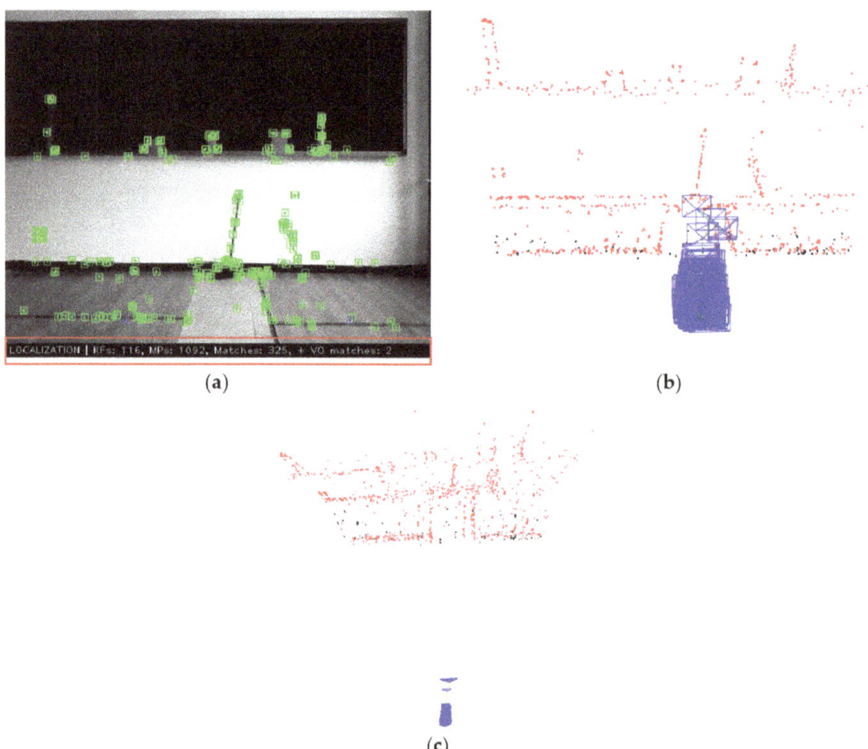

Figure 11. The original mapping effect. (**a**) Current RGB image frame after map data has been stabilized; (**b**) sparse point-cloud map with the same angle-of-view as the camera; (**c**) vertical-view angle of the map.

After the completion of the above experiment, we changed the environment and then performed the test of restore mapping according to the experimental design. The results are shown in Figure 12.

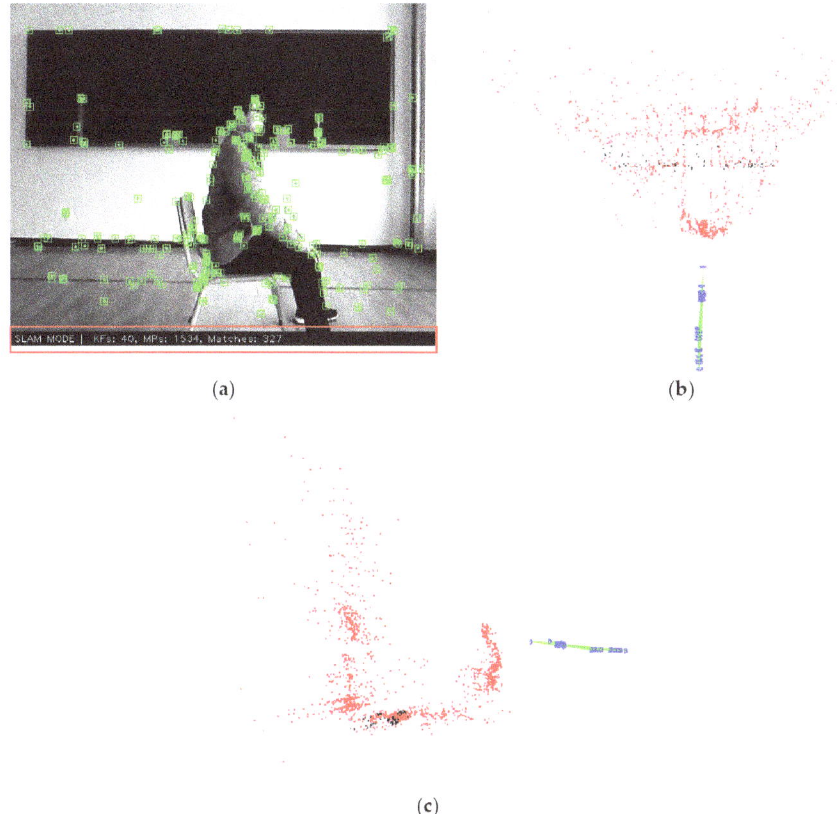

Figure 12. The map status after restore mapping. (**a**) Current RGB frame after restore mapping; (**b**) vertical-view angle of the map; (**c**) side-view angle of the map.

Figure 12 shows the restore mapping status after the environment change. Figure 12a is an RGB frame during mapping, and the image shows that the system has detected the newly added features of the person and the chair. The element data of the map marked by the red box at the bottom of the image has changed significantly. The number of map points continues to increase, but the number of key frames dropped sharply, which is because the environment has changed and the system is continuously updating keyframes. Figure 11b,c are the top view and side view of the map after completion of the construction, respectively. The image shows that some new map points were added in the blank area in Figure 11c that are consistent with the environmental changes, and the outline of the person sitting on the chair (near the side of the camera) can be found in the side view. In addition, in contrast to Figure 11c, the new mapping trajectory also changed according to the movement of the robot.

The analysis of the above experimental results shows that the method proposed in this paper can continue to construct the changing part of the environment after a change in the environment, so as to be applicable to the new environmental structure. In other words, the method proposed in this paper has strong robustness against changes to the environment.

6. Discussion and Conclusions

This paper proposed an improved ORB-SLAM2 system for improving the problems of the original system in terms of startup speed, tracking accuracy, and map reuse. Firstly, we proposed a binary-based vocabulary storage method to improve the startup speed of the ORB-SLAM2 system. Then, we proposed an improved vocabulary training algorithm to train small-scale vocabularies for specific environments for improving system location accuracy. Finally, we proposed the offline map construction method and the rapid relocation method of a robot, which enables the system to quickly restore the location tracking state based on an offline map, avoiding repeated mapping and changing the working mode of the original ORB-SLAM2 system. In addition, when the environment changes, the improved system is able to restore mapping to reconstruct the changing part of the environment. Through experiments and results analysis, we proved that the method proposed in this paper could achieve the expected goal, and the improved ORB-SLAM2 system can be better applied to mobile robots. However, the proposed methods in this paper also have limitations. Firstly, since we do not have much related research, the improved system cannot provide a dense point-cloud map. In addition, our Binary-based Vocabulary Storage method can provide a binary vocabulary for the system, but from the perspective of readability, this is not very user-friendly; thus, if researchers need to read vocabulary information directly, our method may not be applicable.

In future studies, we will focus on the rapid dense map construction of the visual SLAM and the semantic SLAM. We hope to promote the diversification of the functionality of the visual SLAM to match the complex task requirements of the intelligent mobile robot.

Author Contributions: Conceptualization, G.Y. and Z.C.; methodology, G.Y. and Z.C.; software, Z.C., Y.L.; validation, Z.C., Y.L. and Z.S.; formal analysis, Z.C.; investigation, Y.L., Z.S.; resources, G.Y.; data curation, Z.C., Y.L.; writing—original draft preparation, Z.C.; writing—review and editing, G.Y.; supervision, G.Y.

Funding: This work is supported by the National Natural Science Foundation of China under Grant Nos. 61863005 and 91746116, the Science and Technology Foundation of Guizhou Province under grant PTRC[2018]5702, [2017]5788, [2018]5781, ZDZX[2013]6020, and LH[2016]7433. Science and Technology Foundation of Guizhou Province in 2019.

Conflicts of Interest: The authors declare no conflicts of interest.

References

1. Ball, D.; Upcroft, B.; Wyeth, G.; Corke, P.; English, A.; Ross, P.; Patten, T.; Fitch, R.; Sukkarieh, S.; Bate, A. Vision-based Obstacle Detection and Navigation for an Agricultural Robot. *J. Field Robot.* **2016**, *33*, 1107–1130. [CrossRef]
2. Ran, L.; Zhang, Y.; Zhang, Q.; Yang, T. Convolutional Neural Network-Based Robot Navigation Using Uncalibrated Spherical Images. *Sensors* **2017**, *17*, 1341. [CrossRef] [PubMed]
3. Dissanayake, G.; Durrant-Whyte, H.; Bailey, T. A computationally efficient solution to the simultaneous localisation and map building (SLAM) problem. *IEEE Trans. Robert* **2013**, *17*, 229–241. [CrossRef]
4. Hess, W.; Kohler, D.; Rapp, H.; Andor, D. Real-time loop closure in 2D LIDAR SLAM. In Proceedings of the IEEE International Conference on Robotics and Automation, Stockholm, Sweden, 16–21 May 2016; pp. 1271–1278. [CrossRef]
5. Kerl, C.; Sturm, J.; Cremers, D. Dense visual SLAM for RGB-D cameras. In Proceedings of the IEEE/RSJ International Conference on Intelligent Robots and Systems, Tokyo, Japan, 3–7 November 2014; pp. 2100–2106. [CrossRef]
6. Mur-Artal, R.; Montiel, J.M.M.; Tardós, J.D. ORB-SLAM: A Versatile and Accurate Monocular SLAM System. *IEEE Trans. Robot.* **2015**, *31*, 1147–1163. [CrossRef]
7. Mur-Artal, R.; Tardos, J.D. ORB-SLAM2: An Open-Source SLAM System for Monocular, Stereo, and RGB-D Cameras. *IEEE Trans. Robot.* **2017**, *33*, 1255–1262. [CrossRef]
8. Davison, A.J.; Reid, I.D.; Molton, N.D.; Stasse, O. MonoSLAM: Real-time single camera SLAM. *IEEE Trans. Pattern Anal. Mach. Intell.* **2007**, *29*, 1052. [CrossRef] [PubMed]

9. Bailey, T.; Nieto, J.; Guivant, J.; Stevens, M.; Nebot, E. Consistency of the EKF-SLAM Algorithm. In Proceedings of the 2006 IEEE/RSJ International Conference on Intelligent Robots and Systems, Beijing, China, 9–15 October 2006; pp. 3562–3568. [CrossRef]
10. Martinezcantin, R.; Castellanos, J.A. Unscented SLAM for large-scale outdoor environments. In Proceedings of the IEEE/RSJ International Conference on Intelligent Robots and Systems, Edmonton, AB, Canada, 2–6 October 2005; pp. 3427–3432. [CrossRef]
11. Holmes, S.; Klein, G.; Murray, D.W. A Square Root Unscented Kalman Filter for visual monoSLAM. In Proceedings of the IEEE International Conference on Robotics and Automation, Pasadena, CA, USA, 19–23 May 2008; pp. 3710–3716. [CrossRef]
12. Sim, R.; Elinas, P.; Griffin, M.; Shyr, A.; Little, J.J. Design and analysis of a framework for real-time vision-based SLAM using Rao-Blackwellised particle filters. In Proceedings of the The Canadian Conference on Computer and Robot Vision, Quebec, QC, Canada, 7–9 June 2006; p. 21. [CrossRef]
13. Klein, G.; Murray, D. Parallel Tracking and Mapping for Small AR Workspaces. In Proceedings of the IEEE and ACM International Symposium on Mixed and Augmented Reality, Nara, Japan, 13–16 November 2007; pp. 1–10. [CrossRef]
14. Engel, J.; Schöps, T.; Cremers, D. LSD-SLAM: Large-Scale Direct Monocular SLAM. In *Computer Vision—ECCV 2014*; Fleet, D., Pajdla, T., Schiele, B., Tuytelaars, T., Eds.; Lecture Notes in Computer Science; Springer: Cham, Switzerland, 2014; Volume 8690, pp. 834–849.
15. Brand, C.; Schuster, M.J.; Hirschmuller, H.; Suppa, M. Stereo-vision based obstacle mapping for indoor/outdoor SLAM. In Proceedings of the IEEE/RSJ International Conference on Intelligent Robots and Systems, Chicago, IL, USA, 14–18 September 2014; pp. 1846–1853. [CrossRef]
16. Lee, Y.J.; Song, J.B. Visual SLAM in indoor environments using autonomous detection and registration of objects. In Proceedings of the IEEE International Conference on Multisensor Fusion and Integration for Intelligent Systems, Seoul, Korea, 20–22 August 2008; pp. 671–676. [CrossRef]
17. Lv, Q.; Lin, H.; Wang, G.; Wei, H.; Wang, Y. ORB-SLAM-based tracing and 3D reconstruction for robot using Kinect 2.0. In Proceedings of the Control and Decision Conference, Chongqing, China, 28–30 May 2017; pp. 3319–3324. [CrossRef]
18. Hornung, A.; Wurm, K.M.; Bennewitz, M.; Stachniss, C.; Burgard, W. OctoMap: An efficient probabilistic 3D mapping framework based on octrees. *Auton. Robot.* **2013**, *34*, 189–206. [CrossRef]
19. Huajie, W.Z.L.D. A SLAM Method Based on Inertial/Magnetic Sensors and Monocular Vision Fusion. *Robot* **2018**, 1–9. [CrossRef]
20. Caldato, B.A.C.; Achilles Filho, R.; Castanho, J.E.C. ORB-ODOM: Stereo and odometer sensor fusion for simultaneous localization and mapping. In Proceedings of the 2017 Latin American Robotics Symposium (LARS) and 2017 Brazilian Symposium on Robotics (SBR), Curitiba, Brazil, 8–11 November 2017; pp. 1–5. [CrossRef]
21. Zeng, F.; Zeng, W.; Gan, Y. ORB-SLAM2 with 6DOF Motion. In Proceedings of the IEEE International Conference on Image, Vision and Computing (ICIVC), Chongqing, China, 27–29 June 2018; pp. 556–559. [CrossRef]
22. Wang, S.; Yue, J.; Dong, Y.; Shen, R.; Zhang, X. Real-time Omnidirectional Visual SLAM with Semi-Dense Mapping. In Proceedings of the IEEE Intelligent Vehicles Symposium (IV), Changshu, China, 26–30 June 2018; pp. 695–700. [CrossRef]
23. Wu, Y.; Hu, Z. PnP Problem Revisited. *J. Math. Imaging Vis.* **2006**, *24*, 131–141. [CrossRef]
24. Kneip, L.; Scaramuzza, D.; Siegwart, R. A novel parametrization of the perspective-three-point problem for a direct computation of absolute camera position and orientation. In Proceedings of the CVPR, Providence, RI, USA, 20–25 June 2011; Volume 42, pp. 2969–2976. [CrossRef]
25. Abdel-Aziz, Y.I.; Karara, H.M.; Hauck, M. Direct Linear Transformation from Comparator Coordinates into Object Space Coordinates in Close-Range Photogrammetry. *Photogramm. Eng. Remote Sens.* **2015**, *81*, 103–107. [CrossRef]
26. Lepetit, V.; Moreno-Noguer, F.; Fua, P. EPnP: An Accurate O (n) Solution to the P n P Problem. *Int. J. Comput. Vis.* **2009**, *81*, 155–166. [CrossRef]
27. Yang, J.; Jiang, Y.G.; Hauptmann, A.G.; Ngo, C.W. Evaluating bag-of-visual-words representations in scene classification. In Proceedings of the International Workshop on Multimedia Information Retrieval, Augsburg, Bavaria, Germany, 24–29 September 2007; pp. 197–206. [CrossRef]

28. Schnabel, R.; Wahl, R.; Klein, R. Efficient RANSAC for Point-Cloud Shape Detection. *Comput. Graph. Forum* **2010**, *26*, 214–226. [CrossRef]
29. Kümmerle, R.; Grisetti, G.; Strasdat, H.; Konolige, K.; Burgard, W. G2O: A general framework for graph optimization. In Proceedings of the IEEE International Conference on Robotics and Automation, Shanghai, China, 9–13 May 2011; pp. 3607–3613. [CrossRef]
30. Galvez-Lopez, D.; Tardos, J.D. Bags of Binary Words for Fast Place Recognition in Image Sequences. *IEEE Trans. Robot.* **2012**, *28*, 1188–1197. [CrossRef]
31. Arthur, D.; Vassilvitskii, S. K-Means++: The Advantages of Careful Seeding. In Proceedings of the Eighteenth Annual ACM-SIAM Symposium on Discrete Algorithms, New Orleans, LA, USA, 7–9 January 2007; pp. 1027–1035. [CrossRef]
32. Wu, H.C.; Luk, R.W.P.; Wong, K.F.; Kwok, K.L. Interpreting TF-IDF term weights as making relevance decisions. *ACM Trans. Inf. Syst.* **2008**, *26*, 55–59. [CrossRef]
33. Yang, G.; Yang, J.; Sheng, W.; Fef, J.; Li, S. Convolutional Neural Network-Based Embarrassing Situation Detection under Camera for Social Robot in Smart Homes. *Sensors* **2018**, *18*, 1530. [CrossRef] [PubMed]
34. Sturm, J.; Engelhard, N.; Endres, F.; Burgard, W. A benchmark for the evaluation of RGB-D SLAM systems. In Proceedings of the IEEE/RSJ International Conference on Intelligent Robots and Systems, Vilamoura, Portugal, 7–12 October 2012; pp. 573–580. [CrossRef]

Article

A Precise and Robust Segmentation-Based Lidar Localization System for Automated Urban Driving

Hang Liu [1], Qin Ye [1,*], Hairui Wang [2], Liang Chen [3] and Jian Yang [3]

1 College of Surveying and Geo-Informatics, Tongji University, Shanghai 200092, China; liuhang@tongji.edu.cn (H.L.)
2 Beijing Momenta Technology Company Limited, Beijing 100190, China; hairui@momenta.ai
3 School of Computer Science and Engineering, Nanjing University of Science and Technology, Nanjing 210094, China; liangchen@njust.edu.cn (L.C.); csjyang@njust.edu.cn (J.Y.)
* Correspondence: yeqin@tongji.edu.cn; Tel.: +86-1391-786-6679

Received: 28 April 2019; Accepted: 3 June 2019; Published: 4 June 2019

Abstract: Real-time and high-precision localization information is vital for many modules of unmanned vehicles. At present, a high-cost RTK (Real Time Kinematic) and IMU (Integrated Measurement Unit) integrated navigation system is often used, but its accuracy cannot meet the requirements and even fails in many scenes. In order to reduce the costs and improve the localization accuracy and stability, we propose a precise and robust segmentation-based Lidar (Light Detection and Ranging) localization system aided with MEMS (Micro-Electro-Mechanical System) IMU and designed for high level autonomous driving. Firstly, we extracted features from the online frame using a series of proposed efficient low-level semantic segmentation-based multiple types feature extraction algorithms, including ground, road-curb, edge, and surface. Next, we matched the adjacent frames in Lidar odometry module and matched the current frame with the dynamically loaded pre-build feature point cloud map in Lidar localization module based on the extracted features to precisely estimate the 6DoF (Degree of Freedom) pose, through the proposed priori information considered category matching algorithm and multi-group-step L-M (Levenberg-Marquardt) optimization algorithm. Finally, the lidar localization results were fused with MEMS IMU data through a state-error Kalman filter to produce smoother and more accurate localization information at a high frequency of 200Hz. The proposed localization system can achieve 3~5 cm in position and 0.05~0.1° in orientation RMS (Root Mean Square) accuracy and outperform previous state-of-the-art systems. The robustness and adaptability have been verified with localization testing data more than 1000 Km in various challenging scenes, including congested urban roads, narrow tunnels, textureless highways, and rain-like harsh weather.

Keywords: Lidar localization system; unmanned vehicle; segmentation-based feature extraction; category matching; multi-group-step L-M optimization; map management

1. Introduction

Localization is one of the most basic and core technologies of unmanned vehicles. Precise and real-time localization service is needed in many modules, including behavior decision, motion planning and feedback control. At present, a high-cost GNSS (Global Navigation Satellite System) and IMU (Integrated Measurement Unit) integrated navigation system [1] is mostly used in unmanned vehicles. These two complementary systems solve the shortcomings of the low frequency of GNSS and the integration drift of IMU [2–4]. The accuracy of single-point positioning technology is low, about 5~10 m, which cannot meet the needs of unmanned vehicles. The real-time carrier-phase based differential GNSS technology can eliminate satellite orbit and block errors, tropospheric and ionospheric delays to achieve centimeter-level localization accuracy, i.e. RTK (Real Time Kinematic) technology [5,6].

However, the differential signal cannot be received everywhere and the accuracy decreases with the distance from the base station. The GNSS position information is prone to jump because of signal occlusion in urban canyons and the multipath effect in large area of flat and smooth ground or water scenes [7]. In addition, the high-cost GNSS receiver and IMU modules, coupled with the paid real-time differential services, make the cost of this localization solution very high. Considering the above factors, it is necessary to design a more accurate, stable, and economical localization system before the large-scale landing of automatic driving technology.

Lidar (Light Detection and Ranging) can obtain the 3D point cloud of the scene by the multiple rotating laser beams. We can obtain accurate 6DoF (Degree of Freedom) pose in the global coordinate system by matching the online frame with the priori map, it can be used as a better input than RTK in INS (Integrated Navigation System), which can work stably in almost all scenes [8–10]. Hence, how to achieve more accurate and efficient matching is particularly important in a Lidar-based localization system.

Here, we propose a precise and robust segmentation-based Lidar localization system with MEMS (Micro-Electro-Mechanical System) IMU aided. The main contributions of this paper are summarized as follows:

1. A novel efficient low-level semantic segmentation-based feature extraction algorithm is designed to extract multiple types of stable features from the online frames, including ground, road-curb, edge, and surface. They ensure accurate pose estimation for frame-frame and frame-map matching.
2. A priori information considered category matching method and a multi-group-step L-M [11] (Levenberg Marquardt) optimization algorithm are proposed, which can avoid most of the mismatching to improve the accuracy, and increase the efficiency by reducing the Jacobian matrix dimension.
3. An efficient priori map management is presented, the map is stored in tiles with overlap and the local map surrounding the vehicle is dynamically loaded to save the computation resource.
4. A complex and complete vehicle localization system has been accomplished by integrating all modules reasonably, which can provide high-accuracy and real-time localization service for high-level unmanned vehicles in various challenging scenes and harsh weather.

The remainder of this paper is organized as follows: Section 2 briefly reviews the popular SLAM (Simultaneous Localization and Mapping) system and the Lidar localization technologies fusion with other sensors. Section 3 presents the framework of the proposed segmentation-based Lidar localization system and describes the algorithms used in each module. Section 4 demonstrates the performance qualitatively and quantitatively in various challenging scenes. Finally, we conclude our works and discuss the future research directions in Section 5.

2. Related Works

The problem of Lidar localization has been a popular research topic in recent years, which has evolved from SLAM technology [12]. The task of SLAM is to estimate the pose and simultaneously build a point cloud map with vision-based [13–16] or Lidar-based [17–20] methods.

Raul et al. present ORB-SLAM2 [21,22], a complete feature-based visual SLAM system for monocular, stereo, and RGB-D (Color-Depth) cameras, including map reuse, DBoW2-based [23] loop closing, and relocalization capabilities. Gao et al. [24] presents an extension of Direct Spares Odometry [25] to monocular visual SLAM system with loop closure detection and pose-graph optimization. Although vision-based methods have many advantages in loop-closure detection, their sensitivity to illumination and viewpoint change may make such capabilities unreliable if used as the sole navigation sensor.

Lidar can directly obtain fine detailed 3D information of a wide range of scenes even at night. A low-drift and real-time lidar odometry and mapping (LOAM) method is proposed in [17,26], they divide the complex problem into two algorithms, one algorithm performs odometry at a high-frequency but at low fidelity to estimate velocity, and another runs at a lower order of magnitude frequency for fine matching and registration of the point cloud. LOAM's resulting accuracy is best achieved by a Lidar-only estimation method on the KITTI [27] odometry benchmark site.

On the basis of LOAM, LeGO-LOAM [28] is a lightweight and ground-optimized LOAM for pose estimation of UGVs (Unmanned Ground Vehicle) in complex environments with variable terrain, it is lightweight, as real-time pose estimation and mapping can be achieved on an embedded system. Point cloud segmentation is performed to discard points that may represent unreliable features after ground separation and integrates the ability of loop closures based on ICP (Iterative Closest Point) [29] to correct motion estimation drift. IMLS-SLAM [30] proposes a scan-to-model matching with implicit moving least squares surface representation method, which can achieve a global drift of only 0.69%.

Although there are many excellent back-end optimization and loop-closure detection algorithms, the pose estimation drift will accumulate continuously in long-term large-scale scene with the SLAM scheme and the resulting error will be too large to meet the needs of unmanned vehicles. Therefore, it is necessary to establish a high-precision map in advance and realize accurate localization through matching online frames with the priori map [31–35].

Ryan W [36] exploits the structure in the environment to perform scan match with Gaussian mixture maps, which are a collection of Gaussian mixture over the z-height distribution. They achieve real-time performance by developing a novel branch-and-bound, a multiresolution approach that makes use of rasterized lookup tables of these Gaussian mixtures. Levinson J [37] yields substantial improvements in vehicle localization based on a previous work [31], including higher precision, the ability to learn and improve maps over time, and increased robustness to environment changed and dynamic obstacles. They model the environment as a spatial grid where every cell is represented as its own gaussian distribution over remittance values and then the Bayesian inference is able to preferentially weight parts of the map most likely to be stationary and of consistent angular reflectivity, thereby reducing uncertainty and catastrophic errors.

In the Lidar-only localization system, the output frequency of localization information is too low to meet the needs of high-speed unmanned vehicles, and the trajectory is not smooth enough. Hence, some studies focus on the fusion of more information from multiple type sensors to obtain more reliable localization results [38–40]. An integrated GNSS, INS, and Lidar-SLAM positioning method for highly accurate forest stem mapping is proposed in [39], the heading angles and velocities extracted from GNSS and INS are used to improve the positioning accuracy of the SLAM solution. In [41], they develop a localization system that adaptively uses information from complementary sensors such as RTK, Lidar, and IMU to achieve high localization accuracy and resilience in challenging scenes, Lidar intensity and altitude cues are used instead of 3D geometry to improve the accuracy and robustness, and an error-state Kalman filter is applied to fuse the localization measurements from different sources with novel uncertainty estimation, it can achieve relatively high accuracy at 5~10 cm RMS (Root Mean Square). This method relies on the intensity information of point cloud, but it is not easy to calibrate the Lidar intensity, and each Lidar has its own differences. In addition, its localization accuracy still needs to be improved.

3. Methodology

3.1. System Overview

The architecture overview and the complete step-by-step algorithm flow of the proposed Lidar localization system are shown in Figure 1 and Algorithm 1. The input of the system includes the raw online cloud, raw MEMS IMU data as well as the prior map dataset, and output real-time accurate 6 DoF pose information.

The system can be divided into five main modules:

(1) Feature Extraction: we organized the raw online point cloud and eliminate the dynamic objects within the road, and then extracted ground, road-curb, surface, and edge features through a series of

proposed efficient low-level semantic segmentation algorithms. The following cloud matching process is based on the extracted features.

(2) Lidar Odometry: frame-frame matching was performed to obtain the ego-motion between two adjacent frames. The result was used as the initial value of frame-map matching in Lidar localization module.

(3) Local Map Load: the local prior feature map is loaded dynamically based on the current vehicle localization information.

(4) Lidar Localization: the current frame was matched with the loaded local map to obtain an accurate global pose and pushback the result into the filter.

(5) Error-state Kalman Filter: we established an error-state Kalman filter that uses Lidar localization result and IMU raw data as measurement input and output high-frequency precise 6 DoF pose. After the filter was initialized, the filter result was used as the initial value of frame-map matching instead of Lidar odometry. The Lidar odometry node lost its meaning and was shutdown to save computing resources.

Furthermore, in order to balance the computing resources and ensure the localization accuracy, each node ran at different frequencies (Figure 1) to form the complete segmentation-based Lidar localization system. The detailed algorithm principle of each modules will be introduced in the following sections.

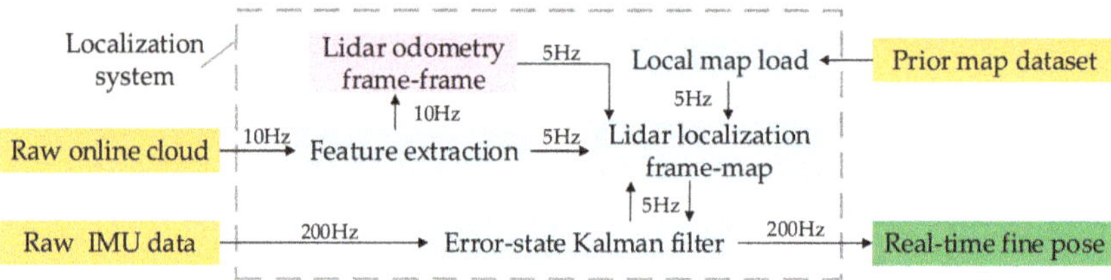

Figure 1. Overview of the proposed Lidar localization system architecture. IMU: Integrated Navigation Unit; Lidar: Light Detection and Ranging

Algorithm 1. Lidar Localization System

Input: prior map M, online point cloud C at 10 Hz, IMU data I at 200 Hz.
Output: fine pose T^a.

establish the error-state Kalman filter F, and pushback the IMU data I_n into F;
for frame C_i **do**
delete the dynamic object C_i^d within the road;
based on low-level semantic segmentation, extract the feature point cloud road C_i^r, road_curb C_i^c, surface C_i^s, and edge C_i^e from the remaining point cloud in turn;
if F initialized **then**
calculate the transform ΔT between the C_{i-1} and C_i from IMU odometry queue;
else
get ΔT by category matching C_{i-1} with C_i;
end if
calculate the rough pose $T_i^r = T_{i-1}^a * \Delta T$;
dynamically update and load the local map M_i;
category match C_i with M_i, and use T_i^r as the initial value to get the fine pose T_i^a by multi-group-step L-M optimization;
pushback T_i^a into F;
end for
calculate the final fine pose T_n^a using F, and output at 200 Hz;
return T_n^a;

3.2. Lidar Feature Map Management

The high precision Lidar localization system depends on a priori map. Therefore, it was important to design an effective method to store and load the map of large-scale scenes. Here, our priori map separately stored the extracted multiple types of feature point clouds and the specific feature extraction methods are demonstrated in 3.3. In order to manage a large-scale point cloud map, we extended the two dimensions into three-dimensional tiles to store the map, each tile was identified by (i, j, k), which indicate the index in x, y, and z directions. In addition, the relationship between the point cloud file path and the tile index was also stored for conveniently finding and loading the specified tiles.

In the localization system, the currently visible area was calculated according to the vehicle position information as well as considering the view angle and measurement range of Lidar. We then loaded the tiles in the area to form the current local map. When loading map data, the K-d tree of the point cloud in each tile was also established to prepare for finding the nearest neighbors of the point cloud matching process in Lidar localization module. To avoid wasting computing resources of re-loading the point cloud and reestablishing the K-d tree, we retained the loaded tiles, only loaded the newly added, and removed the unused when the vehicle position was updated.

The map was stored by tangent tiles in [41], which causes the K-d tree at the tile edge unable to reflect the real nearest neighbor relationship and produces an incorrect result when searching for the nearest points. Here, we proposed an overlapped tile to store the map. Figure 2 is the 2D representation of the relation between tangent tiles (blue) and overlapped tile (green). On the basis of keeping the center of the tile unchanged, the overlapped tile expanded leaf size so that the adjacent tiles are overlapped. Therefore, the K-d tree at the edge (red) of the original tile was ensured to be correct. The overlap does not need to be too large, thus wasting storage space and computing resources, as long as the correctness at the edge (red) can be ensured. Here, we set the overlap as 6 m and the effectiveness had been proved by experience. It should be noted that there was no relationship between overlap and leaf size, and we did not need to adjust overlap when the leaf size was changed in using.

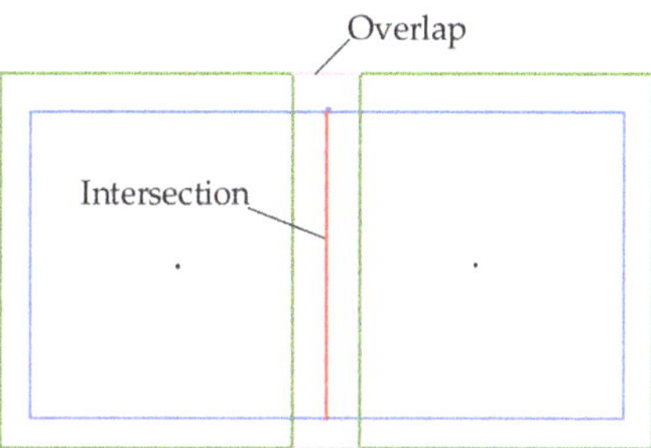

Figure 2. The 2D representation of the relation between tangent tiles and overlapped tiles.

Figure 3 shows the scope of the test area in this study, which had a size of about 3 × 4 km; (b) is the tiled point cloud map of ground feature and the tiles are given random color. The leaf size of each tile was 50 m and with 6 m overlap between two adjacent tiles. The map accuracy was the premise of the localization system accuracy. We used a total station to measure some obvious feature points on the spot and found the same points in the generated point cloud map to evaluate the map accuracy, which can achieve centimeter level and meet the requirement of Lidar-based localization system.

Figure 3. The scope of the test area. (**a**) The test roads in satellite image. (**b**) Tiled point cloud map.

3.3. Segmentation-Based Feature Extraction

As each frame data contained millions of points, the computation was too large to satisfy real-time processing if all points were used for pose estimation [42–44]. In addition, there were many dynamic objects in the road scene, which were noise in the registration process, and reduced the registration accuracy or even lead to failure [45]. In this study, we first eliminated the dynamic objects within the road by the pre-built RoadROI (Road Region of Interest), which was generated through ground segmentation and region growing. In the following, we separately introduced a series of proposed low-level semantic segmentation-based feature extraction algorithms, including ground, road-curb, surface, and edge.

3.3.1. Ground

Lidar acquired data by continuously rotating laser scans and publishes them as small packets. These packets were collected for the time of the sensor revolution in what we called a frame; the time was 100 ms for Velodyne VLP-32C. Before running our algorithms, we first inertially corrected the frame distortion caused by the vehicle's motion through IMU or odometry information [46].

Then, the point cloud was projected onto a cylinder and expanded into a two-dimensional range image, called an organized point cloud. Each row of the range image represented a laser line, which had the same vertical angle θ_v; each column represented the points of different scan lines at the same azimuth angle θ_a. Therefore, we exploited the clearly defined neighborhood relations directly in the 2D range image without additionally computing resource to build K-d tree, and the 2D range image had more flexibility in use. For example, we easily obtained a point in specified laser line or column and it offered many conveniences for point cloud segmentation. The subsequent segmentation-based feature extraction algorithms were all implemented on the basis of the organized point cloud.

The ground extraction was the primary task of rule-based point cloud segmentation [47], calculating complex normal difference and gradient, then performing region growing to extract ground from a Lidar frame. This occupied too much computing resources, which made it unsuited for such a complex and real-time system. An image of when the laser scanning was on a flat ground surface can be seen in Figure 4. The differences in the z direction was far less than the x, y directions between two adjacent points in the same column (the two points in the purple box of Figure 4) and the former one was close to 0. Considering such a geometric feature, we defined a term α_i to represent the vertical angle between two adjacent points in the same column:

$$\alpha_i = tan^{-1}\left(\frac{\delta^c_{z,i}}{\sqrt{\delta^c_{x,i} * \delta^c_{x,i} + \delta^c_{y,i} * \delta^c_{y,i}}}\right) \tag{1}$$

where $\delta^c_{x,i}$, $\delta^c_{y,i}$ and $\delta^c_{z,i}$ represents the difference in x, y, z direction between two adjacent points in the cth column. We traversed all points in m rows below the range image to calculate α_i, because only these laser lines can scan to the ground and $m = 17$ for Velodyne VLP-32C. When $\alpha_i < 2.5°$ and $\delta^c_{z,i} < 0.05$, they are the ground points. Furthermore, the vertical angle plays a more important role than the elevation difference in ground segmentation. The threshold was relatively loose, so the algorithm can still perform well at the transitions between flat and sloped terrain. There was no difference for the algorithm when the vehicle was on a flat surface or slope, because the point cloud is in the Lidar coordinate system, which was relatively parallel to the ground. The segmentation results contain not only road surface but also flat ground, where the flat ground points played the same role as road surface in optimizing pose, so they were also needed in our localization system.

3.3.2. Road-Curb

We solved pose using the constraints provided by feature points, while the stable edge and surface features were fewer in textureless scenes, such as a highway where the constraints were fewer, which increased the possibility of matching failure in such scenes. Road-curb was also a kind of structurally stable feature, which provided the constraints of multiple DoF. As shown in Figure 4 [48], road-curb has many spatial features different from the ground surface. For example, the evaluation of road-curb changes sharply while the ground surface points are smooth and continuous. Based on these features, the road-curb detection method is proposed.

We first calculated the theoretical threshold based on the spatial features to separate road-curb from ground surface. $\delta^l_x, \delta^l_y, \delta^l_z$ represents the difference in x, y, z direction between two adjacent points of ground surface in the lth laser line (the two points in the red box of Figure 4), and it is defined by

$$\delta^l_x = h_s \cdot cot\theta^l_v \cdot (1 - cos\theta_h) \quad (2)$$

$$\delta^l_y = h_s \cdot cot\theta^l_v \cdot sin\theta_h \quad (3)$$

where h_s is the sensor height, θ^l_v is the vertical angle of the lth laser line, and θ_h is the angle resolution of the sensor, which is set to 0.2° for Velodyne VLP-32C. If the ground surface satisfies the assumption that it is completely smooth, then $\delta^l_z = 0$. While considering the measurement error and the slight roughness of the ground surface, here we set $\delta^l_z = 0.003m$. All points of the ground surface were then traversed to calculate the difference $\delta^l_{x,i}, \delta^l_{y,i}, \delta^l_{z,i}$ in x, y, z direction between two adjacent points in the lth laser line. If $\delta^l_{x,i} > \delta^l_x, \delta^l_{y,i} > \delta^l_y, \delta^l_{z,i} > \delta^l_z$, the points were road-curb candidate points. There was some noise in the candidate points, considering that the road-curb was an approximately straight line, and we iteratively extracted all road-curb lines from the candidate points.

The above algorithm has a hypothesis that the road-curb is approximately parallel to the x- or y- axis of the Lidar coordinate system. The hypothesis was destroyed when the vehicle was turning. Here, we determined the vehicle turning state based on θ_{yaw} using localization information and then used the accumulated turning angle to invert the point cloud to satisfy the hypothesis. Finally, the road-curb was detected with high precision and recalled in various scenes such as going straight and turning.

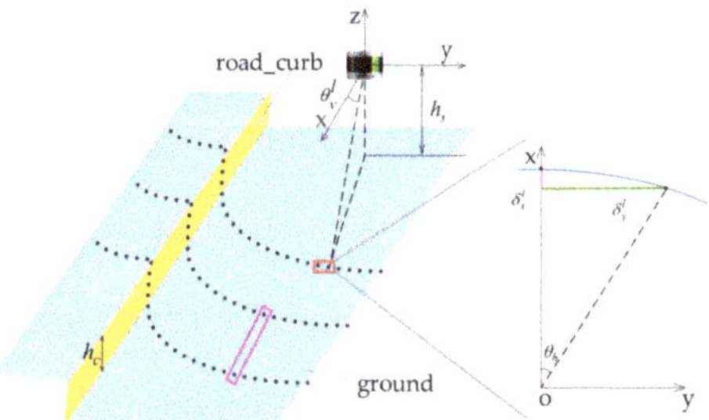

Figure 4. Ground and road-curb scene description.

3.3.3. Surface

There were also other surface features expect the extracted ground plane, including architecture walls, traffic signs, billboards, etc. According to the scanning characteristics of a horizontally installed Lidar, when the laser line was scanned in a plane, its three-dimensional coordinate values were smooth, continuous, and symmetrical. For a sparse Lidar point cloud of a single frame, when the laser line scanned the distant wall surface, although it was only a single line, but it should be defined as a surface feature instead of edge. In [26], the smoothness of a point is defined, which is used to distinguish surface and edge, but it cannot perform well in a complex urban road environment, and the extraction results are confusing. Here, we define a more concise term $\boldsymbol{P}$ that can achieve better results to describe the planeness of the local surface:

$$\boldsymbol{P} = \left(\boldsymbol{I}^l_{c-1} + \boldsymbol{I}^l_c + \boldsymbol{I}^l_{c+1}\right)/3 - \boldsymbol{I}^l_c \tag{4}$$

where $\boldsymbol{I}^l_{c-1}, \boldsymbol{I}^l_c, \boldsymbol{I}^l_{c+1}$ represent three adjacent points in the lth rows of the range image. When $\boldsymbol{P}_x, \boldsymbol{P}_y, \boldsymbol{P}_z$ are all less than the threshold, the point is the candidate surface point. The threshold is set as 0.03m here and it is proportional to the horizontal resolution of LiDAR, which is 0.02° for Velodyne VLP-32C. All the points of expected ground feature points are then traversed to obtain the candidate surface points set, and to eliminate the discontinuous candidate points to get the final surface feature point cloud with high quality.

3.3.4. Edge

In the road scene, the stable edge feature included light poles, tree trunks, architecture ridges, etc., all of which are vertical. The consistency of the organized point cloud in the column direction was worse than the row direction. Using an extraction method similar to the surface, which assumes the edge feature points are in the same column of the range image, we could not precisely extract edge features. A more effective approach is presented here.

The edge feature extraction process is shown in Figure 5. First, the remaining points after removing the ground and surface points were clustered, since the traditional Euclidean distance clustering method required a lot of computation resources. Here, we adopted a grid-based method to project point cloud into XOY plane, which is composed by x-axis and y-axis of the Lidar coordinate system. Then, we performed region growing according to the plane distance, which clustered the point cloud quickly and effectively [49]. Each cluster was then fitted with a line using the RANSAC (Random Sample Consensus) method to obtain the points set and normalized line parameters. When the line was approximately parallel to the Z axis, or, in other words, the z in line parameters were close to 1, and x, y were close to 0, the points set was an edge feature point cloud.

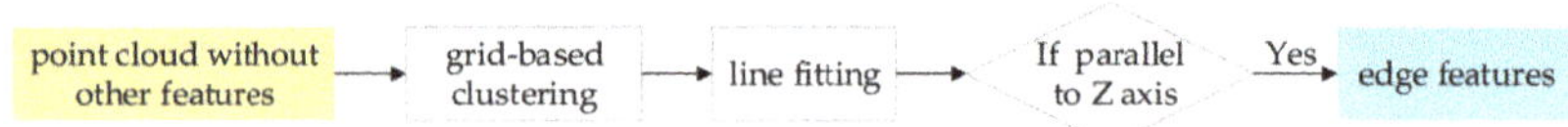

Figure 5. Edge feature extraction process.

3.4. *Priori Information Considered Category Matching*

Based on the extracted feature points, frame-frame matching was performed to estimate the sensor ego-motion between two consecutive frames in Lidar odometry module and frame-map matching was performed to obtain the accurate global localization information in the Lidar localization module. The processing of the two modules was almost the same in Sections 3.4 and 3.5, except for a few differences in the data association part. The details of the processing will be introduced in the following paragraphs.

We extended the matching strategy of point-to-line and point-to-plane in [17], which was more reasonable and robust than the traditional point-to-point for the sparse Lidar data. For the sake of brevity, the detailed procedures of point-to-line and point-to-plane matching can be found in [17]. Here, we propose a priori information considered category matching method to improve the accuracy and efficiency.

(1) Category Matching: In Lidar odometry and Lidar localization modules, we needed to find the corresponding features for points in the current frame C_i from the last frame C_{i-1} and the current dynamically loaded local map M_i respectively. In frame-map matching, the tile index in the map of each point in C_i was calculated and then the k-nearest neighbor points np^k in the correspondence tile are found. There were four categories of feature point clouds extracted by the same method in C_i, C_{i-1}, and M_i, and they are maintained separately. Thus, we only found correspondences in the same category, because the structure of the feature points was stable. Therefore, it could be extracted to the same category feature repeatedly in different frames. For example, for the edge features in C_i^e, we found its corresponding points in C_{i-1}^e and M_i^e points set. This method can not only improve the efficiency by reducing the potential candidate points, but also improve the accuracy of data association.

(2) Consider priori information: Before calculating the distance of point-to-line and point-to-plane, the equations of the line and plane need to be calculated. The line and plane equations can be expressed with a direction vector $d\left(d_x, d_y, d_z\right)$ or normal vector $n\left(n_x, n_y, n_z\right)$ and a point in it, where the point can be obtained by nearest neighbor search, and the key and difficulty is the solution of the vector. In the frame-frame matching of [17] and [28], they first find the two or three nearest points in different laser lines, then use Formula (5) and (6) to calculate the direction and normal vector, respectively.

$$d = np^0 - np^1 \tag{5}$$

$$n = \left(np^0 - np^1\right) \times \left(np^2 - np^1\right) \tag{6}$$

In the frame-map matching, they obtain the vector by performing the PCA (Principal Component Analysis) with the five nearest points and the computation was relatively high. Furthermore, the vector could not precisely express the local spatial geometry because of the sparseness of Lidar data. Here, we considered the priori information of the extracted features. For example, the direction of edge features was vertical and the plane of the ground features parallel to the Lidar coordinate system's XOY plane. Therefore, when using the point-to-line matching method for edge features, we directly assigned the direction vector to $d(0,0,1)$ to form a line equation after finding the nearest point. For the simplicity of expression, Table 1 shows the match method, the needed nearest points numbers, the calculation method of direction or normal vector for different feature types in frame-frame, and frame-map matching. The matching methods are the same for frame-frame and frame-map, but the nearest points number and the calculation method of direction or normal vector are different. The computation complexity of the three methods for vector calculation, PCA, direct assignment and formula (5)(6) decrease in turn, but the accuracy is increase. By fully considering the priori information to simplify the solution of direction or normal vector and improving the accuracy of data association, the accuracy and efficiency of pose estimation can be improved.

Table 1. The data association method of different types of feature points. PCA: Principal Component Analysis

Feature Type	Match Method	Nearest Points Num		Direction/Normal Vector	
		Frame-Frame	Frame-Map	Frame-Frame	Frame-Map
edge	point-to-line	1	1	(0,0,1)	(0,0,1)
road-curb	point-to-line	3	5	Formula (5)	PCA
ground	point-to-plane	1	1	(0,0,1)	(0,0,1)
surface	point-to-plane	2	5	Formula (6)	PCA

3.5. Multi-Group-Step L-M Optimization

The distance of all correspondences of the feature point cloud was calculated by point-to-line and point-to-plane distance formulas. Therefore, the Lidar motion was recovered by minimizing the overall distances. The formula for optimizing the pose using the gradient-descent-based L-M method is deduced in detail in [26], where optimize 6DoF $\{t_x, t_y, t_z, \theta_{roll}, \theta_{pitch}, \theta_{yaw}\}$ with all the points together. This results in a higher dimension Jacobian matrix j and an increase in the computational resource consumption. In addition, the ground points have no benefit for optimization $\{t_x, t_y, \theta_{yaw}\}$, or even reduce accuracy.

Due to the different spatial characteristics of each type of feature points, there was a significant difference in their ability to constrain the variables in the pose. For example, the ground points have a strong constrain on $\{t_z, \theta_{roll}, \theta_{pitch}\}$, but were invalid for other variables. Furthermore, the edge points were good at constraining $\{t_x, t_y\}$, but they do not make sense for $\{t_z\}$. Here, we propose a multi-group-step optimization method; Figure 6 shows the pipeline. It was successively optimized by four steps of edge, road-curb, ground, and surface, wherein each step optimizes different variables (shown in the box bottom) that depended on the feature type, and the optimized result of each step was used as the initial value of the next step, better initial values reduced the likelihood of falling into a local optimum. After a four-step optimization, we can obtain a more accurate pose from the initial coarse pose. In addition, since the Jacobian matrix dimension in each step was relatively lower, the computation time was reduced. It should be noted that the optimization was orderly, the edge feature with the least constraints was used for optimization first, and the surface feature with the most constraints was used for optimization last; experiments verify that the best results can be obtained in this order. The Multi-group-step L-M optimization method can not only reduce the computing resource but also improve the accuracy of pose estimation.

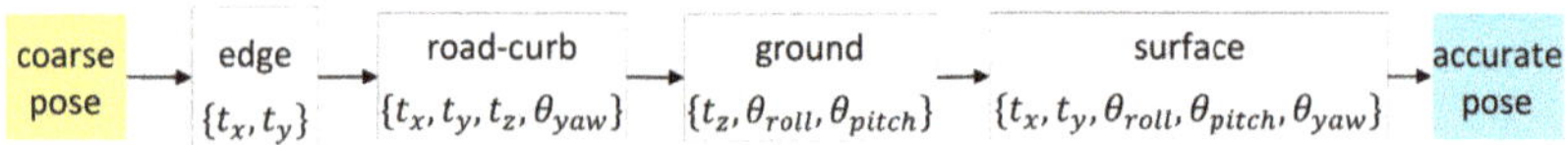

Figure 6. The pipeline of Multi-group-step L-M optimization.

3.6. Lidar and IMU Fusion

The output frequency of the pure Lidar localization system only reached 10 Hz at most, which could not meet the needs of unmanned vehicles running at high speed. IMU can produce a high frequency triaxial acceleration and rotation rate and it is sufficiently accurate to provide robust state estimations between Lidar measurements. IMU and Lidar are two complementary sensors and we can obtain more accurate and smooth real-time localization results after fusing their information.

Here, we used an error-state Kalman filter fusion framework to fuse the Lidar and MEMS IMU measurements. The error-state was the difference between the estimated state and truth state, including position, orientation, velocity, gyroscopes biases, and accelerometers biases. With the MEMS IMU data, the SINS (strap-down inertial navigation system) was used to predict the position, orientation, and velocity as a prediction model in the Kalman filter propagation phase by integrating the specific force measured by the accelerometer and the rotation rate measured by the gyroscope. With Lidar pose measurements that estimated by Lidar frame-map matching, we update the error-state Kalman Filter's state and then used the error-state to correct the SINS state as an update model in the Kalman filter. The fusion framework is similar to [41] and the input and output difference are shown in Table 2. The proposed system's input is Lidar localization results and MEMS IMU data, dose not rely on RTK. The output includes 6DoF pose $\{t_x, t_y, t_z, \theta_{roll}, \theta_{pitch}, \theta_{yaw}\}$, velocity, accelerometers, and gyroscopes bias. After the fusion of Lidar localization information and MEMS IMU data, the system produced a smoother, high-frequency and high-accuracy 6DoF pose, providing localization service for other modules of unmanned vehicles.

Table 2. The filter input and output difference between [41] and ours. RTK: Real Time Kinematic; DoF: Degree of Freedom.

Method	Input	Output
[41]	RTK, Lidar Localization, IMU	2D position, altitude, velocity, accelerometers and gyroscopes bias
Ours	Lidar Localization, IMU	6DoF pose, velocity, accelerometers and gyroscopes bias

4. Experiment Results and Discussions

4.1. Hardware System and Evaluation Method

The proposed system was validated on a computer with a 3.6GHz i7-8700 CPU processor, 16GB memory, and GPU was not used in the experiment. A Velodyne VLP-32C was horizontally installed on the car roof to collect LiDAR data. The VLP-32C measurement range was up to 180 m with an accuracy of ±3 cm, it had 32 laser channels with a 30° vertical FOV (Field of View) and 360° horizontal FOV, and it could obtain data up to 10 Hz. The MEMS IMU(ADIS16465) was installed under LiDAR and it included a triaxial digital gyroscope and triaxial digital accelerometer. In addition, the unmanned vehicle was equipped with a high-cost and high-precision RTK/IMU INS(XW-GI7660) for evaluation. The INS achieved 0.02 m in position and 0.01° in orientation RMS accuracy.

In order to evaluate the localization accuracy quantificationally, the ground truth of the vehicle motion trajectories should be established. We have designed two methods for different scenes, which with good and bad GNSS signals respectively. **Online evaluation mode:** The ground truth was generated through the INS, but it could only be used in an open scene with a good GNSS signal and the accuracy declined rapidly when entering the areas with poor GNSS signal such as urban canyons. **Offline evaluation mode:** The base station was set up at the control point to post-process the original RTK/IMU integrated navigation data and some Lidar SLAM technologies were added, including scan-match, loop closure, and global pose-graph optimization, we finally got more accurate trajectories as the ground truth. Actually, that was also the pipeline of Lidar mapping.

We calculated the absolute 6 DoF errors, where the units were expressed in meters for translation and in degrees for rotation. Then we statistically calculated the 1σ, 2σ, and 3σ of a data section, which indicated 68.26%, 95.4%, and 99.73% of the errors was less than the value respectively, and the 1σ was the RMS, and each data section was about 5 km.

The errors have been transformed from the world coordinate system to car coordinate system, in which x and y denote the vehicle forward and transverse direction respectively. The localization system performance can be expressed more clearly.

4.2. Segmentation-Based Feature Extraction Result Analysis

The feature extraction module plays a vital role in the proposed localization system. Hence, we first demonstrated the proposed multiple segmentation-based feature extraction algorithms in various scenes before analyzing the localization performance. If we quantificationally evaluated the feature extraction algorithms' performance by calculating the detection rate of each type of feature, a dataset with detailed manual labeling was needed, which required substantial of labor and time. At present, our data were not labeled and there was no such public data set available for use. Therefore, we qualitatively evaluated the segmentation algorithms using some representative scenes. In Figures 7–11, the gray points are the original point cloud of a frame and the blue, red, green, and purple points were the extracted ground, road-curb, surface, and edge features respectively.

As shown in the figures, most of the features can be detected, and the detail is explained in the following paragraphs.

Figure 7 shows the ground segmentation results in different scenes. Figure 7a is a scene where vehicles pass through the crossroads, and the ground in all directions can be completely detected. For some challenging cases, for example, Figure 7b representing a congested urban road with severe occlusion, Figure 7c representing a highway ramp with slope, the algorithm is robust, and can still achieve the same performance as the normal scene. The detection results include not only the road surface but also the flat ground.

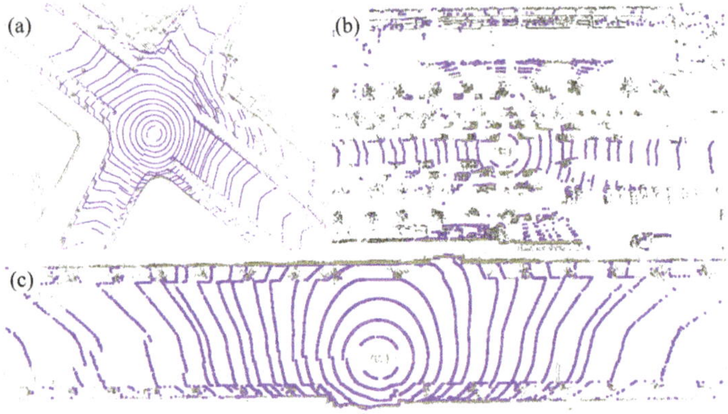

Figure 7. The ground segmentation results in various scenes. (**a**) Crossroads scene; (**b**) congested urban road scene; (**c**) highway ramp with slope scene.

Figure 8 shows the road-curb segmentation result in some representative scenes with different difficulty levels to demonstrate the wide adaptability of the proposed algorithm. Figure 8a is the simple straight road scene, where the multiple road-curb line on both sides of the road can be detected steadily. Figure 8b shows the vehicle going straight through the intersection, the road-curb of the two roads parallel and perpendicular to the vehicle travel direction can all be detected. For the most challenging turn scenes, the algorithm can still work by adding the reverse rotation point cloud processing, Figure 8c,d are left-turn and right-turn cases respectively.

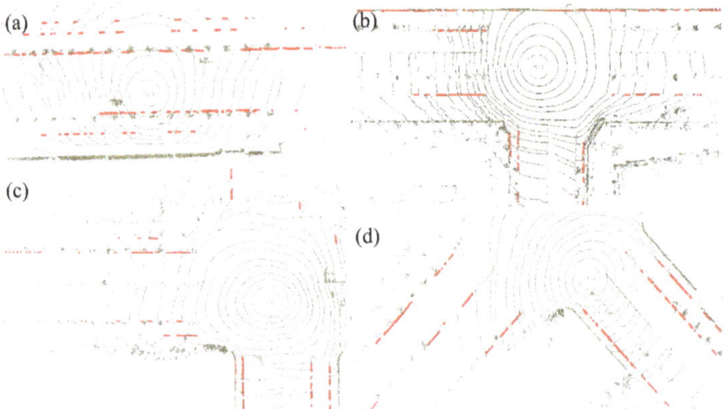

Figure 8. The road-curb segmentation results in various scenes. (**a**) Straight scene; (**b**) straight through the intersection scene; (**c**) left-turn scene; (**d**) right-turn scene.

Figure 9 shows the surface segmentation results, including architecture walls, traffic signs, billboards, etc. For sparse multi-channel Lidar data, these high and distant horizontal lines should still be defined as surface features. In addition, it can eliminate most of the scattered leaves and other unreliable points, which should be considered as noisy, as they are unlikely to be seen in two consecutive frames.

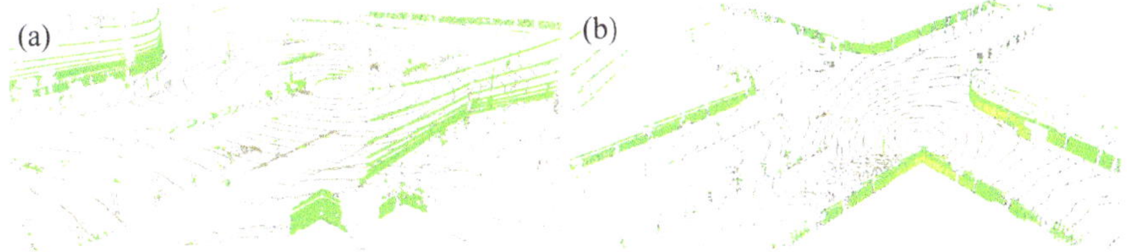

Figure 9. The surface segmentation results in various scenes; (**a**) and (**b**) are two different scenes.

The edge segmentation results are shown in Figure 10. It can extract almost all the rod-like objects at a very long range, such as lamp poles, traffic poles, tree trunks, and the vertical lines of architecture. There may be a few false negative points, for example, multiple columns laser points scan on a lamp pole that is very close to the vehicle, the lamp pole will be detected as surface instead of edge feature. Such a case is rare, and compared with [26] and [28], our precision and recall are far higher, and their extracted edge and surface features are confusing.

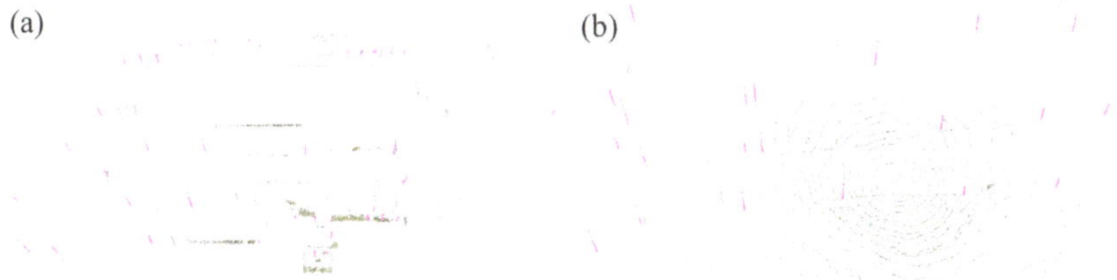

Figure 10. The edge segmentation results in various scenes; (**a**) and (**b**) are two different scenes.

We propose a series of efficient low-level semantic segmentation algorithms for sparse Lidar point cloud by making full use of the spatial geometric features. They are not only robust for all kinds of challenging scenes but also very lightweight with low time and space complexity, the time consumption of each frame is only about 35ms. The algorithms can work stably on online frames. Figure 11 shows the performance in three scenes, it can be seen that the algorithms can eliminate dynamic objects, remove the cluttered tree leaves points from the raw point cloud, and segment them into four structured semantic point clouds, including ground, road-curb, edge, and surface. The segmentation results are downsampled by the voxel grid method, and then fewer feature points are used for matching.

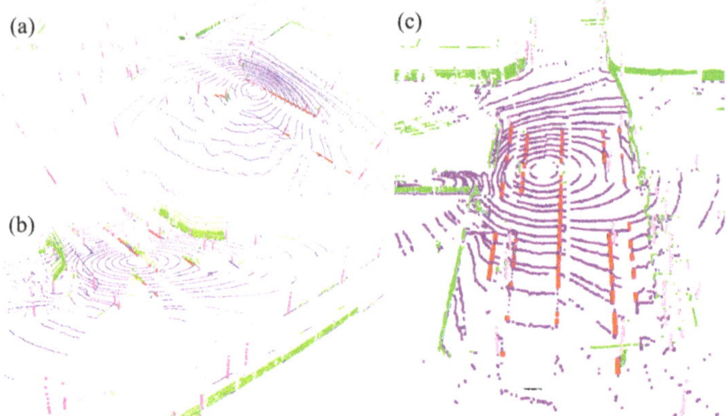

Figure 11. The efficient low-level semantic segmentation-based feature extraction results; (**a**), (**b**), and (**c**) are three different scenes.

4.3. Qualitative and Quantitative Analysis of Localization Accuracy

Our Lidar-based localization system was deployed in several unmanned vehicles and conducted daily localization tests to verify the stability of the system. The main test area is shown in Figure 3 and the total test mileage is more than 1000 Km. Here, we selected three typical sections of data of scenes with varying difficulty to evaluate the performance qualitatively and quantitatively. S1 is a typical urban road scene, S2 is a crowed urban road scene in rainy day, and S3 is a textless highway scene.

The difficulty of S2 was that a large number of dynamic vehicles point cloud in crowded urban roads were noise for the localization and block the view of the Lidar. In addition, harsh weather, such as rain and fog, had similar effects on Lidar data, which reduced the Lidar measurement range and created some noisy points; this phenomenon was more serious in rainy days. Therefore, we selected a section data of rainy day with worse quality for comparison. While the system first removed the most dynamic objects and the segmentation-based feature extraction algorithms had a strong suppression effect on noise, the limited point cloud information was fully used to determine a high-precision pose.

In S3, there were almost no stable surface features in the highway scene, such as Figure 10b, the constraints were too few, causing potential fail in pose estimation. The edge and road-curb could be extracted to ensure the number of constrains and the system work stably with the proposed matching and optimization algorithms.

The segmentation-based feature point extraction method (Section 3.3) extracted the most structurally stable points in the scene and provided enough basis for point cloud matching. The proposed priori information considered category matching method (Section 3.4) added the filter of category matching, which effectively reduced mismatching and improved the correctness of data association. Furthermore, in the proposed multi-group-step L-M optimization method (Section 3.5), we fully considered the characteristics of each features and analyze which variables of pose were constrained by them, and we divided the six pose six variables into groups and solved them step by step. All of these improvements ensured that the system worked steadily and precisely. Each online frame matched well with the priori map and the estimated trajectory closely coincides with the ground truth. The computational complexity was small enough to meet the need for real-time processing and no frame skipping occurs.

We show the quantitative results in Table 3, where the units are expressed in meters for translation and in degrees for rotation, and the evaluation methods of which have been expounded in 4.1. Reasonably, the translation and rotation errors in all directions of S1 were better than S2 and S3, but they were still on the same level. In addition, most errors of S2 were slightly less than S3. This shows that textless is more challenging than noise for the proposed Lidar localization system, which was due to the strong suppression of noise by the proposed feature extraction method. In order to calculate a more accurate pose, we needed to optimize to make full use of the limited information in the textless scene to calculate a more accurate pose. The RMS errors of translation achieved 3~5 cm and the maximum errors of t_x, t_y were less than 15 cm. The RMS errors of rotation are less than 0.1° and mainly in θ_{yaw}, and even though θ_{yaw} was larger than θ_{roll} and θ_{pitch}, it was small enough to be used. In fact, the changes in θ_{roll} and θ_{pitch} were very small when the vehicle travels. In the existing reports [41], is the most complete and accurate state-of-the-art Lidar-based localization system, where the t_x, t_y achieved 5~10 cm RMS accuracy and t_x is about 10 cm RMS. Our system outperformed the previous state-of-the-art systems and the localization error was reduced by half.

Table 3. Pose errors in the three different scenes.

Scene	X			Y			Z		
	1σ	2σ	3σ	1σ	2σ	3σ	1σ	2σ	3σ
S1	0.034	0.065	0.090	0.030	0.064	0.100	0.042	0.124	0.201
S2	0.040	0.099	0.132	0.035	0.067	0.118	0.048	0.127	0.218
S3	0.048	0.080	0.125	0.040	0.069	0.100	0.046	0.134	0.211
Scene	**Roll**			**Pitch**			**Yaw**		
	1σ	2σ	3σ	1σ	2σ	3σ	1σ	2σ	3σ
S1	0.018	0.037	0.066	0.019	0.034	0.042	0.083	0.173	0.208
S2	0.022	0.049	0.073	0.019	0.041	0.082	0.080	0.186	0.275
S3	0.057	0.089	0.100	0.018	0.034	0.042	0.100	0.201	0.227

Figure 12 is the localization error curves of a data piece with about 2600 s, which include almost all kinds of scenes, such as typical urban road, congested urban roads, narrow tunnels, and textureless highways. Figure 12a,b represent translation and rotation errors respectively. The fluctuation of the t_z and θ_{yaw} are larger than t_x, t_y and $\theta_{roll}, \theta_{pitch}$, while they are within the accuracy requirements of planning, control and other modules in unmanned vehicles. In fact, the t_z error also confused some other factor that the ground truth in t_z is not accurate enough as the RTK accuracy in the z direction is lower than x and y.

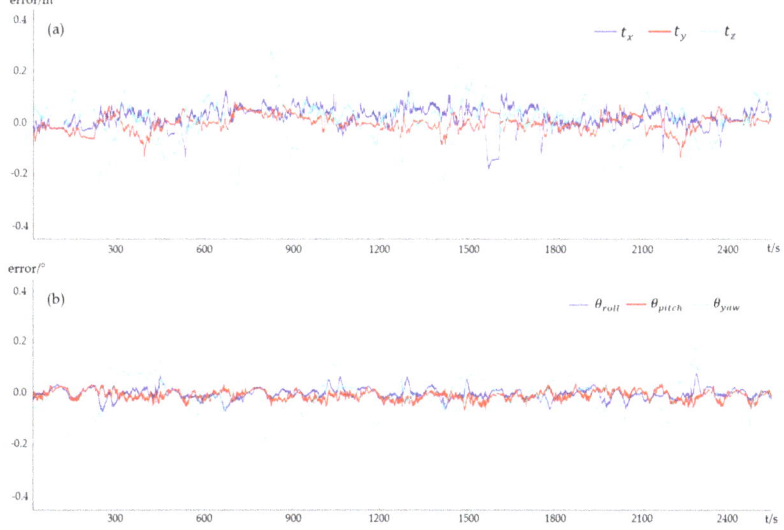

Figure 12. Pose error curves. (**a**) Position error curves; (**b**) orientation error curves.

5. Conclusions

We have proposed a complex and complete segmentation-based Lidar localization system, designed for high level autonomous driving applications, which can not only reduce the cost of localization, but also improve the accuracy and stability. The implementation of the precise and robust real-time localization system benefits from a series of optimization improvements, including the prior map storage management and dynamic loading method, the efficient low-level semantic segmentation-based multiple type feature point clouds extraction algorithm, the priori information considered category matching data association method, the multi-group-step L-M iterative optimization algorithm, the fusion with MEMS IMU data through a state-error Kalman filter, and the reasonable and effective coupling of multiple modules in the system framework.

Our localization system can produce a stable and precise 6DoF pose at 200Hz, achieve 3~5 cm in position and 0.05~0.1° in orientation RMS accuracy; the accuracy outperforms previous state-of-the-art systems. These results are sufficient to meet the automatic driving requirements in control and planning modules. The system can achieve almost identical performance in various challenging scenes, including congested urban roads, narrow tunnels, and textureless highways. In rain, fog, and other harsh weather, even though the Lidar range decrease and the noise increase, the system can still work normally. Our system has been deployed in multiple unmanned vehicles and large-scale test verifications have been conducted every day. The robustness and adaptability have been verified with more than 1000 Km localization testing data.

As the proposed Lidar localization system relies on the prior map dataset, if there are many large modifications in the scene, such as the architecture demolition or construction, the accuracy will decrease or the localization may even fail. Hence, methods to update the map dataset automatically, efficiently and cheaply will be one of the researches focus in future work. In addition, using the model-to-model data association method instead of the used point-to-line or point-to-plane methods can further reduce the computation complexity of the system, which is also a research topic of future interests.

Author Contributions: Conceptualization, H.L., Q.Y. and H.W.; Data curation, H.L., L.C. and J.Y.; Formal analysis, H.L. and H.W.; Funding acquisition, Q.Y.; Investigation, H.L., Q.Y. and H.W.; Methodology, Q.Y., H.L. and H.W.; Project administration, Q.Y.; Resources, H.L., L.C. and J.Y.; Software, H.L. and H.W.; Supervision, H.L.; Validation, H.L., H.W., L.C. and J.Y.; Visualization, H.L.; Writing—original draft, H.L.; Writing—review & editing, H.L. and Q.Y., J.Y.

Funding: This research was funded by the National Natural Science Foundation of China under grant number 41771480.

Acknowledgments: The authors would like to gratefully acknowledge Beijing Momenta Technology Company Limited who provided us the data support.

Conflicts of Interest: The authors declare no conflict of interest.

References

1. Groves, P.D. *Principles of GNSS, Inertial, and Multisensor Integrated Navigation Systems*; Artech House: Fitchburg, MA, USA, 2013.
2. Falco, G.; Pini, M.; Marucco, G. Loose and tight GNSS/INS integrations: Comparison of performance assessed in real urban scenarios. *Sensors* **2017**, *17*, 255. [CrossRef] [PubMed]
3. Zhang, X.; Miao, L.; Shao, H. Tracking architecture based on dual-filter with state feedback and its application in ultra-tight GPS/INS integration. *Sensors* **2016**, *16*, 627. [CrossRef]
4. Wei, W.; Zongyu, L.; Rongrong, X. INS/GPS/Pseudolite integrated navigation for land vehicle in urban canyon environments. In Proceedings of the IEEE Conference on Cybernetics and Intelligent Systems, Singapore, 1–3 December 2004; pp. 1183–1186.
5. Woo, R.; Yang, E.J.; Seo, D.W. A Fuzzy-Innovation-Based Adaptive Kalman Filter for Enhanced Vehicle Positioning in Dense Urban Environments. *Sensors* **2019**, *19*, 1142. [CrossRef] [PubMed]
6. Choi, K.; Suhr, J.; Jung, H. FAST Pre-Filtering-Based Real Time Road Sign Detection for Low-Cost Vehicle Localization. *Sensors* **2018**, *18*, 3590. [CrossRef] [PubMed]
7. Wang, D.; Xu, X.; Zhu, Y. A Novel Hybrid of a Fading Filter and an Extreme Learning Machine for GPS/INS during GPS Outages. *Sensors* **2018**, *18*, 3863. [CrossRef] [PubMed]

8. Van Brummelen, J.; O'Brien, M.; Gruyer, D.; Najjaran, H. Autonomous vehicle perception: The technology of today and tomorrow. *Transp. Res. Part C Emerg. Technol.* **2018**, *89*, 384–406. [CrossRef]
9. Castorena, J.; Agarwal, S. Ground-edge-based LIDAR localization without a reflectivity calibration for autonomous driving. *IEEE Robot. Autom. Lett.* **2018**, *3*, 344–351. [CrossRef]
10. Gawel, A.; Cieslewski, T.; Dubé, R.; Bosse, M.; Siegwart, R.; Nieto, J. Structure-based vision-laser matching. In Proceedings of the IEEE/RSJ International Conference on Intelligent Robots and Systems (IROS), Daejeon, Korea, 9–14 October 2016; pp. 182–188.
11. Roweis, S. *Levenberg-Marquardt Optimization*; University Of Toronto: Toronto, ON, Canada, 1996.
12. Hsu, C.M.; Shiu, C.W. 3D LiDAR-Based Precision Vehicle Localization with Movable Region Constraints. *Sensors* **2019**, *19*, 942. [CrossRef] [PubMed]
13. Forster, C.; Pizzoli, M.; Scaramuzza, D. SVO: Fast semi-direct monocular visual odometry. In Proceedings of the IEEE international conference on robotics and automation (ICRA), Hong Kong, China, 31 May–7 June 2014; pp. 15–22.
14. Engel, J.; Schöps, T.; Cremers, D. LSD-SLAM: Large-scale direct monocular SLAM. In Proceedings of the European Conference on Computer Vision (ECCV), Zurich, Switzerland, 6–12 September 2014; pp. 834–849.
15. Qin, T.; Li, P.; Shen, S. Relocalization, global optimization and map merging for monocular visual-inertial SLAM. In Proceedings of the IEEE International Conference on Robotics and Automation (ICRA), Brisbane, Australia, 21–26 May 2018; pp. 1197–1204.
16. Weiss, S.; Achtelik, M.W.; Lynen, S.; Chli, M.; Siegwart, R. Real-time onboard visual-inertial state estimation and self-calibration of mavs in unknown environments. In Proceedings of the IEEE International Conference on Robotics and Automation, St. Paul, MN, USA, 14–18 May 2012; pp. 957–964.
17. Ji, Z.; Singh, S. Low-drift and real-time lidar odometry and mapping. *Auton. Robot.* **2017**, *41*, 401–416.
18. Moosmann, F.; Stiller, C. Velodyne slam. In Proceedings of the IEEE Intelligent Vehicles Symposium (IV), Baden-Baden, Germany, 5–9 June 2011; pp. 393–398.
19. Kuramachi, R.; Ohsato, A.; Sasaki, Y.; Mizoguchi, H. G-ICP SLAM: An odometry-free 3D mapping system with robust 6DoF pose estimation. In Proceedings of the IEEE International Conference on Robotics and Biomimetics (ROBIO), Zhuhai, China, 6–9 December 2015; pp. 176–181.
20. Hess, W.; Kohler, D.; Rapp, H.; Andor, D. Real-time loop closure in 2D LIDAR SLAM. In Proceedings of the IEEE International Conference on Robotics and Automation (ICRA), Stockholm, Sweden, 16–21 May 2016; pp. 1271–1278.
21. Mur-Artal, R.; Montiel, J.M.M.; Tardos, J.D. ORB-SLAM: A Versatile and Accurate Monocular SLAM System. *IEEE Trans. Robot.* **2017**, *31*, 1147–1163. [CrossRef]
22. Mur-Artal, R.; Tardos, J.D. ORB-SLAM2: An Open-Source SLAM System for Monocular, Stereo and RGB-D Cameras. *IEEE Trans. Robot.* **2017**, *33*, 1255–1262. [CrossRef]
23. Galvez-Lopez, D.; Tardos, J.D. Bags of Binary Words for Fast Place Recognition in Image Sequences. *IEEE Trans. Robot.* **2012**, *28*, 1188–1197. [CrossRef]
24. Gao, X.; Wang, R.; Demmel, N.; Cremers, D. LDSO: Direct sparse odometry with loop closure. In Proceedings of the IEEE/RSJ International Conference on Intelligent Robots and Systems (IROS), Madrid, Spain, 1–5 October 2018; pp. 2198–2204.
25. Engel, J.; Koltun, V.; Cremers, D. Direct sparse odometry. *IEEE Trans. Pattern Anal. Mach. Intell.* **2018**, *40*, 611–625. [CrossRef] [PubMed]
26. Zhang, J.; Singh, S. LOAM: Lidar Odometry and Mapping in Real-time. In Proceedings of the Robotics: Science and Systems(RSS), Berkeley, CA, USA, 12–16 July 2014; pp. 109–111.
27. Geiger, A.; Lenz, P.; Urtasun, R. Are we ready for autonomous driving? the kitti vision benchmark suite. In Proceedings of the IEEE Conference on Computer Vision and Pattern Recognition, Providence, RI, USA, 16–21 June 2012; pp. 3354–3361.
28. Shan, T.; Englot, B. LeGO-LOAM: Lightweight and Ground-Optimized Lidar Odometry and Mapping on Variable Terrain. In Proceedings of the IEEE/RSJ International Conference on Intelligent Robots and Systems (IROS), Madrid, Spain, 1–5 October 2018; pp. 4758–4765.
29. Besl, P.J.; McKay, N.D. Method for registration of 3-D shapes. In Proceedings of the Sensor Fusion IV: Control Paradigms and Data Structures, Boston, MA, USA, 12–15 November 1992; pp. 586–607.
30. Deschaud, J.E. IMLS-SLAM: Scan-to-model matching based on 3D data. In Proceedings of the IEEE International Conference on Robotics and Automation (ICRA), Brisbane, Australia, 21–26 May 2018; pp. 2480–2485.

31. Levinson, J.; Montemerlo, M.; Thrun, S. Map-based precision vehicle localization in urban environments. In Proceedings of the Robotics: Science and Systems (RSS), Atlanta, GA, USA, 27–30 June 2007; p. 1.
32. Yoneda, K.; Tehrani, H.; Ogawa, T.; Hukuyama, N.; Mita, S. Lidar scan feature for localization with highly precise 3-D map. In Proceedings of the IEEE Intelligent Vehicles Symposium, Ypsilanti, MI, USA, 8–11 June 2014; pp. 1345–1350.
33. Baldwin, I.; Newman, P. Laser-only road-vehicle localization with dual 2d push-broom lidars and 3d priors. In Proceedings of the 2012 IEEE/RSJ International Conference on Intelligent Robots and Systems, Vilamoura, Portugal, 7–12 October; pp. 2490–2497.
34. Chong, Z.J.; Qin, B.; Bandyopadhyay, T.; Ang, M.H.; Frazzoli, E.; Rus, D. Synthetic 2d lidar for precise vehicle localization in 3d urban environment. In Proceedings of the IEEE International Conference on Robotics and Automation, Karlsruhe, Germany, 6–10 May 2013; pp. 1554–1559.
35. Wolcott, R.W.; Eustice, R.M. Robust LIDAR localization using multiresolution Gaussian mixture maps for autonomous driving. *Int. J. Robot. Res.* **2017**, *36*, 292–319. [CrossRef]
36. Wolcott, R.W.; Eustice, R.M. Fast LIDAR localization using multiresolution Gaussian mixture maps. In Proceedings of the IEEE International Conference on Robotics and Automation (ICRA), Lijiang, China, 8–10 August 2015; pp. 2814–2821.
37. Levinson, J.; Thrun, S. Robust vehicle localization in urban environments using probabilistic maps. In Proceedings of the IEEE International Conference on Robotics and Automation, Anchorage, AK, USA, 4–8 May 2010; pp. 4372–4378.
38. Gao, Y.; Liu, S.; Atia, M.; Noureldin, A. INS/GPS/LiDAR integrated navigation system for urban and indoor environments using hybrid scan matching algorithm. *Sensors* **2015**, *15*, 23286–23302. [CrossRef] [PubMed]
39. Qian, C.; Liu, H.; Tang, J.; Chen, Y.; Kaartinen, H.; Kukko, A.; Zhu, L.; Liang, X.; Chen, L.; Hyyppä, J. An integrated GNSS/INS/LiDAR-SLAM positioning method for highly accurate forest stem mapping. *Remote Sens. Lett.* **2017**, *9*, 3. [CrossRef]
40. Klein, I.; Filin, S. LiDAR and INS fusion in periods of GPS outages for mobile laser scanning mapping systems. In Proceedings of the ISPRS Calgary 2011 Workshop, Calgary, AB, Canada, 9–31 August 2011.
41. Wan, G.; Yang, X.; Cai, R.; Li, H.; Zhou, Y.; Wang, H.; Song, S. Robust and precise vehicle localization based on multi-sensor fusion in diverse city scenes. In Proceedings of the IEEE International Conference on Robotics and Automation (ICRA), Brisbane, Australia, 21–26 May 2018; pp. 4670–4677.
42. Li, J.; Zhong, R.; Hu, Q.; Ai, M. Feature-based laser scan matching and its application for indoor mapping. *Sensors* **2016**, *16*, 1265. [CrossRef] [PubMed]
43. Grant, W.S.; Voorhies, R.C.; Itti, L. Efficient Velodyne SLAM with point and plane features. *Auton. Robot.* **2018**, 1–18. [CrossRef]
44. Ravankar, A.; Ravankar, A.A.; Hoshino, Y.; Emaru, T.; Kobayashi, Y. On a hopping-points svd and hough transform-based line detection algorithm for robot localization and mapping. *Int. J. Adv. Robot. Syst.* **2016**, *13*, 98. [CrossRef]
45. Yu, C.; Liu, Z.; Liu, X.J.; Xie, F.; Yang, Y.; Wei, Q.; Fei, Q. DS-SLAM: A Semantic Visual SLAM towards Dynamic Environments. In Proceedings of the IEEE/RSJ International Conference on Intelligent Robots and Systems (IROS), Madrid, Spain, 1–5 October 2018; pp. 1168–1174.
46. Bogoslavskyi, I.; Stachniss, C. Fast range image-based segmentation of sparse 3D laser scans for online operation. In Proceedings of the IEEE/RSJ International Conference on Intelligent Robots and Systems (IROS), Daejeon, Korea, 9–14 October 2016; pp. 163–169.
47. Na, K.; Byun, J.; Roh, M.; Seo, B. The ground segmentation of 3D LIDAR point cloud with the optimized region merging. In Proceedings of the International Conference on Connected Vehicles and Expo (ICCVE), Las Vegas, NV, USA, 2–6 December 2013; pp. 445–450.
48. Zhang, Y.; Wang, J.; Wang, X.; Dolan, J.M. Road-segmentation-based curb detection method for self-driving via a 3D-LiDAR sensor. *IEEE Trans. Intell. Transp. Syst.* **2018**, *19*, 3981–3991. [CrossRef]
49. Xu, Y.; Wei, Y.; Hoegner, L.; Stilla, U. Segmentation of building roofs from airborne LiDAR point clouds using robust voxel-based region growing. *Remote Sens. Lett.* **2017**, *8*, 1062–1071. [CrossRef]

Article

Design, Calibration, and Evaluation of a Backpack Indoor Mobile Mapping System

Samer Karam [1,*], George Vosselman [1], Michael Peter [2], Siavash Hosseinyalamdary [1] and Ville Lehtola [1]

1 Department of Earth Observation Science, Faculty ITC, University of Twente, 7514 AE Enschede, The Netherlands; george.vosselman@utwente.nl (G.V.); s.hosseinyalamdary@utwente.nl (S.H.); v.v.lehtola@utwente.nl (V.L.)

2 Independent researcher, 46397 Bocholt, Germany; michael-peter@windowslive.com

* Correspondence: s.karam@utwente.nl; Tel.: +31-53-4894577

Received: 26 February 2019; Accepted: 8 April 2019; Published: 13 April 2019

Abstract: Indoor mobile mapping systems are important for a wide range of applications starting from disaster management to straightforward indoor navigation. This paper presents the design and performance of a low-cost backpack indoor mobile mapping system (ITC-IMMS) that utilizes a combination of laser range-finders (LRFs) to fully recover the 3D building model based on a feature-based simultaneous localization and mapping (SLAM) algorithm. Specifically, we use robust planar features. These are advantageous, because oftentimes the final representation of the indoor environment is wanted in a planar form, and oftentimes the walls in an indoor environment physically have planar shapes. In order to understand the potential accuracy of our indoor models and to assess the system's ability to capture the geometry of indoor environments, we develop novel evaluation techniques. In contrast to the state-of-the-art evaluation methods that rely on ground truth data, our evaluation methods can check the internal consistency of the reconstructed map in the absence of any ground truth data. Additionally, the external consistency can be verified with the often available as-planned state map of the building. The results demonstrate that our backpack system can capture the geometry of the test areas with angle errors typically below 1.5° and errors in wall thickness around 1 cm. An optimal configuration for the sensors is determined through a set of experiments that makes use of the developed evaluation techniques.

Keywords: IMMS; indoor mapping; MLS; mobile laser scanning; SLAM; point clouds; 2D laser scanner; 2D laser range-finder; LiDAR; LRF; sensors configurations

1. Introduction

Accurate measurement and representation of indoor environments have attracted a large scientific interest because of the multitude of potential applications [1–5]. In particular, the use of indoor mobile mapping systems (IMMS) has shown promise in indoor data collection. Indoor spaces are satellite-denied environments, so it is an obvious choice to map them using relative positioning techniques, i.e., simultaneous localization and mapping (SLAM). A typical IMMS utilizes multiple sensors, e.g., laser scanners, inertial measurement units (IMU) and/or cameras, to capture the indoor environment. The sensors are attached onto a mobile platform that can be a pushcart, a robot, or human-carriable equipment [6–10]. Laser scanners are used to measure the geometry, cameras are used to measure the texturing, and IMUs are used to estimate the changes in orientation of the scanner for SLAM purposes. The reason behind this use of the sensors is that RGB camera-based visual SLAM algorithms are extremely sensitive to lighting conditions, and fail in textureless spots, which are common in indoor environments. In turn, depth cameras (or RGB-D cameras) employed to alleviate this shortcoming have a very short range, which is insufficient for large indoor spaces.

Multiple human-carriable systems that employ laser scanners have been developed [10–14]. This is not surprising, as easily carriable equipment is widely applicable, e.g., unlike pushcarts, the carriable equipment can be taken up and down the stairs, and because laser scanners are the best sensors in capturing indoor geometry as discussed earlier. This group of mobile mapping systems is further divided into hand-held and backpack systems. Lehtola et al. [4] identify the state-of-the-art of these types. For hand-held commercial systems, Kaarta Stencil and ZEB1 REVO arguably present the current best on the market. For backpack systems, there is the Leica Pegasus [13].

The IMMS are quite different from each other. This is because when using relative positioning, the physical scanner platform and the employed data association method are intertwined. Therefore, advances oftentimes cannot be and are not incremental, since changing the hardware has an impact on the software and vice versa, and it can sometimes be advantageous if both the hardware and the software are re-designed.

In this paper, therefore, we introduce the design and the performance of our triple-2D-LRF (laser range finder) backpack system that is capable of outputting 3D indoor models from a 6 degree-of-freedom (6DOF) trajectory. Notably, this work differs from the previous triple-2D-LRF configuration state-of-the-art [11,14,15] by employing two LRFs in slanted angles. Using slanted angles appears as a minor detail but turns out to be a quite fundamental issue. Specifically, it allows for combining the scan lines from the three 2D LRFs to form a quasi-3D point subset in the local platform coordinates that can then be robustly matched against a planar feature in the world coordinates. In other words, slanting the LRFs enables the use of robust planar features for SLAM-based data association and measurements of all three LRFs are used simultaneously for an integral estimation of the backpack pose, planes, calibration and relative sensor orientations. Studying the use of planar features is advantageous for two reasons. First, oftentimes the final representation of the indoor environment is wanted in a planar form and formulating the use of planes already into the SLAM-algorithm is therefore motivated. Second, a typical wall in an indoor environment physically has a planar shape.

As a second contribution, we present alternative evaluation techniques for assessing the performance of IMMSs. The proposed evaluation techniques estimate the reconstruction accuracy and quality even in the absence of a ground truth model. Here, in contrast to previous works [4,16–19] that employ 3D ground truth data, the proposed methods uses 2D information in form of architectural constraints, i.e., the perpendicularity and parallelism of walls, or if available, floor plans. Furthermore, the proposed evaluation methods are utilized to find practical optima for the slanted LRFs angles.

This paper is organized as follows. Section 2 presents an overview of the previously developed human-carriable IMMSs and the state-of-the-art for evaluation methods on generated maps. Section 3 describes the design of our backpack system and the planar-feature SLAM method, based on the earlier works in [5] and [20]. The calibration process of the mounted LRF is explained in Section 4. We elaborate the strategy of the registration process for LRFs in Section 5. We also present the proposed techniques to evaluate the system performance in Section 6, as partly introduced in [5]. In Section 7, we show all implemented experiments that lead to the optimal configuration of the system. The paper draws conclusions in Section 8.

2. Related Works

Human-carriable systems can be divided into two categories: hand-held systems and backpack systems. After discussing the literature on these, we shall outline the literature on evaluation methods.

2.1. Hand-Held Systems

Hand-held systems offer more flexibility because theoretically anywhere the operator can walk, the system can map. Examples of hand-held systems include ZEB1 from 3D Laser Mapping/CSIRO and Viametris iMS2D.

ZEB1 consists of a laser range-finder (Hokuyo UTM-30LX with 30 m range) and an inertial measurement unit (IMU, a MicroStrain 3DM-GX2) mounted on a passive linkage mechanism [6].

The system is based on the 6DOF SLAM algorithm that was developed to work with the capricious movement of the sensor. To operate ZEB1, it must be gently oscillated by the operator towards and away from him or her with a connection to the IMU to provide a solution.

In comparison with other IMMS systems, ZEB1 has accessibility characteristics that allows it to map most of the areas in indoor environments, including stairwells. On the other hand, the performance of the device is acceptable only under specific conditions. For example, ZEB1 is not suitable for some environments in which the motion is not observable because the areas are featureless, large or open. Furthermore, the proposed SLAM algorithm will struggle if the oscillation of the sensor head stops for more than a few seconds.

In the recent years, GeoSLAM has developed the mobile kinematic laser scanner ZEB-REVO as a commercial system for the measurement and mapping of multi-level 3D environments [21]. It is also handheld, but the LRF is rotated on a fixed pole instead of irregular motion on a spring.

iMS2D is a handheld scanner released by Viametris in 2016 for 2D indoor scanning [22]. It comprises simply a 2D Hokuyo laser range-finder and fisheye camera.

2.2. Backpack Mapping Systems

These are instruments that are carried by a human operator. The key characteristic of this kind of system is that they have a non-zero pitch and roll.

Naikal et al. [11] mounted three LRFs (Hokuyo URG-04LX) orthogonally to each other together with a camera on a backpack platform. They aim to retrieve 6DOF localization in 3D space by integrating two processes. In the first one, the transformation is estimated by applying the visual odometry technique and in the second one, the rotation angles are estimated from the three scanners by applying the scan-matching algorithm. In later work by the same group, [14] added one more 2D scanner and two IMUs (HG9900 and InterSense) to the backpack system.

The overall goal of their work was to estimate the trajectory the system follows during mapping. To achieve this goal, they developed four algorithms, which depend mainly on scan-matching, to retrieve the 6DOF pose translation of the system over time. The proposed framework is quite similar to that of the 6DOF scan-matching process in the SLAM approach of [2].

The proposed algorithms have tradeoffs in terms of performance depending on the building's environment. For instance, those algorithms that rely on planar floor assumptions provide more precise results only in the case of planar floor availability in the captured area. In addition, the algorithms lack a systematic filter that optimally combines the sensors' measurements, such as a Kalman filter.

In other work by the same group, Liu et al. [15] replaced the yaw scanner (Hokuyo URG-04LX) by the Hokuyo UTM-30LX and added three cameras to the backpack platform. They used the previously developed algorithms [14] to estimate the system's trajectory based on integrating the laser and IMU data. Each of the sensors is used independently to estimate one or more parameters of system's pose (x, y, z, roll, pitch, yaw) over time, e.g., the z value is estimated from the pitch scanner while x, y, and yaw values are estimated from the yaw scanner. The left pose parameters, namely, roll and pitch, are estimated using the InterSense IMU. Since the camera is approximately synchronized with the scanners, Liu et al. estimate the pose of each image by nearest-neighbor interpolation of the pose parameters in order to texture the 3D model. Since the sensor rotation is determined independently for each of the three axes and not in an integrated manner, they need to assume that each scanner keeps scanning in the same plane over time, but that is unrealistic because of human operator motion. Therefore, this assumption will reflect negatively on the accuracy in the case of backpack rotation.

Kim [23] presented an approach for 3D positioning of a previously developed backpack system [11] in an indoor environment and also for generating point clouds of this environment using Rao-Blackwellized Particle Filter (RBPF)-based SLAM algorithm. The system consists of five LRFs (Hokuyo UTM-30LX), two IMUs (HG9900 and InterSense) and two fisheye cameras (GRAS-14S5C). In contrast to the other approaches which use all the data, Kim's approach identifies and incorporates the most credible data from each range finder. For localizing the system in an indoor environment, the cumulative shifts of the system over time are computed from yaw and pitch range finders using scan-matching techniques. Similar to [15], the TORO optimizer is used to minimize the accumulated positional error over time and solve the loop closure problem. Next, the point cloud is generated from roll range finders, restructured using a plane reconstruction algorithm, and textured using captured images. To avoid an expected misalignment in the case of the complex indoor environment, two 2D SLAM algorithms are proposed and integrated. The first one is to localize the system in the z-axis direction, and the second one for xy localization. The role of orientation sensor (InterSense IMU) is to measure roll and pitch angles to correct the measurements of the pitch and roll scanners and thus increase the accuracy of the scanner-based localization method. Only data from the roll scanner are used to generate the point cloud, while the pitch and yaw scanners will be responsible for 3D localization. In contrast to the yaw scanner, which scans in a plane parallel to the floor and helps to determine the xy location, a pitch scanner scans in a plane perpendicular to the floor and provides the third dimension (z) of the location. Thus, the localization algorithm may fail in case of discontinuities between consecutive walls or transparent objects, such as windows.

Wen et al. [9] developed an indoor backpack mobile mapping system consisting of three 2D LRFs (Hokuyo UTM-30LX) and one IMU (Xsens MTi-10). The system configuration consists of one LRF mounted horizontally while the other two are vertical. A 2D map of the building is constructed by a particle filter-based 2D SLAM using data from the horizontal range finder and then applying the rotations captured by the IMU to obtain a 3D pose of the system and thus a 3D map of the building. At the same time, the two vertical LRFs are responsible for creating 3D point clouds.

Filgueira et al. [24] presented a backpack mapping system constructed from a 3D LiDAR and an IMU for indoor data acquisition. The LiDAR is Velodyne VLP-16 that provides 360° horizontal coverage and 30° vertical coverage. The SLAM algorithm utilizes the combination of two algorithms proposed in [25] for indoor and outdoor positioning and mapping adapted for handling Velodyne's data. The iterative closest point algorithm (ICP) was used for data association. The system is tested using the Faro Focus 3D scanner as a ground truth in two indoor environments with different characteristics. In later work by the same group, Lagüela et al. [26] made some adjustments in the design of the system such as increasing the height of the Velodyne to avoid the occlusions that might occur because of the operator's body. Moreover, they mounted two webcams in the system for inspection purposes. In order to analyze the performance of the recent version of their backpack system, they did a comparison not only with the static scanner (Faro Focus 3D) as before, but also with the ZEB-REVO scanner.

Blaser et al. [10] proposed a wearable indoor mapping platform (BIMAGE) to provide 3D image-based data for indoor infrastructure management. The platform is mounted by a panoramic camera (FLIR Ladybug5), IMU and two Velodyne VLP-16 (horizontal, vertical). Subsequent camera-based georeferencing is used to improve the camera positions provided by LiDAR-based SLAM.

2.3. Evaluation Methods

Various evaluation strategies have been proposed to investigate the performance of the state-of-the-art IMMSs and quantify the quality of resulting point clouds. The most common strategy is a point cloud to point cloud (pc2pc) comparison after registering both clouds to the ground truth coordinate system, typically using CloudCompare software [9,16,27]. While Thomson et al. [16] investigated the earlier Viametris i-MMS and ZEB1 systems using TLS (Faro Focus3D) as ground truth, Maboudi et al. [17] tested the later generations of Zebedee and Viametris (iMS3D and ZEB-Revo) using TLS (Leica P20) as ground truth. In addition to the pc2pc comparison, Maboudi et al. [17] they compared the building information model's (BIM) geometry derived from the tested systems to that derived from TLS. In later work [18], three additional analyses are proposed, namely, points-to-planes distance, target-to-target distance and model-based evaluation. In a broader assessment process, Lehtola et al. [4] proposed metrics to evaluate the full point cloud of eight state-of-the-art IMMSs against the point cloud of two TLSs (Leica P40, Faro Focus3D). Tran et al. [19] provided comparison metrics for the evaluation of 3D planar representations of indoor environments. Specifically, if a 3D planar reference model is given, the completeness, correctness, and accuracy of the obtained model can be estimated against it.

3. Backpack System ITC-IMMS

3.1. System Description

Due to the limited use and problems experienced by the previous indoor mapping systems, we developed our own indoor mobile mapping system shown in Figure 1. Our aim is to combine the proven accuracy of 2D SLAM-based trajectory estimates of push-cart systems with the flexibility of 3D hand-held or backpack systems. The system design has been proposed in [20] and is now implemented, optimised, calibrated, and evaluated. This backpack system consists of three LRFs (Hokuyo UTM-30LX) which are all utilized for a 3D (6 DOF) SLAM. In contrast to available 3D laser scanners, we try to keep the system design less expensive by only making use of these simple LRFs.

Figure 1. The laptop used and the backpack system mounted by three LRFs S_0 (Top), S_1 (left), and S_2 (right) fitted with markers.

The ranging noise according to our LRF's specifications is ±30 mm for [0.1 10] m range and ±50 mm for [10 30] m range [28]. This gives the Hokuyo UTM-30LX a key advantage over the range camera (Kinect) in capturing data inside large buildings such as airports where the dimensions of interior areas usually exceed 10 m.

The top LRF (here referred to as S_0) is mounted on the top of backpack system and it is approximately horizontal while the other two LRFs (S_1, S_2) are mounted to the right and left of the top one and are rotated around the moving direction (as in the i-MMS) as well as around the operator's shoulder axis. These two rotation axes are perpendicular to each other as shown in Figure 1. To find the optimal values for the rotation angles, we conducted experiments that will be described in Section 7. There are two objectives for the rotation of the range finders: First, the laser scanning covers surfaces perpendicular to the moving direction e.g., walls both behind and in front of the system, and second, it eases the association of points on new scan lines to previously seen walls. In the case where the scan lines would intersect walls vertically, a strong data association is not guaranteed when walking around corners or through doors. The field of view of the LRFs is limited to 270°, and accordingly, there will be a 90° gap in each scanline. In order to cover all walls as well as possible, the two range finders (S_1, S_2) are rotated around their axes such that their gaps (shadow areas) are directed towards the floor and the ceiling, respectively [20]. A laptop running Ubuntu 16.04.X and the robot operation system (ROS) is used to communicate with all mounted sensors during data capture.

3.2. Coordinate Systems

The proposed mapping system is a multi-sensor system and each one of the three mounted sensors has its own coordinate system. Next to the aforementioned sensor's coordinate system, there are two additional coordinate systems: the frame (backpack) and model (local world) coordinate system.

To integrate the data of the three LRF sensors, coordinates in their individual coordinate systems must be transformed into a unified coordinate system, which is termed the "frame coordinate system (f)". We adopt the sensor coordinate system of S_0 as the frame coordinate system. Assuming all sensors are rigidly mounted on the frame, the sensor coordinate systems of S_1 and S_2 are registered in this frame coordinate system using six transformation parameters, namely, three rotation parameters $(\omega_{s_i}, \varphi_{s_i}, \kappa_{s_i})$ and three translation parameters $(dX_{s_i},\ dY_{s_i},\ dZ_{s_i})$. These parameters are determined in the registration process described in Section 5.

Since the frame coordinate system is attached to a moving backpack system, a fixed coordinate system should be defined as a reference and a space in which the final indoor model will be described. This fixed coordinate system is termed the "model coordinate system (m)". This model coordinate system is assumed to be the frame coordinate system at the start point of the trajectory. As long as the frame coordinate system is moving in 3D space, it is registered in the model coordinate system using six transformation parameters over time (t), namely, three rotation parameters $(\omega_f(t), \varphi_f(t), \kappa_f(t))$ and three translation parameters $(dX_f(t),\ dY_f(t),\ dZ_f(t))$. Those changing parameters that originate from the 6DOF SLAM algorithm are explained in more detail in the next section.

3.3. 6DOF SLAM

We defined a feature-based SLAM algorithm in which the range observations of all three scanners contribute to the integral estimation of all six pose parameters. The starting point for the SLAM is the association of newly measured points to already estimated planes in the indoor environment. The six pose parameters are modeled as a function over time using B-splines. The planes are simply defined by a normal vector (n) and distance to the origin (d) in the model coordinate system. For a point X_m in the model coordinate system it should therefore hold that:

$$nX_m - d = 0, \tag{1}$$

As the laser scanners after registration provide point coordinates X_f in the frame coordinate system, we write $X_m = R(t)\ X_f + v(t)$ to transform a point X_f in the frame coordinate system to point X_m in the model coordinate system by a rotation $R(t)$ and a translation $v(t)$. Substituting X_m in Equation (1) provides the observation equation

$$E\left\{n\left[R(t)X_f + v(t)\right] - d\right\} = 0, \tag{2}$$

The trajectory of the frame, as well as the rotations, is modelled by B-splines as a function of time (t). For instance, roll ω is formulated as follow:

$$\omega(t) = \sum\nolimits_i \alpha_{\omega,i}.B_i(t), \tag{3}$$

where $\alpha_{\omega,i}$ is the spline coefficient for ω to be estimated on interval i.

The model coordinate system is defined based on the first scans of the three scanners. Since there is no information about the system speed yet, the rotation and translation defined during establishing the model coordinate system are used to predict the orientation and translation of the system over the time interval of the first two scanlines using a constant local spline. Later, more data will be captured by the LRFs. Then, for pose parameter prediction, the local spline estimation is implemented using the data of only three to four scanlines of each of the laser scanners. The locally estimated splines are linearly extrapolated to obtain a prediction of the frame pose over the time interval of the next scanline acquisition.

After segmenting the next scanline using a line segmentation procedure [29], a test on a distance threshold is used to decide whether a segment should be associated with an already reconstructed plane or be used to instantiate a new plane need. Currently, only horizontal and vertical planes are used. After setting up the corresponding observation equations, the pose parameters are estimated for the next time interval. After processing the whole dataset with locally defined spline functions, one integral adjustment estimates spline coefficients for the whole trajectory as well as all parameters of planes in the model coordinate system.

4. Calibration Process

In this research the term "calibration" refers to the estimation of biases in the raw range data acquired by every single LRF, in our case the Hokuyo UTM-30LX. The calibration of the laser range finders is needed to optimise the quality of the reconstructed point cloud.

4.1. Calibration Facility

For carrying out the calibration process, a classroom in ITC faculty building was selected as a calibration facility. The room has an almost rectangular shape with white walls and is of a suitable size. The reference data were captured by tape measurements.

4.2. Calibration

Equation (4) formulates the relationship between the coordinates in the LRF sensor system (X_s, Y_s), and the model system (X_m, Y_m), and Equation (5) describes the known location of a wall in the model system (distance d, orientation θ). All relationships are in 2D as we assume the LRF to be scanning perpendicular to the walls.

$$\begin{pmatrix} X_m \\ Y_m \end{pmatrix} = \begin{pmatrix} cos\beta & sin\beta \\ -sin\beta & cos\beta \end{pmatrix} \begin{pmatrix} X_s \\ Y_s \end{pmatrix} + \begin{pmatrix} X_0 \\ Y_0 \end{pmatrix}, \tag{4}$$

$$X_m cos\theta + Y_m sin\theta - d = 0, \tag{5}$$

where β is the rotation of the LRF, and (X_0, Y_0) represent the location of the LRF in the local model coordinate system.

Each indoor environment is decorated differently, and the surface materials are different, e.g., on walls. The surface material properties, e.g., color, brightness, and smoothness, impact the range measurements to a small degree [30]. This change in surface properties to the range measurements is compensated with the calibration of Equation (6). In our calibration model we use a scale factor (λ_r) and offset (Δr) for the range measurements as well as a scale factor (λ_α) for the scanning direction. The coordinates in LRF sensor system are obtained from the observed polar coordinates (range r, scanning direction α).

$$\hat{r} = \lambda_r r + \Delta r, \tag{6}$$

$$\begin{pmatrix} X_s \\ Y_s \end{pmatrix} = \begin{pmatrix} cos\lambda_\alpha\alpha & sin\lambda_\alpha\alpha \\ -sin\lambda_\alpha\alpha & cos\lambda_\alpha\alpha \end{pmatrix} \begin{pmatrix} \hat{r} \\ 0 \end{pmatrix}, \tag{7}$$

Equations (4)–(7) are combined to obtain a single equation with the pose of the LRF (X_0, Y_0, β) and the calibration parameters (λ_r, Δr, λ_α) as the unknown parameters. The LRF to be calibrated is put on different locations in the calibration room with different rotations to optimise the estimability of the calibration parameters. After a warming up period, the data of a few scan lines per pose are used to estimate all pose and calibration parameters. Points of those scan lines were manually labelled with the index number of the corresponding wall. Estimated range offsets and range scale factors were typically below 4 cm and 0.8%, respectively, whereas the estimated angle scale error is usually below 0.7%. After calibration, the remaining residuals between the points and the wall planes show a root mean square value below 1 cm. This is clearly better than the noise level specified by the manufacturer (3 cm).

4.3. Self-Calibration

Similar to self-calibration in the photogrammetric bundle adjustment, it is feasible to include the estimation of the sensor calibration parameters of all three LRFs in the SLAM process. In absence of a reference (tape) measurement in the SLAM procedure, we can, however, not estimate the range scale factors of all LRFs as the scale of the resulting point cloud would then be undetermined. Hence, we fix the range scale of the top LRF (S_0) to the value obtained in the calibration room and include the remaining eight calibration parameters as additional unknowns to the SLAM equations.

5. Relative Sensor Registration

To accurately fuse data from the three LRF sensors, their coordinate systems must be registered to a common reference system. This requires the estimation of the relative pose of the LRFs with respect to each other. We adopt the sensor coordinate system of the horizontal LRF as the backpack frame coordinate system and register as accurately as possible the two slanted LRFs with respect to this system. The registration is performed in two steps: marker registration and fine registration. These processes do not require a room with known dimensions, but the data should be captured in a specific way as described in the following two paragraphs.

5.1. Initial Registration

The registration method is based on 3D tracking technology for markers (see Figure 1) attached to the heads of the mounted LRFs to achieve an approximate registration. We make use of the 'ar_track_alvar' package [31], which is a robot operation system (ROS) wrapper for Alvar, an open source library for virtual and augmented reality (AR) marker tracking. As the laptop's webcam is involved in the registration process, it also needs to be calibrated using another ROS package. As the rotation and translation between the markers and the LRF sensor coordinate systems can be determined to a few mm and degree, the relative marker positions estimated with the ROS package can be used to infer approximate values for the parameters of the relative registration of the three LRFs.

5.2. Fine Registration

The goal of this process is to refine the approximate values for the registration parameters obtained during the previous approach and assumed to be acceptably accurate. This fine registration imposes two constraints on the captured sensor data. As the indoor environment usually contains large planes, the first constraint used is co-planarity of three line segments on the same plane simultaneously scanned by the three LRFs [32,33]. The second constraint is inferred from the perpendicularity of two observed planes [33].

The data collection is carried out with the backpack (ITC-IMMS) on the back of the operator. In order to estimate all registration parameters, the planar surfaces should be observed by the backpack system with different orientations. Therefore, first the operator stands inside a suitable area, in which the previous constraints are applicable, and starts capturing data while bending forward and sideward (right and left). Then, the operator rotates by 90° and bends again in the same way. These rotation and bending steps are repeated until the operator is back at the initial orientation.

The captured data pass through a series of processing steps before being subject to the registration's constraints. Firstly, the scanlines from each LRF are segmented by a line segmentation algorithm [29] and transformed to a frame system using the approximated parameters. Next, the all pairs of nearly co-planar line segments captured by two different LRFs are collected.

To define the aforementioned constraints, two types of observation equations are formulated, namely, perpendicularity and coplanarity. Let us denote l_j^i as the direction vector of the segment, where i refers to a plane (A, B) and j refers to one of the LRFs (0, 1, 2). The relative transformations of S_1 and S_2 with respect to S_0 are described by 12 parameters, a 3D rotation ($R_1(\omega_1, \phi_1, \kappa_1)$ and $R_2(\omega_2, \phi_2, \kappa_2)$) and a 3D translation ($T_1$ and T_2) for each.

If two planes A and B are perpendicular it will hold that:

$$\left(l_0^A \times R_1 l_1^A\right) \cdot \left(l_0^B \times R_1 l_1^B\right) = 0, \tag{8}$$

where: $l_0^A \times R_1 l_1^A$ is the normal vector of plane A expressed in the coordinate system of S_0, and $l_0^B \times R_1 l_1^B$ is the normal vector of plane B expressed in the coordinate system of S_0. The unknowns in this equation are the rotation angles of S_1 in R_1.

In a common coordinate system, the two direction vectors of the line segments as well as the vector connecting the midpoints $\left(p_0^i, p_1^i, p_2^i\right)$ must be coplanar. Taking the coordinate system of S_0, the coplanarity equation for plane A can be formulated as follows:

$$\left(l_0^A \times R_1 l_1^A\right) \cdot \left(p_0^A - R_1 p_1^A - T_1\right) = 0, \tag{9}$$

As three-line segments could be recorded with three different LRFs in each of three perpendicular planes, the data captured at a single pose of the mapping system would yield nine independent coplanarity equations and three independent perpendicularity equations. Thus, this would already provide sufficient observations to estimate all 12 registration parameters. However, to increase the reliability of the estimation we use a much larger number of equations with data of different poses of the backpack captured according to the described bending procedure. The scanning frequency of the Hokuyo used is 40 Hz; therefore, after one minute, each LRF records 2400 scanlines thereby leading to a very large number of observation equations.

Using the available approximated values from marker registration and after linearizing the formulated equations, an accurate estimate of the transformation parameters can be obtained by applying a least-squares estimation. The standard deviations of the estimated parameters are around 1 mm for the translations and around 0.05° for the rotation angles.

5.3. Self-Registration

In analogy to the self-calibration for the intrinsic sensor parameters of the LRFs, we can also extend the SLAM equations with the parameters describing the relative poses of the LRFs. We refer to this registration approach as the self-registration. Approximate values of the 12 registration parameters are obtained from either the initial registration or fine registration described above. When the sensor data are captured with a good variation in the rotations of the backpack with respect to the surrounding walls, ceiling and floor, as realized by the bending procedure, all 12 self-registration parameters can be estimated well as part of the overall estimation of all pose spline coefficients and plane parameters. This is not the case when the backpack IMLS is used in a normal mode when the operator walks upright through a building. In that case the top LRF, scanning in an approximately horizontal plane, will only capture vertical walls. As a consequence, the vertical offset between this LRF and the other two cannot be estimated. In this scenario the self-registration is restricted to 11 parameters.

6. SLAM Performance Measurements and Results

This section elaborates the methodology to evaluate indoor laser scanning point clouds described in our previous work [5] with some additions. Moreover, the measurements taken by our mapping system (ITC-IMMS) are processed by applying this methodology to investigate the performance of the 6DOF SLAM and to assess the capability of ITC-IMMS of capturing the true geometry of building interiors and preserve an accurate positioning when moving from one room to another.

6.1. Dataset

The dataset used is collected by ITC-IMMS at the University of Braunschweig, Germany. The scanned area shows a distinct office environment that has many windows and doors leading to rooms. Due to renovation work, the rooms were nearly empty. On the one side, this allows an easier identification of planar surfaces. On the other side, the number of surfaces is relatively small and a missed surface may have a larger impact on the estimability of the pose spline coefficients. The generated point cloud and the reconstructed planes are shown in Figure 2. About 73 million points were captured during a 9-minute walk through the rooms.

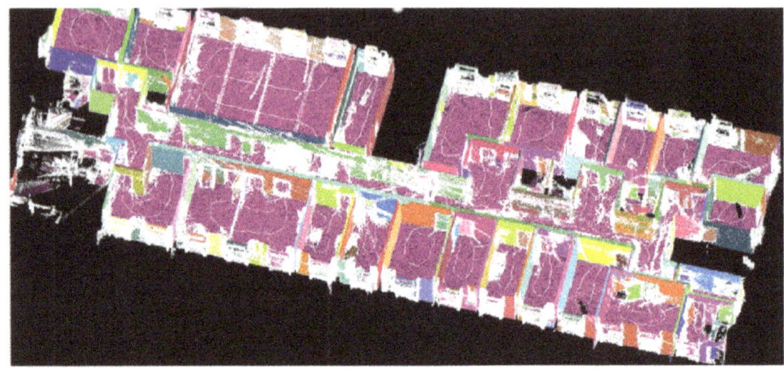

(a)

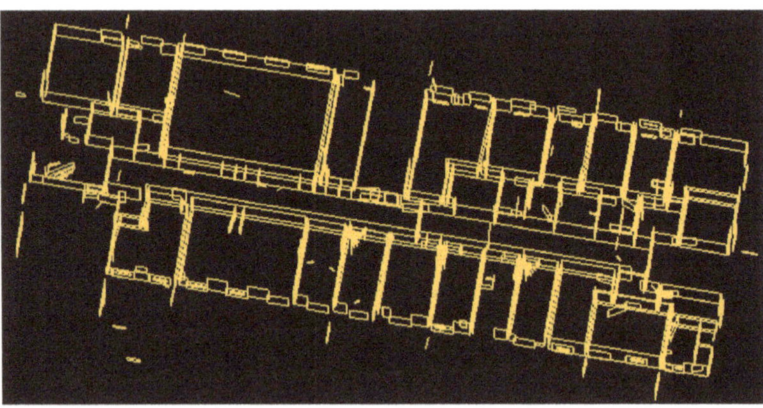

(b)

Figure 2. ITC-IMMS outputs. (**a**) The generated point cloud (colors show plane association) with the trajectory followed (white). (**b**) The reconstructed planes.

Point to Plane Association (Data Association)

The point is assigned to the closest plane if its distance to this plane is lower than 20 cm. Of the 73 million points, 53 million points were associated to 503 planes during the SLAM and used to estimate a total of 27880 pose spline coefficients and plane parameters. The distribution of the residual distance from the point to its associated plane is shown in Figure 3a. In total, 97% of the points have residuals below 3 cm and the RMS value of these residuals is only 1.3 cm. This means that the method is self-consistent. The RMS value, however, does not adequately represent the overall quality of the dataset. Further quality measures are therefore developed and used in the next section.

Figure 3b shows a top view of the generated point cloud, where points are colored based on their respective residuals. The data association rule which assigns points to planes experiences problems when two distinct planes are too close to each other. For example, a door that is wide open and thus close to the wall, or a door that is only slightly open and thus close to the other wall are typical causes for this behaviour. Moreover, there can be dynamic noise, for instance, if a door is opening while the data is being captured. Clear examples of both problematic cases are highlighted in Figure 3b. The problem resulting from merging a door with a nearby wall can be seen inside the orange dashed rectangle.

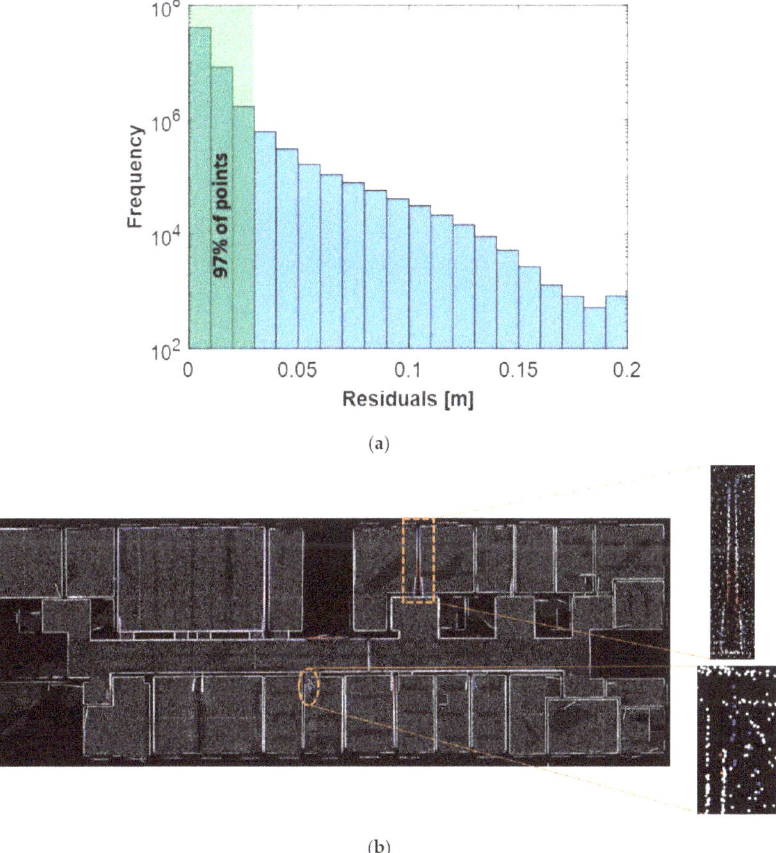

(a)

(b)

Figure 3. The residuals between the points and the estimated planes. (**a**) Frequency of the residuals with a logarithmic scale for the y-axis and a linear scale for the x-axis. (**b**) Top view of the generated point cloud. All white points have residuals below 3 cm and points with larger residuals are marked with either red or blue color, depending on the sign. The orange dashed rectangle marks an example of a plane representing a door being merged with another plane, which represents a wall that is near the door. The orange dashed oval surrounds an opening door.

6.2. Evaluation Techniques

As SLAM-based point clouds usually suffer from registration errors because of the dead-reckoning nature of SLAM algorithms, the performance of the mapping system and the accuracy of the provided results, which needs to be analyzed. While most current evaluation methods rely on the availability of reference data, we develop several techniques to investigate the mapping system in the absence of an accurate ground truth model. The proposed techniques take advantage of regularities in wall configurations to check how well the rooms are connected, and thus how well the environment is reconstructed.

Since most buildings have a floor plan (though often outdated), we utilize that as an external information source to check the quality of the generated indoor model, but without relying on an accurate registration of the point cloud to the floor plan. We classify the developed techniques into three independent groups: (1) techniques using architectural constraints; (2) techniques using a floor plan; and (3) completeness techniques.

To simplify the process and because the permanent structure of man-made indoor environments mainly consists of planar and vertical structures, the first two groups make use of 2D edges derived from such structures. As our feature-based SLAM outputs both point clouds and 3D reconstructed planes, the 2D edges are derived from the projection of the vertical planes onto the XY-plane, as presented in our previous work [5]. We address the third group of evaluation techniques in the study of the optimal configuration described in Section 7.

6.2.1. Evaluation Using Architectural Constraints

We make use of the predominant characteristics in indoor man-made environments, namely, perpendicularity and parallelism of walls, to investigate the ability of our mapping system to capture the true geometry of the mapped environment. Two sides of a particular wall are parallel and two neighbouring walls in a room are usually perpendicular. Thus, the corresponding reconstructed pairs of planes resulting from the indoor mapping should be both parallel or perpendicular as well. Nearly perpendicular pairs of planes with nearby endpoints are labelled as perpendicular edges. Nearly parallel planes at a short distance and with opposite normal vector directions are labelled as parallel edges. Moreover, we make a histogram of the estimated wall thickness. As most walls will have the same thickness in reality, we expect a clear peak in this histogram. The angles between the planes at opposite wall sides and the wall thickness histogram provide a good impression of the ability of the mapping system to maintain an accurate positioning when moving from one room to another.

We assume two walls to be perpendicular when the angle between their 2D edges in the XY plane is between 85° and 95° and their end points, that are close to the intersection point and should represent the corner point, are within 30 cm. Furthermore, the angle between the parallel edges should be below 5° and the distance between them should not exceed 30 cm. The results of the Braunschweig data are shown in Figure 4a,b. In addition, we compute the angle error as the deviation from the perfect parallelism (0°) and perpendicularity (90°) and build histograms of these errors as shown in Figure 4c,d. In reality walls are, of course, never constructed perfectly parallel or perpendicular to other walls, but the deviation from this is expected to be an order of magnitude smaller than the deviations observed in the reconstructed model.

The results show that the angle between two sides of a wall is determined less accurately than that between two perpendicular planes in the same room. This is consistent with the expected performance of SLAM algorithms, as the two walls sides are not seen at the same point of time. Moreover, we note high percentages in the above histograms in bins where the angles deviate by more than 2.5° from their expected values of 0° and 90°. By tracking the source of these high percentages, we found that they mainly originate from incorrectly reconstructed planes, such as open doors. In addition, the measurements of walls' thickness demonstrate that there are two standard types of walls in the building and the standard deviation of the thickness is around 1 cm.

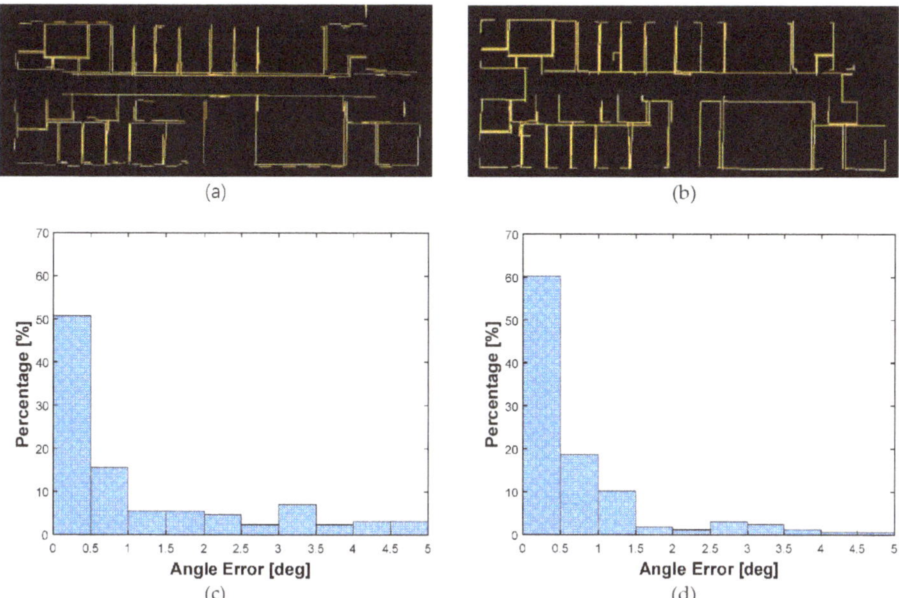

Figure 4. The results of the architectural constraints method. (**a**) All pairs of parallel edges. (**b**) All pairs of perpendicular edges. (**c**) Percentages of angle errors between parallel edges in the range [0°, 5°]. (**d**) Percentages of angle errors between perpendicular edges in the range [85°, 95°].

6.2.2. Evaluation Using a Floor Plan

Nowadays, many buildings have 2D floor plans reflecting the as-planned state from before construction. We investigate the feasibility of using a simple 2D floor plan in analyzing the accuracy of the reconstructed model.

Transformation: As the 2D edges derived from our SLAM-based point clouds and those in the floor plan (see Figure 5) are in two different coordinate systems, we have to register them in the same coordinate system for valid comparison. We use a 2D similarity transformation and estimate the transformation parameters based on a number of manually selected corresponding points. The main goal of this transformation is to identify correspondences between the edges extracted from the point cloud and those in the floor plan. We do not estimate residual distances or angles between an edge in the point cloud and an edge in the floor plan, because we want to keep the comparison process independent of the chosen coordinate systems and quality of the registration. Therefore, we only compare the angles and distances between edges or points extracted from the point cloud to the angles and distances between the corresponding edges or points in the floor plan. The left image in Figure 5 shows the digitized floor plan of the scanned floor.

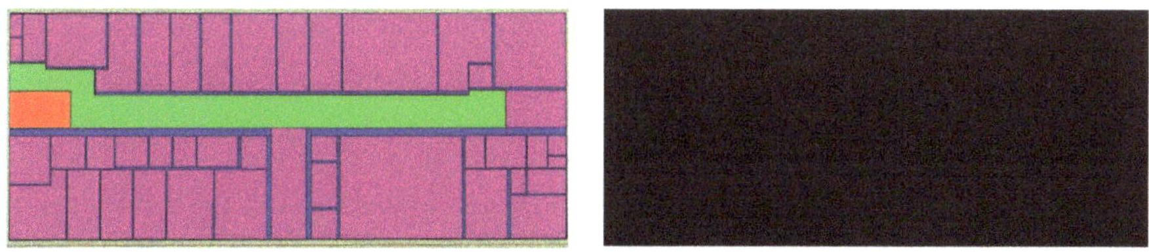

Figure 5. The digitized floor plan (left) and point cloud-based edges for Braunschweig data (right).

Edge Matching: When both sets of edges are registered in the same coordinate system, we start matching the corresponding edges. Firstly, we detect all point cloud-based edges that are expected to belong to a room in the floor plan using a buffer around the room polygon, termed a polygon-buffer. Secondly, we choose which of the detected edges most probably represents the side of a wall in that room using another buffer (30 cm width) around each of the room's edges, termed an edge-buffer as well as the normal vector direction to avoid confusion with edges of the opposite wall side.

In contrast to other objects, walls usually are reconstructed as large and more reliable planes in the SLAM process. Therefore, a further selection based on height information is implemented to keep only edges that most probably belong to walls. We implement the filtering process room by room in order to estimate the floor and ceiling height for each room separately and exclude non-wall edges based on their height. An edge is classified as a window-edge and removed if the corresponding plane is not connected to either the floor or ceiling, and if its height is less than 2 m. Similarly, an edge is classified as a door-edge and removed if the corresponding plane is connected to the floor and its height is less than 2.2 m. Thus, the remaining edges $\boldsymbol{E_{PC}}$, that most probably represent walls in the building, form what we term the PC-based map. Figure 6 shows the resulting PC-based map consists of 144 edges matched to the floor plan edges $\boldsymbol{E_F}$.

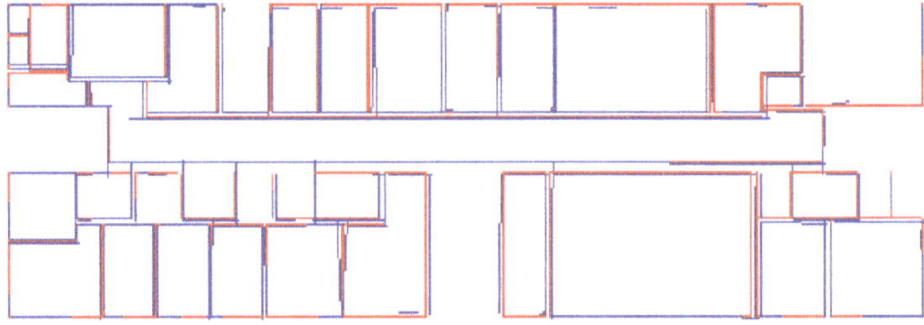

Figure 6. The final point cloud edges E_{PC} (blue) that match the floor plan edges E_F (red).

Analysis: The final step is to pair edges from both edge sets ($\boldsymbol{E_{PC}}$, $\boldsymbol{E_F}$) and form a set of tuples of matched edges. Based on these set of edges, we perform the statistical computations needed to check the accuracy of the PC-based map, and thus the accuracy of generated point clouds.

a) Error in angle in relation to distance

We want to study the impact of distance on the angle errors. Let $\boldsymbol{E_{PC}}$ and $\boldsymbol{E_F}$ be edge sets extracted from the point cloud and floor plan, respectively. Let $(e_{pc}, e_F)_i$ be pairs of matched edges where $i = 1, 2, \ldots, n$ and n is the number of pairs. We pick the ith pair of edges $(e_{pc}, e_F)_i$ and compute the angles $(\alpha_{pc}, \alpha_F)_{ij}$ and distances $(d_{mf})_{ij}$ with respect to all other pairs of edges $(e_{pc}, e_F)_j$ where $(\alpha_F)_{ij}$ is the angle between $(e_F)_i$ and $(e_F)_j$, $\left(\alpha_{pc}\right)_{ij}$ is the angle between $(e_{pc})_i$ and $(e_{pc})_j$, $(d_{mf})_{ij}$ is the distance between midpoints of $(e_F)_i$ and $(e_F)_j$, and $j = i+1, i+2, \ldots, n$. For each pair of edges, we compute the difference between the angle in the point cloud and angle in the floor plan: $(\Delta\alpha = \alpha_{pc} - \alpha_F)_{ij}$. Hence, we obtain $n(n-1)/2$ angle differences and the corresponding distances between the edges $\left(\Delta\alpha, d_{mf}\right)$. We compute these values for the Braunschweig data where 10296 pairs of edges are examined to obtain the results displayed in Figure 7.

The results presented in Figure 7a,b demonstrate that the errors in the angle between point cloud edges are small; approximately 81% are in the range [−1°, 1°]. The two remarkable small peaks in Figure 7b around ±3° refer to some larger errors which may belong to only a few poorly reconstructed planes. Overall, as can be seen in Figure 7a, the distance between edges has no impact on the error in the angle between them.

To identify the poorly estimated outlier edges, we construct Figure 8 in which all edges pairs that have an angle error of 3° or more are presented. The pattern in this figure clearly indicates which edges are mostly involved in edge pairs with large angle errors. We observed five outlier edges and excluded them from the computations in order to obtain a better picture of the potential quality of the system; see Figure 7c,d. Table 1 shows standard deviation values and the number of edge pairs that are involved in the computations for both cases, both before and after excluding outlier edges. We can see that the removal of the outlier edges leads to a 25% decrease in the estimated standard deviation.

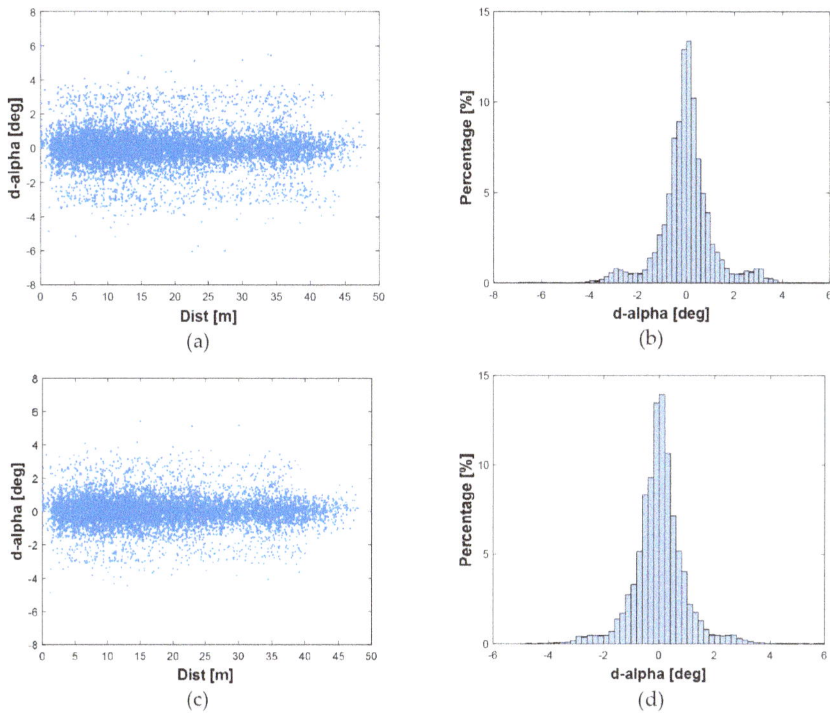

Figure 7. (**a**) Errors in angle as relation of distance. (**b**) Histogram of the percentages of errors. (**c**) Errors in angle as a function of distance after excluding outlier edges. (**d**) Histogram of the percentages of errors after excluding outlier edges.

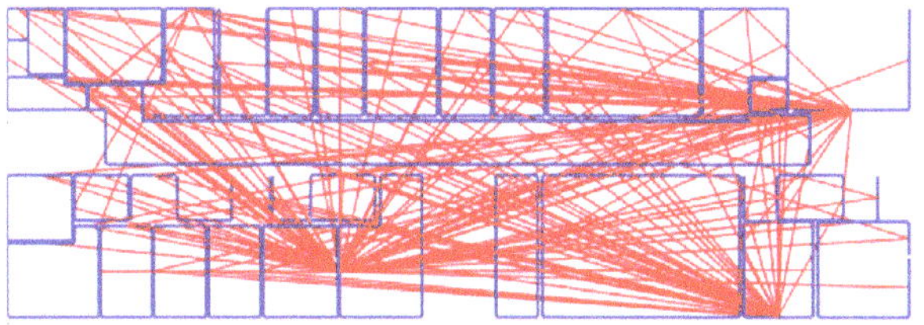

Figure 8. Red lines for edge pairs (blue) that have an angle error of 3° or more.

Table 1. Values of mean, standard deviation, and the number of edges pairs that are involved in the computations for both cases, before and after excluding outlier edges.

	Before	After
Mean	0.01°	0.00°
Std Dev.	1.14°	0.85°
Edge-Pairs Nr	10296	9591

Since the perpendicular and parallel edges are already labelled (Section 3.2), we also computed these values for each type of edge separately. We found that the error in the angle over distance is not related to the attitude of one edge to other. Moreover, we studied the impact of time on the angle errors as the 3D planes are reconstructed over time through applying the SLAM algorithm. However, the results show no relation between time and angle errors. The reason for this is that the operator returned to the same corridor during the data capturing in Braunschweig which in turn leads to frequent loop closures that prevent the errors from accumulating.

b) Error in distance in relation to distance:

Besides the previous computations of angle errors based on the pairs of edges, we compute the distance errors based on pairs of edges' end points. However, because the point cloud-based map is usually incomplete, we find corner points by intersecting the neighbouring edges. We utilize the topology of the floor plan and intersect edges from $\boldsymbol{E_{PC}}$ if their matched floor plan edges e_F are connected.

Let P_{pc} and P_F be intersection points obtained from the floor plan and the point cloud, respectively. Let $(p_{pc}, p_F)_i$ be pairs of points where $i = 1,2, \ldots, n$ and n is the number of pairs. We pick the ith pair of points $(p_{pc}, p_F)_i$ and compute the distances $(d_{pc}, d_f)_{ij}$ with respect to all other pairs of points $(p_{pc}, p_F)_j$ where $(d_f)_{ij}$ is the distance between $(p_F)_i$ and $(p_F)_j$, $(d_{pc})_{ij}$ is the distance between $(p_{pc})_i$ and $(p_{pc})_j$, and j = $i+1, i+2, \ldots, n$. Next, the error in the distances $(\Delta d = d_f - d_{pc})_{ij}$ will be plotted against the distances $(d_f)_{ij}$ to check if the error in distance depends on the distance between floor plan points. Computing the error in distance in this way will remove the systematic error that would otherwise result from errors in the transformation.

From the data used, we obtain 128 corner points leading to 8128 pairs of points involved in the distance errors computation as function of distance. Figure 9 shows that the errors in distance are sometimes quite large (~40 cm). The source of such error is not necessarily the mapping system or the proposed SLAM algorithm, but it could be also the outdated floor plan used. We noted some differences in the width of some walls between the floor plan and the realised construction.

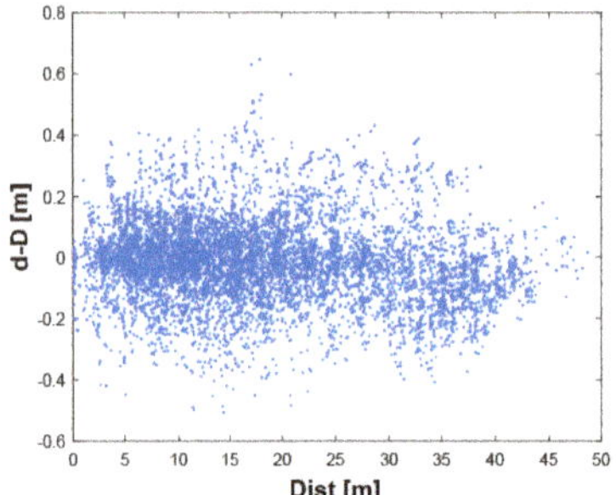

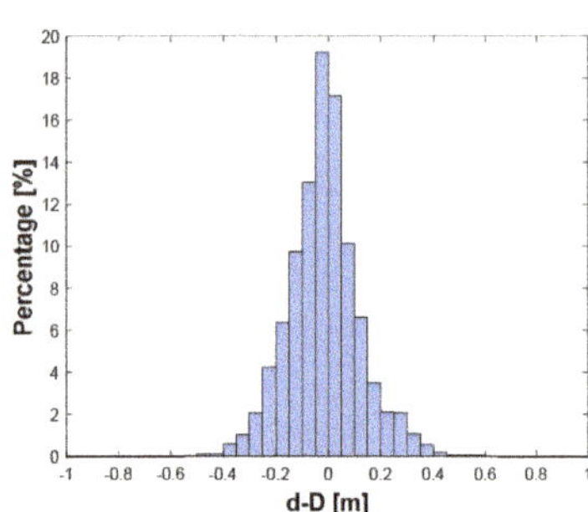

Figure 9. Errors in distance in relation to the distance (left) and Histogram of the percentages of errors (right).

We carried out an analysis similar to that shown in Figure 8 to identify the poorly reconstructed corners (outlier points) using the distance between p_{pc} and p_F. The comparison of the results before and after excluding these outlier points does not show a significant improvement. However, it is not possible to draw the conclusion that these errors are caused by the ground truth model used or the mapping system.

7. Determining Optimal Configuration

As our system is equipped with several sensors, we utilize the proposed evaluation techniques in the previous section to find the optimal configuration.

7.1. Studied Configurations

It is important to avoid occlusion and to acquire sufficient geometrical information of the building to be mapped. As our mapping system is composed of three 2D sensors, we seek the optimal mount of these sensors (LRFs) on the backpack by making use of the proposed evaluation methods above. The optimal sensor configuration is defined through an experimental comparison of different configurations.

The experiments were conducted in an indoor office environment at our university. The three selected rooms, with a corridor in-between, were captured by the system for several possible configurations. A set of criteria was used to select the studied configurations.

In order to see all walls around the system, S_0 is always be horizontal and above the operator's head, and only rotatable around its rotational axis. As the Hokuyo LRF has a 270° field of view, we rotate S_0 around its axis in such a way that the shadow area (gap) points to the left or right side of the operator in order to achieve a good coverage of the surfaces both behind and in front of the system.

In contrast to the top LRF, the left and right LRFs (S_1, S_2) have three rotational degrees of freedom around three axes: the operator's moving direction (X_f), the operator's shoulder axis $\left(Y_f\right)$, and the LRF's rotational axis (see Figure 1). However, some points need to be considered in these rotations. For a good data association, it is better to mount S_1 and S_2 in a way that they scan in two different planes. The oblique scanlines provide a good coverage of walls behind and in front of the system and ensure the overlap with old data when passing through doors and corners. Also, this geometry of the scanlines provides sufficient observations that strengthen the system of equations and make the pose estimation process more robust [21].

Nevertheless, we found empirically that a small angle of rotation around the shoulder axis leads to a better coverage of the surfaces around the system, while a wide angle of rotation may lead to a loss of coverage of the floor and ceiling. Thus, this reduces the estimability of the system's movement in z direction. To avoid occlusion by the operator's body, we should have forward-slanted scanlines.

Based on the aforementioned criteria, a set of orientation configurations for the two LRFs as listed in Table 2 was tested. Specifically, the table lists the rotations of LRFs S_1 and S_2, namely, θ_1 and θ_2, around the moving direction (X_f) and the shoulder axis (Y_f), respectively. In addition, the LRFs are rotated around their axis to ensure that the gap of one points to the floor and the gap of other points to the ceiling. This provides as many observations as possible on all surfaces (walls, floor, and ceiling) in order to position the backpack.

Table 2. Tested configurations are defined by the first five columns (explained in Section 7.1) and the results from the three distinct evaluation techniques are listed in the last five columns (explained in Section 7.2). The resulting values are highlighted with colors, as follows. For RMSE: < 0.80° green, [0.80°, 1°] yellow, and ≥ 1° red. For λ expressed in percentages: [85, 100] green, [70, 85] yellow, and ≤ 70 red. The last column represents the completeness state of the point cloud in three different cases: existence of large gaps, existence of small gaps, or almost no gaps.

Orientation Configuration of LRFs	θ_1 [deg]		θ_2 [deg]		Arch. Constraints		Floor Plan		Gap
	S_1	S_2	S_1	S_2	RMSE [deg]	λ %	RMSE [deg]	λ %	Size
1	20 [1]	20 [1]	20 [2]	20 [2]	1.57	70	0.79	67	Small
2	30	30	30	30	0.81	85	0.40	85	No
3	40	40	40	40	0.94	83	0.56	76	No
4	15	20	30	30	0.94	84	0.57	73	Small
5	30	30	20	20	0.76	90	0.51	79	No
6	20	20	40	40	0.54	95	0.71	83	Small
7	40	40	20	20	0.77	86	0.57	74	No

Table 2. *Cont.*

Orientation Configuration of LRFs	θ_1 [deg]		θ_2 [deg]		Arch. Constraints		Floor Plan		Gap
	S_1	S_2	S_1	S_2	RMSE [deg]	λ %	RMSE [deg]	λ %	Size
8	20	20	30	30	1.38	83	0.82	72	Small
9	0	20	20	90	1.25	81	0.76	70	Large
10	0	30	30	90	0.65	99	0.54	87	Large
11	0	40	40	90	0.73	91	0.52	80	Large

[1] The LRF is rotated 20° around the moving direction and 20° is the angle between its scanline plane and the horizontal plane $(XY)_f$. [2] The LRF is rotated 70° (i.e., 90° − 20°) around the shoulder axis, but 20° is the angle between its scanline plane and the vertical plane $(YZ)_f$.

7.2. *Experimental Comparison of Configurations*

For each suggested configuration, we find the relative orientation of the slanted LRFs (S_1, S_2) with respect to the horizontal LRF (S_0) with the registration procedures explained in Section 5. Next, we scan the test area with all possible configurations and each time, we obtain a dataset of point cloud and 3D reconstructed planes.

7.2.1. Accuracy

Evaluation Using Architectural Constraints: The test area consists of rectangular rooms as can be seen in Figure 10. Regarding the analysis of the perpendicularity, the basic approach explained in Section 6.2.1 is extended to involve all walls that should be perpendicular to each other in the building and not only the neighboring walls in a room. Similarly, the algorithm looks for all walls that should be parallel to each other in the building and not only both sides of a wall. Next, we compute the Root Mean Square Error (RMSE) of the computed angles to estimate the deviations from the perfect perpendicularity and parallelism, respectively.

Evaluation Using the Floor Plan: The floor plan of the test area is available as a 2D CAD drawing; see Figure 10. The evaluation of errors in angles and distances is conducted as described in Section 6.2.2 for the Braunschweig dataset.

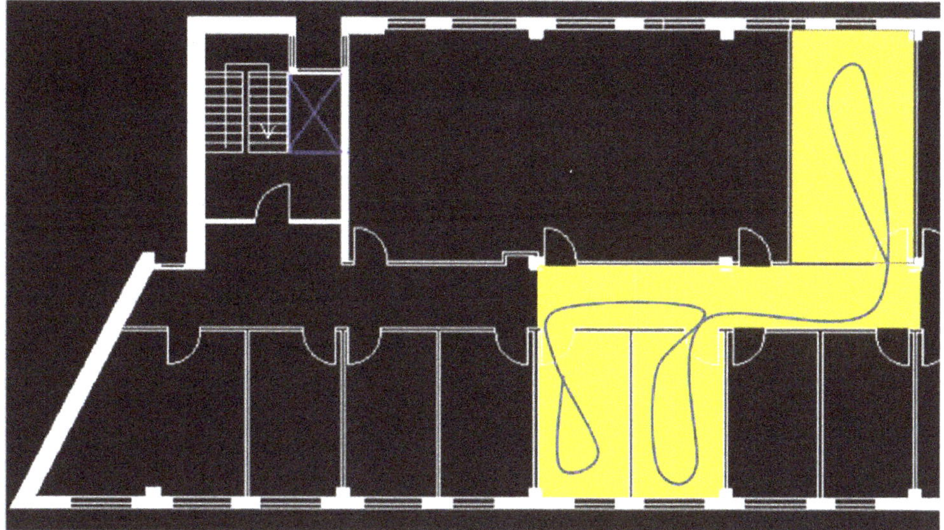

Figure 10. Part of the 2D CAD drawing of the 3rd floor in the Citadel building with highlighted scanning area (yellow) and the trajectory followed (blue).

All the resulting statistical values for both architectural constraints and using the floor plan are listed in Table 2. The comparison between the listed configurations is done using these values which reflect the reconstruction accuracy. The sixth column presents the sum of RMSE values computed in the two evaluation techniques introduced before (perpendicularity, parallelism) for each configuration (see Section 6.2.1). Also, for each configuration, we count the points that have an angle error of less than 1° and divide this amount by the total number of points, yielding a rate of angle errors λ. We take the average rate in these two techniques (perpendicularity, parallelism) and present the results in the seventh column. Similarly, the eighth and ninth columns in Table 2 demonstrate the computed RMSE and λ computed for each configuration with the availability of floor plan. These statistical values are computed in order to get an overall impression of the reconstruction accuracy of each configuration, and thus it helps in decision making on the optimal configuration.

The sixth and seventh columns in the Table 2 show that the configurations (1, 8, 9) have clearly larger deviations than the other configurations, therefore being less accurate in capturing the geometry of the building interiors. The evaluation using the floor plan, as shown in the eighth and ninth columns, confirms this conclusion. RMSE values for parallelism indicate that the configurations (2–7, 10, 11) are the most accurate in maintaining an accurate localization when moving from one room to another. The common characteristic of these configurations is that they have at least one of the two rotation angles in the range of 30° to 40°.

7.2.2. Completeness of Data Capturing

We visually inspect the completeness of the captured data in which walls, floor, and ceiling are recorded. The process relies on a simulated point cloud representing a scan of a virtual corridor (loop) and generated for each configuration. Figure 11a shows the 3D model of this virtual area and Figure 11b the point cloud of configuration 6 where the colors relate to the three different LRFs and the white polyline represents the followed trajectory. The simulated point cloud generating process assumes the operator walks around the corridor in an anti-clockwise direction starting from the middle of the corridor.

The analysis process is based on a set of point clouds corresponding to all suggested configurations to compare the areas covered by points and find which configuration provides the better coverage. We want to investigate whether a more accurate configuration in geometry reconstruction provides a more complete point cloud of the scene.

Overall, configurations (1–8) appear to give a good coverage. However, the configurations (2, 3, 5, 7) provide the most complete point cloud of the scene, while the LRFs with the other configurations miss the lower/upper part of the wall in the scanning geometry when the system turns around the corner. An example of a configuration (6) that results in an incomplete point cloud is shown in Figure 11b. This configuration misses a part of the wall close to the corner (see Figure 11c). Figure 11d shows the data recorded in one scan line of each LRF in the configuration 6. The tenth column in Table 2 demonstrates that the configurations (9, 10, 11) have the largest gaps in their point clouds. The common characteristic of these configurations is that only one of the slanted LRFs is scanning the walls both right and left of the system.

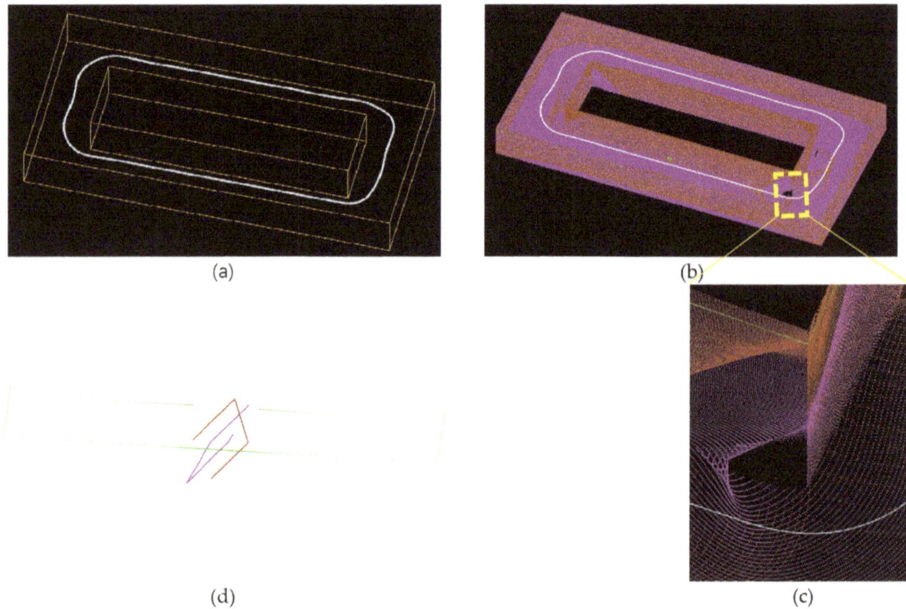

Figure 11. Simulation data. (**a**) 3D model of the virtual corridor used as a test area with the trajectory followed (white). (**b**) The resulting point cloud of the test area for the configuration 6 without ceiling's points. The yellow dashed rectangle shows an example of a gap on the wall around the corner (**c**). (**d**) The simulated geometry of LRFs' scanlines for the configuration 6 in which the colors relate to the three different LRFs (S_0 *green*, S_1 *purple*, S_2 *red*).

7.3. Discussion of Configuration Experiments

The results of Section 7.1 and Section 7.2 lead to the conclusion that the more accurate configuration in geometry reconstruction does not necessarily provides a more complete point cloud of the scene, and vice versa. Although the configurations (10, 11) show a better performance than configurations (2, 8) in terms of the reconstruction accuracy, they provide a less complete point cloud.

Finally, our system has the top LRF mounted horizontally and on a level that the environment is not occluded by the operator's head. The shadow area of this LRF is pointed to the walls to the left or right of the operator. We discovered that the other two LRFs should be scanning the surfaces parallel to the moving direction e.g., walls both right and left of the system. Also, we found out that these LRFs should be rotated not only about the shoulder axis (Y_f), but at least one of them should also be rotated around the moving direction (X_f) by an angle in the range of 30° to 40°. Moreover, the results revealed that determining where the LRFs' data gap is pointing at plays a pivotal role in the completeness of the resulting point clouds. The best coverage is achieved when the gap of one slanted LRF is aimed at the floor and the gap of other is aimed at the ceiling.

8. Conclusions

We presented the design, calibration and registration methods, and performance analysis of a multi-sensor backpack indoor mobile mapping system (ITC-IMMS). We have proposed and presented several evaluation techniques for the investigation of the system's ability to acquire geometric information of an interior environment. Evaluations can also be performed when there is no ground truth model or only a floor plan available. If the floor plan is outdated, this will usually surface as a large error in the evaluation and can therefore be identified as outlier. The results on the Braunschweig data showed some differences in the width of some walls between the floor plan and the realized construction. Such changed walls can then be removed from the map. The proposed evaluation methods are not limited to our mapping system.

The experimental results showed the ability of ITC-IMMS to map an office building with an angle error within 1.5° between its planar surfaces, and the precision in generating the width of wall is around 1 cm. Although we did not consider the errors in the outdated floor plan, the point cloud-based map shows a good internal consistency.

We have carried out an experimental comparison of selected configurations to find the best configuration by studying the properties of 3D planes and point clouds reconstructed with these configurations. The selection of the optimal sensor configuration was built in terms of data occlusion, the success of the algorithm, and the accuracy and completeness of the resulting map and point cloud. In order to see all walls around the system, we left the top LRF mounted horizontally in the optimal configuration for the backpack system and on a level that the environment cannot be occluded by the operator's head. To achieve a good coverage of the surfaces both behind and in front of the system, it must be rotated around its axis to locate the shadow area on the walls to the left or right of the operator. The other two LRFs should be scanning in a plane perpendicular to the moving direction. Also, they should be rotated not only about the shoulder axis, but at least one of them should also be rotated about the moving direction by an angle in the range of 30° to 40°. The gap of one should be pointed at the floor, while the gap of other should be pointed at the ceiling. In this way, the system achieves improved coverage of the environment and ensures a good data association when passing through doors and corners, and thus has a robust estimation of the plane and pose parameters.

Nevertheless, the analysis of the system's performance may be slightly different in another indoor environment with much larger or smaller spaces. For such environments we would then need to repeat the experiments of Section 7, but we do not expect this will be necessary for many buildings.

In the near future, we plan to expand the scope of application of the current system and SLAM algorithm to include more complex situations such as staircases and fancy architecture (e.g., slanted walls, round walls, non-horizontal floor). To do that, we will integrate IMU data in the local pose estimation and as consequence, we can use higher order splines to predict future poses. We anticipate that this integration will lead to a better hypothesis generation of planar structures and an optimal estimation for the whole trajectory.

Author Contributions: S.K. and G.V. jointly developed and tested the algorithms and wrote the paper. G.V. implemented the registration and calibration methods. M.P. implemented the segmentation algorithm and contributed to the backpack construction, marker calibration, and data collection. This research was done while M.P. was working at the Faculty of Geo-Information Science and Earth Observation, University of Twente. All the co-authors, M.P., S.H., and V.L. Lehtola supervised the work and reviewed the paper drafts. Data curation, S.K.; Investigation, S.K.; Methodology, S.K. and G.V.; Project administration, G.V.; Software, S.K., G.V. and M.P.; Supervision, G.V., M.P., S.H. and V.L.; Visualization, S.K.; Writing original draft, S.K.; Writing review & editing, G.V., M.P., S.H. and V.L.

Funding: This research received no external funding.

Acknowledgments: The authors would like to thank Gerben te Riet o/g Scholten from Robotics and Mechatronics department at Twente University for building the prototype system. Also, the authors wish to thank -Ing. Markus Gerke for the opportunity to collect the data used in the experiments in the building of TU Braunschweig, Germany.

Conflicts of Interest: The authors declare no conflict of interest.

References

1. Biber, P.; Andreasson, H.; Duckett, T.; Schilling, A. 3D modeling of indoor environments by a mobile robot with a laser scanner and panoramic camera. In Proceedings of the 2004 IEEE/RSJ International Conference on Intelligent Robots and Systems (IROS), Las Vegas, NV, USA, 28 September–2 October 2004; Volume 4, pp. 3430–3435.
2. Borrmann, D.; Elseberg, J.; Lingemann, K.; Nüchter, A.; Hertzberg, J. Globally consistent 3D mapping with scan matching. *Rob. Auton. Syst.* **2008**, *56*, 130–142. [CrossRef]
3. Henry, P.; Krainin, M.; Herbst, E.; Ren, X.; Fox, D. RGB-D Mapping: Using Depth Cameras for Dense 3D Modeling of Indoor Environments. In *Experimental Robotics, Springer Tracts in Advanced Robotics*; Springer: Berlin/Heidelberg, Germany, 2014; Volume 79, pp. 477–491.
4. Lehtola, V.V.; Kaartinen, H.; Nüchter, A.; Kaijaluoto, R.; Kukko, A.; Litkey, P.; Honkavaara, E.; Rosnell, T.; Vaaja, M.T.; Virtanen, J.P.; et al. Comparison of the selected state-of-the-art 3D indoor scanning and point cloud generation methods. *Remote Sens.* **2017**, *9*, 796. [CrossRef]
5. Karam, S.; Peter, M.; Hosseinyalamdary, S.; Vosselman, G. An Evaluation Pipeline for Indoor Laser Scanning Point Clouds. *ISPRS Ann. Photogramm. Remote Sens. Spat. Inf. Sci.* **2018**, *4*, 85–92. [CrossRef]
6. Bosse, M.; Zlot, R.; Flick, P. Zebedee: Design of a spring-mounted 3-D range sensor with application to mobile mapping. *IEEE Trans. Robot.* **2012**, *28*, 1104–1119. [CrossRef]
7. Viametris iMS3D-VIAMETRIS. Available online: http://www.viametris.com/products/ims3d/ (accessed on 20 November 2018).
8. Trimble Applanix: TIMMS Indoor Mapping. Available online: https://www.applanix.com/products/timms-indoor-mapping.htm (accessed on 20 November 2018).
9. Wen, C.; Pan, S.; Wang, C.; Li, J. An Indoor Backpack System for 2-D and 3-D Mapping of Building Interiors. *IEEE Geosci. Remote Sens. Lett.* **2016**, *13*, 992–996. [CrossRef]
10. Blaser, S.; Cavegn, S.; Nebiker, S. Development of a portable high performance mobile mapping system using the robot operation system. *ISPRS Ann. Photogramm. Remote Sens. Spat. Inf. Sci.* **2018**, *4*, 13–20. [CrossRef]
11. Naikal, N.; Kua, J.; Chen, G.; Zakhor, A. Image Augmented Laser Scan Matching for Indoor Dead Reckoning. In Proceedings of the 2009 IEEE/RSJ International Conference on Intelligent Robots and Systems, St. Louis, MO, USA, 10–15 October 2009; pp. 4134–4141.
12. Lehtola, V.V.; Virtanen, J.-P.; Vaaja, M.T.; Hyyppä, H.; Nüchter, A. Localization of a mobile laser scanner via dimensional reduction. *ISPRS J. Photogramm. Remote Sens.* **2016**, *121*, 48–59. [CrossRef]
13. Leica Geosystems. Leica Pegasus: Backpack. Available online: https://leica-geosystems.com/ (accessed on 4 February 2019).
14. Chen, G.; Kua, J.; Shum, S.; Naikal, N.; Carlberg, M.; Zakhor, A. Indoor localization algorithms for a human-operated backpack system. In *Proceedings 3D Data Processing, Visualization, and Transmission*; University of California: Berkeley, CA, USA, 2010; pp. 15–17.
15. Liu, T.; Carlberg, M.; Chen, G.; Chen, J.; Kua, J.; Zakhor, A. Indoor localization and visualization using a human-operated backpack system. In Proceedings of the 2010 International Conference on Indoor Positioning and Indoor Navigation, Zurich, Switzerland, 15–17 September 2010; pp. 1–10.
16. Thomson, C.; Apostolopoulos, G.; Backes, D.; Boehm, J. Mobile Laser Scanning for Indoor Modelling. *ISPRS Ann. Photogramm. Remote Sens. Spat. Inf. Sci.* **2013**, *2*, 289–293. [CrossRef]
17. Maboudi, M.; Bánhidi, D.; Gerke, M. Evaluation of Indoor Mobile Mapping Systems. In Proceedings of the GFaI Workshop 3D North East 2017, Berlin, Germany, 7–8 December 2017; pp. 125–134.
18. Maboudi, M.; Bánhidi, D.; Gerke, M. Investigation of geometric performance of an indoor mobile mapping system. *ISPRS Arch. Photogramm. Remote Sens. Spat. Inf. Sci.* **2018**, *42*, 637–642. [CrossRef]

19. Tran, H.; Khoshelham, K.; Kealy, A. Geometric comparison and quality evaluation of 3D models of indoor environments. *ISPRS J. Photogramm. Remote Sens.* **2019**, *149*, 29–39. [CrossRef]
20. Vosselman, G. Design of an indoor mapping system using three 2D laser scanners and 6 DOF SLAM. *ISPRS Ann. Photogramm. Remote Sens. Spat. Inf. Sci.* **2014**, *2*, 173–179. [CrossRef]
21. GeoSLAM GeoSLAM—The Experts in 'Go-Anywhere' 3D Mobile Mapping Technology. Available online: https://geoslam.com/ (accessed on 25 November 2018).
22. VIAMETRIS iMS2D-VIAMETRIS. Available online: http://www.viametris.com/products/ims2d/ (accessed on 4 December 2018).
23. Kim, B.K. Indoor localization and point cloud generation for building interior modeling. In Proceedings of the IEEE RO-MAN, Gyeongju, Korea, 26–29 August 2013; pp. 186–191.
24. Filgueira, A.; Arias, P.; Bueno, M. Novel inspection system, backpack-based, for 3D modelling of indoor scenes. In Proceedings of the 2016 International Conference on Indoor Positioning and Indoor Navigation (IPIN), Alcalá Henares, Spain, 4–7 October 2016; pp. 4–7.
25. Zhang, J.; Singh, S. LOAM: Lidar Odometry and Mapping in Real-Time. In Proceedings of the Robotics: Science and Systems Foundation, Berkeley, CA, USA, 12–16 July 2014. [CrossRef]
26. Lagüela, S.; Dorado, I.; Gesto, M.; Arias, P.; Gonz, D.; Lorenzo, H. Behavior analysis of novel wearable indoor mapping system based on 3D-SLAM. *Sensors* **2018**, *18*, 766. [CrossRef] [PubMed]
27. Sirmacek, B.; Shen, Y.; Lindenbergh, R.; Zlatanova, S.; Diakite, A. Comparison of ZEB1 and Leica C10 indoor laser scanning point clouds. *ISPRS Ann. Photogramm. Remote Sens. Spat. Inf. Sci.* **2016**, 143–149. [CrossRef]
28. Hokuyo Ltd. Hokuyo Automatic Co., Ltd. Available online: https://www.hokuyo-aut.jp/ (accessed on 4 December 2018).
29. Peter, M.; Jafri, S.R.U.N.; Vosselman, G. Line segmentation of 2D laser scanner point clouds for indoor SLAM based on a range of residuals. *ISPRS Ann. Photogramm. Remote Sens. Spat. Inf. Sci.* **2017**, *4*, 363–369. [CrossRef]
30. Park, C.-S.; Kim, D.; You, B.-J.; Oh, S.-R. Characterization of the Hokuyo UBG-04LX-F01 2D laser rangefinder. In Proceedings of the 19th International Symposium in Robot and Human Interactive Communication, Viareggio, Italy, 13–15 September 2010; pp. 385–390.
31. Niekum, S. ar_track_alvar-ROS Wiki. 2013. Available online: http://wiki.ros.org/ar_track_alvar (accessed on 25 November 2018).
32. Fernández-Moral, E.; Arévalo, V.; González-Jiménez, J. Extrinsic calibration of a set of 2D laser rangefinders. In Proceedings of the 2015 IEEE International Conference on Robotics and Automation (ICRA), Seattle, WA, USA, 26–30 May 2015; pp. 2098–2104. [CrossRef]
33. Choi, D.G.; Bok, Y.; Kim, J.S.; Kweon, I.S. Extrinsic calibration of 2D laser sensors. In Proceedings of the 2014 IEEE International Conference on Robotics and Automation (ICRA), Hong Kong, China, 31 May–7 June 2014; pp. 3027–3033. [CrossRef]

Article

Automatic Extrinsic Self-Calibration of Mobile Mapping Systems Based on Geometric 3D Features

Markus Hillemann [1,2,*], Martin Weinmann [1], Markus S. Mueller [1] and Boris Jutzi [1]

1 Institute of Photogrammetry and Remote Sensing, Karlsruhe Institute of Technology, 76131 Karlsruhe, Germany
2 Fraunhofer Institute of Optronics, System Technologies and Image Exploitation IOSB, 76275 Ettlingen, Germany
* Correspondence: markus.hillemann@kit.edu

Received: 24 July 2019; Accepted: 17 August 2019; Published: 20 August 2019

Abstract: Mobile Mapping is an efficient technology to acquire spatial data of the environment. The spatial data is fundamental for applications in crisis management, civil engineering or autonomous driving. The extrinsic calibration of the Mobile Mapping System is a decisive factor that affects the quality of the spatial data. Many existing extrinsic calibration approaches require the use of artificial targets in a time-consuming calibration procedure. Moreover, they are usually designed for a specific combination of sensors and are, thus, not universally applicable. We introduce a novel extrinsic self-calibration algorithm, which is fully automatic and completely data-driven. The fundamental assumption of the self-calibration is that the calibration parameters are estimated the best when the derived point cloud represents the real physical circumstances the best. The cost function we use to evaluate this is based on geometric features which rely on the 3D structure tensor derived from the local neighborhood of each point. We compare different cost functions based on geometric features and a cost function based on the Rényi quadratic entropy to evaluate the suitability for the self-calibration. Furthermore, we perform tests of the self-calibration on synthetic and two different real datasets. The real datasets differ in terms of the environment, the scale and the utilized sensors. We show that the self-calibration is able to extrinsically calibrate Mobile Mapping Systems with different combinations of mapping and pose estimation sensors such as a 2D laser scanner to a Motion Capture System and a 3D laser scanner to a stereo camera and ORB-SLAM2. For the first dataset, the parameters estimated by our self-calibration lead to a more accurate point cloud than two comparative approaches. For the second dataset, which has been acquired via a vehicle-based mobile mapping, our self-calibration achieves comparable results to a manually refined reference calibration, while it is universally applicable and fully automated.

Keywords: mobile mapping; laser scanning; self-calibration; 3D point clouds; geometric features

---

1. Introduction

Mobile Mapping is an efficient technology to acquire spatial data of the environment. The spatial data is further processed to 3D city models, building models or models of indoor environments which are nowadays an essential data source for applications in crisis management, civil engineering or autonomous driving. Depending on the scale of the environment, mobile platforms like airplanes, cars, Unmanned Areal Vehicles (UAVs) or mapping backpacks are considered. The mobile platform is typically equipped with one or more mapping sensors. A widely utilized mapping sensor is a laser scanner, also known as Light Detection And Ranging (LiDAR) sensor, since it acquires accurate and dense spatial data of the environment in form of a 3D point cloud.

The quality of the 3D point cloud captured with a Mobile Mapping System (MMS) is limited by the accuracy of the mapping sensor itself and mainly three more components: The estimation of the *pose* of the MMS, the *intrinsic calibration* of the individual sensors and the *extrinsic calibration* of the MMS.

The *pose* describes the position and orientation of the MMS with respect to a superordinate coordinate frame. It is used to register a single scan to a common point cloud. The pose is estimated in a specific frame which we call the navigation frame. The pose estimation task can be solved by using a technology that observes the sensor system, like for example the Global Navigation Satellite System (GNSS), or by using on-board sensors and an odometry- or Simultaneous Localization and Mapping (SLAM)-algorithm.

The *intrinsic calibration* is the process of estimating the interior calibration parameters of the individual sensors. In case of a laser scanner, these are for example a range finder offset and beam direction influences. As the impact of the interior calibration parameters on the accuracy of the point cloud is typically lower than the impact of the other mentioned components, the interior calibration parameters of mapping sensors are often neglected. However, for highest accuracy requirements, the interior calibration parameters must be taken into account.

The objective of the *extrinsic calibration* of a MMS is to find a rigid transformation from the navigation frame to the frame of the mapping sensor. The extrinsic calibration is also known as determining the *relative pose* between the two sensors.

For the sake of clarity, in this work we focus on estimating the extrinsic calibration parameters. For simplicity, we will call this process the calibration from now on.

Figure 1 visualizes the effect of an inaccurate calibration on the basis of a point cloud resulting from Mobile Mapping within an indoor scenario. Due to the use of a line scanning device as mapping sensor and heterogeneous movements of the sensors, the points are not uniformly sampled. The point cloud on top is generated using an insufficient calibration with an error of five degrees in the pitch angle. This leads to a systematically distorted point cloud. In contrast, the point cloud at the bottom is calibrated using a target calibration approach for cameras and laser scanners [1–3].

Figure 1. Point cloud as a result of a mobile mapping in an indoor environment. Due to the use of a line scanning device and non-uniform movements of the mobile mapping platform, the points are inhomogeneously sampled. The color of each point illustrates its height. Top: Inaccurate guess of the extrinsic calibration parameters. Bottom: Calibrated using the Robust Automatic Detection in Laser Of Calibration Chessboards (RADLOCC) [1–3], which is a target calibration approach for cameras and laser scanners.

In general, the calibration approaches can be arranged into three categories: (i) Calibration approaches which are performed in the laboratory, (ii) calibration approaches which use specific, artificial targets and (iii) self-calibrations. Self-calibrations are the most practical and time-saving approaches, as it is possible to calibrate the system based on the data that shall be collected anyways. There is no need to prepare the environment before data collection, to design artificial targets or to use additional sensors in self-calibration processes. Moreover, self-calibrations can be more accurate because changes of the calibration parameters during data collection can be taken into account.

In this paper, we propose a self-calibration approach which is based on geometric features. Therefore, we also provide a brief introduction to these features.

1.1. Geometric Features

Geometric features are often used to automatically analyze 3D point clouds. The most popular task in this context is classification [4,5], but geometric features have also been used for coarsely registering 3D point clouds [6] and retrieving objects in 3D point clouds [7]. Particularly, the geometric features derived from eigenvalues of the 3D structure tensor have proven to be descriptive [4]. As these features represent a specific property of the local neighborhood by a single value, they are rather intuitive. The three features of linearity, planarity and sphericity enable for example to quickly identify the primary dimensionality within a local neighborhood.

In this paper, we use such geometric features to evaluate the quality of a point cloud and thus to evaluate the calibration of a MMS with a laser scanner and a pose estimation sensor.

1.2. Contributions and Structure

With this paper, we introduce a novel extrinsic self-calibration approach for Mobile Mapping Systems. In this context, we also analyze the suitability of different cost functions for the extrinsic self-calibration task. The MATLAB code of our self-calibration framework is publicly available (https://github.com/markushillemann/FeatCalibr).

Our self-calibration approach combines all of the following advantages simultaneously: (i) the estimation of all extrinsic calibration parameters between an arbitrary pose estimation sensor and a mapping sensor, (ii) a universal solution suitable for different kinds of mapping sensors (e.g., 2D and 3D laser scanners), (iii) no assumptions about the environment and no requirements for artificial targets or additional sensors, (iv) a solution robust to initial calibration errors due to a multi-scale approach.

The paper is organized as follows: Section 2 provides an overview of related work in the field of extrinsic self-calibration of MMSs and in the field of geometric point cloud features. The methodology of our calibration approach, which is based on a nonlinear least squares optimization of the calibration parameters, is presented in Section 3. Section 4 describes the synthetic and real datasets and the experiments that have been performed on these datasets. We examine if some of the potentially usable geometric features are better suited for the use as cost function for our self-calibration than others and compare the results to a self-calibration which uses a cost function from the literature [8]. Furthermore, we focus on questions about the applicability to different sensors and environments and prove the suitability to real sensor data. In Section 5, we elaborate on the results of the experiments and derive under which conditions the self-calibration achieves good results. Finally, Section 6 provides concluding remarks as well as an outlook on future work.

2. Related Work

As the use of geometric features in terms of a self-calibration is rare, we split the related work section. Section 2.1 presents the related work in terms of relevant calibration approaches. These approaches usually distinguish between two types of mapping sensors: 2D and 3D laser scanners. 2D laser scanners measure the range to the illuminated surface while scanning the surface linearly. Therefore, 2D laser scanners are also known as line laser scanners. The scanning is usually performed sequentially, or more rarely synchronously, by utilizing linearly arranged fiber optics. An example of a 2D laser scanner is the *Hokuyo UTM-30LX-EW* (Hokuyo Automatic Co., Ltd., Osaka, Japan). 3D laser scanners are in principle 2D laser scanners combined with a rotating platform for scanning perpendicular to the previous mentioned scanning direction. An example of a 3D laser scanner is the *Velodyne HDL-64E* (Velodyne Lidar, Inc., San Jose, CA 95138 USA).

The literature offers many calibration approaches which require a 3D laser scanner. As our approach is also suited to estimate the extrinsic calibration parameters of 2D laser scanners and as this is the more restrictive sensor, we focus this section on approaches that can handle both types of sensors. As geometric features are a fundamental component of our calibration approach, we also present the related work of these features in Section 2.2. Most of the previous works on geometric features are linked to classification tasks.

2.1. Extrinsic Calibration of Mobile Mapping Systems

Many previous works handle the extrinsic calibration of MMSs with specific combination of sensors. Especially for the extrinsic calibration of MMSs with a laser scanner and a camera, numerous approaches have been presented in literature. There are target calibrations that require special calibration objects like triangles, folding patterns, cubes or more complex calibration objects consisting of multiple connected planes [9–15] and widely used approaches with the need of planar checkerboards [1–3,16–18]. The latter approaches are practical because the utilized checkerboards can also be used to perform the intrinsic calibration of the camera. All of these methods are designed on the idea of minimizing the distance of points to the surface of the calibration target. The calibration parameters can be carried out based on corresponding observations of the two sensors. However, these approaches work exclusively with the specific combination of sensors from a laser scanner and a camera. Moreover, they require an at least partially overlapping field of view of the laser scanner and the camera. Calibration methods which work with arbitrary pose estimation sensors and which are independent of the position and orientation of the sensors with respect to each other are rarely addressed in the literature.

Using GNSS/INS for pose estimation and a laser scanner as mapping sensor, it is not possible to exploit corresponding observations. This sensor combination is typical for vehicle-based mobile mapping. The GNSS/INS system is utilized for the registration and georeferencing of the acquired point clouds. Using this sensor combination for a calibration is equivalent to using an arbitrary six degrees of freedom (6-DOF) pose estimation sensor. Thus, calibration methods that are similar to the presented method are often connected to these fields of use.

Several calibration approaches exist that consider artificial targets. These artificial targets can be ground control points signalized with retro-reflecting material [19], reference geometries like planar surfaces [20,21] or more special geometries like a half-circle surface [22]. The artificial targets must either be arranged in a specific way [20], or the positions or even the 6-DOF poses have to be provided [21]. Moreover, the artificial targets have to be identified in the point cloud which can be time-consuming and error-prone.

Closer to our work is the *rigorous approach* [23]. It is, like our approach, not restricted to GNSS/INS-derived pose but adaptable to any kind of 6-DOF pose estimation sensor. The basic assumption of the rigorous approach is that the point cloud contains groups of points which can be conditioned to lie on surfaces of a known form, particularly on a plane. The position and orientation of these planes are estimated together with the rotation parameters of the extrinsic calibration and one parameter for a constant range finder offset in a Gauss-Helmert adjustment model. The residual function they use is, straightforwardly, formulated as the sum of the distances of each selected point to its corresponding plane. In contrast to our approach, the translation parameters of the extrinsic calibration are assumed to be known and must therefore be determined separately. Moreover, the form of the surfaces has to be determined and adapted manually in the algorithm if the calibration environment does not fit the model assumption. Finally, the identification of the points lying on the specific surfaces is an additional, time-consuming processing step.

The *quasi-rigorous approach* [24] is another choice for extrinsic calibration that does not rely on artificial targets. This approach requires laser scans and associated poses of the sensor system, as well as initial calibration parameters. The quasi-rigorous approach uses at least two overlapping laser scans to optimize the initial calibration parameters. Similar to the approaches described so far, the quasi-rigorous approach minimizes the sum of distances of a point from the first laser scan to a corresponding plane from the second laser scan, i.e., a point-to-plane metric is used. This corresponding plane is a triangular patch which is derived from a Triangulated Irregular Network (TIN) generation of the second laser scan. Consequently, there is no need to use artificial targets. In contrast to the quasi-rigorous approach, we do not use a point-to-plane metric for minimization, but geometric 3D features. Registration approaches based on Iterative Closest Point (ICP) [25,26] are known for not being robust to initial misalignments. Applied to the calibration, this means that the initial calibration parameters already have to be known very precisely so that the optimization converges to the global minimum.

Recently, the quasi-rigorous approach has successfully been used for UAV-based mobile mapping [27]. As it is important for the interpretation of the results from our real data experiments in Section 4.2.2, we briefly summarize some details of this work: For a typical UAV-based mobile mapping, it is possible to robustly estimate five of the six parameters of the calibration solely based on the flight data. Therefore, the scene has to have a horizontal and vertical expansion and the scene has to be flown over several times in different directions and altitudes. However, the vertical component of the translation, can only be estimated using vertical control, i.e., in form of horizontal planar patches with known altitude. For a more detailed discussion about the minimal requirements on the poses or the environment in context of calibration, the reader is referred to the quasi-rigorous approach [27].

A self-calibration approach that has no assumptions about the environment has been presented by [8]. Instead, a cost function which is based on the Rényi quadratic entropy (*RQE*) is minimized to optimize initial calibration parameters. The cost function is defined as

$$f_{RQE} = -\sum_{i=1}^{N}\sum_{j=1}^{N} G(X_i - X_j, \Sigma), \tag{1}$$

where $X_i, X_j \in \mathbb{R}^3$ are points of the point cloud, N is the number of points and $G(\mu, \Sigma)$ is a Gaussian function with mean μ and covariance Σ. A derivation of the cost function is not repeated here and the reader is referred to [8]. For the covariance $\Sigma = \sigma^2 I$ the authors use an isotropic kernel I and the same variance σ^2 for each dimension. This reduces the tuning parameters of the cost function to a single parameter σ. It is suggested to choose this tuning parameter close to the expected accuracy of a captured 3D point. The RQE can be interpreted as a measure of the "crispness" of the point cloud and depends on pairwise point distances and a tuning parameter σ. The authors calibrate a system consisting of three 2D laser scanners mounted on a rotating platform.

The self-calibration based on the *RQE* has also been used to calibrate vehicle-based MMSs [28]. In a first step, the authors extrinsically calibrate a MMS with a 3D laser scanner and a GNSS/INS unit. In a second step, the resulting point cloud is exploited to calibrate additional 2D laser scanners. This is done by maximizing the similarity between the point cloud resulting from the first step and the point cloud from the 2D laser scanners. To evaluate the similarity between these two point clouds, they use the Jensen-Rényi Divergence [29]. In essence, this is equivalent to finding the entropy contribution of the cross-terms between the two point clouds. This approach has also been used to calibrate the *Oxford RobotCar* which also includes 3D laser scanners, 2D laser scanners and a GNSS/INS unit [30].

The main advantage of the self-calibration based on the *RQE* [8] is that it has no assumptions about the environment. In contrast, other approaches assume for example planes at specific positions, the existence of surfaces of a known form or a 3D model of the environment. Consequently, it is also more practical than the other approaches. Our self-calibration approach shares these advantages. In contrast to the self-calibration based on the *RQE*, we use different cost functions for optimization and evaluate their suitability for self-calibration. We use a neighborhood with a fixed size for calculation of the cost function instead of taking all point pairs into account which reduces the computational complexity. Moreover, we present results of a multi-scale approach that increases the radius of convergence and improves the accuracy of the self-calibration.

2.2. Geometric Features

To describe the local characteristics around a considered point $\mathbf{X}_i \in \mathbb{R}^3$, a diversity of geometric features may be utilized. Among these geometric features, particularly the ones derived from the 3D structure tensor $\mathbf{S}_i \in \mathbb{R}^{3\times3}$ are widely used. Thereby, the 3D structure tensor represents a 3D covariance matrix calculated on the basis of the 3D coordinates of those points within the local neighborhood $\mathcal{N}_i$ of $\mathbf{X}_i$. The three eigenvalues of this 3D covariance matrix indicate the extent of a 3D ellipsoid along its principal axes. Denoting the eigenvalues by $\lambda_{1,i}, \lambda_{2,i}, \lambda_{3,i} \in \mathbb{R}$ with $\lambda_{1,i} \geq \lambda_{2,i} \geq \lambda_{3,i} \geq 0$, the following structures may easily be recognized:

- A linear (1D) structure is given for $\lambda_{1,i} \gg \lambda_{2,i}, \lambda_{3,i}$, since the respective points in the local neighborhood are mainly spread along one principal axis.
- A planar (2D) structure is given for $\lambda_{1,i}, \lambda_{2,i} \gg \lambda_{3,i}$, since the considered points spread within a plane spanned by two principal axes.
- A volumetric (3D) structure is given for $\lambda_{1,i} \approx \lambda_{2,i} \approx \lambda_{3,i}$, since the considered points are similarly spread in all directions.

An extension of this has been presented with the analytical consideration of the eigenvalues for describing the local point cloud characteristics with respect to a set of specific shape primitives [31]. Instead of such an analysis with respect to specified parametric models, the local structure around $\mathbf{X}_i$ may also be described via geometric features derived from the eigenvalues of the 3D structure tensor without assuming a parametric model. In this regard, geometric features that are also known as local 3D shape features [32,33] are commonly used which are rather intuitive and represent a specific property of the local neighborhood by a single value. Thus, these features are interpretable, but recent investigations on the accuracy and robustness of such geometric features reveal that some of them are more susceptible to discretization and noise whereas others are more robust [34].

Other commonly used geometric features are represented by angular characteristics [35], point distances and height differences [36], a variety of low-level geometric 3D and 2D features [4], and moments and height features [5]. Besides the diversity of interpretable geometric features, there are also more sophisticated geometric features like spin image descriptors [37], 3D shape context descriptors [7], Signature of Histograms of OrienTations (SHOT) descriptors [38,39], shape distributions [40] or point feature histograms [41]. These are typically derived by sampling specific properties within the local neighborhood of $\mathbf{X}_i$, e.g., in form of histograms, so that the resulting feature vectors describing the local structure are of higher dimension and single entries are hardly interpretable.

In our work, we assume that self-calibration is performed best when the derived point cloud represents real physical circumstances the best. Such real physical circumstances in turn can be described by interpretable geometric features with a single value. Consequently, we focus on the use of a metric based on interpretable geometric features, also known as local 3D shape features [32,33], to evaluate when real physical circumstances are represented the best.

3. Methodology

First, we introduce the principle of the self-calibration approach and give a general overview of our methodology. Then, we present the relevant processing steps in more detail and explain important design choices.

The fundamental assumption of the self-calibration is that the calibration parameters are estimated the best when the derived point cloud represents the real physical circumstances the best and, therefore, when the point cloud has the best quality. To evaluate the quality, we use a cost function which is based on a geometric feature. This geometric feature relies on the 3D structure tensor derived from the local neighborhood of each point. This means that local characteristics of the point cloud are considered which are described by the extent of the neighborhood around each point in three perpendicular directions. By iteratively optimizing the calibration parameters in a nonlinear least squares optimization the cost function is minimized and thereby the point cloud quality is maximized.

Figure 2 shows the workflow of the self-calibration and serves as a general overview of the method. The inputs to the self-calibration are laser scans and corresponding poses of the pose estimation sensor as well as initial calibration parameters and an initial grid size for a voxel grid filter. The basis of the calibration is a nonlinear least squares optimization, that is represented by the box in the middle of Figure 2. Each step inside this box is performed with each iteration of the optimization. First, the initial point cloud is calculated with the given input data. Next, this point cloud is downsampled using a voxel grid filter. For each point and its k nearest neighbors in the downsampled point cloud, the value of the selected geometric feature is determined. To make the algorithm more robust to outliers, we then filter the points based on the value of the geometric feature. Further, a cost function $\mathcal{R}$ which is based on the feature values of the remaining points is evaluated. This cost function is minimized in a nonlinear least squares optimization. Consequently, the initial calibration parameters are optimized such that the cost function is minimized. The optimization stops if the change of the calibration parameters is less than a given threshold θ. To improve the calibration result and increase the radius of convergence of the approach, we optimize recursively in multiple scales. Thus, we refine the scale by reducing the size of the voxel grid after the optimization has stopped and perform consecutive optimizations. We record the final calibration parameters after the optimization has run through a specified number of scales.

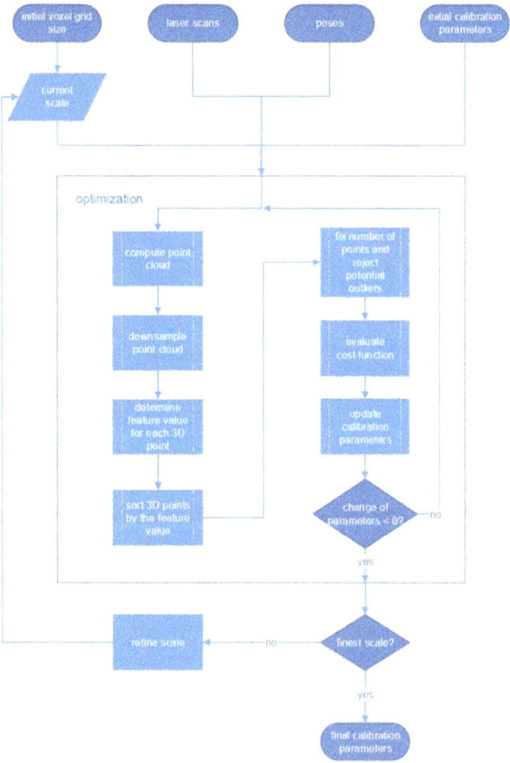

Figure 2. Workflow of the self-calibration. Besides the scan points and associated poses, initial calibration parameters and an initial voxel grid size are the inputs to a nonlinear least squares optimization. This optimization is performed recursively with multiple scales to be robust as well as accurate. It estimates the calibration parameters by minimizing an cost function which is based on a selected geometric feature.

3.1. Input and Output Parameters

As depicted in Figure 2 there are four categories of input parameters for the self-calibration approach:

1. The first category consists of a single parameter that represents the initial size a of a voxel grid filter. The size is defined by the edge length of each voxel. Section 3.2.2 gives more details on the downsampling.
2. The second category contains the measurements of the mapping sensor. For each timestep t the mapping sensor measures a point or a point set ${}^{m}\mathbf{X}_t$ in the Cartesian mapping frame m. The set of point sets for each utilized time step is the input to the self-calibration.
3. The third category contains the measurements of the pose estimation sensor, analogous to the second category. For each timestep t the pose estimation sensor provides a 6-DOF pose ${}^{w}_{n}\mathbf{M}_t$ that is associated to a single point set. The 6-DOF pose ${}^{w}_{n}\mathbf{M}_t$ represents the rigid transformation from the navigation frame n to the world frame w. For this simplified formal description, synchronized sensors and the interpolation of movements during the scanning are assumed.
4. The fourth category consists of the six initial calibration parameters. The initial calibration matrix ${}^{n}_{m}\mathbf{C}$ is a reparametrization of these parameters. In Section 4.1.2 we investigate how accurate the initial calibration parameters must be for the self-calibration to be applicable.

The third category, namely the poses, can best be controlled by the operator of the MMS. Therefore some considerations follow how the MMS should be moved so that the result of the calibration is accurate. As mentioned, local characteristics of the neighborhood around each point are considered to quantify the quality of the point cloud. To ensure that the effects of inaccurate calibration parameters can be quantified with this metric, the same objects must be captured from different angles. However, this does not mean that identical physical points must be captured from different angles. This is one characteristic that distinguishes the presented self-calibration approach fundamentally from procedures that minimize the difference between multiple measurements of ground control points (like for example [19]) or that are based on the direct comparison of two point clouds (like for example [24,27]). In principle, the 6-DOF poses should show large variations for all parameters in a confined space to capture the same objects from different angles.

The above mentioned parameters are the input to a nonlinear least squares optimization. This nonlinear least squared optimization iteratively optimizes the initial calibration parameters. Thereby, the output parameters of the self-calibration approach are optimized calibration parameters of the final calibration matrix ${}^{n}_{m}\hat{\mathbf{C}}$.

3.2. Optimization

As the nonlinear least squares optimization is the basis of the calibration, we describe each step in more detail in this section.

3.2.1. Computation of the Point Cloud $\mathcal{P}$

Given the point set ${}^{m}\mathbf{X}_t$ in the Cartesian mapping frame m. The points ${}^{w}\mathbf{X}_t$ in the world frame w are then computed by following the transformation sequence:

$$ {}^{w}\mathbf{X}_t = {}^{w}_{n}\mathbf{M}_t \;\; {}^{n}_{m}\mathbf{C} \;\; {}^{m}\mathbf{X}_t, \tag{2} $$

where ${}^{w}_{n}\mathbf{M}_t$ is the pose at time t and ${}^{n}_{m}\mathbf{C}$ is the calibration matrix, which we assume to be constant in the selected period of the data acquisition.

Then the point cloud $\mathcal{P}$ is the set of laser scans in the common world frame:

$$ \mathcal{P} = \{ {}^{w}\mathbf{X}_t \,|\, t \,\epsilon\, \{1, ..., T\}\}, \tag{3} $$

where T is the number of time steps or poses and associated laser scans.

3.2.2. Downsampling

In principle, when working with scanning devices, the acquired point cloud $\mathcal{P}$ is inhomogeneously sampled. Close regions show a higher density than far regions. Furthermore, a non-uniform movement of the mobile mapping platform effects the homogeneity of the sampling. To overcome these effects, it is common practice to apply a voxel grid filter $f_d(x, a)$ [5], where x is the input point cloud and a is the edge length of each voxel. After applying the voxel grid filter, the downsampled point cloud $\mathcal{P}' = f_d(\mathcal{P}, a)$ has a homogeneous sampling. The applied voxel grid filter f_d approximates all the points in a voxel by the centroid of these points. Thereby, the downsampled point cloud $\mathcal{P}'$ approximates the geometry of the original point cloud $\mathcal{P}$ well and even small changes of the geometry are observable. This is important for the iterative calibration approach. In contrast, the approximation of each voxel by a random point within the voxel would be more efficient, but less accurate as small changes would not be observable. The downsampling has two additional benefits: Firstly, it can remove small objects in the point cloud that might disturb the calibration process. This mainly depends on the size of the used voxels and the size of the objects. Secondly, it accelerates the calibration process, because the selected geometric feature has to be computed for less points in coarse scales. However, the downsampling also raises a new challenge: The number of points in the point cloud may vary at each iteration of the least squares optimization. To be able to use a standard optimization algorithm anyway, we need a workaround which is described in Section 3.2.4.

3.2.3. Determination of the Geometric Feature

Given the spatial 3D coordinates corresponding to all points within the local neighborhood $\mathcal{N}_i$ of a considered point $\mathbf{X}_i$, we calculate the 3D structure tensor and its eigenvalues (cf. Section 2.2). Thereby, we define the neighborhood $\mathcal{N}_i$ as the 50 neighbor points with the smallest Euclidean distance to the considered point $\mathbf{X}_i$. The eigenvalues are used to derive a set of geometric features which comprises the features of linearity $f_{L,i}$, planarity $f_{P,i}$, sphericity $f_{S,i}$, omnivariance $f_{O,i}$, anisotropy $f_{A,i}$, eigenentropy $f_{E,i}$ and change of curvature $f_{C,i}$ [32,33]. These geometric features are defined as follows:

$$f_{L,i}(\mathbf{X}_i, \mathcal{N}_i) = \frac{\lambda_{1,i} - \lambda_{2,i}}{\lambda_{1,i}} \tag{4}$$

$$f_{P,i}(\mathbf{X}_i, \mathcal{N}_i) = \frac{\lambda_{2,i} - \lambda_{3,i}}{\lambda_{1,i}} \tag{5}$$

$$f_{S,i}(\mathbf{X}_i, \mathcal{N}_i) = \frac{\lambda_{3,i}}{\lambda_{1,i}} \tag{6}$$

$$f_{O,i}(\mathbf{X}_i, \mathcal{N}_i) = \sqrt[3]{\lambda_{1,i}\lambda_{2,i}\lambda_{3,i}} \tag{7}$$

$$f_{A,i}(\mathbf{X}_i, \mathcal{N}_i) = \frac{\lambda_{1,i} - \lambda_{3,i}}{\lambda_{1,i}} \tag{8}$$

$$f_{E,i}(\mathbf{X}_i, \mathcal{N}_i) = -\sum_{j=1}^{3} \lambda_{j,i} \ln\left(\lambda_{j,i}\right) \tag{9}$$

$$f_{C,i}(\mathbf{X}_i, \mathcal{N}_i) = \frac{\lambda_{3,i}}{\lambda_{1,i} + \lambda_{2,i} + \lambda_{3,i}} \tag{10}$$

For the special case $\lambda_1 = \lambda_2 = \lambda_3 = 0$, the features linearity (4), planarity (5), sphericity (6), anisotropy (8) and change of curvature (10) are not defined. This special case occurs when all points in the considered point set have identical coordinates. In practice there is sensor noise that prevents this.

To determine the calibration parameters, we minimize an cost function which is based on a selected feature. Selecting the feature of linearity, for example, would lead to the assumption that the calibration is estimated best, if linear structures in the point cloud are the sharpest. This would be a restricted assumption because usually linear structures are not predominant in point clouds.

In man-made environments, planar structures like the ground, the ceiling, walls, tables, etc. arise more prevalent. Thus, the feature of planarity would be a more obvious choice. Maximizing the planarity in a point cloud is a similar basic assumption as minimizing the distance of points to their corresponding, detected or estimated planes (cf. Section 2.1).

If continuous surfaces in the point cloud, like lines, planes, cylinders, etc. shall be exploited, the change of curvature, namely the variation of a point along the surface normal [42], may be chosen. So in this case, the assumption would be that the calibration is estimated best when the change of curvature over the whole scene is the smallest. For example, if lines in the point cloud are sharp and points corresponding to a plane have small perpendicular distances to this plane.

However, choosing features like sphericity, omnivariance or eigenentropy is the most generic. By minimizing these features, the sharpness of the cloud is maximized. In contrast to all existing calibration approaches mentioned in Section 2.1 except for the self-calibration based on the *RQE* [8,28] and the quasi-rigorous approach [24], this approach does not assume the existence of geometries with a specific, predefined form. In other words, selecting one of these features does not imply model assumptions about the environment.

In Section 5.1, we discuss the question of which feature is best suited for the self-calibration in more detail based on simulations (cf. Section 4.1.1) and experiments on real data (cf. Section 4.2.1).

3.2.4. Computation of the Cost Function and Parameter Estimation

As mentioned above, we aim to maximize the point cloud quality, by minimizing a cost function $\mathcal{R}$ which is based on the selected geometric feature. Because some geometric features should have large values for the point cloud to be accurate, and others small values, the cost function $\mathcal{R}$ depends on the feature selection. In case of the features of sphericity, omnivariance, eigenentropy and change of curvature small values lead to a more accurate point cloud, while in case of the features of linearity, planarity and anisotropy large values lead to a better quality of the point cloud. For unification, we redefine the geometric features as follows:

$$
\begin{aligned}
g_{L,i}(\mathbf{X}_i,\mathcal{N}_i) &= 1 - f_{L,i}(\mathbf{X}_i,\mathcal{N}_i) && (11)\\
g_{P,i}(\mathbf{X}_i,\mathcal{N}_i) &= 1 - f_{P,i}(\mathbf{X}_i,\mathcal{N}_i) && (12)\\
g_{S,i}(\mathbf{X}_i,\mathcal{N}_i) &= f_{S,i}(\mathbf{X}_i,\mathcal{N}_i) && (13)\\
g_{O,i}(\mathbf{X}_i,\mathcal{N}_i) &= f_{O,i}(\mathbf{X}_i,\mathcal{N}_i) && (14)\\
g_{A,i}(\mathbf{X}_i,\mathcal{N}_i) &= 1 - f_{A,i}(\mathbf{X}_i,\mathcal{N}_i) && (15)\\
g_{E,i}(\mathbf{X}_i,\mathcal{N}_i) &= f_{E,i}(\mathbf{X}_i,\mathcal{N}_i) && (16)\\
g_{C,i}(\mathbf{X}_i,\mathcal{N}_i) &= f_{C,i}(\mathbf{X}_i,\mathcal{N}_i), && (17)
\end{aligned}
$$

where $g_{F,i}(\mathbf{X}_i,\mathcal{N}_i) \epsilon [0,1]$ is the redefined geometric feature with $F \epsilon \{L,P,S,O,A,E,C\}$. As $1 - f_{A,i}(\mathbf{X}_i,\mathcal{N}_i)$ is the same as $f_S(\mathbf{X}_i,\mathcal{N}_i)$, we ignore the feature of anisotropy in the following.

To summarize the processing steps explained so far, the calculation of the cost function to be minimized is described in the following. Given the measured points ${}^m\mathbf{X}_t$ in the Cartesian mapping frame m, the poses ${}^w_n\mathbf{M}_t$ and the initial calibration matrix ${}^n_m\mathbf{C}$, the point cloud $\mathcal{P}$ is determined by inserting Equation (2) in (3):

$$\mathcal{P} = \{ {}^w_n\mathbf{M}_t \; {}^n_m\mathbf{C} \; {}^m\mathbf{X}_t \mid t \epsilon \{1,...,T\}\}. \tag{18}$$

The calibration matrix ${}^n_m\mathbf{C}$ is a reparametrization of the six calibration parameters which are optimized. Then, the downsampled point cloud

$$\mathcal{P}' = f_d(\mathcal{P},a) \tag{19}$$

is calculated. Consequently, the cost function to be minimized is

$$\mathcal{R} = \sum_{i=1}^{N'} (g_{F,i}(\mathbf{X}_i, \mathcal{N}_i))^2, \text{ with } \mathbf{X}_i \in \mathcal{P}', \tag{20}$$

where $g_{F,i}(\mathbf{X}_i, \mathcal{N}_i)$ is the value of the selected, redefined geometric feature and N' is the number of points in the downsampled point cloud $\mathcal{P}'$.

As mentioned in Section 3.2.2, we need a workaround to ensure that the number of considered points is the same for each iteration of the optimization algorithm. Therefore, we sort the points by their feature value and use a subset with the lower ζ percent of the values. We choose a smaller subset for the first scale (ζ_0), because the number of points in the point cloud may vary strongly at the beginning of the optimization, depending on the initial calibration. Usually, the point cloud which results from a low-quality calibration is less compact than from the optimal calibration. This results in a considerably higher number of points in the downsampled point cloud, due to the equidistant nature of the voxel grid filter.

Sorting the points and rejecting potential outliers has another advantage over random rejections for some of the features: It makes the algorithm more robust to environments that do not perfectly fit the assumptions about spatial surface characteristics that some of the features imply. A real environment does not exclusively consist of e.g., planar objects. There is vegetation outdoors and there are houseplants indoors which can lead to large values for the selected redefined feature. So, the calibration methodology has to be robust to scattering objects in the point cloud. In unfavorable situations, the remaining points still contain scattering objects with large feature values. To estimate the parameters robustly anyway, we use an M-estimator [43] in an iteratively re-weighted least squares optimization to down-weight the points with large feature values. We choose the Huber-estimator

$$\rho_k(x) = \begin{cases} \frac{1}{2}x^2 & \text{for } x < k \\ k(|x| - \frac{1}{2}k) & \text{otherwise} \end{cases}, \tag{21}$$

where k is the tuning constant [43].

Thus, the final cost function for the optimization becomes

$$\mathcal{R} = \sum_{i=1}^{L} \rho_k(g_{F,i}(\mathbf{X}_i, \mathcal{N}_i)), \text{ with } \mathbf{X}_i \in \mathcal{P}', \tag{22}$$

where $L = \lceil \zeta M \rceil$ is the number of points in the subset with $\zeta = \zeta_0$ for the first scale and $\mathbf{X}_i \in \mathcal{P}'$ is a point in the transformed, downsampled point cloud $\mathcal{P}'$ (c.f. Equations (18)–(20)). However, using the cost function in Equation (20) for the subset of points should be sufficiently robust in most cases.

This cost function is minimized in order to estimate the six parameters of the calibration, which represent the translation and the rotation between the sensors. As we want to express the rotation with the minimal number of parameters, we choose the axis-angle representation as parametrization of the rotational part of the calibration. In contrast to Euler and quaternion representations, this parametrization is well-suited for unconstrained optimization algorithms. The derivatives needed by the optimization are determined numerically by using the difference quotient. For this purpose, the cost function has to be evaluated at least once for each parameter to be estimated. This is the most time-consuming processing step depending on the number of points N in the point cloud. The iteratively re-weighted least squares optimization should stop as soon as the parameters no longer change significantly. Thus, a threshold on the change of parameters is applied.

3.3. Multi-Scale Approach

The sampling of the point cloud has a large impact on the accuracy and robustness of the calibration (cf. Section 4.1.3). If the sampling is very fine and the initial calibration parameters are inaccurate, the optimization likely converges to a local minimum. If the sampling is coarse, the calibration is more robust, but less accurate. Thus, we decided to choose a multi-scale approach.

We define the scale of the point cloud by the size of the voxel grid, which is used for downsampling. Every time the optimization has converged to a solution, we refine the scale of the point cloud by reducing the size of the voxel grid filter. Through the successive refinement of the point cloud, the optimization takes smaller and smaller areas into account because the geometric features are determined based on a fixed number of neighbors for each scale. For coarse samplings, the optimization roughly minimizes the feature value determined at large-scale structures like the ground and walls. For finer scales, it minimizes the feature value at even smaller structures. Hence, we can achieve the robustness of a coarse scale and the accuracy of a fine scale. The calibration finally ends after a fixed number of scales.

4. Experiments

In Section 4.1 the synthetic datasets and the experiments performed on these datasets are introduced. These experiments focus on analyzing how suitable the different cost functions are for the self-calibration task, how accurate the initial parameters have to be known and how the multi-scale approach performs in comparison with the approach that uses a single scale. Subsequently, Section 4.2 describes the real datasets and the experiments with the calibration performed on these data. These experiments show the applicability of the calibration to real data and different sensor setups.

Additionally, for each dataset, we compare the results of our self-calibration approach to the self-calibration based on the *RQE*. Therefore, we use our workflow and exchange the cost function by the cost function based on Equation (1). The main difference to the cost function presented by [8] is that we use a neighborhood $\mathcal{N}_i$ for each point in the cloud instead of evaluating the distances from every point to every other point in the point cloud. This is conform to a suggestion of [8] and can be justified by the fact that large distances have a very small impact on the cost function. Due to the use of neighborhoods, the impact of the tuning constant σ decreases. However, to take the tuning parameter σ into account, we always test different σ values from optimistic guesses of the point accuracy up to pessimistic guesses and show the best calibration results.

4.1. Synthetic Data

To generate the synthetic data, we use ROS (http://www.ros.org/) (Robot Operating System) and Gazebo (http://gazebosim.org/). With this software it is possible to simulate arbitrary environments, sensors, sensor configurations and trajectories. Our synthetic application area is a closed room with an extent of 10 m × 10 m × 5 m. We simulate a 2D laser scanner just like the *Hokuyo UTM-30LX-EW*, as this is one of the most frequently used mapping sensors for reconstruction tasks in small-scale environments. This laser scanner has a field-of-view of 270°, scans 1080 points per line and captures data in ranges from 10 cm up to 30 m. The range measurements are simulated without noise for this analysis. The laser scanner is rigidly mounted on a carrier platform. This platform moves to random poses inside of the simulated environment in order to scan the surroundings from different positions and view angles. With this setup, we simulate ten different mobile mappings of the synthetic environment. Each of these mobile mappings is represented by 100 system poses and associated scan lines. Before extrinsically calibrating the MMS, we falsify the true calibration parameters by 5° for each angle (initial rotation parameters) and 5 cm in each direction (initial translation parameters). As we know the exact parameters of the calibration on simulated data, we show directly the deviations from the ground truth. For convenience, we present the results of the calibrations using the following metrics based on the L2-norm $||\cdot||_2$:

$$\text{Translation error: } \Delta d = ||\hat{\mathbf{t}} - \tilde{\mathbf{t}}||_2, \tag{23}$$

$$\text{Rotation error: } \Delta r = ||\hat{\mathbf{r}} - \tilde{\mathbf{r}}||_2, \tag{24}$$

where $\hat{\mathbf{t}} = \begin{pmatrix} \hat{t_x} & \hat{t_y} & \hat{t_z} \end{pmatrix}^T$ is the estimated and $\tilde{\mathbf{t}} = \begin{pmatrix} \tilde{t_x} & \tilde{t_y} & \tilde{t_z} \end{pmatrix}$ the true translation vector and $\hat{\mathbf{r}} = \begin{pmatrix} \hat{\phi} & \hat{\theta} & \hat{\psi} \end{pmatrix}^T$ is the estimated and $\tilde{\mathbf{r}} = \begin{pmatrix} \tilde{\phi} & \tilde{\theta} & \tilde{\psi} \end{pmatrix}$ the true rotation vector. So, Δd is the deviation of estimated and true distance between the laser scanner and the navigation frame and Δr is the deviation of estimated and true rotation.

4.1.1. Suitability of the Cost Functions

To find out which cost function is best suited for self-calibration, we perform calibrations with each cost function for all ten mobile mappings. Additionally to the presented geometric features, we compare the results to the self-calibration based on the RQE. Figure 3 shows the results of the calibrations in terms of translation and rotation errors.

The calibration based on the feature of linearity results in parameters, which are very close to the initial calibration parameters. The corresponding errors are significantly larger than the errors of the calibrations and are not shown for display reasons. The calibrations based on the remaining features result in smaller translation and rotation errors compared to the calibration based on RQE. For the features of omnivariance and eigenentropy, the median error is the smallest. To test the suitability of the cost functions under realistic conditions with sensor noise, we compare the results of the calibrations based on the different geometric features again with one of the real datasets in Section 4.2.1.

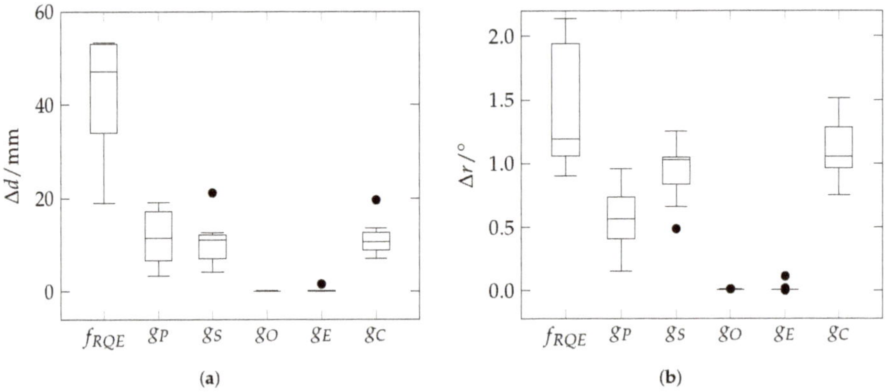

Figure 3. Calibration results depending on the f_{RQE} and the features of planarity (g_P), sphericity (g_S), omnivariance (g_O), eigenentropy (g_E) and change of curvature (g_C). (**a**) shows the translation error, (**b**) the rotation error. The feature of linearity has turned out to be unsuitable for the calibration approach and is not shown for display reasons. The features of omnivariance and eigenentropy clearly outperform f_{RQE} and all other features.

4.1.2. Radius of Convergence

Each optimization needs reliable initial calibration parameters to converge to the global minimum. An important question for an operator is if it is necessary to measure the initial calibration parameters beforehand with additional sensors or if a rough guess might be accurate enough. In order to estimate the radius of convergence, simulations are carried out with varying initial calibration parameters.

Figure 4 shows the calibration error plotted against the error of the initial calibration parameters. Figure 4a shows the results for a simulated initial translation error Δd_0. For initial translation errors up to 2.2 m the final calibration error is clearly below 1 mm and 0.01°. This means that even with a deviation of two meters from the true distance of the sensors, the calibration is still able to converge close to the global optimum. For larger initial translation errors, the final calibration error considerably increases.

For the results in Figure 4b, an initial rotation error from one to 40° is simulated. The rotation error of the calibrations is below 0.01° for initial rotation errors up to 30°. The translation error up to this mark is below 1 mm. For worse initial rotation parameters, the rotation error clearly grows, since the optimization is then no longer able to find the global minimum. The translation error is small for this simulation, as the initial translation has not been falsified.

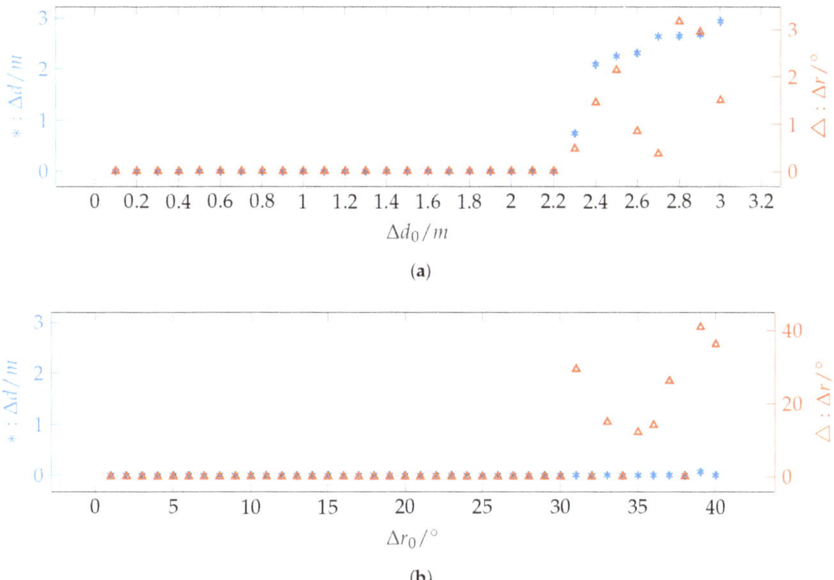

Figure 4. Required accuracy of the initial calibration parameters ((**a**): Initial translation, (**b**): Initial rotation). Up to an initial translation error of 2.2*m* or an initial rotation error of 31° the results are in the range of sub-millimeters and sub-degrees. More inaccurate values for the initial calibration parameters mean that the optimization no longer converges to the global minimum and the errors clearly grow.

4.1.3. Single-Scale vs. Multi-Scale

In order to support the statements from Section 3.3, we compare the calibration results based on different scales (see Figure 5) in terms of robustness against initial calibration errors. As mentioned before, the optimization needs good initial calibration parameters to converge to the global minimum. An important question to an operator might be if it is necessary to measure the calibration parameters beforehand with additional sensors or if a rough guess might be accurate enough. In order to estimate the radius of convergence, simulations were carried out with varying initial calibration translation errors Δd_0 and varying initial rotation errors Δr_0. Since the feature of omnivariance achieved the best calibration results, this investigation was carried out with this feature.

The coarse-scale and the fine-scale approach do not refine the scale iteratively, but use one fixed scale. The fine scale provides a close to the original but still homogeneous sampling of the point cloud. The multi-scale approach starts with the coarse scale and refines it to the fine scale in two iterations.

For this analysis, we add Gaussian noise to the ranges and to the poses, as we want to test the robustness of the single-scale approaches and the multi-scale approach under more realistic conditions. With a Visual-Inertial SLAM and the laser scanner used in the experiments in mind, we add a noise of 2 cm and 0.1° to the poses and a noise of 3 cm to the ranges. However, adding Gaussian noise to all sensor measurements is still a simplification of the reality. For convenience, we only show the translation error for varying initial translation errors and the rotation error for varying initial rotation errors.

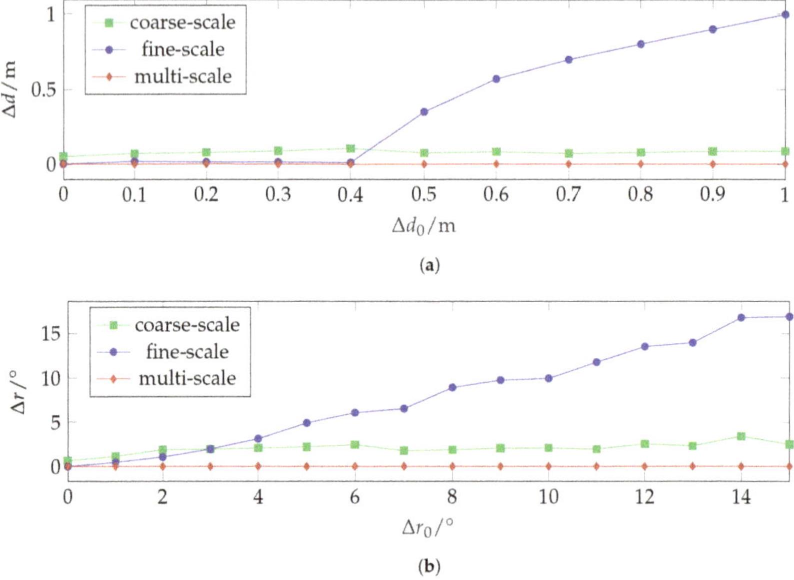

Figure 5. Single-scale vs. multi-scale calibration. (**a**) shows the translation error after performing the calibration depending on the initial translation error. (**b**) shows the rotation error after performing the calibration depending on the initial rotation error. The coarse-scale approach is more robust to large initial calibration errors than the fine-scale approach. However, the fine-scale approach is more accurate for small initial calibration errors. The multi-scale approach combines the advantages of both single-scale approaches.

The experiments show that the coarse-scale approach is less accurate than the fine-scale approach for small initial calibration errors. However, for large initial calibration errors, it achieves better results compared to the fine-scale approach. The multi-scale approach always achieves the best results.

4.2. Real Data

To test our calibration approach on real data, we use two datasets with different characteristics. The first one is a self-collected dataset, where a hand-held 2D laser scanner is moved in a small-scale indoor laboratory environment. A Motion Capture System (MCS) records the poses. We refer to this dataset as the small-scale indoor dataset. The second dataset is a part of the well-known, publicly available (http://www.cvlibs.net/datasets/kitti/eval_odometry.php) KITTI dataset [44]. This dataset is a large-scale outdoor vehicle-based mobile mapping dataset. It consists of the data of a GNSS/INS unit, a 3D laser scanner and a stereo camera system captured in different environments.

Please note, that we use the same parameters for each dataset. Only the amount of data (poses and scan points) was adjusted for each dataset, in order to limit the runtime of the calibration.

4.2.1. Small-Scale Indoor Dataset

The mapping sensor used for this real-data experiment is the 2D laser scanner *Hokuyo UTM-30LX-EW* with the same specifications as described in Section 4.1. The range measurement accuracy of this laser scanner is specified with ±30 mm for ranges up to 10 m and ±50 mm for ranges from 10 m to 30 m. The poses are determined with the MCS *Optitrack Prime 17 W*. The MCS tracks the positions of retro-reflective markers by triangulation with sub-millimeter accuracy. If at least three markers are connected rigidly, it is possible to determine the 6-DOF pose of this set of markers which is also called a rigid body. So, the objective of the calibration in this case is to find the transformation between the mapping frame and the frame of the rigid body which is rigidly attached to the laser scanner.

The environment used for the self-calibration is a laboratory with dimensions of 8 m × 5 m × 2.5 m. The room contains objects like cupboards, desks, chairs and monitors. The captured data consists of 2618 different poses and their corresponding scan lines, which results in a point cloud with 2.8 million points.

As it is not possible to provide exact ground truth for the calibration task, we compare our self-calibration qualitatively to a target calibration (TC). This TC is based on the unified intrinsic and extrinsic calibration of a MMS with a multi-camera-system and a laser scanner (UCalMiCel) [18]. The TC is a two-step approach. The first step is based on RADLOCC [1–3] and estimates the transformation between the laser scanner and the camera. The second step is based on accurate measurements of the MCS and determines the transformation between the camera and the rigid body which is rigidly mounted to the camera. For more details about the TC, the reader is referred to the work of [18]. However, since the target calibration is two-step, it has more potential sources of error compared to our self-calibration which directly determines the transformation between the laser scanner and the rigid body. As mentioned, the results are additionally compared to the self-calibration based on RQE.

Concerning the initial calibration parameters, we use the parameters which are estimated by the TC and falsify these parameters about 5° for each rotation parameter and 5 cm for each translation parameter.

As mentioned in Section 2.2, some of the features are more robust to discretization and noise than others [34]. Therefore, we additionally verify the simulation results by comparing the various features also on real data. Figures 6 and 7 depict all resulting point clouds. The point clouds in the left column are computed with the parameters resulting from the target calibration. For better comparability these point clouds are colorized based on the feature values used for the associated self-calibrations. This means that the point clouds in the left column do not differ apart from their color display. The point clouds in the right column show the results of the self-calibrations colored by means of the respective feature. All feature values have been scaled to values between zero and one.

Figure 6d,f show that the parameters estimated by the self-calibration based on the features of linearity and planarity are inaccurate. For these cases, the optimization converges to a local minimum. In all other cases, the point clouds look basically similar to the target calibration, which means that self-calibration is possible with these features. Furthermore, all point clouds resulting from the self-calibrations except for the one in Figure 6b look more compact and sharper than the point cloud based on the TC. This can be seen especially at the top right and bottom left corners of the room. The geometry of the point clouds which result from the calibrations based on the features of omnivariance and eigenentropy look even sharper. However, it is hard to rank the results of the different features based on visual inspection. Thus, we use two additional quantitative metrics to judge the performance of the features.

Firstly, we compare the feature values computed on the point clouds. Table 1 shows the medians of the feature values computed based on the point clouds resulting from different calibrations. Except for the median of feature of linearity, the medians of the feature values are the smallest for the self-calibration based on omnivariance. Furthermore, all self-calibrations except for the calibrations based on linearity and planarity lead to lower values of all features compared to the target calibration and the calibration based on RQE. This supports the visual impression of Figures 6 and 7.

Secondly, we evaluate the calibration results with a further metric which is independent of the geometric features. For this, we consider the mean distances of points to corresponding planes which are fitted to different planar regions of the point clouds. In other words, we evaluate the average thickness of planar structures in the point cloud. We use six planar structures, one on the floor (Plane 1), one on the ceiling (Plane 2) and one at each wall of the room (Planes 3–6). For this purpose, we define regions of interest for the planar structures, fit planes to the points in these regions of interest and compute the mean point-to-plane distances. Table 2 shows these mean point-to-plane distances. Again, all self-calibrations except for g_L and g_P achieve better results as the TC and the calibration based on RQE. For five of the six planes, g_O achieves the lowest mean point-to-plane distances.

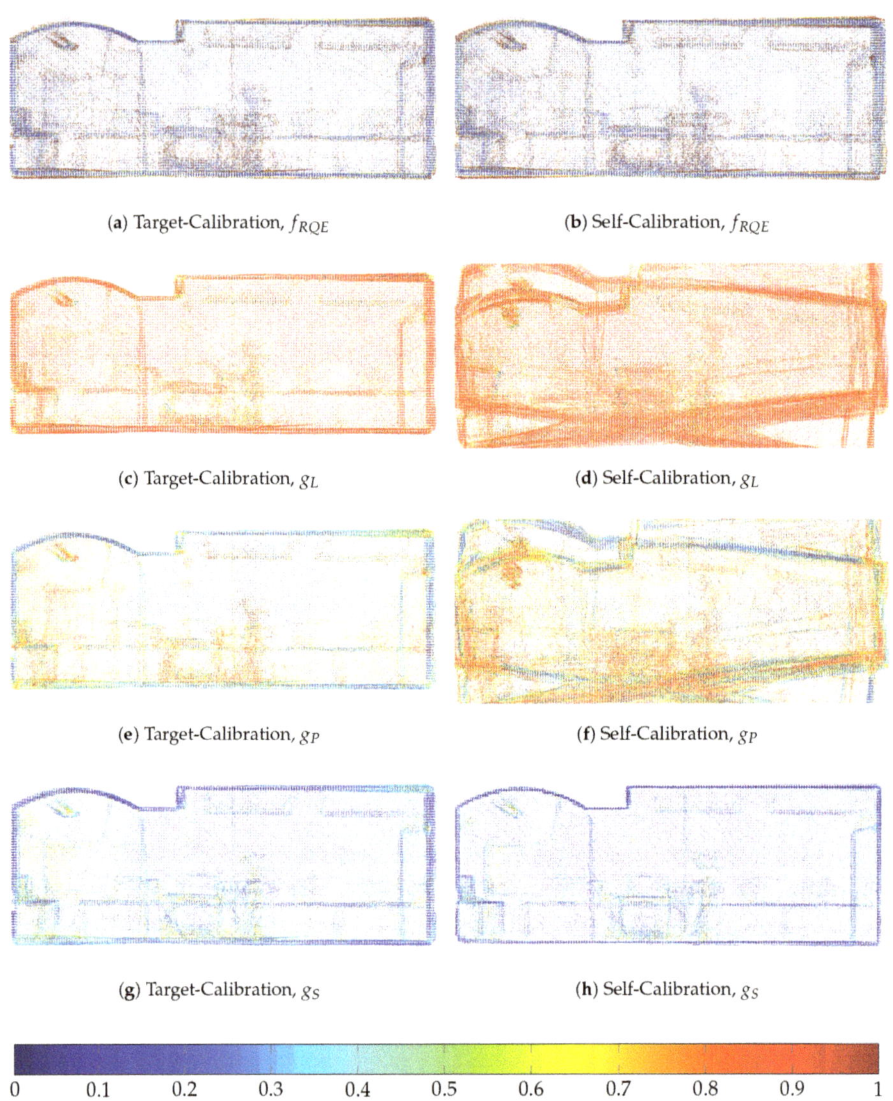

(**a**) Target-Calibration, f_{RQE} (**b**) Self-Calibration, f_{RQE}

(**c**) Target-Calibration, g_L (**d**) Self-Calibration, g_L

(**e**) Target-Calibration, g_P (**f**) Self-Calibration, g_P

(**g**) Target-Calibration, g_S (**h**) Self-Calibration, g_S

Figure 6. Real data results on the small-scale indoor dataset. The self-calibration based on the RQE achieves comparable results to the TC. The self-calibrations based on the features of linearity g_L and planarity g_P failed. The self-calibration based on the feature of sphericity g_S, however, results in a sharper point cloud than the target calibration.

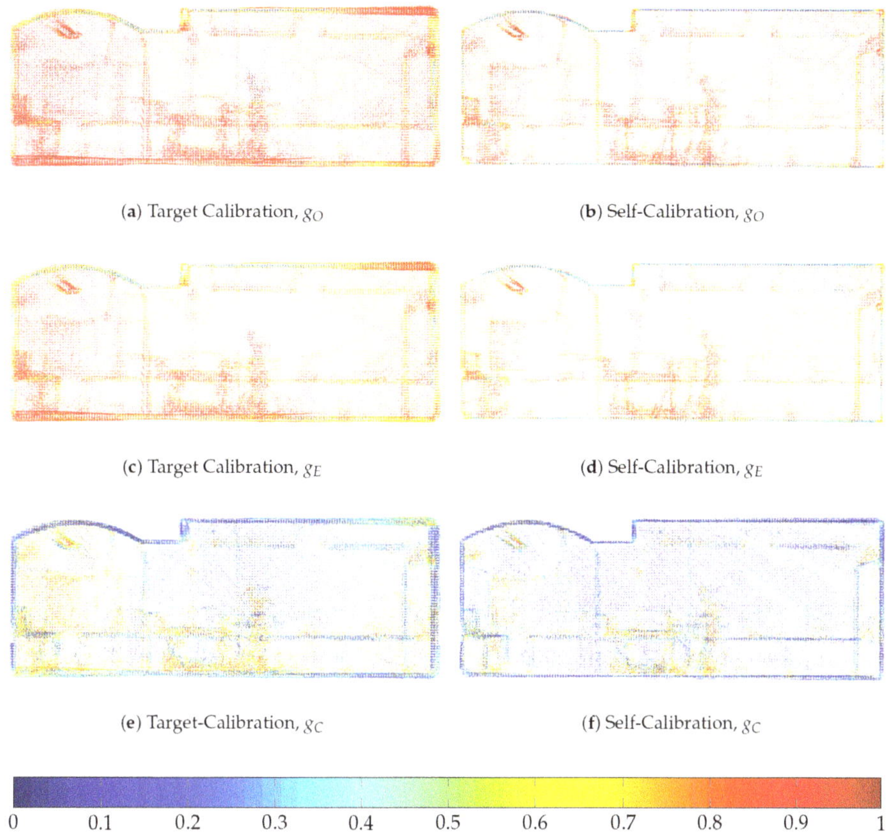

(**a**) Target Calibration, g_O (**b**) Self-Calibration, g_O

(**c**) Target Calibration, g_E (**d**) Self-Calibration, g_E

(**e**) Target-Calibration, g_C (**f**) Self-Calibration, g_C

Figure 7. Real data results on the small-scale indoor dataset. The self-calibrations based on the features of omnivariance g_O, eigenentropy g_E and change of curvature g_C lead to sharper point clouds compared to the target calibration.

Table 1. Comparison of the median of the RQE (med(RQE)) and the redefined features (med(g_L), ..., med(g_C)) for different calibration approaches. TC is the target calibration based on RADLOCC. RQE is the self-calibration based on the RQE. g_L, ..., g_C are the self-calibrations based on the corresponding feature. Lower feature values are better. The lowest feature value in each column is marked in bold. X: Calibration failed. The calibration based on the RQE leads to slightly higher values than the TC. The self-calibrations based on g_S, g_O, g_E and g_C outperform the TC and the calibration based on RQE. Omnivariance g_O achieves the best results.

Approach	med(f_{RQE})	med(g_L)	med(g_P)	med(g_S)	med(g_O)	med(g_E)	med(g_C)
TC	0.161	0.842	0.389	0.215	0.275	0.734	0.106
f_{RQE}	0.205	0.845	0.450	0.268	0.288	0.787	0.129
g_L	X	X	X	X	X	X	X
g_P	X	X	X	X	X	X	X
g_S	0.092	**0.826**	0.281	0.087	0.217	0.546	0.047
g_O	**0.082**	0.830	**0.264**	**0.072**	**0.205**	**0.514**	**0.039**
g_E	0.088	0.828	0.271	0.079	0.211	0.529	0.042
g_C	0.095	**0.826**	0.283	0.090	0.220	0.552	0.048

Table 2. Mean point-to-plane distances for different parts of the point cloud (in mm). X: Calibration failed. The smallest point-to-plane distance in each column is marked in bold. Again, all self-calibrations except for g_L and g_P achieve better results than the TC and the calibration based on RQE. For five of the six planes, the omnivariance g_O achieves the lowest mean distances.

Approach	Plane 1	Plane 2	Plane 3	Plane 4	Plane 5	Plane 6
TC	14.1	21.7	13.6	15.1	14.5	22.8
f_{RQE}	14.3	33.0	17.0	25.0	28.4	26.3
g_L	X	X	X	X	X	X
g_P	X	X	X	X	X	X
g_S	7.1	11.4	9.5	7.5	**9.9**	6.7
g_O	**6.3**	**8.3**	**7.7**	**7.4**	10.7	**6.6**
g_E	10.2	**8.3**	8.1	9.3	10.1	8.0
g_C	8.6	11.2	7.8	7.9	11.6	9.7

Finally, Figure 8 shows the calibration results of the TC, the results of the self-calibration based on the RQE and the self-calibration based on the feature of omnivariance for the small-scale indoor dataset. The parameters are separated into their single components instead of using the L2-norm, to visualize whether the precision varies significantly. Figure 8a–c show the translation parameters, Figure 8d–f show the rotation errors. For the TC, only a single set of parameters is available without information about the precision. The medians of multiple calibrations based on the RQE and also based on the feature of omnivariance differ significantly from the results of the target calibration. The precision of the results based on the feature omnivariance, is higher than the precision of the calibration based on RQE for each parameter. The self-calibration based on the feature of omnivariance estimates each parameter with a precision of a few millimeters and millidegrees, respectively.

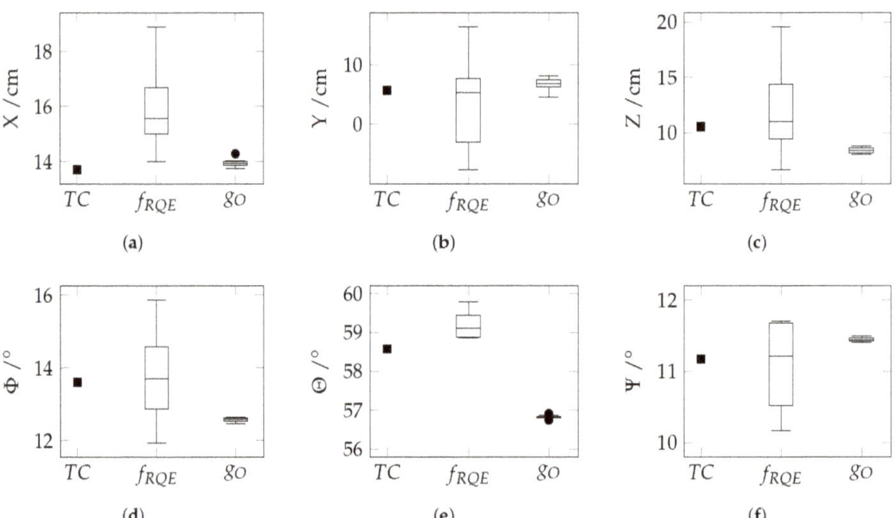

Figure 8. Calibration results of the different approaches for the small-scale indoor dataset. (**a–c**) show the translation parameters, (**d–f**) show the rotation parameters. For the TC, only a single set of parameters is available without information about the precision. The medians of multiple calibrations differ significantly from the results of the target-calibration. The self-calibration based on omnivariance g_O estimates all parameters with higher precision compared to the calibration based on RQE.

4.2.2. Large-Scale Outdoor Dataset

This experiment shows the calibration results on a part of the publicly available KITTI dataset [44]. The KITTI dataset is a large-scale outdoor vehicle-based mobile mapping dataset. The dataset includes the data of a GNSS/INS unit, a 3D laser scanner and a stereo camera system.

To test the calibration approach with a more inaccurate pose estimation sensor compared to the highly accurate MCS which is used for the small-scale indoor dataset, we use the images of the stereo camera system in a visual SLAM framework instead of the highly accurate GNSS/INS unit for this analysis. The visual SLAM framework we use is called ORB-SLAM2 [45]. Thus, in this case, the extrinsic calibration aims to determine the transformation between a 3D laser scanner and a stereo camera system.

The purpose of this experiment is to evaluate the robustness of the proposed extrinsic self-calibration approach in terms of different environments and sensors. The environment mainly differs from the environment in the small-scale indoor data set by larger dimensions of the point cloud, a larger number of points and a larger percentage of vegetation. The sensors differ from those used in the small-scale indoor data set in terms of the operating principle of the pose estimation sensor, the number of scanned lines per rotation of the laser scanner, and the accuracies of both sensors.

The laser scanner is a *Velodyne HDL-64E*. It has a measurement range of 120 m and a range measurement accuracy of 5 cm (1σ). It has a vertical field of view of 26.8° and a horizontal field of view of 360°. The vertical angular resolution is 0.4° and the horizontal angular resolution is 0.09°. The laser scanner measures up to approximately one million points per second.

Each single camera of the stereo camera system has a 1/2″ CCD sensor without color filters, 1.4 megapixel, a pixel size of 3.75 µm and a global shutter. For the pose estimation by ORB-SLAM2 on the KITTI dataset, we assume a translation accuracy of 1.15% and a rotation accuracy of 0.0027° [46].

We use the first of 22 sequences of the KITTI dataset. In order to accelerate the calibration process, we use a subset of this sequence and randomly downsample the laser scanner points. However, to ensure an accurate pose-estimation of ORB-SLAM2, we wait for the first loop closure. Consequently, we use every third of the first 1650 poses and associated laser scans.

We compare the estimated parameters against the parameters of the reference calibration (RC), which is provided with the dataset. We falsify the values of the RC about 5° for each rotation angle and 5 cm for each translation parameter and use these values as initial calibration parameters. We compare the RC qualitatively and quantitatively with the optimized parameters after performing the self-calibration.

The RC is a two-step semi-automatic procedure. In the first step, coarse calibration parameters are determined based on the automatic single-shot calibration of range and camera sensors utilizing checkerboards [47]. Their algorithm proceeds a segmentation to identify the checkerboard planes in the data of the range sensor. In the following, transformation hypotheses are generated by random plane associations. The best hypotheses are then refined and verified. A final non-maxima suppression step determines all acceptable solutions. In the second step, the result of this automatic calibration is refined manually. Therefore, the number of disparity outliers is minimized jointly with the reprojection error of manually selected correspondences [44].

The accuracy of the calibration results of the automatic step is given in [47]. The median rotation error without Gaussian noise is given with approximately 10° with a 25th percentile of approximately 1° and a 75th percentile of approximately 18°. The median translation accuracy is given with approximately 0.2 m, the 25th percentile with a few centimeters and the 75th percentile with approximately 0.5 m. The accuracy of the manually refined calibration (after the second step) is not mentioned.

Figure 9 shows the point clouds associated with this experiment. The color of each point represents its height. Figure 9a,c,e show an overview of the point clouds computed with the parameters of the RC, the self-calibration based on the *RQE* and the self-calibration based on the feature of omnivariance, respectively. Figure 9b,d,f show detailed top views of these point clouds.

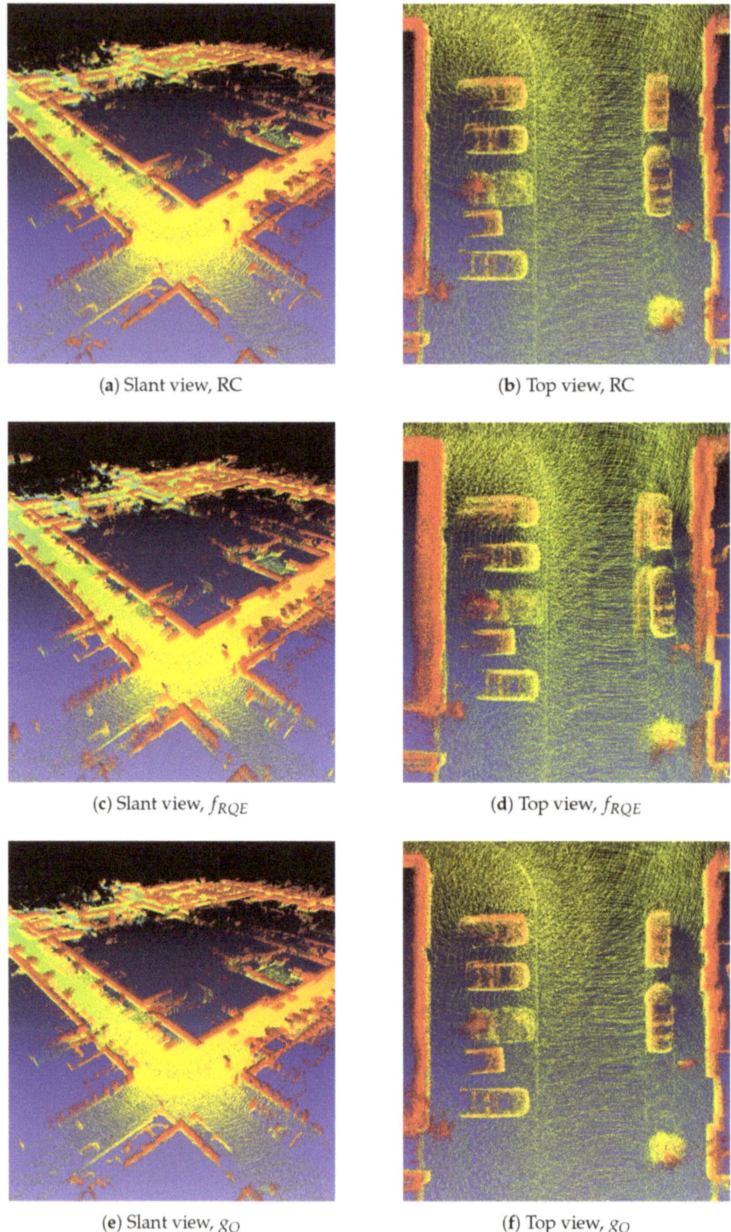

(a) Slant view, RC (b) Top view, RC

(c) Slant view, f_{RQE} (d) Top view, f_{RQE}

(e) Slant view, g_O (f) Top view, g_O

Figure 9. Results on the publicly available KITTI dataset [44]. The color of each point depicts its height. Both calibrations improved the initial calibration parameters such that objects and basic structures are identifiable in the point clouds. Moreover, the RC and the calibration based on the feature of omnivariance lead to very similar point clouds. The calibration based on the f_{RQE}, however, results in a more noisy point cloud, which can be seen especially at the vertical walls and at artifacts next to the two cars on the right.

Both calibrations improved the initial calibration parameters significantly such that objects and basic structures are identifiable in the point clouds. The RC and the calibration based on the feature of omnivariance lead to very similar point clouds. They only differ in details but quality differences are difficult to verify based on visual inspection. The calibration based on the f_{RQE}, however, results in a more noisy point cloud, which can be seen especially at the vertical walls and the artifacts next to the two cars on the right.

Table 3 shows mean point-to-plane distances for different parts of the point cloud in analogy to Section 4.2.1. Planes 1–3 are excerpts from the street level at different positions, Planes 4–6 are excerpts from walls of different buildings. *RC* is the calibration performed by [44] including the manual refinement. f_{RQE} is the self-calibration based on the *RQE* and g_O is our self-calibration based on the feature of omnivariance. The differences between the mean point-to-plane distances are in the range of a few centimeters for most planes. However, for three out of six planes the RC achieves the best results. For the other three planes the self-calibration based on the feature of omnivariance achieves the best results. This supports the visual impressions of Figure 9 that the results of the *RC* and g_O are of similar quality whereas the quality of f_{RQE} is lower.

Finally, Figure 10 shows the results of the RC, the self-calibration based on the *RQE* and our self-calibration based on the the feature omnivariance for the large-scale outdoor dataset. Figure 10a–c show the translation parameters and Figure 10d–f shows the rotation parameters. The calibration parameters must be interpreted with respect to the orientation of the navigation frame and the mapping frame. For the large-scale outdoor dataset, the *X*-axis of the navigation frame points to the right, the *Y*-axis to the bottom and the *Z*-axis to the front. For the mapping frame, the *X*-axis points to the left, the *Y*-axis to the top and the *Z*-axis to the front.

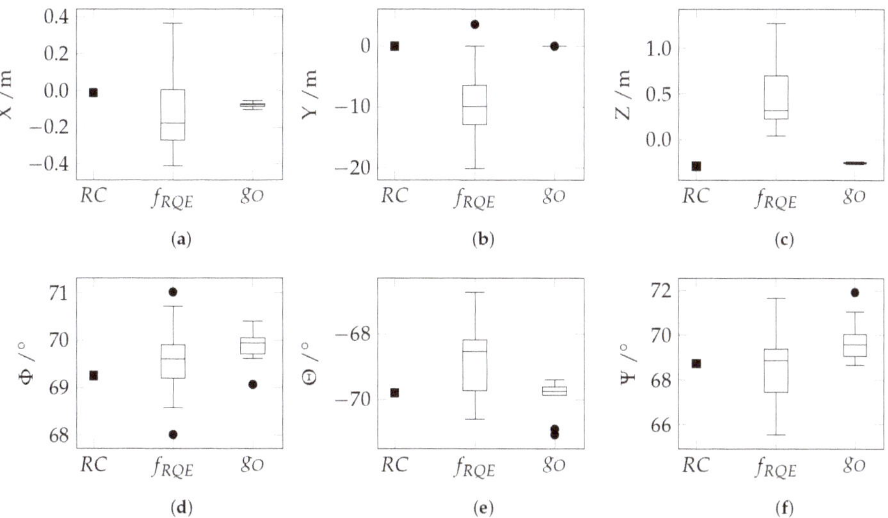

Figure 10. Calibration results of the reference calibration (RC), the self-calibration based on the *RQE* and our self-calibration based on the the feature of omnivariance for the large-scale outdoor dataset. (**a**–**c**) show the translation parameters, (**d**–**f**) show the rotation parameters. The estimation of the vertical component *Y* of the translation fails for the self-calibration based on *RQE*. The estimation of the calibration parameters is more precise for the self-calibration based on the omnivariance g_O, however, Φ and Ψ is estimated with a larger deviation from the reference calibration.

Table 3. Mean point-to-plane distances for different parts of the point cloud of the large-scale outdoor dataset (in cm). The smallest point-to-plane distance in each column is marked in bold. For three out of six planes the RC achieves the best results. For the other three planes the self-calibration based on the feature of omnivariance achieves the best results.

Approach	Plane 1	Plane 2	Plane 3	Plane 4	Plane 5	Plane 6
RC	5.4	**3.0**	5.4	**8.3**	**8.2**	16.4
f_{RQE}	11.4	10.3	7.9	8.6	10.0	23.1
g_O	**5.0**	4.5	**4.1**	8.7	9.4	**13.3**

The estimation of the vertical component Y of the translation fails for the self-calibration based on RQE as the error is about 10 m for this parameter. Nevertheless, objects like cars are identifiable in the resulting point cloud shown in Figure 9d. The reason is that a calibration error of -10 m for the vertical component causes the entire point cloud to be shifted 10 m upwards without affecting the position of one point relative to another. This point is discussed in more detail in Section 5.4.1. The estimation of the translation parameters is more precise as well as more accurate for the calibration based on the feature of omnivariance.

The result of the rotation parameters is more precise for the self-calibration based on the omnivariance, however, Φ and Ψ are estimated with a larger deviation from the reference calibration. As the self-calibration approach based on the RQE is not precise, a single calibration is likely to be less accurate than the result shown in Figure 10. However, the median of multiple calibrations shows errors of less than 1° for all rotation parameters.

5. Discussion

In the following, we discuss the results of each performed experiment in detail. First, we discuss the results of the investigations concerning the different cost functions, the radius of convergence, and the multi-scale approach. Subsequently, we discuss the experiments with real data in more detail.

5.1. Suitability of the Cost Function

The experiments in Section 4 have shown that it is possible to optimize initial calibration parameters with the self-calibration approach and most investigated cost functions. The cost functions based on linearity and planarity have proven to be unsuitable for the self-calibration. However, some of the cost functions lead to significantly better results than others. The experiments show that the cost functions based on the features omnivariance and eigenentropy outperform the other cost functions including the one based on the RQE. These results could be shown for synthetic as well as for real data.

When comparing Equations (4)–(10), it is noticeable that, in contrast to all other features, no quotient is formed from the eigenvalues for the features omnivariance and eigenentropy. In these cases, no ratio between multiple eigenvalues is taken into account. Instead, the omnivariance becomes small if at least one of the three eigenvalues is small. This ensures that linear and planar structures in the point cloud as well as isolated points lead to small values of the omnivariance. Moreover, a point that belongs to two perpendicular planes like the edge of a room also leads to a small value of the omnivariance. In contrast, structures which cause all three eigenvalues to be considerably larger than zero result in large values of the omnivariance. So disordered structures which result for example from wrong calibrations or from vegetation are penalized. The eigenentropy behaves similar to the omnivariance. However, structures that deviate from an ideal linear or planar structure are penalized more severely. So the feature eigenentropy is supposed to be less suited to calibrate inaccurate sensors with noisy measurements. The omnivariance and the eigenentropy lead to a more generic approach compared to other features and are, thus, supposed to be suitable for more kinds of environments. This is supposed to be the reason that these features are more suitable for self-calibration. Hence, it means that the only restriction is that the environment should not predominantly consist of regions where each eigenvalue of the structure tensor is large like for vegetation. In this case, the cost function is supposed to have a small gradient such that the optimization likely converges to a local minimum. Another reason might be, that the features omnivariance and eigenentropy have proven to be robust to noise, whereas the features linearity and planarity are very sensitive to noise [34].

Compared to the cost function based on the feature of omnivariance, the cost function based on the RQE leads to less accurate and less precise results. The cost function based on the RQE represents the sum over the weighted exponentiated distances between all possible points in the specified neighborhood [8]. Smaller distances are weighted higher. The distances are computed along the coordinate axis. The cost is therefore not only dependent on the extent of the considered points, but also on the orientation of the points with respect to the coordinate axes. The omnivariance essentially represents the volume of the 3D ellipsoid around the considered points oriented along its principal axes, since the eigenvalues can be interpreted as the extent of the 3D ellipsoid along the principal axes. Consequently, the extent is not determined along the coordinate axes, but along the principal axes of the considered points. This cost function is therefore independent of the orientation of the considered points with respect to the coordinate axes.

5.2. Radius of Convergence

Figure 4 shows that radius of convergence of the self-calibration is with more than 2 m and 30° impressively large. Consequently, the approach is robust against large initial translation and rotation errors. It is therefore sufficient in most cases to roughly estimate the initial translation and initial rotation parameters. Although the distortion of the point cloud increases for large initial calibration errors, the experiments show that approximating the angles between the sensors, measuring it with simple instruments or using the design drawings is accurate enough for the optimization to converge to the global minimum. It is also possible to compute the initial point cloud with Equation (2), tune the parameters manually until basic structures are identifiable and perform the optimization afterwards to obtain accurate calibration parameters.

5.3. Single-Scale vs. Multi-Scale

Figure 5a shows, that the fine-scale approach is robust to initial translation errors and achieves accurate results. However, for large initial translation errors, the estimation of the translation parameters fails. The coarse-scale approach is more robust to initial translation errors, but not as accurate as the fine-scale approach. The multi-scale approach achieves the most accurate results and is robust concerning the tested initial translation errors.

In contrast, Figure 5b shows that the fine-scale approach is not very robust to initial rotation errors. The calibration achieves less accurate results than the coarse-scale approach for initial rotation errors of more than 3° in this simulation. Even if the exact numbers of this investigation refer only to the given sensor noise, Figure 5b shows that the coarse-scale approach achieves better calibration results from a certain size of initial rotation error. The multi-scale approach also achieves the most accurate results for the tested initial rotation errors.

The coarse-scale approach is more robust to large initial calibration errors than the fine-scale approach. However, the fine-scale approach, is more accurate for small initial calibration errors. The multi-scale approach combines the advantages of both approaches, which means it is robust as well as accurate, i.e., if the initial rotation parameters are not known in advance, the multi-scale approach should be chosen. The only disadvantage of the multi-scale approach is an increased runtime. However, as the calibration is usually not time-critical, the self-calibration should be performed with multiple scales.

5.4. Real data

The real-data experiments show that the self-calibration is in principle applicable to different scenarios such as different sensor combinations (2D laser scanner + MCS and 3D laser scanner + a stereo camera together with ORB-SLAM2) and different environments (small-scale indoor and large-scale outdoor). The experiments with real data also show that the self-calibration approach is sufficiently robust against objects with high scattering characteristics like vegetation. This robustness is achieved by filtering or down-weighting the residuals that have large values.

5.4.1. Small-Scale Indoor Dataset

The sensor setup used for the small-scale indoor dataset is the most challenging for calibration. In contrast to 3D laser scanners, where two different 3D scans can be registered using well-known techniques like the ICP algorithm, the registration of lines of a 2D laser scanner in 3D is not possible solely based on the scanning data. Thus, a wide range of approaches is not applicable. However, we can compare our results to a state-of-the-art target calibration, which is based on RADLOCC and described in more detail by [18]. However, this target calibration can not be seen as ground truth as its result has proven to be less accurate than some of the self-calibrations.

Figures 6–8 and Tables 1 and 2 show, that the parameters estimated by our self-calibration lead to a sharper point cloud than the parameters that are estimated by the target calibration. However, this is not a proof of outperforming the reference calibration in general. In contrast to the target calibration, our self-calibration is universally applicable, completely data-driven, and does not require a calibration target.

In comparison to the self-calibration based on the RQE, the self-calibration based on the feature of omnivariance leads to a sharper point cloud as well. This can be seen particularly at the qualitative comparison of Figures 6f and 7b and the quantitative comparison in Table 2.

5.4.2. Large-Scale Outdoor Dataset

For the vehicle-based large-scale outdoor dataset, the result of our self-calibration is very similar to the reference calibration. However, neither the qualitative evaluation which is based on the appearance of the point cloud nor the quantitative evaluations are sufficient to assess which calibration is better. Anyway, the experiment with this dataset has proven the applicability of our self-calibration to vehicle-based mobile mapping with a 3D laser scanner and a stereo camera together with ORB-SLAM2 for pose estimation. Although the characteristics of the environment and the sensors are very different from the small-scale indoor dataset, the self-calibration improves the initial calibration parameters. The experiment thus shows the robustness of the self-calibration against these factors. In particular, the experiment also shows the suitability for a pose estimation approach that utilizes on-board sensors. However, systematic errors in pose estimation may have a negative impact on self-calibrations in general and should therefore be reduced to a minimum. The more inaccurate pose estimation and measurements of the laser scanner are supposed to be the main reasons that the results for the large-scale outdoor dataset are not as accurate as for the small-scale indoor dataset.

Compared to the automatic single-shot calibration of range and camera sensors [47] that was performed as the first step of the reference calibration of this dataset, our self-calibration is more accurate. The translation parameters are estimated with comparable precision. However, the estimation of the rotation parameters, is with approximately a factor of 10 much more accurate for our approach. One reason for this might be that our approach makes use of the stereo camera system, whereas the reference calibration is based on the measurements of a single camera. Another reason might be that our approach uses much more observations due to the nature of a self-calibration approach. Compared to the second step of the reference calibration, namely after manual refinement, the quality of our self-calibration is comparable.

The self-calibration based on the *RQE* again leads to less precise results compared to the self-calibration based on the feature of omnivariance. However, using the medians of multiple calibrations leads to point clouds with acceptable quality for most use cases.

6. Conclusions and Outlook

In this article, we propose a novel self-calibration approach for extrinsic calibration of Mobile Mapping Systems that include a mapping sensor and any kind of pose estimation sensor. Since it is a self-calibration, there is no need to involve calibration targets or manipulate the environment. Therefore, it can be performed based on the data that shall be collected for the 3D mobile mapping task. The approach uses geometric features that are computed based on the local neighborhood around each point of the cloud.

The experiments with synthetic data as well as with two different real datasets show the applicability to several sensor combinations in different environments. The investigations of the suitability of different cost functions demonstrate that the cost function based on the feature of omnivariance lead to more accurate results than the other investigated cost functions, including the cost function based on the Rényi quadratic entropy [8]. The self-calibration approach based on this feature has no assumptions about the environment. Coarse initial calibration parameters are assumed to be known. However, as the experiments with synthetic data show, the initial calibration parameters only need to be known approximately and can be roughly guessed in many cases, since the radius of convergence of the self-calibration is impressively large. Further, the experiments show that the use of a multi-scale approach considerably improves the calibration results. In contrast to single-scale approaches, the multi-scale approach is more robust as well as more accurate.

For all experiments with real sensor data, the accuracy of the calibration is about 1° for the rotation parameters. The accuracy of the translation parameters depends on the utilized sensors. For accurate sensors like in the small-scale indoor dataset, the accuracy of the translation is in the range of a few millimeters.

Future work focuses on the impact of different environments on the self-calibration. Further, it is theoretically possible to extend the presented self-calibration approach so that intrinsic parameters like a range offset of a laser scanner are estimated together with the extrinsic parameters.

In the current implementation, each point of the point cloud is used multiple times due to the neighborhood search. This is a disadvantage in terms of computational performance. In principle, there is potential here to make the approach more efficient. This could allow to use the basic ideas of our approach for on-line self-calibrations.

Author Contributions: Conceptualization, M.H., M.W., M.S.M. and B.J.; Formal analysis, M.H. and M.W.; Investigation, M.H.; Methodology, M.H. and B.J.; Project administration, B.J.; Software, M.H.; Supervision, B.J.; Validation, M.W. and M.S.M.; Visualization, M.H.; Writing—original draft, M.H. and M.W.; Writing—review & editing, M.H., M.W., M.S.M. and B.J.

Funding: This research received no external funding.

Acknowledgments: We thank Jochen Meidow for helpful discussions, which have contributed to a deeper understanding of the topic. We also thank Marcus Hebel and Joachim Gehrung for the preparation of the *MLS 1—TUM City Campus dataset* [48] so that it could be used for testing purposes in the context of our work. We acknowledge support by the KIT-Publication Fund of the Karlsruhe Institute of Technology.

Conflicts of Interest: The authors declare no conflict of interest.

References

1. Zhang, Q.; Pless, R. Extrinsic Calibration of a Camera and Laser Range Finder (Improves Camera Calibration). In Proceedings of the IEEE/RSJ International Conference on Intelligent Robots and Systems (IROS), Sendai, Japan, 28 September–2 October 2004; Volume 3, pp. 2301–2306.
2. Kassir, A.; Peynot, T. Reliable automatic camera-laser calibration. In Proceedings of the 2010 Australasian Conference on Robotics and Automation (ARAA), Brisbane, Australia, 1–3 December 2010.
3. Vasconcelos, F.; Barreto, J.P.; Nunes, U. A Minimal Solution for the Extrinsic Calibration of a Camera and a Laser-Rangefinder. *IEEE Trans. Pattern Anal. Mach. Intell.* **2012**, *34*, 2097–2107. [CrossRef] [PubMed]

4. Weinmann, M.; Jutzi, B.; Hinz, S.; Mallet, C. Semantic point cloud interpretation based on optimal neighborhoods, relevant features and efficient classifiers. *ISPRS J. Photogramm. Remote Sens.* **2015**, *105*, 286–304. [CrossRef]
5. Hackel, T.; Wegner, J.D.; Schindler, K. Fast semantic segmentation of 3D point clouds with strongly varying density. *ISPRS Ann. Photogramm. Remote Sens. Spat. Inf. Sci.* **2016**, *3*, 177–184.
6. Bueno, M.; Bosché, F.; González-Jorge, H.; Martínez-Sánchez, J.; Arias, P. 4-Plane congruent sets for automatic registration of as-is 3D point clouds with 3D BIM models. *Autom. Constr.* **2018**, *89*, 120–134. [CrossRef]
7. Frome, A.; Huber, D.; Kolluri, R.; Bülow, T.; Malik, J. Recognizing objects in range data using regional point descriptors. In Proceedings of the European Conference on Computer Vision, Prague, Czech Republic, 11–14 May 2004; pp. III:224–III:237.
8. Sheehan, M.; Harrison, A.; Newman, P. Self-calibration for a 3D laser. *Int. J. Robot. Res.* **2012**, *31*, 675–687. [CrossRef]
9. Kong, J.; Yan, L.; Liu, J.; Huang, Q.; Ding, X. Improved Accurate Extrinsic Calibration Algorithm of Camera and Two-dimensional Laser Scanner. *J. Multimed.* **2013**, *8-6*, 777–783. [CrossRef]
10. Yu, L.; Peng, M.; You, Z.; Guo, Z.; Tan, P.; Zhou, K. Separated Calibration of a Camera and a Laser Rangefinder for Robotic Heterogeneous Sensors. *J. Adv. Robot. Syst.* **2013**, *10-10*, 367–377. [CrossRef]
11. Hu, Z.; Li, Y.; Li, N.; Zhao, B. Extrinsic Calibration of 2-D Laser Rangefinder and Camera From Single Shot Based on Minimal Solution. *IEEE Trans. Instrum. Meas.* **2016**, *65*, 915–929. [CrossRef]
12. Sim, S.; Sock, J.; Kwak, K. Indirect Correspondence-Based Robust Extrinsic Calibration of LiDAR and Camera. *Sensors* **2016**, *16*, 933. [CrossRef]
13. Dong, W.; Isler, V. A novel method for the extrinsic calibration of a 2-D laser-rangefinder & a camera. In Proceedings of the 2017 IEEE International Conference on Robotics and Automation (ICRA), Singapore, 29 May–3 June 2017; pp. 5104–5109.
14. Li, N.; Hu, Z.; Zhao, B. Flexible extrinsic calibration of a camera and a two-dimensional laser rangefinder with a folding pattern. *Appl. Opt.* **2016**, *55*, 2270. [CrossRef]
15. Chen, Z.; Yang, X.; Zhang, C.; Jiang, S. Extrinsic calibration of a laser range finder and a camera based on the automatic detection of line feature. In Proceedings of the 2016 9th International Congress on Image and Signal Processing, BioMedical Engineering and Informatics (CISP-BMEI), Datong, China, 15–17 October 2016; pp. 448–453.
16. Zhou, L. A New Minimal Solution for the Extrinsic Calibration of a 2D LIDAR and a Camera Using Three Plane-Line Correspondences. *IEEE Sens. J.* **2014**, *14*, 442–454. [CrossRef]
17. Tulsuk, P.; Srestasathiern, P.; Ruchanurucks, M.; Phatrapornnant, T.; Nagahashi, H. A novel method for extrinsic parameters estimation between a single-line scan LiDAR and a camera. In Proceedings of the 2014 IEEE Intelligent Vehicles Symposium, Dearborn, MI, USA, 8–11 June 2014; pp. 781–786.
18. Hillemann, M.; Jutzi, B. UCalMiCeL–Unified intrinsic and extrinsic calibration of a Multi-Camera-System and a Laserscanner. *ISPRS Ann. Photogramm. Remote Sens. Spat. Inf. Sci.* **2017**, *4*, 17–24. [CrossRef]
19. Talaya, J.; Alamus, R.; Bosch, E.; Serra, A.; Kornus, W.; Baron, A. Integration of a terrestrial laser scanner with GPS/IMU orientation sensors. In Proceedings of the XXth ISPRS Congress, Istanbul, Turkey, 12–23 July 2004; Volume 35-B5, pp. 1049–1055.
20. Gräfe, G. High precision kinematic surveying with laser scanners. *J. Appl. Geod.* **2007**, *1-4*, 185–199. [CrossRef]
21. Heinz, E.; Eling, C.; Wieland, M.; Klingbeil, L.; Kuhlmann, H. Development, calibration and evaluation of a portable and direct georeferenced laser scanning system for kinematic 3D mapping. *J. Appl. Geod.* **2015**, *9-4*, 227–243. [CrossRef]
22. Jung, J.; Kim, J.; Yoon, S.; Kim, S.; Cho, H.; Kim, C.; Heo, J. Bore-sight calibration of multiple laser range finders for kinematic 3D laser scanning systems. *Sensors* **2015**, *15*, 10292–10314. [CrossRef] [PubMed]
23. Skaloud, J.; Lichti, D. Rigorous approach to bore-sight self-calibration in airborne laser scanning. *ISPRS J. Photogramm. Remote Sens.* **2006**, *61*, 47–59. [CrossRef]
24. Habib, A.F.; Kersting, A.P.; Shaker, A.; Yan, W.Y. Geometric calibration and radiometric correction of LiDAR data and their impact on the quality of derived products. *Sensors* **2011**, *11*, 9069–9097. [CrossRef]
25. Besl, P.; McKay, N.D. A method for registration of 3-D shapes. *IEEE Trans. Pattern Anal. Mach. Intell.* **1992**, *14*, 239–256. [CrossRef]

26. Chen, Y.; Medioni, G. Object modelling by registration of multiple range images. *Image Vis. Comput.* **1992**, *10*, 145–155. [CrossRef]
27. Ravi, R.; Shamseldin, T.; Elbahnasawy, M.; Lin, Y.J.; Habib, A. Bias Impact Analysis and Calibration of UAV-Based Mobile LiDAR System with Spinning Multi-Beam Laser Scanner. *Appl. Sci.* **2018**, *8*, 297. [CrossRef]
28. Maddern, W.; Harrison, A.; Newman, P. Lost in translation (and rotation): Rapid extrinsic calibration for 2D and 3D LIDARs. In Proceedings of the 2012 IEEE International Conference on Proceedings of the Robotics and Automation (ICRA), Saint Paul, MN, USA, 14–18 May 2012; pp. 3096–3102.
29. Wang, F.; Syeda-Mahmood, T.; Vemuri, B.C.; Beymer, D.; Rangarajan, A. Closed-form Jensen-Renyi divergence for mixture of Gaussians and applications to group-wise shape registration. In Proceedings of the International Conference on Medical Image Computing and Computer-Assisted Intervention, London, UK, 20–24 September 2009; pp. 648–655.
30. Maddern, W.; Pascoe, G.; Linegar, C.; Newman, P. 1 year, 1000 km: The Oxford RobotCar dataset. *Int. J. Robot. Res.* **2017**, *36*, 3–15. [CrossRef]
31. Jutzi, B.; Gross, H. Nearest neighbour classification on laser point clouds to gain object structures from buildings. *Int. Arch. Photogramm. Remote Sens. Spat. Inf. Sci.* **2009**, *38*, 4–7.
32. West, K.F.; Webb, B.N.; Lersch, J.R.; Pothier, S.; Triscari, J.M.; Iverson, A.E. Context-driven automated target detection in 3-D data. In *Automatic Target Recognition XIV*; International Society for Optics and Photonics: Orlando, FL, USA, 2004; Volume 5426; pp. 133–143.
33. Pauly, M.; Keiser, R.; Gross, M. Multi-scale feature extraction on point-sampled surfaces. *Comput. Graph. Forum* **2003**, *22*, 281–289. [CrossRef]
34. Dittrich, A.; Weinmann, M.; Hinz, S. Analytical and numerical investigations on the accuracy and robustness of geometric features extracted from 3D point cloud data. *ISPRS J. Photogramm. Remote Sens.* **2017**, *126*, 195–208. [CrossRef]
35. Munoz, D.; Bagnell, J.A.; Vandapel, N.; Hebert, M. Contextual classification with functional max-margin Markov networks. In Proceedings of the IEEE Conference on Computer Vision and Pattern Recognition, Miami, FL, USA, 20–25 June 2009; pp. 975–982.
36. Waldhauser, C.; Hochreiter, R.; Otepka, J.; Pfeifer, N.; Ghuffar, S.; Korzeniowska, K.; Wagner, G. Automated classification of airborne laser scanning point clouds. *Solving Computationally Expensive Engineering Problems: Methods and Applications*; Koziel, S., Leifsson, L., Yang, X.-S., Eds.; Springer: Cham, Switzerland, 2014; pp. 269–292.
37. Johnson, A.E.; Hebert, M. Using spin images for efficient object recognition in cluttered 3d scenes. *IEEE Trans. Pattern Anal. Mach. Intell.* **1999**, *21*, 433–449. [CrossRef]
38. Tombari, F.; Salti, S.; Di Stefano, L. Unique signatures of histograms for local surface description. In Proceedings of the European Conference on Computer Vision, Heraklion, Greece, 5–11 September 2010; pp. III:356–III:369.
39. Salti, S.; Tombari, F.; Di Stefano, L. SHOT: unique signatures of histograms for surface and texture description. *Comput. Vis. Image Underst.* **2014**, *125*, 251–264. [CrossRef]
40. Osada, R.; Funkhouser, T.; Chazelle, B.; Dobkin, D. Shape distributions. *ACM Trans. Graph.* **2002**, *21*, 807–832. [CrossRef]
41. Rusu, R.B.; Blodow, N.; Beetz, M. Fast point feature histograms (FPFH) for 3D registration. In Proceedings of the IEEE International Conference on Robotics and Automation, Kobe, Japan, 12–17 May 2009; pp. 3212–3217.
42. Rusu, R.B. Semantic 3D object maps for everyday manipulation in human living environments. *Künstl. Intell. (KI)* **2010**, *24*, 345–348. [CrossRef]
43. Huber, P.J. Robust statistics. In *International Encyclopedia of Statistical Science*; Springer: Cham, Switzerland, 2011; pp. 1248–1251.
44. Geiger, A.; Lenz, P.; Urtasun, R. Are we ready for autonomous driving? The KITTI vision benchmark suite. In Proceedings of the 2012 IEEE Conference on Proceedings of the Computer Vision and Pattern Recognition (CVPR), Providence, RI, USA, 16–21 June 2012; pp. 3354–3361.
45. Mur-Artal, R.; Tardós, J.D. ORB-SLAM2: An open-source SLAM system for monocular, stereo, and RGB-D cameras. *IEEE Trans. Robot.* **2017**, *33*, 1255–1262. [CrossRef]
46. Geiger, A.; Lenz, P.; Stiller, C.; Urtasun, R. Visual Odometry/SLAM Evaluation 2012. Available online: http://www.cvlibs.net/datasets/kitti/eval_odometry.php (accessed on 23 July 2019).

47. Geiger, A.; Moosmann, F.; Car, O.; Schuster, B. Automatic camera and range sensor calibration using a single shot. In Proceedings of the International Conference on Robotics and Automation (ICRA), Saint Paul, MN, USA, 14–18 May 2012; pp. 3936–3943.
48. Gehrung, J.; Hebel, M.; Arens, M.; Stilla, U. An Approach to Extract Moving Objects from MLS Data Using a Volumetric Background Representation. *ISPRS Ann. Photogramm. Remote Sens. Spat. Inf. Sci.* **2017**, *4*, 107–114. [CrossRef]

Article

A Flexible Architecture for Extracting Metro Tunnel Cross Sections from Terrestrial Laser Scanning Point Clouds

Zhen Cao [1], Dong Chen [1,*], Yufeng Shi [1], Zhenxin Zhang [2], Fengxiang Jin [3], Ting Yun [4], Sheng Xu [4], Zhizhong Kang [5] and Liqiang Zhang [6]

1. College of Civil Engineering, Nanjing Forestry University, Nanjing 210037, China; caozhennjsg@gmail.com (Z.C.); yfshi@njfu.edu.cn (Y.S.)
2. Advanced Innovation Center for Imaging Technology, Capital Normal University, Beijing 100048, China; zhangzhx@cnu.edu.cn
3. School of Surveying and Geo-Informatics, Shandong Jianzhu University, Jinan 250101, China; fxjin@sdjzu.edu.cn
4. College of Information Science and Technology, Nanjing Forestry University, Nanjing 210037, China; njyunting@gmail.com (T.Y.); xusheng404@gmail.com (S.X.)
5. School of Land Science and Technology, China University of Geosciences, Beijing 100083, China; zzkang@cugb.edu.cn
6. The State Key Laboratory of Remote Sensing Science, Faculty of Geographical Science, Beijing Normal University, Beijing 100875, China; zhanglq@bnu.edu.cn

* Correspondence: chendong@njfu.edu.cn

Received: 5 January 2019; Accepted: 29 January 2019; Published: 1 February 2019

Abstract: This paper presents a novel framework to extract metro tunnel cross sections (profiles) from Terrestrial Laser Scanning point clouds. The entire framework consists of two steps: tunnel central axis extraction and cross section determination. In tunnel central extraction, we propose a slice-based method to obtain an initial central axis, which is further divided into linear and nonlinear circular segments by an enhanced Random Sample Consensus (RANSAC) tunnel axis segmentation algorithm. This algorithm transforms the problem of hybrid linear and nonlinear segment extraction into a sole segmentation of linear elements defined at the tangent space rather than raw data space, significantly simplifying the tunnel axis segmentation. The extracted axis segments are then provided as input to the step of the cross section determination which generates the coarse cross-sectional points by intersecting a series of straight lines that rotate orthogonally around the tunnel axis with their local fitted quadric surface, i.e., cylindrical surface. These generated profile points are further refined and densified via solving a constrained nonlinear least squares problem. Our experiments on Nanjing metro tunnel show that the cross sectional fitting error is only 1.69 mm. Compared with the designed radius of the metro tunnel, the RMSE (Root Mean Square Error) of extracted cross sections' radii only keeps 1.60 mm. We also test our algorithm on another metro tunnel in Shanghai, and the results show that the RMSE of radii only keeps 4.60 mm which is superior to a state-of-the-art method of 6.00 mm. Apart from the accurate geometry, our approach can maintain the correct topology among cross sections, thereby guaranteeing the production of geometric tunnel model without crack defects. Moreover, we prove that our algorithm is insensitive to the missing data and point density.

Keywords: terrestrial laser scanning; tunnel central axis; tunnel cross section; enhanced RANSAC; quadric fitting; constrained nonlinear least-squares problem

1. Introduction

Tunnel displacement and deformation monitoring are critical factors for engineers to evaluate the tunnel health and safety. Some indices such as the cross section's chord [1], convergence and dislocations [2], tunnel axis's settlement [2], and tunnel model's clearance [3] are important measures to evaluate the stability of tunnels. At the initial stage, some traditional methods are used to monitor the tunnel's health. The instruments such as tape extensometer [4] and total station [5] are extensively used for high precision deformation monitoring. Although they have a very high precision, these traditional ground surveying measurements are time-consuming, labor-intensive and costly. In some cases, they fail to get the deformation because sparsely measured points are not dense enough to evaluate the overall tunnel. Apart from the above geodetic surveys, an alternative method based on digital photogrammetry has been proposed because more measured points along a profile can be extracted efficiently. However, many of the acquired images easily suffer from occlusions, low contrast and unfavorable perspectives, thereby making automatic image matching and understanding harder. For example, by combining stereo images and laser-lit spots, Wang et al. [5,6] propose a profile-image method to measure cross sections. Although this method captures the densified points along a profile more than traditional geodetic surveys, it is restricted by insufficient lighting conditions in an actual tunnel. We recommend the readers to the recent survey [7] to get the review of the state-of-the-art methods and applications in tunnel inspections based on photogrammetric and computer vision techniques. Recently, the Terrestrial Laser Scanning (TLS) technique has been developed to efficiently and accurately collect three-dimensional point clouds from the reflected objects. Since the advantages of TLS technique include being highly accurate, efficient, automatic and has a lack of contact, the TLS technique has been proved to be an effective way to monitor the deformation of engineering structures, particularly in the fields of tunnel management [8,9] and deformation analysis [10–12].

Based on TLS point clouds, tunnel deformation monitoring techniques can be generally categorized into profile-based methods and model-based methods. The profile-based method tries to analyze the tunnel's deformation by using a series of thin profiles directly derived from TLS tunnel point clouds. Numerous published papers along this line [1,13–15] demonstrate the accuracy and efficiency of this type of method. For example, Han et al. [13] propose an automated and efficient method to estimate the tunnel centerline and cross-sectional plane. In their method, the centerline is first estimated by skeletonizing a binary image produced by projecting three-dimensional point clouds onto a horizontal plane. Based on the tunnel boundary points and the centerline, the cross-sectional planes are estimated and further adjusted, and the final cross sections are generated by projecting the nearby points to the adjusted planes. Although the authors improve the efficiency of surveying and data processing, they cannot extract the continuous profiles because the parametric equation of tunnel centerline is not provided. In addition, this method is sensitive to the non-lining points, i.e., pipes and equipment attached to the lining. To overcome the drawback regarding the above centerline, the methods in [1,3] use multiple models, i.e., straight line, transition curve and curve to commonly fit the different segments of tunnel centerline, and then adopt the global least-squares adjustment to maintain the consistency between adjacent fitting models. To cope with missing data, Kang et al. [1] recover the blank areas of tunnel profiles with a surficial interpolation algorithm. To obtain the accurate direction of the cross-sectional planes, the authors in [14] optimize the initial cross sections by means of twice adjustments using the total least squares method and Rodrigues' rotation formula, respectively.

The critical step for profile-based methods is to obtain the real cross-sectional points. To this end, some methods focus on how to eliminate non-lining points (accessories) from the raw point clouds. Generally, two types of methods are used to remove non-lining points. The first method is based on the geometric characteristics of the tunnel. More specifically, it is assumed that the tunnel can be fitted by the circular model [1,11,12,14,15] and then uses the circular filtering to filter cables and other equipment in the tunnel. The second method separates liner and accessories by combining TLS geometric and radiometric information. For example, the results in [16] demonstrate that the corrected intensity information is an effective physical criterion and a complementary data source to remove accessories that cannot be eliminated by sole geometric information. In another method, Yoon et al. [8] use the radiometric characteristics for Mobile Laser Scanning (MLS) data to filter the accessories from the tunnel raw data, which cannot be directly used for processing the TLS point clouds because not only the scattering property of an object but also other variables, especially distance and incidence angle effects will influence backscattered energy in TLS data [3]. To address this deficiency, Tan et al. [17] further investigate the effect of distance on the intensity of TLS data and propose a new correction method for different TLS scanners to overcome the deficiency of directly using the original intensity.

Although the acquisition of actual geometric parameters for inspection of tunnel is efficient, accurate and has flexible implementation from TLS point clouds, the profile-based works are more restricted regarding pointwise and piecewise description of the profiles. That is, the profile-based method has some potential risks between the areas of adjacent cross sections, where the slight deformations might not be tracked. To address this issue, the profile-based method has been expanded to an overall evaluation of tunnel models. For example, Fekete et al. [9] reconstruct the models by means of triangulating tunnel point clouds to detect the underbreak and overbreak areas. Dimitrov and Golparvar-Fard [18] use the non-uniform rational B-spine (NURBS) fitting technique to achieve parametric representation of arbitrary 3D geometries from point clouds. Nahangi and Haas [19] present an algorithm for automated quantification of discrepancies for components of assemblies by using skeletonized 3D models and TLS laser scans. Qiu and Cheng [3] propose a novel clearance inspection technique to generate the high-resolution digital surface model of a railway tunnel surface (bare lining) from TLS point clouds. By introducing a high-accuracy interpolation and filtering algorithm, this method can generate bare-lining models without limitation of tunnel cross section shapes. In Ref. [20], the scanned surface of a bored tunnel is approximated with a cylindrical model, and the point-wise deformation analysis is performed by comparing surface patches. Although the tunnel model-based method can detect the minor differences by comparison with the tunnel models generated at different times, this type of method is sensitive to non-lining point clouds, which means the real deformation and displacement can be hidden behind the artifacts of the created tunnel models. In addition, the generation of tunnel models ranging from geometric primitives, e.g., planes, cylinder or spheres, to more complex ones, such as parametric patches and NURBS is time-consuming, which does not meet the requirement of real-time interaction for processing large-scale TLS point clouds.

In this paper, we employ the hybrid concept by combining the profile-based and model-based methods for tunnel deformation monitoring. More specifically, based on the concept of the profile-based method, we propose a novel cross-sectional extraction method, which not only gets the continuous points within each cross section, but also more importantly has the capability to generate continuous profiles over the entire metro tunnel. Utilizing these continuous points and profiles, the entire tunnel model can be easily generated by triangulation. Thus, if we have two-phase TLS tunnel data acquired at different times, we can evaluate the whole differences by the superposition of two phase tunnel models. Given that our work is built on profile-based framework, we explicitly state our original contributions as follows:

- **Tunnel Central Axis Extraction:** we present an algorithm for tracking the tunnel central axis in topologically-ware and geometrically-aware manners. That is, we first propose a slice-based algorithm for tracking the initial central axis, which is further refined by integrating a Minimum Spanning Tree (MST) and Savitzky–Golay smoothing.
- **Tunnel Central Axis Segmentation:** we propose an enhancement to the traditional Random Sample Consensus (RANSAC) algorithm. The enhanced algorithm translates the problem of hybrid linear and nonlinear segment extraction into a sole segmentation of linear elements defined at the tangent space rather than raw data space, significantly simplifying segmentation problem.
- **Tunnel Profile Determination:** we introduce a high-accuracy interpolation and filtering algorithm for extracting continuous tunnel profiles over the entire tunnel. The generated profile points are determined by using interpolation, and further densified via solving a constrained nonlinear least squares formulation.

This paper is organized as follows: Section 1 reviews profile-based and model-based methods in tunnel deformation. Section 2 describes the detailed methodology including tunnel centerline extraction and cross section determination. In Section 3, the experimental dataset, the performance evaluation results of geometric and topological accuracy based on actual TLS tunnel data are presented, analyzed, and discussed. Finally, Section 4 concludes the paper along with a few suggestions for future research topics.

2. Methodology

2.1. Tunnel Axis Extraction via the Slice-based Method

The tunnel axis determines the tunnel's position and orientation, and meanwhile reflects the distribution of the tunnel points in three-dimensional space. The tunnel axis is also a basis for extraction of a cross-sectional plane at a specific location because the tunnel's central axis needs to be orthogonal to all of the cross sections. To accurately extract the tunnel axis, in this paper, we introduce a slice-based method, which transforms the problem of extracting a three-dimensional tunnel axis into extracting a two-dimensional axis problem at two projected spaces, i.e., xy- and xz-planes. That is, we use two-axis equations in two-dimensional space to commonly represent the tunnel axis equation in three-dimensional space.

To get the tunnel axis in the projected xy-plane, the projected point clouds are first divided into a series of slices at equal intervals along the direction of x-axis. In each slice, the minimum and maximum values on y-axis are determined. The mean point calculated by averaging the minimum and maximum points is regarded as an axis point within the corresponding slice. All of the mean points constitute the complete tunnels' axis points in the projected xy-plane. As shown in Figure 1a, the point clouds are partitioned into a series of slices by the dash lines at a given interval δ. The point $\bar{p}_i$ is an estimated tunnel axis point calculated by averaging the points p_i^{max} and p_i^{min} within i-th slice. Using the above method, all of the points denoted in red are extracted, representing the final tunnel axis points in the projected xy-plane. Similarly, the tunnel axis' points are extracted from the projected xz-plane as evident in Figure 1b.

It should be noted that the calculated axis points in the beginning and ending areas might deviate from the real position of a tunnel axis due to irregular shapes of the raw data. As these pseudo axis points indicated by red ovals in Figure 1 undoubtedly weaken the accuracy of the extracted tunnel axis, they are eliminated using a local curvature measure. After that, the axis points in the xy-plane are fitted by a fifth degree polynomial regression model, i.e., $f_{xy}(x) = \sum_{i=0}^{n} a_i x^i$. Similarly, axis points in the xz-plane are fitted by a linear regression model, i.e., $f_{xz}(x) = ax + b$ (see Section 2.2.3). Therefore, the tunnel axis in three-dimensional data space can be represented through joining these two models together:

$$\mathcal{L} = \begin{cases} y = f_{xy}(x), \\ z = f_{xz}(x). \end{cases} \tag{1}$$

Compared with the designed location of the tunnel axis, the height of the tunnel axis in the xz-plane has been raised and is usually higher than the theoretical (designed) value. This discrepancy is that, when we scan tunnel point clouds, the track laying has been finished, thereby raising the bottom of a tunnel. Their geometric relationships can be vividly shown in Figure 2, from which the extracted axis point o in red is higher than a center of real tunnel, denoted by a black point o'.

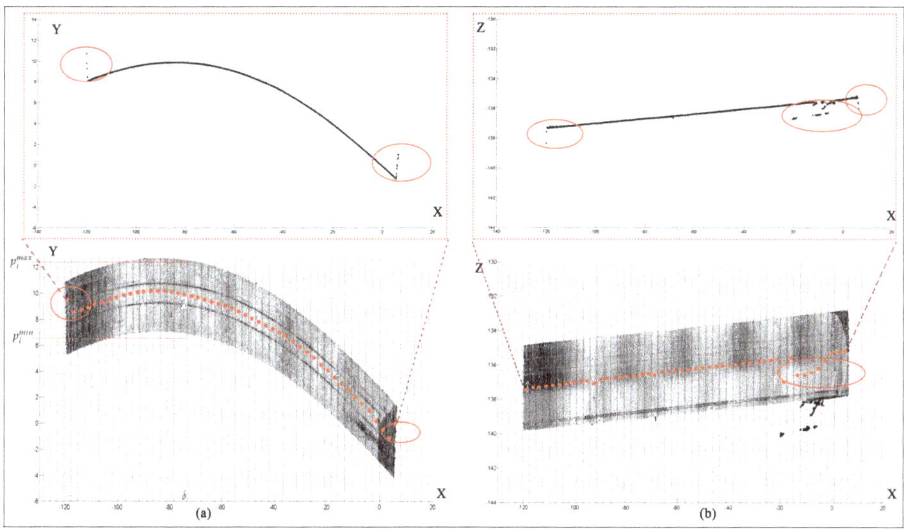

Figure 1. Extraction of tunnel axis in the two projected spaces. (**a**) tunnel axis points in *xy*-plane; (**b**) tunnel axis points in the *xz*-plane. The gray points represent the raw tunnel point clouds and the red points represent the calculated tunnel axis points.

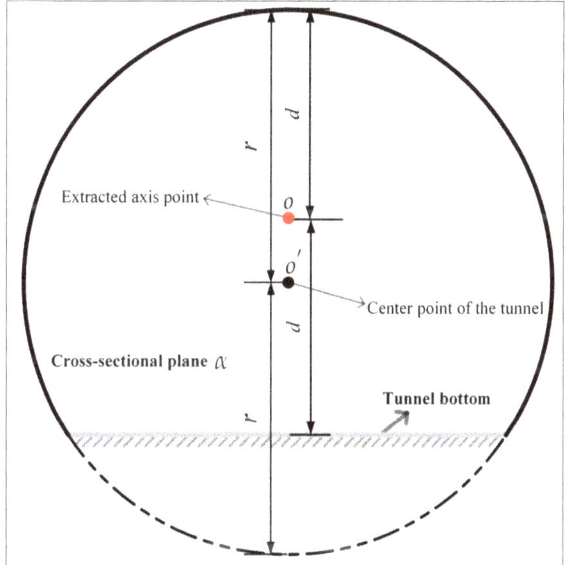

Figure 2. Geometric relationships between the designed and the extracted axes.

2.2. Tunnel Axis Segmentation

According to the planning requirements regarding tunnel safety, the tunnel generally includes linear parts and nonlinear/circular segments with fixed curvature [3]. Once the tunnel axis has been extracted, it should be further segmented into linear and nonlinear segments. For the linear segments, we refit them by the linear least squares regression. However, we use a high order polynomial model to fit the nonlinear segments. The advantage of axis segmentation is twofold: (1) The tangent and directional vectors corresponding to circular and linear segments can be easily determined. This means the axis direction for any point along the axis can be estimated, which prepares the conditions for the subsequent extraction of cross sections as the tunnel axis should be orthogonal to the cross sections. (2) The axis segmentation can help us to achieve the tunnel's geometric representation at multiple levels of details (LoDs). More specifically, for linear segments, a few tunnel cross sections are enough to fit/represent the tunnel geometry. In contrast, for circular segments or the transition zones between linear and circular segments, the high density of cross sections are needed to achieve an accurate geometric representation. Through tunnel axis segmentation, we can enhance the flexibility of geometric representation via using the concept of LoD representation. Meanwhile, we can also make a good balance between the tunnel's geometric accuracy and compactness via using a few tunnel cross sections.

The proposed axis segmentation algorithm consists of three stages: (1) We first restore the topological relationships of tunnel axis points using MST. (2) The sequence axis points are then provided as input to the smoothing step which uses the Savitzky–Golay (SG) [21] filtering algorithm to refine the axis. (3) In the final step, the axis has been divided into linear and nonlinear segments using an enhanced RANSAC [22] algorithm, which works at the tangent space rather than the data space defined in the traditional sense.

2.2.1. Constructing MST of Point Clouds along the Tunnel Axis

The initial extracted axis points in Section 2.1 are unorganized and without topology. We need to restore the topological relationships, i.e., clock-wise or counterclockwise order of these points. To this end, we first construct the MST, from which the weight on any edge is the Euclidean distance between two points. After the completion of the MST, the maximum depth of a subtree is traversed via the depth-first searching (DFS) algorithm. This subtree implicitly includes the topological relationship among axis points, i.e., clockwise or counterclockwise sequence of axis points. As shown in Figure 3a, the central axis of a tunnel can be well recognized via undirected graph MST although it has a certain degree of noises along the central axis. After DFS traversing, in Figure 3b, the sequence of the central axis points connected by the green lines is well maintained. Apart from restoring topology of axis points, it can be observed in Figure 3b, our MST-based traversing strategy has the capability to remove a certain degree of noises which deviates far from the tunnel axis to some extent, thereby refining the axis points.

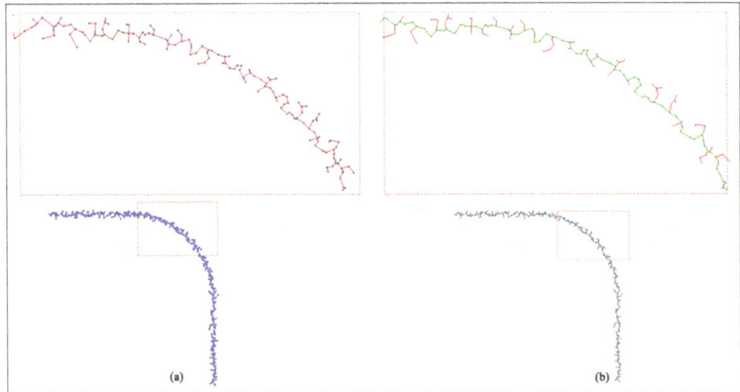

Figure 3. Restoring the topological relationship among axis points via MST (Minimum Spanning Tree). (**a**) the axis point clouds are organized by an undirected graph MST; (**b**) the maximum depth of a subtree is recognized and overlaid on the MST. The blue points present the initial axis points. The red and green lines represent the established MST and the extracted axis, respectively.

2.2.2. Smoothing Central Axis of a Tunnel

To obtain more accurate axis segments, the central axis acquired in Section 2.2.1 needs to be further smoothed to depress the noises. Fortunately, we find that the SG smoothing filtering is an ideal tool to strike a balance between maintenance of axis details and depression of noise. Essentially, this algorithm performs a local least squares polynomial regression of degree k over at least $k + 1$ points.

Through SG smoothing, the linear and nonlinear patterns/characteristics of the central axis become more obvious and it can be easily recognized in the subsequent segmentation algorithm. As shown in Figure 4, we compare the distribution of points along the central axis in the tangent space, i.e., (d, θ) plane. Parameter d represents the sum of the lengths of any two connected points along a central axis limited from the beginning point to the current processing point, and θ represents the angle between the tangent to the current points and the horizontal axis defined in the tangent plane. Obviously, a linear segment of a tunnel is expected to be transformed into a horizontal linear segments in the tangent space. However, a circular segment with a fixed curvature is anticipated to be transformed into a sloping linear segment because the angle between tangents at endpoints of the circular arc is proportional to the length of the arc. In practice, even though we perform the axis segmentation in tangent plane, the axis points contain too many details and "zigzag" noises because irregularly spaced axis point measurements can blur the horizontal and the sloping linear characteristics in tangent space.

We use two types of the geometric tunnel axis based on synthetic points to prove the applicability of the SG algorithm. We adopt the synthetic data rather than the real tunnel point clouds because the Nanjing tunnel dataset used in our paper has a small curvature, which directly results in the curvature of the extracted tunnel axis is not prominent in the projected xy-plane. Because of this, the real tunnel point clouds cannot fully test the applicability of the SG filtering algorithm and its influences for the follow-up tunnel axis segmentation algorithm. As shown in Figure 4a, two kinds of geometric tunnels axis points without noise are transformed into tangent plane, where the linear features are prominent. However, after adding a certain degree of noise, the linear distribution patterns in (d, θ) space are unrecognizable as proved in Figure 4b. After further performing the SG smoothing, the linear distribution of axis points in Figure 4c becomes notable again although it is not as prominent as the situation in Figure 4a.

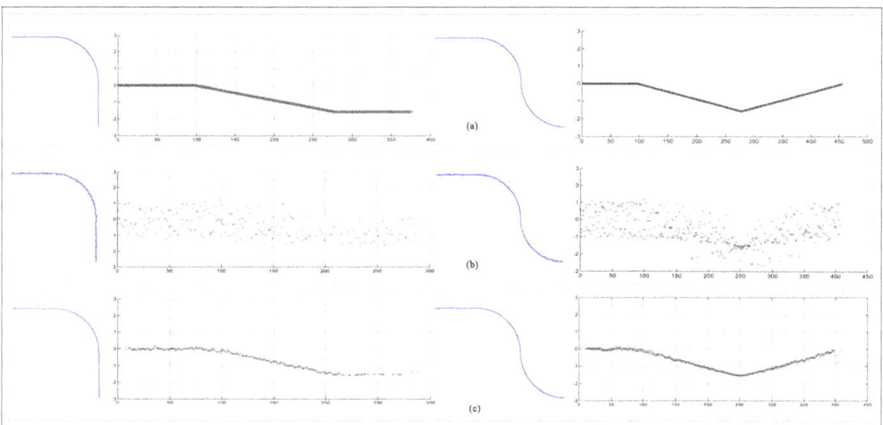

Figure 4. Comparisons of linear distributions of axis points in tangent space before and after SG (Savitzky-Golay) smoothing. (**a**) the ideal tunnel axis points without noise in their data and tangent spaces; (**b**) some degree of Gaussian noises have been added into the axis point clouds; (**c**) the distribution of axis points in data and tangent spaces after SG smoothing.

2.2.3. Segmentation of the Tunnel Axis

We employ an enhanced RANSAC algorithm to partition the central tunnel axis into linear and circular segments. If the conventional RANSAC algorithm is simply used to simultaneously recognize linear and circular segments in data space, over-segmentation often occurs. For example, a circular segment with low curvature might be mistakenly segmented as multiple linear segments. To solve this problem, we transform the question of detecting the linear and circular segments from the original data space into the question of only detecting linear segments in the transformed tangent space, significantly reducing the complexity of segmentation of the tunnel axis. An enhanced RANSAC algorithm is described as follows:

(a) The central axis point clouds are transformed into tangent space (d, θ).
(b) An optimal linear segment is first detected from (d, θ) space, and then the axis point clouds are divided into inliers and outliers.
(c) We determine the extent of inliers in tangent space. If any point from outliers is inside the extent of the inliers, the corresponding point is removed from the outliers.
(d) The point clouds from the remaining outliers are provided as input. The steps from (b) to (c) are executed repeatedly until all of the remaining outliers have been processed.
(e) The sequence of linear elements extracted in tangent space is restored according to the sequence of their middle points. Hence, we obtain anchor points by circulating around pairwise neighboring linear segments. More precisely, if a linear segment $\mathcal{S}_i$ is a neighbor with a linear primitive $\mathcal{S}_j$, there is an anchor point whose position is exactly at the intersection of $\mathcal{S}_i$ and $\mathcal{S}_j$.
(f) Using these anchor points, the linear and circular segments in data space are simultaneously determined, since the axis points have a one-to-one mapping relationship between data and tangent spaces.

It should be aware that we only need two parameters to execute the enhanced RANSAC algorithm: (1) the probability P that at least one of the selected axis points does not contain an outlier, and (2) the Euclidean distance ϵ from a point to the hypothetical linear model that determines the number of points consistent with the linear model in the tangent space. In our case, we set parameter P to 0.99, which ensures the optimal linear model within a specific number of iterations controlled by P can be obtained. The other parameter ϵ is determined by the definition of the tangent space. We set ϵ to 0.01. It is to be noted that the parameter of ϵ in the tangent space is dimensionless although its unit is consistent with the unit of raw point clouds in original data space. Figures 5 and 6 show the tunnel axis segmentation results with regard to synthetic and real data. It can be seen that the proposed method can effectively segment the linear and nonlinear axis geometry by introducing the tangent space. Meanwhile, thanks to the SG filter, it makes our axis segmentation insensitive to the noises.

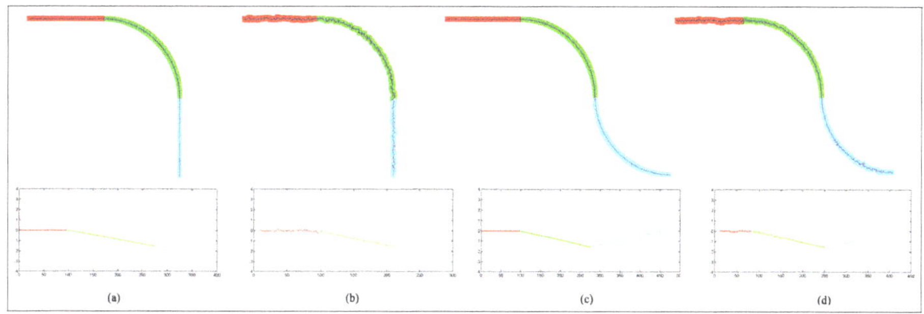

Figure 5. Comparisons of tunnel axis segmentation with and without noise. (**a**,**c**) are results of tunnel axis point clouds without noise; (**b**,**d**) are results of the tunnel axis after adding a certain degree of Gaussian noise.

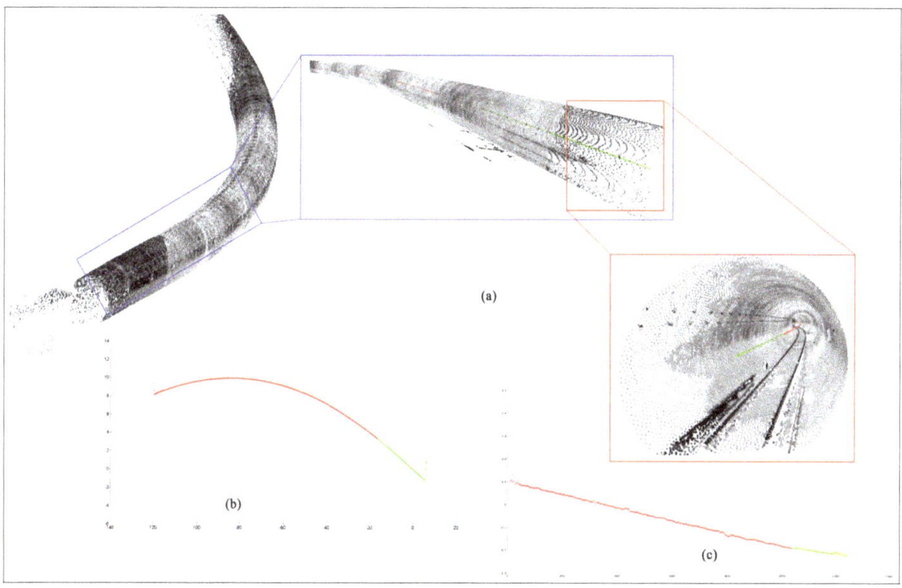

Figure 6. The central tunnel axis segmentation of real data. (**a**) the segmented three-dimensional axis is superimposed onto their associated tunnel point clouds; (**b**,**c**) represent the two-dimensional segmented tunnel axis in data space (xy-plane) and tangent space, respectively.

After segmentation, we use the linear regression model to fit the linear axis model. The polynomial regression model is employed to fit the axis points for nonlinear tunnel segments. The details are described below:

Generally, the polynomial of degree n can be written into:

$$y = a_0 + a_1 x + a_2 x^2 + \cdots + a_n x^n. \tag{2}$$

As the variable x is the observation value, we let $X_1 = x, X_2 = x^2, \ldots, X_n = x^n$. The nonlinear Equation (2) can be transformed into the linear equation below:

$$y = a_0 + a_1 X_1 + a_2 X_2 + \cdots + a_n X_n. \tag{3}$$

We assume that there are m points on the nonlinear axis. Based on Equation (3), these points constitute a set of equations:

$$\begin{cases} y_1 = a_0 + a_1 X_{11} + a_2 X_{12} + \dots a_n X_{1n}, \\ y_2 = a_0 + a_1 X_{21} + a_2 X_{22} + \dots a_n X_{2n}, \\ \dots \\ y_m = a_0 + a_1 X_{m1} + a_2 X_{m2} + \dots a_n X_{mn}. \end{cases} \quad (4)$$

Equation (4) can be further transformed into a matrix equation:

$$Y = BA, \quad (5)$$

where $Y = [y_1, y_2, \dots, y_n]^T$, $B = \begin{bmatrix} 1 & X_{11} & X_{12} & \dots & X_{1n} \\ 1 & X_{21} & X_{22} & \dots & X_{2n} \\ \vdots & \vdots & \vdots & \vdots & \vdots \\ 1 & X_{m1} & X_{m2} & \dots & X_{mn} \end{bmatrix}$, and $A = [a_0, a_1, \dots, a_n]^T$. Y and B are estimated from the observed data. Equation (5) can be solved using the linear least squares techniques:

$$Y + e = (B + E_B)A, \quad (6)$$

where e and E_B are the error matrix of Y and B, respectively. In linear algebra, the coefficients of matrix A can be solved through the singular value decomposition (SVD) [23,24] of an augmented matrix $[BY]$.

2.3. Tunnel Cross Section Extraction

It consists of four steps to extract cross-sectional points at an arbitrary point o along the tunnel axis: (1) We first calculate the cross-sectional plane α at an arbitrary point o on the tunnel axis. (2) We generate a series of straight lines that rotate orthogonally around the tunnel axis within α at a specific rotation interval $\Delta\varphi$. (3) According to each straight line l_i, we determine its associated local tunnel point set, which is further fitted by a cylinder model. That is, l_i can pass through the region determined by the local tunnel point clouds. (4) The cross-sectional points are then produced by intersecting a series of straight lines with their associated cylinder surfaces.

2.3.1. Determination of the Cross-Sectional Plane

Based on Equation (1), we can get a directional vector of an arbitrary point $o = \{o_x, o_y, o_z\}$ on the tunnel axis $\mathcal{L}$, i.e., $\vec{n}_o = \{1, f'_{xy}|_{x=o_x}, f'_{xz}|_{x=o_x}\}$. Obviously, the cross section that passes through o can be determined by $\alpha = \vec{op} \cdot \vec{n}_o$, from which the symbol p is an arbitrary point that is different from a point o on plane α. Symbol "$\cdot$" is a dot product of two vectors. It should be noted that, for a linear segment, $\vec{n}_o$ is a directional vector of the fitted tunnel's axis segment. For a nonlinear segment, $\vec{n}_o$ is a tangent vector of the fitted nonlinear axis segment on point o (see Section 2.2.3).

2.3.2. Determination of a Straight Line Equation

This step aims to partition cross-sectional plane α according to a given rotation interval angle $\Delta\varphi$. After being partitioned, a series of straight lines can be obtained. Actually, the number of straight lines equals the number of cross-sectional points. As shown in Figure 7a, to get an arbitrary straight line equation from α, we select an arbitrary point p from α, and a new vector $\vec{l}$ from a plane α can be determined via $\vec{l} = \vec{op}$. Another different vector $\vec{m}$ from a plane α can also be simultaneously determined via $\vec{m} = \vec{n_o} \times \vec{l}$, where the symbol "$\times$" represents a cross product of two vectors. Based on the two vectors mentioned above $\vec{l}$ and $\vec{m}$, an arbitrary vector $\vec{k}$ on a plane α can be represented by a linear combination of the normalized vectors $\vec{l}$ and $\vec{m}$. That is,

$\vec{k} = (\vec{l}/|\vec{l}|)\cos\varphi + (\vec{m}/|\vec{m}|)\sin\varphi$. A straight line equation, i.e., $\mathcal{L}_k = \frac{x-o_x}{\vec{k}_x} = \frac{y-o_y}{\vec{k}_y} = \frac{z-o_z}{\vec{k}_z}$ is solely determined through vector $\vec{k} = \{\vec{k}_x, \vec{k}_y, \vec{k}_z\}$ and $\vec{o}$.

2.3.3. Determination of Cross-Sectional Points

This step aims to determine all of the cross section points on the plane α. To this end, we first determine the corresponding local tunnel point clouds regarding each straight line $\mathcal{L}_k$. Then, the local tunnel point clouds are fitted by the nonlinear cylinder surface model. The intersection points between $\mathcal{L}_k$ and the cylinder surface are obtained and the potential pseudo intersection point is further removed under the constraint of local tunnel point clouds' data extent.

More precisely, we first get a point q_k from a straight line $\mathcal{L}_k$ and q_k is subject to the constraint $q_k = o + r\vec{k}$. As shown in Figure 7b, it is obvious that all of the q_k from the straight lines constitute the circle with center o and radius r. As we previously mentioned in Section 2.1, the centers o and o' do not overlap, which means that the path of q_k is not consistent with the real tunnel surface. Once we determine each q_k, the neighborhood searching centered at q_k is performed. The local point set $\mathcal{Q}_k$ indicated by green ovals are expected to be obtained. It is noted that the number of points within $\mathcal{Q}_k$ is determined by $\rho \times \mathcal{A}$, from which $\mathcal{A}$ is the area of searching extent and ρ is the density of tunnel point clouds. To improve the computational efficiency of the neighborhood searching, the K-D is established using the tunnel points in advance. For each local point set $\mathcal{Q}_k$, the conventional RANSAC algorithm is used to fit a cylinder surface whose parameters include an axis of a cylinder, an arbitrary point from the axis, and a radius of a cylinder. The intersection points between $\mathcal{L}_k$ and its corresponding cylinder surface can be calculated on the plane α. It should be aware that, for each straight line, two intersection points can be obtained, but one of them is a pseudo point. In this case, the pseudo point should be eliminated under the constraint of local tunnel points' extent. As a result, the real intersection point remains and can be regarded as one of the tunnel's cross-sectional point corresponding to $\mathcal{L}_k$. The above process is executed repeatedly until all of the cross-sectional points on the plane α are calculated.

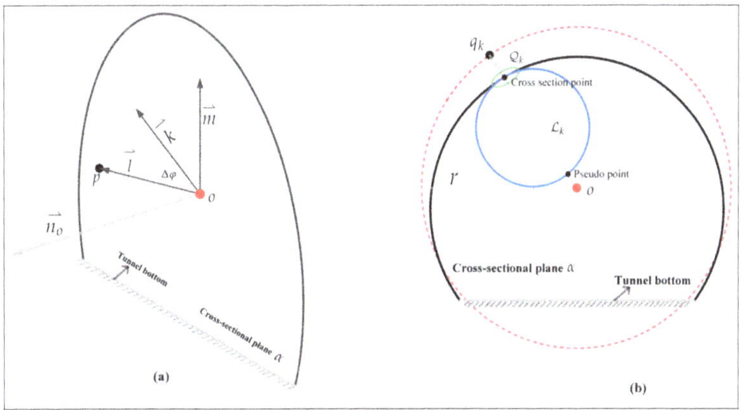

Figure 7. The diagram of determination of cross-sectional planes and points. (**a**) cross-sectional plane determination; (**b**) the geometric relationship for determination of cross-sectional points. Note that the dashed circle in red represents the auxiliary circle centered at axis point o, while the black circle denotes the real tunnel profile shape centered at point o'. The blue circle represents the fitted cylinder using the local point set in $\mathcal{Q}_K$.

In practice, it can be observed that some extracted cross sections have a degree of missing data as it is evident in Figure 8b. That is, some points on the specific cross sections are missing, causing the cross section's points to be incomplete. This deficiency is mainly because the point set Q_k includes outliers and noises, or probably because the ratio of the number of liner points in Q_k is not sufficient to fit a cylinder surface. To further refine the extracted cross sections of a tunnel, we adopt a strategy of "fitting-and-resampling". More specifically, we refit the points of each cross section via the circular model and then get the complete cross-sectional points at a specific rotation interval $\Delta\varphi$ through resampling from the fitted tunnel circles.

To get the parameters of the fitted circle, the least squares regression method under the constraint that the new generated center of the circle must lie on the cross-sectional plane α. We assume the circle center $\bar{o} = (\bar{o}_x, \bar{o}_y, \bar{o}_z)$ with radius $\bar{r}$ on the plane α. The equation of the circle can be expressed below:

$$\begin{gathered} x^2 + y^2 + z^2 = 2x\bar{o}_x + 2y\bar{o}_y + 2z\bar{o}_z + \bar{r}^2 - \bar{o}_x^2 - \bar{o}_y^2 - \bar{o}_z^2, \\ s.t.\ \ a\bar{o}_x + b\bar{o}_y + c\bar{o}_z + d = 0, \end{gathered} \tag{7}$$

where $\{a, b, c\}$ is the normal vector of plane α and d is the constant term of normal plane. In this equation, the terms $x^2 + y^2 + z^2$, $2x$, $2y$ and $2z$ are regarded as observation values, and meanwhile the variables $\bar{o}_x$, $\bar{o}_y$, $\bar{o}_z$ and $\bar{r}^2 - \bar{o}_x^2 - \bar{o}_y^2 - \bar{o}_z^2$ are regarded as estimated values. We assume that the extracted cross section has m observation points. Equation (7) can be written as the matrix form:

$$\begin{gathered} Y + e = (B + E_B)A \\ s.t.\ \ KA = K_0 \end{gathered} \tag{8}$$

where, $Y = \begin{bmatrix} x_1^2 + y_1^2 + z_1^2 \\ x_2^2 + y_2^2 + z_2^2 \\ \vdots \\ x_m^2 + y_m^2 + z_m^2 \end{bmatrix}$, $B = \begin{bmatrix} 2x_1 & 2y_1 & 2z_1 & 1 \\ 2x_2 & 2y_2 & 2z_2 & 1 \\ \vdots & \vdots & \vdots & \vdots \\ 2x_m & 2y_m & 2z_m & 1 \end{bmatrix}$, $A = \begin{bmatrix} \bar{o}_x \\ \bar{o}_y \\ \bar{o}_z \\ \bar{r}^2 - \bar{o}_x^2 - \bar{o}_y^2 - \bar{o}_z^2 \end{bmatrix}$, $K = [a, b, c, 0]$ and $K_0 = -d$. Parameter e and E_B are the error matrices of Y and B. Based on Equation (8), two new matrices $Y_\epsilon = \left[Y^T, K_0\right]^T$ and $B_\epsilon = \left[B^T, K^T\right]^T$ can be generated. Then, we use the SVD technique [23] mentioned in Section 2.2.3 to decompose an augmented matrix $[B_\epsilon, Y_\epsilon]$ to estimate the parameters of the fitted circle defined in matrix A.

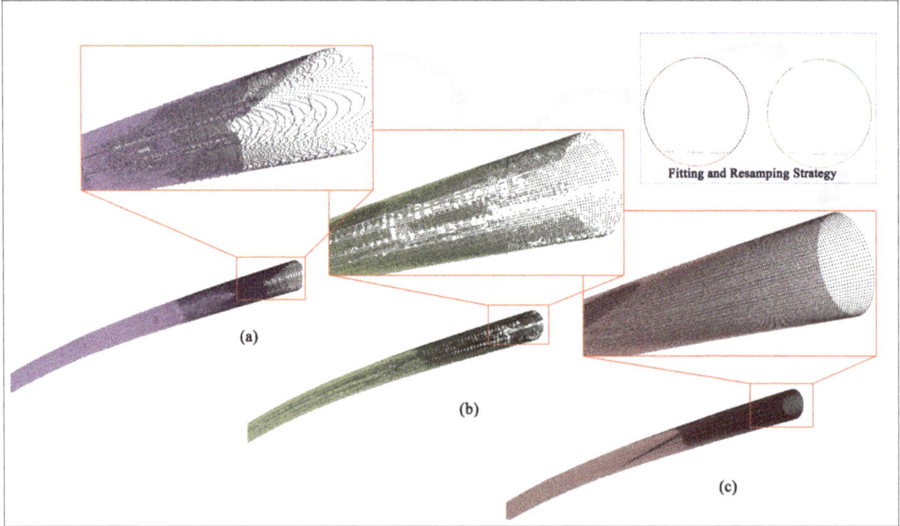

Figure 8. Comparisons of tunnel cross-sectional points before and after refinement. (**a**) the raw tunnel point clouds; (**b**,**c**) are results of tunnel cross-sectional points before and after refinement.

2.3.4. Continuous Extraction of Tunnel Cross Sections

Once we determine the individual tunnel cross section, the same method can be extended to the whole tunnel to extract the continuous tunnel cross sections. More specifically, we can use an integral formula Equation (9) to repeatedly determine the accurate location x_{p_d} of the next point $p_d = \{x_{p_d}, y_{p_d}, z_{p_d}\}$ along the tunnel axis at a regular interval Δd and a starting point $p_s = \{x_{p_s}, y_{p_s}, z_{p_s}\}$:

$$\Delta d = \int_{x_{p_s}}^{x_{p_d}} \sqrt{1 + f'_{xy}(x,y)^2 + f'_{xz}(x,z)^2} d_x. \tag{9}$$

When we get the value of x_{p_d}, we substitute it into Equation (1) to get the other two elements of y_{p_d} and z_{p_d} of point p_d. This procedure executes repeatedly until all of the locations of the tunnel axis have been calculated. At these locations, the method described in Section 2.3.3 is used to extract their associated tunnel cross-sectional points.

3. Performance Evaluation and Discussion

3.1. Dataset Specifications

We use Terrestrial Laser Scanner FARO® Focus^{3D} X 330 (Lake Mary, FL, USA) to scan Nanjing metro line S3 from Tiexin bridge to Chunjiang road. The entire scene has five scans with a total length of 127 m and a total number of 200 million points. All the scans are registered into a common reference frame, i.e., coordinate system based on sphere targets. The entire preprocessing procedure relies on FARO processing and registration software SCENE (SCENE 5.5, FARO Technologies Inc., Lake Mary, FL, USA) (https://www.faro.com/products/product-design/faro-scene/). The range of FARO® X 330 is from 0.6 m to 330 m. We acquire five scans of data where we set the distance between any adjacent scans about 23 m to maintain the point density. The point density, i.e., spacing of point clouds is highly dependent on scan resolution. In the case of the FARO® Focus^{3D} X 330 instrument used herein, the scan resolution is expressed as a fraction and the user can choose among 1/1, 1/2, 1/4, 1/5, 1/8, 1/10, 1/16, 1/20 and 1/32, with higher resolution requiring higher scanning time. In our case, we set 1/4, taking 10–15 min for each scan and having 6.136 mm point density at approximately 10 m. In practice, the optimal resolution setting should make a trade-off between the required density points per square meter and the scan duration [25]. To decrease the missing data and data occlusion, we set the field of view zenith angle range from –60° to 90° and azimuth angle range from 0° to 360°. The range error is defined as a systematic measurement error 10 m and 25 m, and one sigma is 0.3 mm under a constraint of a 90% reflective surface. Through an experiment of a fitting flat indoor facçade, we find that the precision of the scanner is around 1.5 mm at approximately 10 m scanning ranges.

3.2. Geometric Error Evaluation

We evaluate our tunnel cross sections from three aspects including geometry, topology and LoDs. From the perspective of geometry, we analyze the deviations of radii of the extracted cross sections and the chord length based on a series of the central angles, and calculate the nearest Euclidean distance from the cross-sectional points to their nearest raw point clouds.

To evaluate the accuracy of the cross-sectional points, we first analyze the deviations of radii of 1258 cross sections with an interval of 0.1 m and compare them to the designed radius. The mean value and the standard deviation of the radii of these cross sections are 2.7492 m and 1.4 mm, respectively. The statistic results are shown in Figure 9a. Compared with the designed tunnel radius of 2.75 m, the RMSEbetween the extracted radii of tunnel and the designed radius is only 1.6 mm. Considering the influence of the precision of the laser scanner, we select a very flat indoor façade to calibrate the precision of the laser scanner at a distance of 10 m. We fit the flat façade and calculate the mean residual of the fitting error of 1.5 mm between the raw point clouds and the fitted façade. This value is between

1.4 mm and 1.6 mm, which implies that the radii of the extracted cross sections are very stable and accurate compared with the design radius of 2.75 m.

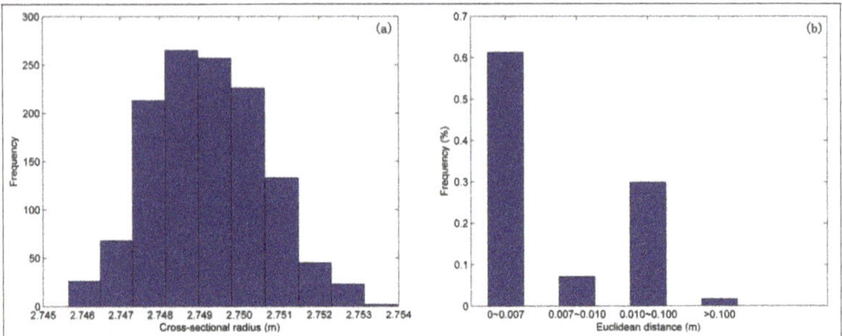

Figure 9. Histogram of cross-sectional radii and minimum Euclidean distance from the cross section points to their nearest raw point clouds. (**a**,**b**) represent the statistics of cross-sectional radius and minimum Euclidean distance, respectively.

We also analyze the minimum Euclidean distance from each cross-sectional point to its nearest raw point clouds to inform the "changed" initial raw point clouds. To this end, we divide the horizontal coordinates of the minimum Euclidean distance shown in Figure 9b into four categories: >0.100 m, 0.010–0.100 m, 0.007–0.010 m and 0–0.007 m. The cross section points from the first category only take up to 2%, which is mainly because of the missing data at the vicinity of tunnel track. This missing data arises from occlusion of the tunnel track having higher elevation than its associated rail slab. Because of missing data, the generated cross-sectional points are far away from its nearest raw point clouds, causing the deviation over 0.1 m, as is evident in the Figure 10a. In Figure 10b, the cross-sectional points from the second category account for 30%, which mainly arises from the tunnel bottom and the bolts used to fix the tunnel linings and other installations, e.g., bolts, pipes and wires. The third category accounts for 7% of the cross-sectional points created. From the distribution in Figure 10c, we can see that these points are scattered far away from their corresponding scanners denoted by the blue points in Figure 10. In these areas, the raw point clouds have low density; therefore, the local fitted quadric surfaces are not as accurate as the fitted surfaces derived from the high density of points. Furthermore, some installations and other equipment attached to the tunnel linings can also result in the large deviations. The last category takes up to 61% of the extracted cross-sectional points. The detailed distribution of those point clouds is shown in Figure 10d, where it can be seen that most of the cross-sectional points are distributed approximately 10 m away from their associated scanner. From FARO® Focus^{3D} X 330 scan data processing and registration software SCENE, we obtain the spacing among point clouds at approximately 10 m is 6.136 mm, which makes it consistent with a maximum of 7 mm and explains why these points are distributed well in the vicinity of the scanner around 10 m.

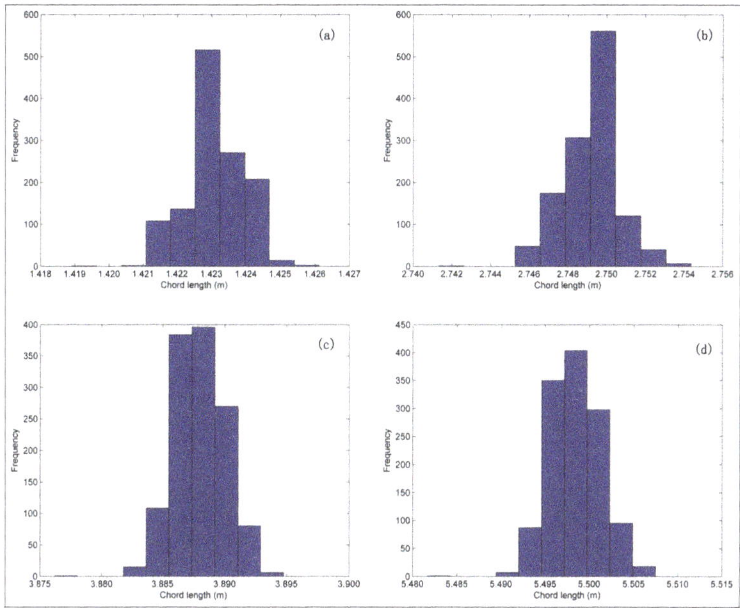

Figure 10. Histogram of chord length. (**a**–**d**) are histograms of different chord lengths with respect to central angle of 30°, 60°, 90° and 180°, respectively.

To further evaluate the geometric accuracy of 1258 cross sections, four types of chord lengths associated with central angles of 30°, 60°, 90° and 180° are compared and evaluated. The histogram and statistics of chord length corresponding to the different angles above are shown in Figure 11 and Table 1. It can be clearly seen that the standard deviations $\mathcal{SD}$ and RMSE with respect to columns Mean and Ref are monotonically increasing, reaching maximum 2.8 mm and 3.2 mm, respectively. This implies that the errors of cross sections become prominent when the chord length gradually increases. Note that the standard deviation of the chord length with a central angle of 180° reaches the maximum value of 2.8 mm, which is exactly twice the number of standard deviations of the radius (1.4 mm). This means that the deviation of chord length with a center angle of 180° is consistent with twice the deviation of the radius. In addition, compared with the references of chord length calculated using the design radius of 2.75 m, the deviation of the chord length is extremely small, just reaching the maximum of 3.2 mm. Therefore, we can make a safe conclusion that our created cross sections have high geometric accuracy by indirectly evaluating four levels of chord lengths.

Table 1. Chord length statistics for 1258 cross sections. The symbols Min, Max, Mean and $\mathcal{SD}$ represent the minimum, maximum, mean and standard deviation of chord length. 'Ref' means the reference (standard) of chord length calculated using the design radius of 2.75 m. '$\mathcal{SD}$' and 'RMSE' denote the standard deviations which are calculated using the numbers in the columns 'Mean' and 'Ref', respectively.

Central Angles	Min	Max	**Mean**	Ref	$\mathcal{SD}$	**RMSE**
30°	1.4189 m	1.4261 m	1.4231 m	1.4235 m	0.0008 m	0.0009 m
60°	2.7414 m	2.7544 m	2.7492 m	2.7500 m	0.0015 m	0.0017 m
90°	3.8762 m	3.8948 m	3.8880 m	3.8891 m	0.0020 m	0.0023 m
180°	5.4817 m	5.5075 m	5.4984 m	5.5000 m	0.0028 m	0.0032 m

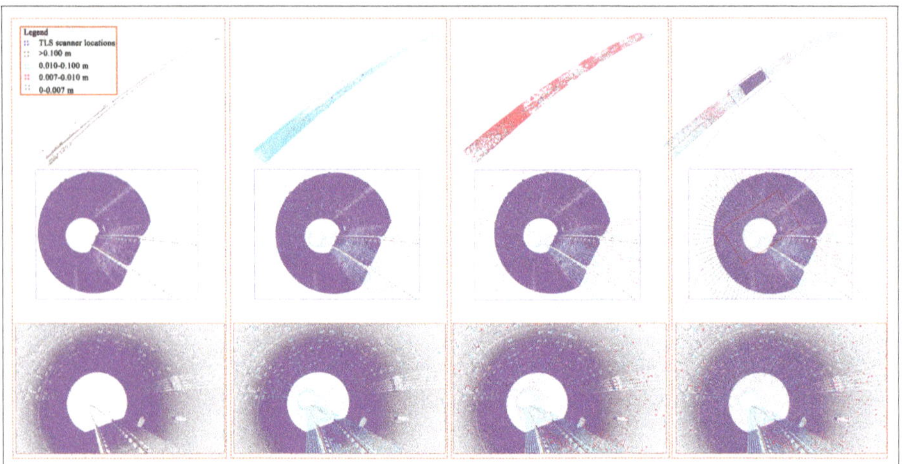

Figure 11. The evaluation of cross-sectional points through the measure of minimum Euclidean distance. The cross-sectional points are rendered by the different colors according to the deviations from their raw point clouds. To clearly show where the deviation arises from, the different categories are gradually superimposed onto their raw point clouds, denoted by purple.

3.3. *Topology Evaluation and LoD Representation*

In the field of computational geometry, topology is a crucial index to determine whether the tunnel can be reconstructed via a series of cross sections. Fortunately, our methodology has the capability to maintain the topology of cross sections in two aspects: (1) For each cross section, the tunnel points are generated by sequentially intersecting a straight line with its local fitted quadric surface. Because of this, the generated points within each cross-sectional plane have topological relationships, i.e., clockwise or counter-clockwise order. Using this type of relation, the geometric shape of a specific cross section can be created. (2) For extraction of multiple cross sections (see Section 2.3.4), we also sequentially generate them along the tunnel axis once given an explicit split interval, thereby forming a series of continuous cross sections. That is, the relationship like the former and later cross sections of the currently processing one is considered, and thus the points from any two arbitrary cross sections have one-to-one correspondence. Using the relationship among multiple cross sections, the entire geometric tunnel shape is expected to be generated. As shown in Figure 12a, the purple triangles shown in the enlarged view are the wire meshes of the tunnel that are waved according to the two topologies above.

Because we can simultaneously maintain the point order within each cross section and the order of multiple cross sections, which lays a solid foundation for LoD representation of the metro tunnel. That is, we can use any arbitrary number of cross sections to achieve the abstraction/representation of metro tunnel. Low levels of LoDs with fewer cross sections are lightweight and suitable for storage, web transmission, acceleration of rendering and other tunnel-related applications. In contrast, although the high level of LoDs are not as compact as the shapes with low level LoDs, the geometric accuracy is relatively high. In practice, selection of the optimal LoD is often determined by factors including data acquisition cost, time, accuracy and labor investment and specific tunnel-related applications. For example, when the tunnel models are used in equipment installation on the tunnel linings, the relatively simple tunnel models generated by the sparse cross sections can meet the requirements. When considering tunnel profile deformation, crack and defect detection, and/or water leakage detection, the most detailed tunnel models with high geometric accuracy are expected to meet particular project needs. Furthermore, in the area of tunnel interior visualization, it often needs tunnel models at different LoDs rather than a single LoD. In this case, a multiple LoDs should be generated on the fly to enhance the users' experience. As shown in Figure 12a–g, the tunnel model is rendered by different LoDs corresponding to profiles of 0.2 m, 0.5 m, 2.0 m, 4.0 m, 6.0 m, 8.0 m and 10.0 m, respectively.

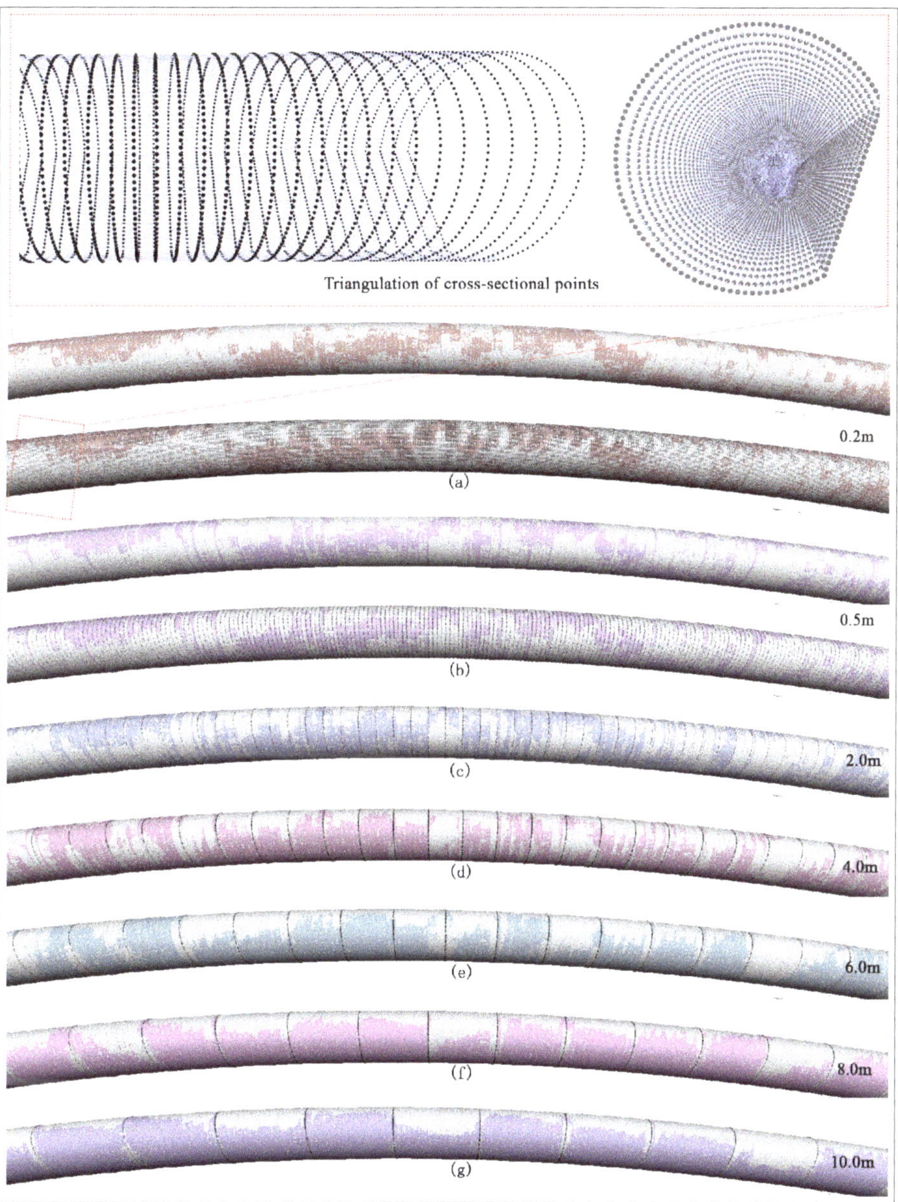

Figure 12. The geometric tunnel representation at different LoDs (Level of Details). (**a**–**g**) are the created tunnel models using the extracted cross sections at the split intervals of 0.2 m, 0.5 m, 2.0 m, 4.0 m, 6.0 m, 8.0 m and 10.0 m. Note that the created tunnel models are superimposed onto their 'reference model' denoted by the gray color. The reference model is produced by a screened Poisson surface reconstruction algorithm through using the raw tunnel point clouds.

To get an optimal split interval Δd, we first define the measures on two tunnel meshes $\mathcal{X}$ and $\mathcal{Y}$ to evaluate the similarity of two geometric shapes. We first use a Monte Carlo sampling method [26] to obtain a set of points over $\mathcal{X}$ and, for each point on $\mathcal{X}$, it searches the closest point (face or edge) on the other mesh $\mathcal{Y}$. By this way, we can calculate the Mean and RMSE between the sampling points on the sampled mesh $\mathcal{X}$ and their closest points on the target mesh $\mathcal{Y}$. These two measures are regarded as the criterion to determine the similarity of two shapes. In our case, $\mathcal{X}$ represents a tunnel model constructed by a series of profiles at a specific interval. A target mesh $\mathcal{Y}$ denotes a tunnel model that is created by the screened Poisson surface reconstruction algorithm [27] using the raw point clouds. As shown in Figure 12a–g, the target tunnel meshes denoted by gray color are superimposed onto a series of sampled tunnel models with different colors. Through the superimposition, we qualitatively evaluate the similarity by visually inspecting the differences of these two shapes. Meanwhile, we also use the two well accepted statistical measures above to make quantitative evaluations. To this end, we first fix the rotation angle interval $\Delta\varphi$ with a specific value to analyze the variations of Mean and RMSE of the generated tunnel models with respect to the split intervals at 0.2 m, 1.0 m, 2.0 m, 4.0 m, 6.0 m, 8.0 m and 10.0 m. The results are shown in Figure 13, and we unexpectedly discover that, for each polyline corresponding to a specific rotation angle, the Mean and RMSE first decrease and then increase with increasing split interval. This means that the high density of profiles does not guarantee producing a tunnel model with high geometric accuracy. Actually, in our scenario, the split interval of 4.0 m can get the tunnel model with the minimum Mean and RMSE errors. On the contrary, we fix the split interval with a specific value and compare geometric variations with regard to the different rotation angle intervals $\Delta\varphi$, i.e., 3°, 6°, 9° and 12°. The results are shown in Figure 14, where we find that the geometric errors are substantially increasing with the increased rotation angles. The bottom polyline in yellow in Figure 14a corresponds to the tunnel model at a 4.0 m split interval. Therefore, we can make a safe conclusion that extraction of profiles at the space of 4.0 m can guarantee the reconstructed tunnel model with the minimum geometric error. However, in practice, the user should totally make a trade-off between the geometric accuracy, compactness and the specific applications to finally determine how many profiles are needed.

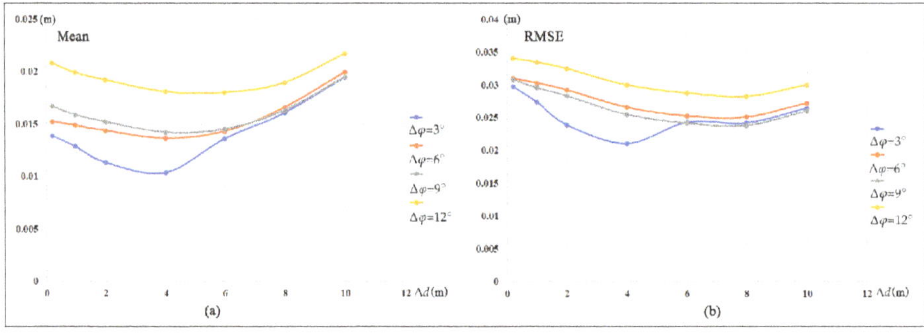

Figure 13. The comparisons of those generated and the reference tunnel model at different Δd and $\Delta\varphi$.

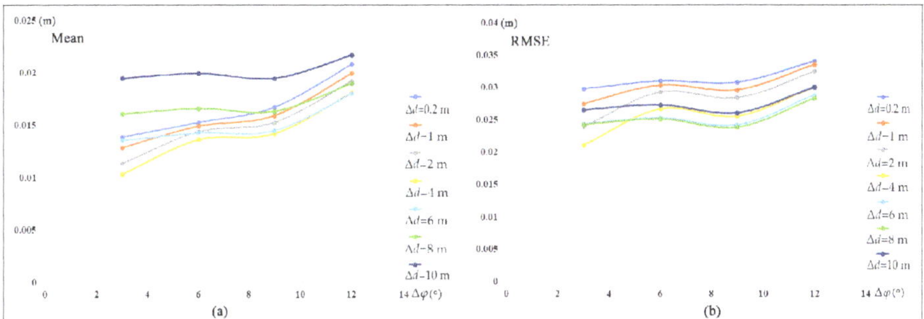

Figure 14. The comparisons of those generated and the reference tunnel model at different $\Delta\varphi$ and Δd.

3.4. Comparison

To further evaluate the accuracy of the proposed method, we compare our algorithm with the algorithm proposed in [1] by using the same dataset, i.e., the metro tunnel in Shanghai, China. The Shanghai dataset is described in Table 2. From the Shanghai dataset, we extract 260 tunnel profiles at the fixed split interval of 0.2 m and compare the variation of the fitted radii with the design radius of 2.75 m. Note that the design radii of the tunnels in the Nanjing and Shanghai datasets are both 2.75 m. The detailed statistics for the radii are shown in Table 3. From this table, the radius difference, i.e., RMSE between the cross sections and the designed value is only 4.6 mm, which outperforms the value of 6 mm estimated in Kang's method [1].

Table 2. Description of Shanghai subway tunnel dataset. #Pts represents the number of points of the Shanghai metro tunnel scan dataset.

Scanner	Scan Angular Resolution	#Pts	Range Accuracy
RIEGL VZ-400	0.046°	2,686,866	± 5 mm

Table 3. Tunnel radius statistics for 260 cross sections from one scan of Shanghai dataset. #Profile represents the number of the extracted cross sections from Shanghai dataset.

#Profile	*Max*	*Min*	Mean	*SD*	RMSE
260	2.7632 m	2.7481 m	2.7538 m	0.0026 m	0.0046 m

Compared with the Nanjing dataset, the range of the zenith angle for acquiring the Shanghai dataset is relatively narrow, causing very severe missing data at the top and bottom of scanner locations, as evident in Figure 15a,b. Although we suffer from very serious missing data, our algorithm can infer the accurate locations of cross section points in these missing data areas by an assumption of circular tunnel shapes, as proved in Figure 15d,e. Moreover, the Shanghai dataset used in our paper only includes one scan data without overlapping point clouds from other scans. Therefore, the point clouds that are far away from the scanner have low point density. As shown in Figure 15c, the distribution of point clouds from one terminal is very sparse. However, in this case, our algorithm is still insensitive to this adverse situation and has the capability to acquire the credible cross sections (see Figure 15f). This is mainly due to the adaption of "self-adaptation" strategy, which dynamically determines the size of local points to finalize the local quadric surface fitting according to point density variation.

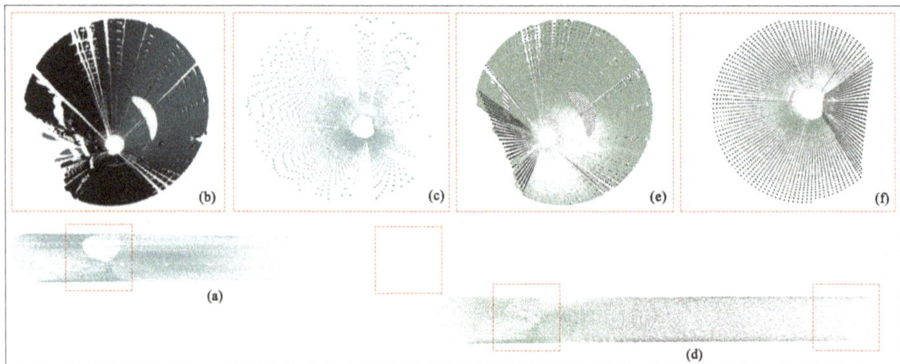

Figure 15. Tunnel profile extraction from one scan of Shanghai metro tunnel dataset. The raw tunnel point clouds are shown in (**a**), where large-area of missing data and severe variations of point density are shown in the enlarged view of (**b**,**c**). The extracted 260 metro tunnel profiles, denoted by the black points, are overlaid onto the raw data. From the zoomed views (**d**–**f**), we can vividly see that the missing data areas are filled by the extracted profiles, and meanwhile the profiles are successfully derived from extremely sparse point clouds.

4. Conclusions

In this paper, we propose an efficient framework to extract tunnel cross sections from the Terrestrial Laser Scanning point clouds. Our framework starts with the extraction of the tunnel axis using a robust slice-based method. To get an accurate position of the tunnel profiles, we further divide the initial axis into linear and nonlinear segments via an enhanced RANSAC algorithm. In contrast with the traditional RANSAC algorithm, our algorithm has been implemented in the tangent space rather than the data space. By this way, the simultaneous linear and nonlinear axis segmentation problem is thus translated into sole linear element segmentation. It does not only reduce the complex of segmentation but also increases the credibility of segmented results. Based on the segmented axis segments, we can fit the linear and nonlinear segments via the linear regression model and the polynomial regression model, respectively. The extracted attitude of the three-dimensional tunnel axis lays a solid foundation for accurate extraction of tunnel cross sections. The initial cross-sectional points are generated by intersecting a sequence of straight lines that rotate orthogonally around the tunnel axis with the corresponding local quadric surfaces. It should be noted that the size of the quadric surface is not a constant but progressive adjustment according to the density of the local point clouds. This advantage helps the algorithm to adapt to the variations of point clouds and makes the algorithm insensitive to the point density. The initial cross-sectional points are further refined and densified via a "fitting-and-resampling" strategy. That is, the coarse cross-section points are first fitted by an assumption of circular tunnel, i.e., using the circular shape/model to fit the cross-sectional points, followed by resampling of the circular shapes. This process significantly enhances the completeness of cross-sectional points, and more importantly makes the proposed algorithm resistant to missing data.

The results on Nanjing dataset show that the RMSE of the fitting accuracy of the cross sections is 1.687 mm. Compared with the design value of 2.75 m, the RMSE of the radii of extracted cross sections is only 1.60 mm. In addition, by trial-and-error experiments, we find that reconstructing the entire metro tunnel model via cross sections evenly spaced at a distance of around 4 m can get the relatively minimum geometric errors. It should be aware that the extracted tunnel profiles have explicit topology information not only existing within each profile but also among multiple profiles. This is beneficial to geometric processing oriented applications such as metro tunnel reconstruction, model editing and retargeting. We also compare our results with the state-of-the-art method proposed by Kang [1] based upon the same dataset, i.e., Shanghai metro tunnel point clouds. It is to be proved that our methodology can cope with large-area missing data and successfully extract the profiles from very sparse and unevenly distributed point clouds. Meanwhile, based on the Shanghai dataset, the RMSEs of extracted cross sections' radii with regard to the design radius of 2.75 m are only 4.6 mm, which is superior to Kang's method of 6 mm.

Although our algorithm achieves promising results on Nanjing and Shanghai datasets, there are some drawbacks and interesting ideas which can be further explored to extend the research reported in this paper. The consistency in the connection areas between the linear and nonlinear axis segments may not be guaranteed. That is to say, the extracted cross sections are not consistent on the transition areas. In future work, the global optimization of axis segments might be expected to relieve and even probably eliminate the deviations in the overlapping parts of adjacent different axis models. In addition, our tunnel extraction algorithm is based on an assumption that the tunnel cross sections can be fitted by the circular shapes; therefore, our algorithm is not applicable to more complex and atypical tunnels. For this problem, we plan to segment a complex tunnel into different parts and then adopt the concept of "hybrid representation" to refit these parts by combining different models, i.e., planar model, circular model, elliptical model, and other free-form models. Furthermore, we find that the track laid in the interior of the tunnel and other installations will undoubtedly influence the accuracy of tunnel profiles. To solve this problem, we plan to first remove these irrelevant objects using the computer vision-based classification and recognition algorithm [28] and then extract the tunnel cross sections from the remaining point clouds.

Author Contributions: Z.C. analyzed the data and wrote the C++ source code. Y.S. and D.C. helped with the project and study design, paper writing, and analysis. Z.Z., F.J., T.Y., S.X., Z.K. and L.Z. helped with field work and data analysis.

Funding: This work was supported by the National Natural Science Foundation of China under Grant No. 31770591 and Grant No. 41701533.

Acknowledgments: The authors would like to thank Zhizhong Kang for providing the Shanghai metro tunnel dataset for comparison.

Conflicts of Interest: The authors declare no conflict of interest.

References

1. Kang, Z.; Zhang, L.; Tuo, L.; Wang, B.; Chen, J. Continuous extraction of subway tunnel cross sections based on terrestrial point clouds. *Remote Sens.* **2014**, *6*, 857–879. [CrossRef]
2. Xie, X.; Lu, X. Development of a 3D modeling algorithm for tunnel deformation monitoring based on terrestrial laser scanning. *Undergr. Space* **2017**, *2*, 16–29. [CrossRef]
3. Qiu, W.; Cheng, Y.J. High-resolution DEM generation of railway tunnel surface using terrestrial laser scanning data for clearance inspection. *J. Comput. Civ. Eng.* **2016**, *31*, 04016045. [CrossRef]
4. Kolymbas, D. *Tunnelling and Tunnel Mechanics: A Rational Approach to Tunnelling*; Springer Science & Business Media: Berlin/Heidelberg, Germany, 2005.
5. Wang, T.T.; Jaw, J.J.; Chang, Y.H.; Jeng, F.S. Application and validation of profile–image method for measuring deformation of tunnel wall. *Tunn. Undergr. Space Technol.* **2009**, *24*, 136–147. [CrossRef]
6. Wang, T.T.; Jaw, J.J.; Hsu, C.H.; Jeng, F.S. Profile-image method for measuring tunnel profile–Improvements and procedures. *Tunn. Undergr. Space Technol.* **2010**, *25*, 78–90. [CrossRef]
7. Attard, L.; Debono, C.J.; Valentino, G.; Di Castro, M. Tunnel inspection using photogrammetric techniques and image processing: A review. *ISPRS J. Photogramm. Remote Sens.* **2018**, *144*, 180–188. [CrossRef]
8. Yoon, J.S.; Sagong, M.; Lee, J.; Lee, K.S. Feature extraction of a concrete tunnel liner from 3D laser scanning data. *Ndt & E Int.* **2009**, *42*, 97–105.
9. Fekete, S.; Diederichs, M.; Lato, M. Geotechnical and operational applications for 3-dimensional laser scanning in drill and blast tunnels. *Tunn. Undergr. Space Technol.* **2010**, *25*, 614–628. [CrossRef]
10. Charbonnier, P.; Chavant, P.; Foucher, P.; Muzet, V.; Prybyla, D.; Perrin, T.; Grussenmeyer, P.; Guillemin, S. Accuracy assessment of a canal-tunnel 3d model by comparing photogrammetry and laserscanning recording techniques. *Int. Arch. Photogramm. Remote Sens. Spat. Inf. Sci.* **2013**, *40*, 171–176. [CrossRef]

11. Walton, G.; Delaloye, D.; Diederichs, M.S. Development of an elliptical fitting algorithm to improve change detection capabilities with applications for deformation monitoring in circular tunnels and shafts. *Tunn. Undergr. Space Technol.* **2014**, *43*, 336–349. [CrossRef]
12. Nuttens, T.; Stal, C.; De Backer, H.; Schotte, K.; Van Bogaert, P.; De Wulf, A. Methodology for the ovalization monitoring of newly built circular train tunnels based on laser scanning: Liefkenshoek Rail Link (Belgium). *Automat. Constr.* **2014**, *43*, 1–9. [CrossRef]
13. Han, S.; Cho, H.; Kim, S.; Jung, J.; Heo, J. Automated and efficient method for extraction of tunnel cross sections using terrestrial laser scanned data. *J. Comput. Civ. Eng.* **2012**, *27*, 274–281. [CrossRef]
14. Cheng, Y.J.; Qiu, W.; Lei, J. Automatic extraction of tunnel lining cross-sections from terrestrial laser scanning point clouds. *Sensors* **2016**, *16*, 1648. [CrossRef] [PubMed]
15. Xu, X.; Yang, H.; Neumann, I. A feature extraction method for deformation analysis of large-scale composite structures based on TLS measurement. *Compos. Struct.* **2018**, *184*, 591–596. [CrossRef]
16. Tan, K.; Cheng, X.; Ju, Q. Combining mobile terrestrial laser scanning geometric and radiometric data to eliminate accessories in circular metro tunnels. *J. Appl. Remote Sens.* **2016**, *10*, 030503. [CrossRef]
17. Tan, K.; Cheng, X.; Ding, X.; Zhang, Q. Intensity data correction for the distance effect in terrestrial laser scanners. *IEEE J. Sel. Top. Appl. Earth Obs. Remote Sens.* **2016**, *9*, 304–312. [CrossRef]
18. Dimitrov, A.; Golparvar-Fard, M. Robust NURBS surface fitting from unorganized 3D point clouds for infrastructure as-built modeling. In Proceedings of the Computing in Civil and Building Engineering, Orlando, FL, USA, 23–25 June 2014; pp. 81–88.
19. Nahangi, M.; Haas, C.T. Skeleton-based discrepancy feedback for automated realignment of industrial assemblies. *Autom. Constr.* **2016**, *61*, 147–161. [CrossRef]
20. Gosliga, R.V.; Lindenbergh, R.; Pfeifer, N. Deformation Analysis of a Bored Tunnel by Means of Terrestrial Laser Scanning. 2006. Available online: https://s3.amazonaws.com/academia.edu.documents/38455846/deformation_analysis_of_a_bored_tunnel_by_means_of_terrestrial_laser_scanning.pdf?AWSAccessKeyId=AKIAIWOWYYGZ2Y53UL3A&Expires=1549004618&Signature=k9CvopqMh2yKDQq15nBgyf2Xbbc%3D&response-content-disposition=inline%3B%20filename%3DDEFORMATION_ANALYSIS_OF_A_BORED_TUNNEL_B.pdf (accessed on 1 February 2019).
21. Savitzky, A.; Golay, M.J.E. Smoothing and Differentiation of Data by Simplified Least Squares Procedures. *Anal. Chem.* **1964**, *36*, 1627–1639. [CrossRef]
22. Fischler, M.A.; Bolles, R.C. Random sample consensus: A paradigm for model fitting with applications to image analysis and automated cartography. *Commun. ACM* **1981**, *24*, 381–395. [CrossRef]
23. Lu, T. Research on the Total Least Squares and Its Application in Surveying Data Processing. Ph.D. Thesis, School of Geodesy and Geomatics, Wuhan University, Wuhan, China, 2010.
24. Golub, G.H.; Van Loan, C.F. An analysis of the total least squares problem. *SIAM J. Numer. Anal.* **1980**, *17*, 883–893. [CrossRef]
25. Kamintzis, J.E.; Jones, J.; Irvine-Fynn, T.; Holt, T.; Bunting, P.; Jennings, S.J.A.; Porter, P.R.; Hubbard, B. Assessing the applicability of terrestrial laser scanning for mapping englacial conduits. *J. Glaciol.* **2018**, *64*, 37–48. [CrossRef]
26. Shapiro, A. Monte Carlo sampling methods. *Handb. Oper. Res. Manag. Sci.* **2003**, *10*, 353–425.
27. Kazhdan, M.; Hoppe, H. Screened poisson surface reconstruction. *ACM Trans. Graph.* **2013**, *32*, 29. [CrossRef]
28. Wang, Z.; Zhang, L.; Fang, T.; Mathiopoulos, P.T.; Tong, X.; Qu, H.; Xiao, Z.; Li, F.; Chen, D. A multiscale and hierarchical feature extraction method for terrestrial laser scanning point cloud classification. *IEEE Trans. Geosci. Remote Sens.* **2015**, *53*, 2409–2425. [CrossRef]

Article

Validation of Portable Mobile Mapping System for Inspection Tasks in Thermal and Fluid–Mechanical Facilities

Manuel Rodríguez-Martín [1,2], Pablo Rodríguez-Gonzálvez [3,*], Esteban Ruiz de Oña Crespo [4] and Diego González-Aguilera [4]

1 Department of Mechanical Engineering, Universidad de Salamanca, 37700 Béjar (Salamanca), Spain; ingmanuel@usal.es

2 Technological Department, Catholic University of Ávila, 05005 Ávila, Spain

3 Department of Mining Technology, Topography and Structures. Universidad de León, 24401 Ponferrada, Spain

4 Department of Cartographic and Land Engineering. Universidad de Salamanca, 05003 Ávila, Spain; estebanrdo@usal.es (E.R.d.O.C.); daguilera@usal.es (D.G.-A.)

* Correspondence: p.rodriguez@unileon.es; Tel.: +34-987-442-055

Received: 30 August 2019; Accepted: 19 September 2019; Published: 20 September 2019

Abstract: The three-dimensional registration of industrial facilities has a great importance for maintenance, inspection, and safety tasks and it is a starting point for new improvements and expansions in the industrial facilities context. In this paper, a comparison between the results obtained using a novel portable mobile mapping system (PMMS) and a static terrestrial laser scanner (TLS), widely used for 3D reconstruction in civil and industrial scenarios, is carried out. This comparison is performed in the context of industrial inspection tasks, specifically in the thermal and fluid-mechanics facilities in a hospital. The comparison addresses the general reconstruction of a machine room, focusing on the quantitative and qualitative analysis of different elements (e.g., valves, regulation systems, burner systems and tanks, etc.). The validation of the PMMS is provided considering the TLS as ground truth and applying a robust statistical analysis. Results come to confirm the suitability of the PMMS to perform inspection tasks in industrial facilities.

Keywords: optical sensors; robust statistical analysis; portable mobile mapping system; handheld; 3D processing; point cloud

1. Introduction

Traditionally, geomatics techniques have been used mainly for the determination and recording of terrain and outdoor scenarios in surveying engineering. However, they also can be applied in industrial context and energy applications [1–4]. In this way, geomatics techniques can be used to generate three-dimensional models of large and complex scenes as nuclear power plants [5], the generation of as-built models in architecture [6,7] or, even, for metrological tasks related with the quality assessment of products [8–11]. Portable mobile mapping systems (PMMSs) allow the generation of dense, geo-referenced, three-dimensional models while the operator is moving though the scene [12]. This novel technology has been developed in the last years becoming the most innovative and emerging technique in surveying tasks [4].

The scope of application of PMMSs is diverse: from cultural heritage constructions [13] and civil structures to industrial sites [14], indoor spaces [15], and natural environments [16]. PMMSs contain two groups of sensors that act in a synchronized way to reconstruct the scene three-dimensionally: navigation and remote sensing modules. The navigation module usually is based on an inertial measurement unit (IMU) and, sometimes, it also can be equipped with a Global Navigation Satellite System (GNSS) receiver. The remote sensing module normally is based on a laser scanning sensor and, sometimes, also different types of cameras. A deep review about this emerging technology and the different typologies of PMMSs is presented in [12].

The PMMS analyzed in this paper is the GeoSLAM Zeb-Revo [17] (from now on, Zeb-Revo). Nowadays, it is one of the most widespread PMMSs, and several investigations have addressed its use for different purposes as natural space management [16], cultural heritage documentation [18], underground documentation [19,20], and disaster analysis [21], among others. In Zeb-Revo, the navigation module is based on an IMU, whereas the remote sensing module is a 2D laser profilometer. The latter is a compact laser scanning range finder which is more compact and light-weight than a traditional 3D laser scanner system and its battery consumption is more optimal [22]. In [23], Zeb-Revo and Leica Pegasus-Backpack were compared with a Z+F Terrestrial Laser Scanner (TLS), providing Zeb-Revo discrepancies around 4 cm. In [22], a robust statistical comparison between Zeb-Revo and Leica Pegasus-Backpack was performed. Although it was demonstrated that both systems performed within manufacturer specifications, Zeb-Revo generally outperformed the manufacturer values despite the presence of outliers.

The adequate distribution of thermal and fluid-mechanical equipment (such as boilers, tanks, heat exchangers, valves, and others) within the machine rooms is an important issue that must be subject of a rigorous study, especially in singular facilities such as hospitals. The tasks of expansion, maintenance, inspection, and improvement of thermal and fluid-mechanical installations require prior knowledge of the elements of the installation as a starting point. Additionally, the criteria for the equipment distribution are normally indicated by international and/or national standards and regulations (e.g., [24,25]). For this reason, meeting the geometric and technical requirements based on standards and regulations is a mandatory task for the facilities engineers. Many of these requirements are based on geometrical criteria such as distance among boilers or other equipment, geometric and volumetric conditions of the rooms, spaces for maintenance tasks, empty spaces for emergency exits, etc. The generation of a three-dimensional model of machine rooms using PMMS is a powerful tool to ensure the quality and efficiency of the service, allowing the registration of the different elements and devices of the installation and their state of conservation, reducing in this manner the risk of possible maintenance problems, making future improvement actions easier such as the expansion of the facilities or improvement actions, all based on accurate, scaled, and duly documented geospatial information. The design and calculation of the facilities must be carried out according to the technical criteria indicated in the standards and regulations, for example, the allowed pipe diameters. The design and calculation of the thermal facilities must be carried out according to the technical criteria that indicate standards [26]. The calculation criteria—for example, the location of valves, the dimensions of the equipment, etc.—must be computed following these indications, being the finished facilities subject to inspection processes also regulated by these standards. Due to this, inspection and documentation tasks are mandatory to certify that the installation meets the standards, consequently, the use of methods that three-dimensionally record these spaces are totally justified. In [4], a method based on PMMS is proposed for the maintenance of machine rooms, providing the automatic extraction of pipes with relative deviation ranges between 0.50% and 11.21% of the mean diameter value.

The aim of this work was to validate the inspection effectiveness of a PMMS, Zeb-Revo, by comparing its results with those obtained with a TLS, Faro Focus 3D, whose survey was used as reference. As a result, many maintenance and inspection tasks could be done in a simpler way and in less time if a PMMS is used instead of a TLS. After this introduction, this paper is structured as follows: Section 2 describes the methodology and materials employed. Experimental results are presented in Section 3, along with the robust statistical analysis. Discussion is detailed in Section 4, whereas the main conclusions are presented at the end of the paper (Section 5).

2. Materials and Methods

A specific methodology for data analysis was stablished to compare the potential of the PMMS, Zeb-Revo, for the evaluation of fluid-mechanical facilities, using a TLS, Faro Focus3D, as ground truth.

2.1. Materials

Zeb-Revo [16] can be considered as the natural evolution of the first version, ZEBedee, which was developed by the CSIRO ICT Center in Brisbane (Australia) [27,28]. This scanning device is based on a 2D profilometer (Hokuyo UTM-30LX) which can be displaced using a platform support. Moreover, Zeb-Revo is equipped with an IMU as navigation module, which is continuously rotating during data acquisition. Then, the device can be equipped with a commercial camera GoPro Hero installed to record the scenario during the data acquisition and, in this manner, the different scanned elements can be easily located on video. It is especially useful because Zeb-Revo does not record radiometry information, so differentiating zones only based on geometry (without color information) is difficult. The 2D laser profilometer is a compact laser scanner, which is more efficient, light-weight, and compact than any TLS used for three-dimensional reconstruction [22].

Due to the continuous rotation of the sensor and the movement of the operator, three-dimensional information (i.e., points) is acquired. The data are stored in a server with a hard disk that is located in a backpack, which is part of the equipment. Specific features of the Zeb-Revo are indicated in Table 1.

Table 1. Technical features of the PMMS, Zeb-Revo.

Parameter	Value
Measurement range (indoor) (m)	30
Measurement range (outdoor) (m)	15
Data capture speed (points/s)	43,200
Accuracy	±0.1%
Relative accuracy	1-3 cm
Field of view	270° × 360°
Operating time (h)	4
Scanner dimensions (mm)	86 × 113 × 470
Weight (kg)	0.85
Rotation frequency (Hz)	0.5

For its part, a TLS, FARO Focus3D 120 (from here, Faro) (Figure 1), was used for the scanning of the reference point cloud (ground truth). This device consists on an infrared laser scanner which takes measures directly using the principle of phase shift in the range of 0.60–120 m at a wavelength of 905 nm. Its field of view covers 320° vertically and 360° horizontally, allowing an angular resolution of 0.009° and a measurement rate of 976,000 points per second, recording radiometric information for each point. The precision provided by the manufacturer is of 2 mm in normal lighting conditions with a beam divergence of 0.19 mrad. This device allows a high accuracy three-dimensional reconstruction, so it was chosen as the ground truth [29].

Furthermore, three reference spheres (14.5 cm of diameter) were required and located in different points of the scene to register the different individual point clouds acquired by Faro and, also, for the subsequent alignment of the point clouds acquired by Zeb-Revo. The spheres acted as reference providing a common Coordinate Reference System (CRS) for the comparison.

2.2. Methods

The developed methodology (Figure 2) was designed specifically for the evaluation of the derived products obtained with the PMMS, Zeb-Revo, specifically oriented to the extraction of the most critical parameters for the inspection of industrial facilities. In addition, the methodology allows the qualitative analysis of in-detail elements from the point cloud generated by the PMMS, Zeb-Revo.

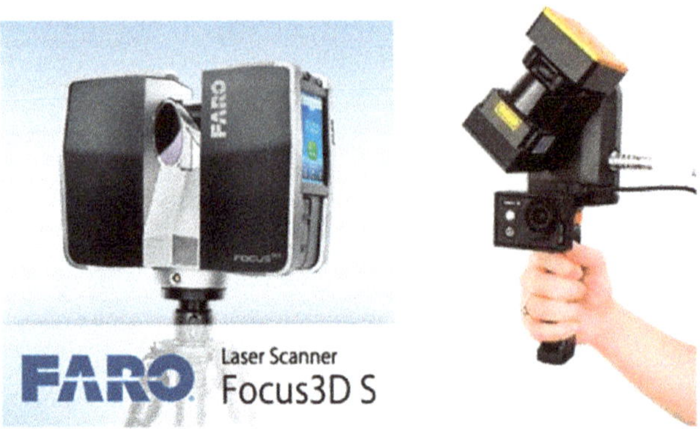

Figure 1. Faro Laser Scanner Focus 3D S (Left) by courtesy of www.faro.com; and GeoSlam Zeb-Revo (Right) by courtesy of https://geoslam.com.

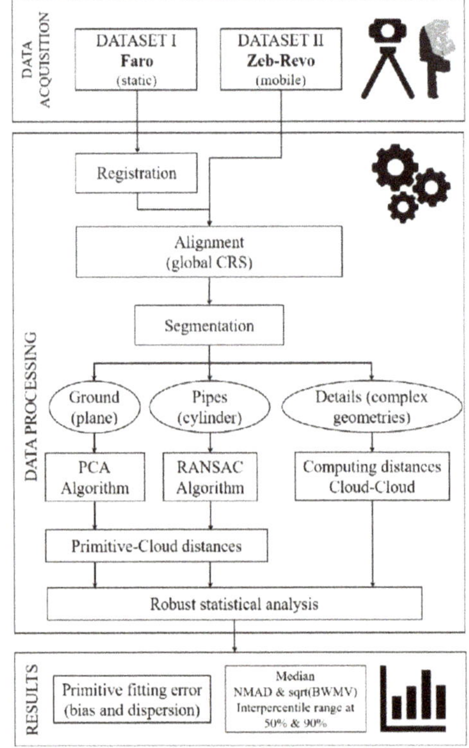

Figure 2. Pipeline for the validation of portable mobile mapping system for inspection tasks in thermal and fluid-mechanical facilities.

2.2.1. Data Acquisition

The first step was the data acquisition which was addressed using both devices (Faro-Dataset I and Zeb-Revo-Dataset II) but, previously, a reference network was set up to analyze the different point clouds generated from Faro and Zeb-Revo in the same CRS. This reference network was designed with spheres attached to different metal parts of the scene, using neodymium magnets to ensure that they remain static during the data acquisition (Figure 3).

Figure 3. Sphere located in a pillar of the machine room (attached by a neodymium magnet).

Regarding the protocol for data acquisition, the PMMS requires to start and finish in the same position of the scene. For this reason, the trajectory planned using Zeb-Revo follows a closed-loop. This trajectory must be established to specifically cover the complete machine room minimizing occlusions. Please note that the error of this kind of PMMS typically increases with time/length of the walk. The maximum data acquisition time provided by the manufacturer (working in continuous mode) is approximate twenty minutes, after which the quality of the reconstruction may be reduced. This time requirement is enough to scan extensive industrial facilities with a single-loop (if that was not enough, several loops could be used to generate different point clouds). Data acquisition protocol contemplates a preliminary inspection of the scene to plan a suitable data acquisition path, detect potentially physical obstacles (e.g., equipment, tools or pipes on the floor, drips, etc.) that would affect the data acquisition and remove them if it is necessary and/or possible.

Once PMMS data (Dataset II) was acquired, the point cloud of the scene was registered in a local CRS. If ground control points are visible in the acquired data, the point clouds can be georeferenced in a global CRS using a rigid Helmert transformation. Alternatively, the individual point clouds can be also referenced in the same local CRS implementing an iterative closest point (ICP) registration [30]. For additional details about Zeb-Revo processing, the reader is referred to [12].

Data acquisition was performed planning the path inside the building (450 m^2) and requiring approximately four minutes. The total track was about 115 m long, so the walking speed was near 0.43 $m{\cdot}s^{-1}$, or 1.6 $km{\cdot}h^{-1}$. The front part of the sensor was always oriented in the direction of advance. A part of the followed trajectory and a 3D view of the reconstruction using the manufacturer's software (GeoSLAM) is shown in Figure 4.

Once the Dataset II using the PMMS was obtained, the TLS, Faro, was used to gather the Dataset I, which was used as ground truth. The data acquisition protocol with the TLS was carefully designed, so that the complete scanning area was covered with the minimum number of scans in order to minimize possible alignment errors. In this way, data acquisition with TLS was implemented from four different positions to ensure that the entire scene was covered. The resulting and aligned point cloud of the machine room was used to validate the dataset obtained with Zeb-Revo.

2.2.2. Data Processing

Once data acquisition protocol for Dataset I (Faro) and Dataset II (Zeb-Revo) was implemented, data processing strategy was applied over the point clouds generated with both laser systems, in order to extract the results. Data processing methodology (Figure 2) was fully oriented to compare the obtained results using Zeb-Revo with respect to the ground truth defined with Faro.

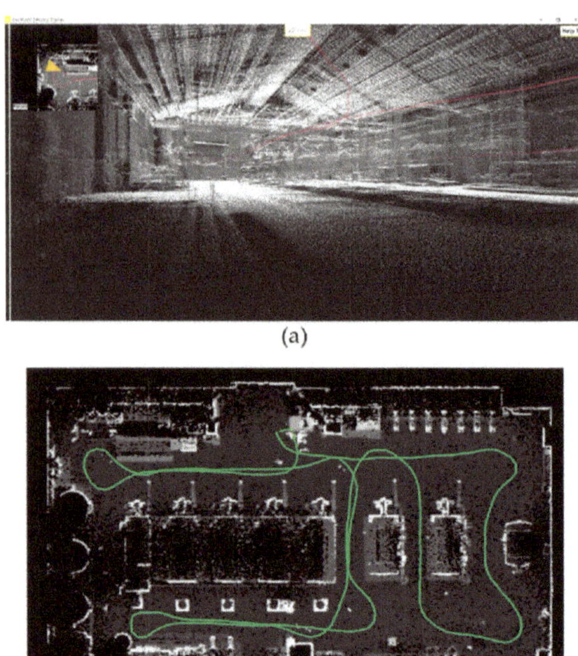

Figure 4. Three-dimensional scene shown by GeoSlam viewer (**a**) and followed trajectory during the data acquisition (green line) (**b**).

Point Clouds Processing: Merging, Alignment, and Segmentation

Dataset I (Faro) was composed by four individual point clouds (Figure 5a) generated from different positions in order to cover the entire scene. Point clouds were merged using the aligning algorithms provided by Faro Scene software. Spheres distributed along the scene were used to stablish plane-to-plane correspondences, which provide the initial approximations for the ICP [31]. As a result, a single point cloud with 165,829,378 points was obtained from the Dataset I (Faro).

Both dataset (Dataset I-Faro, Dataset II-Zeb-Revo) were aligned under the same CRS using the aforementioned spheres. For the alignment process, firstly, the three spheres were segmented for each point cloud and, subsequently, the centers of the spheres were extracted using a RANSAC-based sphere fitting [32]. The standard deviation (SD) resulting from the fitting is shown in Table 2.

In this way, three invariant references (centers of spheres) were obtained with an acceptable fitting error (Table 2), acting as homologous points for each dataset. The average error of the sphere fitting was 3.5 mm and 6.2 mm for the TLS (Dataset I) and the PMMS (Dataset II), respectively. The alignment between both point clouds was implemented by means of CloudCompare [33]. This software provides a registration tool which can align two point clouds if there are at least three corresponding point pairs in both datasets.

The root mean square error (RMSE) of the individual Faro scans was 6.2 mm, whereas the RMSE of the alignment of the PMMS point cloud to the reference system was 12.1 mm. Note that the alignment error between the individual TLS point clouds generates an error propagation in the subsequent analysis—e.g., the cylinder fitting—especially in the small pipes.

Once both datasets (Figure 5) were georeferenced in the same CRS, different regions of interest were used for primitives fitting and quality assessment through a robust statistical analysis (Figure 6).

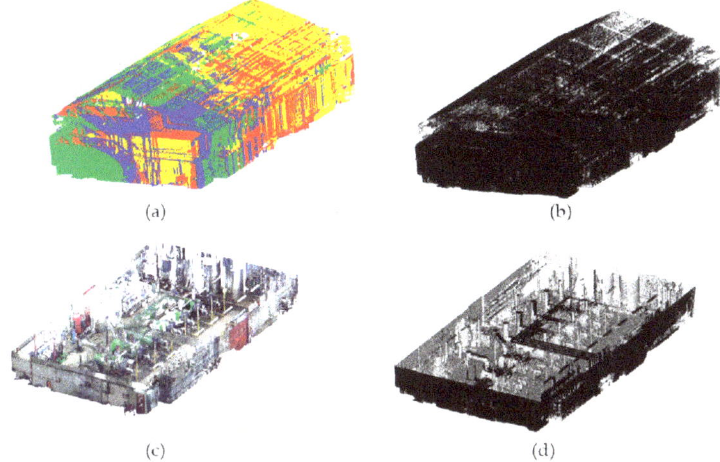

Figure 5. Dataset I: four point clouds obtained and aligned with Faro (**a**). Dataset II: point cloud obtained with Zeb-Revo (**b**). Dataset I: segmented point cloud to appreciate the distribution of facilities using Faro. (**c**). Dataset II: segmented point cloud to appreciate the distribution of facilities using Zeb-Revo (**d**).

Table 2. Standard deviation of the RANSAC-based sphere fitting to align both datasets in a global CRS.

Sphere	SD (m)	
	Zeb-Revo	Faro
1	±0.0064	±0.0041
2	±0.0067	±0.0038
3	±0.0057	±0.0026

The regions chosen for the analysis are the next:

1. Pipes (cylindrical regions): Thermal and fluid-mechanics facilities as the machine rooms normally have numerous pipes with different diameters, typologies and lengths that join the boilers and other equips with the gas connections, with the residual gas evacuations and with the water circuits. For this reason, focusing on inspection tasks, it can be necessary to check if the installed pipes are appropriate considering the design specification criteria for the thermal and fluid-mechanic facilities and the standards and regulations. Furthermore, the registration of pipes is highly useful for maintenance tasks and the parameters of the installed pipes as the diameter must be also considered for future modifications or expansion of the installation. These reasons motivated the choice of diameters as main parameters to be extracted from the point clouds for the identification of the pipes. In this way, three representative pipe regions were chosen for each boiler (Figure 7). These were chosen for having the smallest diameter (e.g., gas pipes which inject the gas into the boiler) and the largest diameter (e.g., water return pipe). This last was evaluated in two different locations: in the connection with the boiler and in the roof (Figure 7a,b). The aim is to control the deviation between the point clouds used to fit the pipes using Zeb-Revo and Faro.
2. Planes referring to the interior enclosures of the machine room: the planes fitted from the vertical walls and the floor are chosen because these regions have interest for the documentation

of volumetric parameters of the rooms, which are relevant for design and maintenance tasks as calculation of ventilation conditions, lighting design, emergency exists, etc. Furthermore, the relative location between the enclosures, the equipment and pipes are criteria to consider for compliance with standards and regulations.

a. Floor plane: One portion of the floor was chosen. The selection of this region was not easy because there were lot of obstacles which prevent its adequate segmentation. This region was selected without obstacles in order to analyze better its deviation (Figure 6).
b. Wall plane: Two portions of vertical walls were chosen (Figure 6). Again, these regions were selected without obstacles for providing a better analysis (Figure 6).

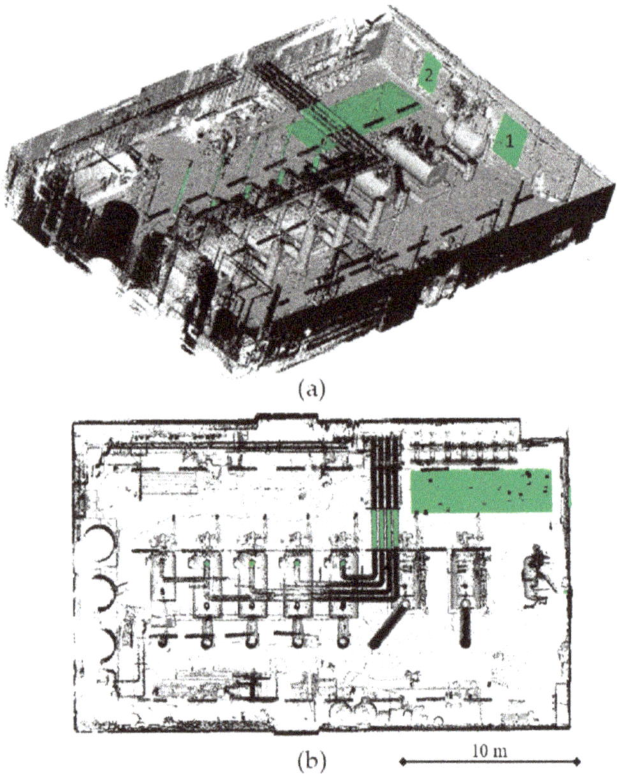

Figure 6. Different regions of interest used to fit the cylinders and planes (marked in green). The wall planes 1 and 2 are numbered in (**a**); while the floor planes and pipes highlighted are also shown in (**b**).

Feature Extraction of Pipes

First, to extract features of pipes a cylinder fitting was applied [32]. Particularly, this algorithm is based on a voting strategy well-known as RANSAC and can be implemented to detect different continuous and idealized geometries from point clouds (e.g., planes, spheres, cylinders, torus, etc.). In this case, the algorithm was applied to fit cylinders from the point clouds acquired in both datasets (Faro, Zeb-Revo). Once the cylinder was fitted, important features as diameters, lengths and a position were obtained. The quality of the adjustment was assessed through the discrepancy of each discrete point with respect to the ideal fitted cylinder.

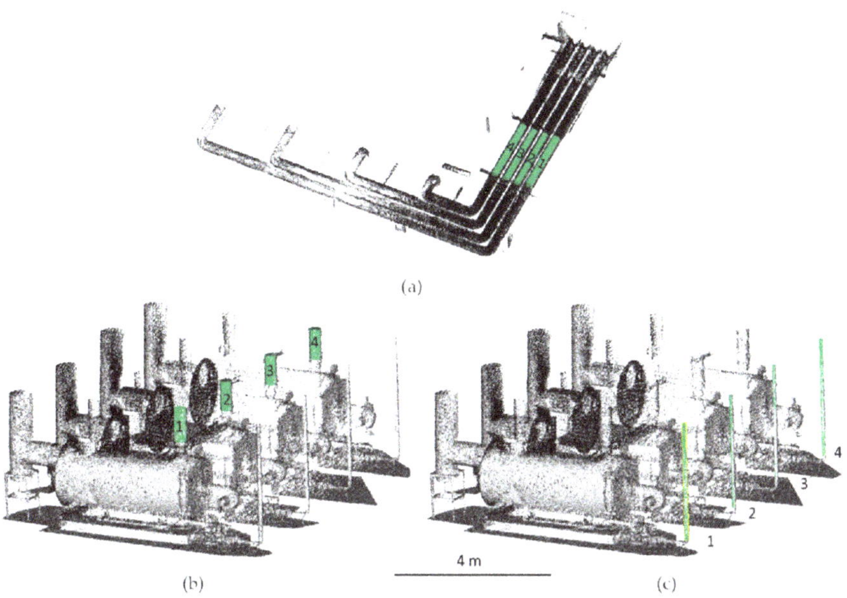

Figure 7. Regions of interest (pipes) used to fit the cylinder primitives: Water return pipe region located in the floor (**a**), water return pipe region located in the boiler (**b**), and gas pipe region (**c**).

Feature Extraction of Floor and Walls

The method to extract the features of the floor and wall regions is based on principal components analysis (PCA) algorithm which has been successfully applied to extract, from photogrammetric point clouds, geometrical features as angles between normal vectors of planes for the misalignment calculation of the steel welded plaques [8,34], as well as to compute the local 3D feature extraction for a machine learning classification of weld beads [35]. Specifically, PCA through the covariance matrix allows us to calculate the main components of the spatial points distribution, using the computation of eigenvectors. For instance, the eigenvector extracted from the smallest eigenvalue (λ_0) indicates the normal direction (v_0) to the fitting plane, and thus the normal direction for the wall and floor planes. Considering a point cloud of n points with coordinates x, y, z where x_m, y_m, z_m are the centroid coordinates, the covariance matrix (2) for each of them is calculated from the matrix of points (**W**). The covariance matrix (**C**) has the values of the variance in the principal diagonal (2). By the diagonalization process of matrix **C**, the eigenvectors of the covariance matrix are obtained, and as a result, the three eigenvalues (λ_1, λ_2, λ_3).

$$\mathbf{W} = \begin{pmatrix} x_1 - x_m & y_1 - y_m & z_1 - z_m \\ \vdots & \vdots & \vdots \\ x_n - x_m & y_n - y_m & z_n - z_m \end{pmatrix} \tag{1}$$

$$\mathbf{C} = \frac{1}{n}\mathbf{W}^T\mathbf{W} = \begin{pmatrix} \sigma_{xx} & \sigma_{yx} & \sigma_{xz} \\ \sigma_{yx} & \sigma_{yy} & \sigma_{yz} \\ \sigma_{zx} & \sigma_{zy} & \sigma_{zz} \end{pmatrix} \tag{2}$$

Using this procedure, planes were fitted from point clouds for both datasets (Faro-Dataset I, Zeb-Revo-Dataset II). The residuals were processed as the distance between each discrete point and the fitted planes. In this way, SD was calculated for the initial evaluation of the quality of the adjustment. Also, the misalignment between the planes was calculated (in mrad) as the angle between the normal vectors.

Primitive Fitting Error and Robust Statistical Analysis

Once planes and cylinders were fitted, a primitive fitting error was provided. The objective was to compare the Zeb-Revo results (Dataset II) with Faro results (Dataset I) used as ground truth. In this respect, point-primitive distances were computed for the elements described in the previous sections (Figures 6 and 7), concretely for the two pipes (Figure 8) and for all the planes. In addition, a robust statistical analysis was applied over the samples to avoid the effect caused by outliers. In this context, several studies [36–38] demonstrated that in the accuracy assessment of data provided by laser scanner systems, as well as photogrammetry, the hypothesis that errors follow a Gaussian distribution is hardly verified. This might be due to the presence of residual system errors, but also to the presence of undesirable scanned objects [22]. In the following analyses non-parametric estimators, the median m, normalized median absolute deviation (NMAD) (3) and the square root of the biweight midvariance (BWMV) (4), were employed. Specifically, the NMAD (3) allows to compare error dispersions from Gaussian samples, since it is normalized by the inverse of the cumulative distribution function of the Gaussian [39]

$$\text{NMAD} = 1.4826 \cdot \text{MAD} \tag{3}$$

$$\text{BWMV} = \frac{n\sum_{i=1}^{n} a_i (x_i - m)^2 \left(1 - U_i^2\right)^4}{\left(\sum_{i=1}^{n} a_i \left(1 - U_i^2\right)\left(1 - 5U_i^2\right)\right)^2} \tag{4}$$

$$a_i = \begin{cases} 1, \; if|U_i| < 1 \\ 0, \; if|U_i| \geq 1 \end{cases} \tag{5}$$

$$U = \frac{x_i - m}{9\text{MAD}} \tag{6}$$

being the median absolute deviation MAD (7) the median (m) of the absolute deviations from the data's median (m_x)

$$\text{MAD} = m(|x_i - m_x|) \tag{7}$$

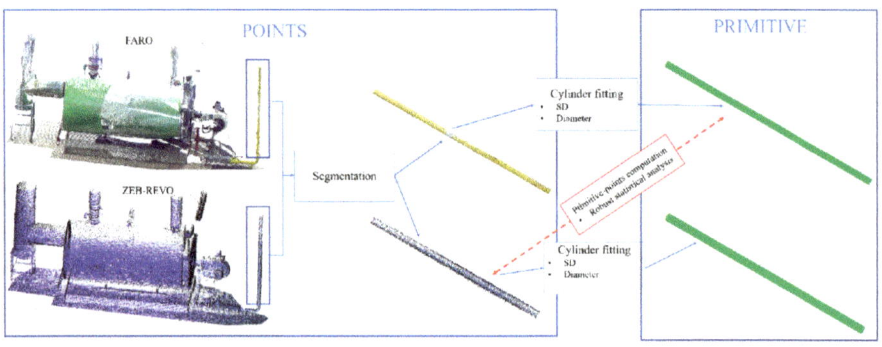

Figure 8. Procedure used to compare the different point clouds in pipes regions. The same procedure is following to compare the plane regions.

Additionally, the interpercentile range (IPR) at 50% and 95% of confidence level were reported. Please note also that due to the asymmetry of the error distribution was not possible to provide a plus-minus range, but an absolute interpercentile range; therefore, the percentile values at 2.5% and 97.5% were reported to outline the aforementioned asymmetry.

3. Results

As consequence of the direct application of the proposed methodology, results were obtained. In this way, a comparison between the cylinders fitted from the two datasets (Faro and Zeb-Revo) is shown in Tables 3–5, whereas the results for the planes fitted (roof and walls) are shown in Table 6. Results from the robust statistical analysis where points from Zeb-Revo were compared with respect to the cylinders and planes fitted from Faro are outlined in Tables 7 and 8.

Table 3. Diameters discrepancies and standard deviations (SD) obtained from the RANSAC cylinder fitting of the pipes: gas pipe regions (Figure 7).

Gas Pipe						
Regions	**Diameter TLS (m)**	**SD TLS (m)**	**Diameter PMMS (m)**	**SD PMMS (m)**	**Discrepancy (m)**	**% Discrepancy**
1	0.078	±0.001	0.097	±0.008	−0.019	24.08%
2	0.076	±0.001	0.094	±0.008	−0.019	24.70%
3	0.075	±0.001	0.093	±0.007	−0.018	24.34%
4	0.077	±0.001	0.088	±0.007	−0.010	13.08%
					−0.016	21.55%

Table 4. Diameters discrepancies and standard deviations (SD) obtained from the RANSAC cylinder fitting of the pipes: water return pipe regions located in the boiler (Figure 7).

Water Return Pipe (Connection in the Boiler)						
Regions	**Diameter TLS (m)**	**SD TLS (m)**	**Diameter PMMS (m)**	**SD PMMS (m)**	**Discrepancy (m)**	**% Discrepancy**
1	0.287	±0.002	0.298	±0.021	−0.011	3.94%
2	0.284	±0.003	0.294	±0.014	−0.010	3.52%
3	0.284	±0.003	0.290	±0.014	−0.006	2.11%
4	0.282	±0.003	0.291	±0.016	−0.009	3.34%
					−0.009	3.23%

Table 5. Diameters discrepancies and standard deviations (SD) obtained from the RANSAC cylinder fitting of the pipes: water return pipe regions located in the roof (Figure 7).

Water Return Pipe (Roof)						
Regions	**Diameter TLS (m)**	**SD TLS (m)**	**Diameter PMMS (m)**	**SD PMMS (m)**	**Discrepancy (m)**	**% Discrepancy**
1	0.285	±0.004	0.275	±0.018	0.010	3.47%
2	0.242	±0.007	0.260	±0.026	−0.018	7.24%
3	0.245	±0.005	0.265	±0.017	−0.020	8.22%
4	0.275	±0.006	0.275	±0.014	−0.000	0.11%
					0.012	4.76%

3.1. Cylinder Fitting for Pipes

The standard deviation (SD) and diameter discrepancies of the cylinders fitted were calculated (Figure 8) (Tables 3–5). Ideally, this discrepancy must be zero but, in this case, it is a useful accuracy measure of the method.

Table 6. Results of the fitting for the vertical planes (walls) and horizontal plane (roof) from point clouds regions (Figure 6).

		Number of Points	Normal Vector Coordinates			SD (m)	Misaligned Angle (Mrad)
			x	y	z		
Floor plane	Faro	3896744	−0.0023	0.0024	1.0000	0.0026	1.334
	Zeb-Revo	120780	−0.0036	0.0027	1.0000	0.0070	
Vertical wall 1	Faro	406026	0.3903	0.9027	−0.0037	0.0019	4.567
	Zeb-Revo	9387	0.3927	0.9197	−0.0033	0.0088	
Vertical wall 2	Faro	443852	0.3935	0.9193	0.0049	0.0017	11.250
	Zeb-Revo	19784	0.3967	0.9179	−0.0058	0.0083	

Table 7. Gaussian vs. Robust statistical analysis: Different regions analyzed (1–4) comparing the discrete points from Zeb-Revo with respect to the fitted surface (cylinder or plane) generated from Faro.

	Gaussian Assessment				Robust Assessment							
	Kurtosis	Skewness	Sample Mean	Sample Deviation	Median	MAD	NMAD	Sqrt (BWMV)	IPR 50%	Perc. 0.025	Perc. 0.975	IPR 95%
Water return pipe (boiler) (2)	6.353	0.618	0.002	0.012	0.001	0.007	0.011	0.012	0.014	−0.020	0.026	0.047
Water return pipe (boiler) (3)	3.694	0.343	0.003	0.014	0.003	0.009	0.014	0.014	0.019	−0.023	0.034	0.057
Gas pipe (2)	2.941	0.202	0.008	0.009	0.007	0.006	0.009	0.009	0.013	−0.009	0.028	0.037
Gas pipe (3)	4.201	−0.043	0.008	0.007	0.008	0.004	0.007	0.007	0.009	−0.006	0.022	0.028
Water return pipe (roof) (3)	32.028	3.213	0.001	0.020	0.000	0.010	0.014	0.015	0.020	−0.030	0.039	0.069
Water return pipe (roof) (4)	3.441	0.047	−0.003	0.014	−0.003	0.009	0.013	0.014	0.018	−0.030	0.026	0.055
Vertical wall (1)	79.303	3.399	−0.027	0.009	−0.027	0.005	0.007	0.008	0.010	−0.042	−0.011	0.031
Vertical wall (2)	4.277	0.184	−0.013	0.008	−0.013	0.005	0.007	0.008	0.010	−0.029	0.005	0.034
Floor plane	11.701	0.146	−0.003	0.007	−0.003	0.004	0.006	0.006	0.008	−0.016	0.011	0.027

Table 8. Summary of bias and dispersion data for the analyzed regions (1–4) comparing the discrete points from Zeb-Revo with respect to the fitted surface (cylinder or plane) generated from Faro.

	Bias	Dispersion					
	Median (mm)	NMAD (mm)	Sqrt (BWMV) (mm)	IPR 50% (mm)	IPR 95% (mm)	Perc 0.025	Perc 0.975
Water return pipe (boiler) (2)	1.4	±10.9	±11.6	14.5	46.5	−20.3	26.3
Water return pipe (boiler) (3)	2.6	±13.7	±14.0	18.6	57.3	−23.2	34.1
Gas pipe (2)	7.5	±9.4	±9.5	12.7	36.6	−9.0	27.6
Gas pipe (3)	8.0	±6.5	±6.8	8.8	28.2	−6.1	22.1
Water return pipe (roof) (3)	−0.4	±14.5	±15.5	19.5	68.8	−29.9	38.9
Water return pipe (roof) (4)	−3.2	±13.4	±13.8	18.2	55.4	−29.7	25.8
Vertical wall (1)	−27.2	±7.1	±7.6	9.7	31.2	−41.8	−10.6
Vertical wall (2)	−12.8	±7.4	±8.0	10.0	34.1	−29.0	5.1
Floor plane	−3.2	±5.9	±6.4	8.0	26.9	−16.3	10.7

The SD for the fitting process for the Dataset I (Faro) was between 1 and 7 mm (being higher in those cylinders located on the roof). These results (Tables 3–5) were considered as reference for comparison (ground truth).

As it was indicated in Tables 3–5, the diameter discrepancies obtained with both sensors (Faro and Zeb-Revo) were between 0.6 and 1.9 cm for all the analyzed pipes, but analyzing discrepancies in relative terms, the deviation between the two diameters of the two cylinders fitted (Faro and Zeb-Revo) was notably higher for small diameter cylinders (21.55%) than for large diameter cylinders (3.2% and 4.8%).

The computation of distances between the point clouds obtained using Zeb-Revo (Dataset II) and the cylinders fitted from the Dataset I (Faro); and the subsequently application of the robust statistical analysis gave some results (Table 7) that can be summarized in the next way:

1. Bias results: the median in the cylinders of bigger diameter is between 2.6 and -3.2 mm, whereas for cylinders of smaller size the median is around 8 mm (Table 7). The latter is compatible with the difference in diameter of approximately 1 cm between the cylinders adjusted for both sensors (Table 3).
2. Dispersion results: Robust analysis using the normalized median absolute deviation (NMAD) and the interpercentile range (IPR) (Tables 7 and 8) shows higher dispersion results for cylinders with larger diameter, always remaining at 1.45 and 0.65 cm, whereas dispersion results are smaller for smaller diameters. The latter and the results of median indicate that there is an initial bias in the cylinders with a small diameter and this error is not due to the dispersion of the data obtained with Zeb-Revo. Please note, that gas pipes shown the highest discrepancies in relative values, but in the present analysis (absolutes values) they are lowest. This behavior is caused by the discrepancy computation which depends inversely proportional to the cylinder diameter.

The robust statistical analysis carried out also discards that the distribution of distances of the points obtained with Zeb-Revo with respect to the cylinders adjusted with Faro follows a Gaussian distribution (on the basis of a visual inspection of a quantile-quantile plot).

3.2. Plane Fitted for Floor and Wall Regions

The standard deviation of the fitting was calculated for each plane (Figure 8). The SD of the fitting is between 1.7 and 2.6 mm for the Dataset I (Faro) and between 7 and 8.8 mm for Dataset II (Zeb-Revo). These results are consistent because the accuracy of Faro is around 2 mm and the expected accuracy of Zeb-Revo is centimetric. The results of Dataset I were considered as reference for comparison for the subsequent robust statistical analysis.

The computation of distances between the cloud of points obtained with Zeb-Revo (Dataset II) and the cylinders fitted from Faro (Dataset I) and the application of the robust statistical analysis gave some results that can be summarized as follows (Table 8):

1. Bias results: The median in vertical planes (walls) was significantly higher than in horizontal plane (floor). It could be the consequence of an initial bias of unknown origin, which, as hypothesis, could be caused by the different surface finish of wall in relation to the floor one, as well, as the relative position and distance regarding the path of the PMMS data acquisition.
2. Dispersion results: Robust statistical analysis with MAD and IPR showed homogeneous dispersion results in all cases (Tables 7 and 8).

3.3. In-Detail Elements: Qualitative Analysis

Different discrete elements have been segmented from point clouds in order to qualitatively assess the level of visualization of these. In Figure 9 is shown a gas pressure regulator system (usually installed in thermal installations) located at the entrance of the boilers. As the reader can see, the difference in the resolution between the two datasets is important but the points of the PMMS are enough for the identification of the valve, because the three elements of the system are distinguishable. Note that both point clouds present outliers.

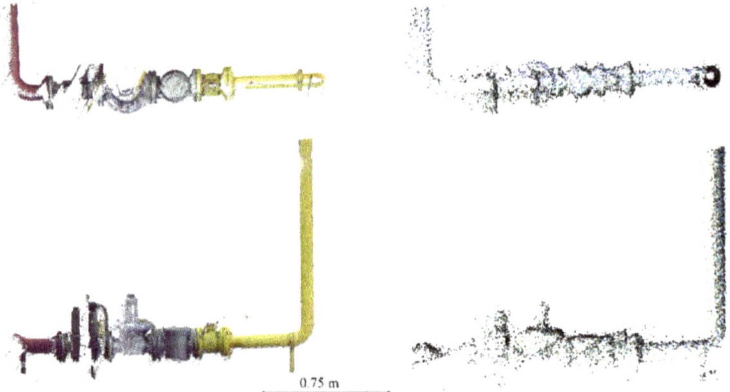

Figure 9. Example of points clouds of the pressure regulation system located at the entrance of the gas in the boilers obtained with Faro (Left) and with Zeb-Revo (Right).

In Figure 10, a cutting valve (usual in most fluid-mechanical installations) is located in the connection to the boiler water pipe. In this, the geometry of the valve is defined in both point clouds, however the final part of the grip is not appreciated in the PMMS point cloud, whereas in the TLS point cloud the valve is fully appreciated. In Figure 11, a boiler burner is shown. In this case, the completeness is higher in the PMMS point cloud but the final part of the grip is not appreciated in the TLS point cloud. This is due to the fact that PMMS takes the data in motion and all the zones are covered, whereas the TLS is static, and the completeness of the models depends on the number of stations of the laser scanner. Finally, a hot water tank is shown in Figure 12. In this case, the same thing happens: the PMMS point cloud is more complete than the TLS point cloud, while this last presents a higher resolution and detail perception.

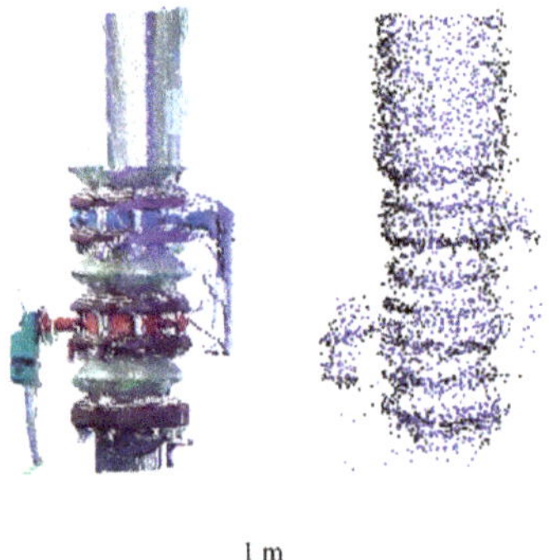

1 m

Figure 10. Example of points clouds of the cutting valve at the entrance of the water in the boilers obtained with Faro (Left) and with Zeb-Revo (Right).

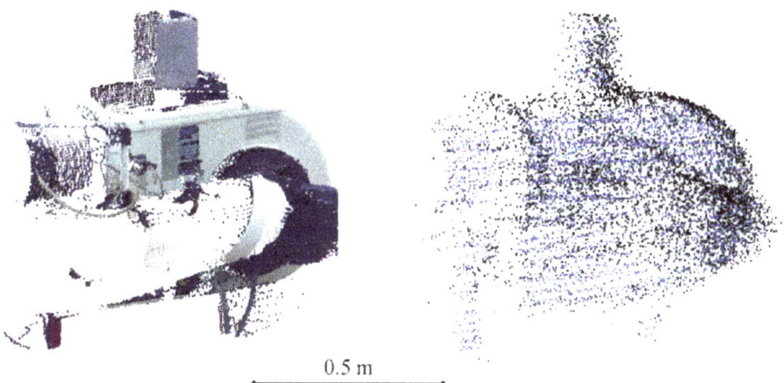

Figure 11. Example of points clouds of the burner system located in the boilers obtained with Faro (Left) and with Zeb-Revo (Right).

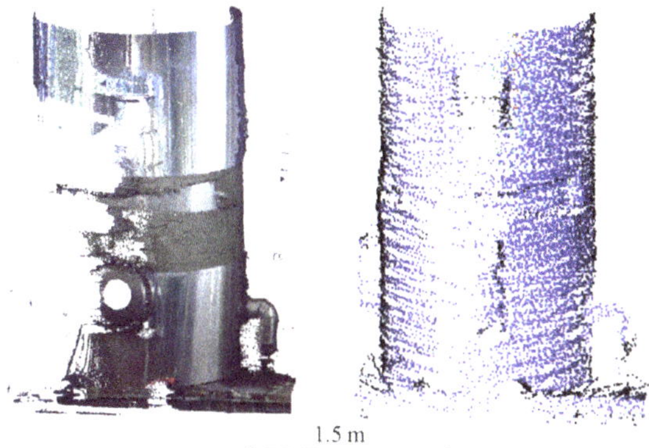

Figure 12. Example of point clouds of the hot water tank obtained with Faro (Left) and with Zeb-Revo (Right).

4. Discussion

Indoor mapping and modeling is probably the main application scenario for PMMS. This becomes even more significant in the specific case of industrial facilities, where the present elements can be reconducted on the basis of parametric geometries and therefore deliver a BIM. For this case, PMMS offers the best compromise between spatial resolution, precision, surveying time, and cost [12].

The potential advantages of PMMS can be summed up on the completeness of the data acquisition thanks to the maneuverability and flexibility of the PMMS in complex scenarios; and the speed of acquisition due to the lack of static stations. Regarding the advantages of the PMMS in relation to the TLS, the key factor is the reduced surveying time. In the present study case, the surveying time for the 4 TLS scans encompassed 28 min and 40 s, whereas the PMMS required only 4 min and 18 s. Ergo, the surveying time was reduced in almost seven times, obtaining at the same a comparable level of completeness.

The results of the primitive fitting are inside the interval expected on the basis of the manufacturer a-priori precision (standard deviation lower than 3 cm). However, in terms of the computed pipe diameters, there was identified a bias. Considering the analysis of the pipes, the bias existing in the smaller pipes is significant. This could be caused by the alignment error during the SLAM processing. Since the precision of the PMMS is about 1 to 3 cm, this error becomes noteworthy for pipes diameters about 8 cm. However, other error sources are not excluded, such as the high curvature of the pipe and/or the surface finish. As a result, this issue will require further investigations. Regarding the water pipes, due to the larger diameter, the relative discrepancy is lower than 5% (Tables 4 and 5). Please note that in the case of the roof water pipes the discrepancies could reach a bit higher values (up to 8.2% as stated in Table 5). This could be caused by the object–sensor distance and the lower completeness due to the limited point of view and small distance among pipes.

The extension of the pipe analysis from the diameter value up to the complete pipe, the PMMS point cloud discrepancies, against the primitive from TLS, decrease for the small pipes (gas), whereas for the larger (water), the discrepancy is zero centered in a ±3 mm interval. Therefore, the PMMS is a suitable sensor for the maintenance, inspection, and safety tasks in fluid-mechanical installation, due to the lack a significant bias, and relative diameter discrepancies lower than 5%. For the gas pipes (diameters lower than 10 cm), the use of the PMMS is limited to general documentation purposes (e.g., location of the pipes).

The static nature of the TLS data acquisition gave place to occlusions in the reference point cloud, but also non-acquisition of parts of the fluid-mechanical installations (e.g., valves, grips). This first issue (occlusions) could not be especially relevant in the primitive fitting due to the careful planning of data acquisition phase; however, the absence of TLS data causes outliers in the point-to-point analysis. So, this discrete analysis requires to take into account the point's neighborhood to avoid mismatches in the discrepancy calculation. Regarding the non-acquisition of small parts, the dynamic nature of the PMMS makes it more suitable than TLS to document industrial facilities.

Regarding the plane analysis, the primitive fitting from PMMS point cloud is compatible with the technical specifications. The derived robust confidence intervals are more symmetrical than the pipes ones. The IPR 95% lower than 35 mm is a very significant results related to the PMMS ability to record this kind of complex environment. For both vertical planes, a significant bias was detected, that could be caused by the surface finish (in relation to the floor plane). In addition, the object-sensor relative position during the data acquisition path (see Figure 4) could be another explanation for the abovementioned bias. The last possible error source is a residual error from the alignment procedure carried out with spheres. This effect is inevitable, but it is diminished by the fact of being both active sensors (the scale parameter is not expected to be significant), remaining only the error from the sphere network design and sphere centroid extraction. However, the centimetric error provided by the PMMS unlike the millimetric error of the TLS, can generate an error in the alignment of the point clouds which justify the bias. The obtention of additional information as the plane misalignment can check the plane orientation, a normally neglected aspect in the analysis of 3D fitting. As expected, the floor plane misalignment is not significant, since it is in the same precision range than the precision construction levels (approximately 0.1 gons). However, for the wall planes it is noticed a significant put-of-plumb condition, with an equivalent displacement of 2 to 4.5 cm for a four meters height (value which exceeds the tolerances of the support plates of the beams used for industrial installations). This angular discrepancy can be attributed to the aforementioned wall's surface finish, and to a lesser extent, the data acquisition path and the residual error on the alignment of the PMMS in relation to the TLS.

Finally, regarding the robust analysis, it can be noted that a slight overestimation of the dispersion values for the Gaussian assessment in relation to the NMAD. In relation to the robust dispersion parameters, NMAD and the square root of BWMV present compatible values. The robust confidence interval provides information related to the asymmetry of the discrepancy sample. When the non-normality is very pronounced (e.g., kurtosis > 10) the Gaussian-derived confidence interval tend to be overestimated. The comparison with the robust interval, for the present study case, reach difference up to 10 mm for the water return pipe 3 (roof).

5. Conclusions

A quantitative and qualitative comparison between the results obtained using a novel PMMS (Zeb-Revo) and the results obtained with a TLS (Faro) for inspection tasks in thermal and fluid-mechanical facilities is provided. Point clouds gathered from both devices were aligned in the same coordinate system using a network of spheres. Segmentation of objects of interest (walls, roofs, floors, and pipes) was implemented to fit ideal geometries useful for the extraction of important features of the facility, as cylinder (pipes) and planes (walls and roof) that allow to extract different features of interest. The different results obtained with the two sensors were compared applying a robust statistical analysis to evaluate the real potential of the PMMS. Additionally, different specific devices of interest—such as valves, regulation systems, tanks, or burners—were segmented from the two datasets to qualitatively show the level of detail which allows Zeb-Revo in comparison with Faro. Globally, the PMMS performance is inside the technical specification provided by the manufacturer, as it is shown by other studies [22].

The deviation in the estimation of the diameter in the analyzed pipes using Zeb-Revo (with respect to the dataset obtained with Faro) is between 9 and 16 mm when a RANSAC fitting methodology is applied. This deviation could be acceptable for inspection tasks of industrial facilities such as the one proposed in [4]. This deviation in relative terms is higher for pipes with smaller diameter (21.55% in an 8 cm pipe), whereas this relative deviation is better for pipes with higher diameter (3.23% and 4.8% in a 28 cm pipe), always remaining around the centimetric precision given by the Zeb-Revo manufacturer.

A robust statistical analysis was applied to compare the deviation of the discrete points acquired from Zeb-Revo with respect the ideal cylinder fitted from Faro. This analysis shows acceptable results of bias and dispersion (NMAD between 6.5 and 14.5 mm), but for those cylinders of smaller diameter, a bias is detected (7.5–8 mm). This bias could be the main cause of the high percentage deviation in the diameter measurement of those pipes and the origin is not completely explained (pipe curvature, surface finish, etc.) as stated in the discussion section.

The studied planes fitted from points for each dataset show geometrical similitude. This is evaluated through the angle between the normal vector of the plane generated from Zeb-Revo and the one generated with Faro. The misaligned angle analysis allowed to identify an angular discrepancy for the wall planes, whereas in the case of the floor plane, the value 1.3 mrad can be considered a very good result. However, while the dispersion results studied with NMAD, square root of BWMV and IPR are suitable for the three planes (NMAD between 5.9 and 7.4 mm), a significant bias is detected in vertical planes (median −27.2 and −12.8 mm).

Finally, the extraction and qualitative analysis of the specific in-detail elements from the two datasets shows the differences in terms of resolution between them. The identification of element is obviously more evident using TLS (also considering the color registration, which is an added value for the definition of these elements). However, the completeness of the PMMS point cloud can be even higher than the dataset obtained from TLS, allowing for visual identification of elements commonly used in the facilities, such as cutting valves, pressure regulation systems, and burner systems.

Future works will address the improvement of the processes of alignment that allows to obtain an answer to the problem of the bias.

Author Contributions: Conceptualization, M.R.-M. and P.R.-G.; Methodology, M.R.-M. and P.R.-G.; Data acquisition, M.R.-M. and P.R.-G.; Validation, M.R.-M., P.R.-G., and E.R.d.O.C.; Formal analysis, M.R.-M. and P.R.-G.; Resources, D.G.-A.; Writing—original draft preparation, M.R.-M., P.R.-G., E.R.d.O.C., and D.G.-A.; Writing—review and editing, M.R.-M., P.R.-G., and D.G.-A.

Funding: This research received no external funding.

Conflicts of Interest: The authors declare no conflict of interest.

References

1. Gonzalez-Aguilera, D.; Del Pozo, S.; Lopez, G.; Rodriguez-Gonzalvez, P. From point cloud to CAD models: Laser and optics geotechnology for the design of electrical substations. *Opt. Laser Technol.* **2012**, *44*, 1384–1392. [CrossRef]
2. Rodríguez-Gonzálvez, P.; Gonzalez-Aguilera, D.; Lopez-Jimenez, G.; Picon-Cabrera, I. Image-based modeling of built environment from an unmanned aerial system. *Autom. Constr.* **2014**, *48*, 44–52. [CrossRef]

3. Caldwell, R. Hull inspection techniques and strategy-remote inspection developments. In Proceedings of the SPE Offshore Europe Conference & Exhibition, Aberdeen, UK, 5–8 September 2017.
4. Rodríguez-Martín, M.; Rodríguez-Gonzálvez, P.; Gonzalez-Aguilera, D.; Nocerino, E. Novel Approach for Three-Dimensional Integral Documentation of Machine Rooms in Hospitals Using Portable Mobile Mapping System. *IEEE Access* **2018**, *6*, 79200–79210. [CrossRef]
5. Hullo, J.F.; Thibault, G.; Boucheny, C.; Dory, F.; Mas, A. Multi-Sensor As-Built Models of Complex Industrial Architectures. *Remote Sens.* **2015**, *7*, 16339–16362. [CrossRef]
6. Pătrăucean, V.; Armeni, I.; Nahangi, M.; Yeung, J.; Brilakis, I.; Haas, C. State of research in automatic as-built modeling. *Adv. Eng. Inform.* **2015**, *29*, 162–171. [CrossRef]
7. Quattrini, R.; Malinverni, E.S.; Clini, P.; Nespeca, R.; Orlietti, E. From TLS to HBIM. High quality semantically-aware 3d modeling of complex architecture. *Int. Arch. Photogramm. Remote Sens. Spat. Inf. Sci.* **2015**, 367–374. [CrossRef]
8. Rodríguez-Martín, M.; Rodríguez-Gonzálvez, P.; Lagüela, S.; González-Aguilera, D. Macro-photogrammetry as a tool for the accurate measurement of three-dimensional misalignment in welding. *Autom. Constr.* **2016**, *71*, 189–197. [CrossRef]
9. Rodríguez-Gonzálvez, P.; Rodríguez-Martín, M.; Ramos, L.F.; González-Aguilera, D. 3D reconstruction methods and quality assessment for visual inspection of welds. *Autom. Constr.* **2017**, *79*, 49–58. [CrossRef]
10. Muhammad, J.; Altun, H.; Abo-Serie, E. Welding seam profiling techniques based on active vision sensing for intelligent robotic welding. *Int. J. Adv. Manuf. Technol.* **2017**, *88*, 127–145. [CrossRef]
11. Shah, H.N.M.; Sulaiman, M.; Shukor, A.Z.; Kamis, Z.; Rahman, A.A. Butt welding joints recognition and location identification by using local thresholding. *Robot. Comput. Integr. Manuf.* **2018**, *51*, 181–188. [CrossRef]
12. Nocerino, E.; Rodríguez-Gonzálvez, P.; Menna, F. Introduction to mobile mapping with portable systems. In *Laser Scanning: An Emerging Technology in Structural Engineering*; CRC Press: Boca Raton, FL, USA, 2019; pp. 37–52.
13. Nocerino, E.; Menna, F.; Toschi, I.; Morabito, D.; Remondino, F.; Rodríguez-Gonzálvez, P. Valorisation of history and landscape for promoting the memory of WWI. *J. Cult. Herit.* **2017**, *29*, 113–122. [CrossRef]
14. Russhakim, N.A.S.; Ariff, M.F.M.; Majid, Z.; Idris, K.M.; Darwin, N.; Abbas, M.A.; Zainuddin, K.; Yusoff, A.R. The Suitability of Terrestrial Laser Scanning for Building Survey and Mapping Applications. *Int. Arch. Photogramm. Remote Sens. Spat. Inf. Sci.* **2019**, *42*, 663–670. [CrossRef]
15. Nikoohemat, S.; Peter, M.; Oude Elberink, S.; Vosselman, G. Exploiting indoor mobile laser scanner trajectories for semantic interpretation of point clouds. *ISPRS Ann. Photogramm. Remote Sens. Spat. Inf. Sci.* **2017**, *2017*, 355–362. [CrossRef]
16. Rodríguez-Gonzálvez, P.; Nocerino, E. Portable Mobile Mapping Systems applied to the Management of Natural Spaces. In *New Developments in Agricultural Research*; Nova Science Publishers: New York, NY, USA, 2019; pp. 1–43.
17. GeoSlam Zeb-Revo. Available online: https://geoslam.com/solutions/zeb-revo/ (accessed on 13 September 2019).
18. Chiabrando, F.; Coletta, C.D.; Sammartano, G.; Spanò, A.; Spreafico, A. "Torino 1911" project: A contribution of a slam-based survey to extensive 3D heritage modeling. *Int. Arch. Photogramm. Remote Sens. Spat. Inf. Sci.* **2018**, 225–234. [CrossRef]
19. Dewez, T.J.B.; Yart, S.; Thuon, Y.; Pannet, P.; Plat, E. Towards cavity-collapse hazard maps with Zeb-Revo handheld laser scanner point clouds. *Photogramm. Rec.* **2017**, *32*, 354–376. [CrossRef]
20. Masiero, A.; Fissore, F.; Guarnieri, A.; Pirotti, F.; Visintini, D.; Vettore, A. Performance Evaluation of Two Indoor Mapping Systems: Low-Cost UWB-Aided Photogrammetry and Backpack Laser Scanning. *Appl. Sci.* **2018**, *8*, 416. [CrossRef]
21. Chiabrando, F.; Sammartano, G.; Spanò, A. A comparison among different optimization levels in 3D multi-sensor models. A test case in emergency context: 2016 Italian earthquake. *Int. Arch. Photogramm. Remote Sens. Spat. Inf. Sci.* **2017**, 155–162. [CrossRef]
22. Nocerino, E.; Menna, F.; Remondino, F.; Toschi, I.; Rodríguez-Gonzálvez, P. Investigation of indoor and outdoor performance of two portable mobile mapping systems. *Proc. SPIE* **2017**, *10332*. [CrossRef]
23. Tucci, G.; Visintini, D.; Bonora, V.; Parisi, E. Examination of Indoor Mobile Mapping Systems in a Diversified Internal/External Test Field. *Appl. Sci.* **2018**, *8*, 401. [CrossRef]

24. AENOR (Spanish Association for Standardisation). Machine Rooms and Gas Fired Self-Contained Apparatus for Heating or Cooling Generation or Cogeneration, Standard UNE 60601:2013. Available online: https://www.aenor.com/normas-y-libros/buscador-de-normas/une/?c=N0052265 (accessed on 12 November 2014).
25. BOE (Official State Gazette of the Government of Spain). *Real Decreto 2060/2008, de 12 de Diciembre, Por el Que se Aprueba el Reglamento de Equipos a Presión y Sus Instrucciones Técnicas Complementarias*; Ministerio de Industria, Turismo y Comercio: Madrid, Spain, 2009; pp. 12297–12388.
26. BOE (Official State Gazette of the Government of Spain). *Real Decreto 1027/2007, de 20 de Julio, Por el Que se Aprueba el Reglamento de Instalaciones Térmicas en Los Edificios*; Ministerio de la Presidencia: Madrid, Spain, 2007; pp. 35931–35984.
27. Bosse, M.; Zlot, R.; Flick, P. Zebedee: Design of a Spring-Mounted 3-D Range Sensor with Application to Mobile Mapping. *IEEE Trans. Robot.* **2012**, *28*, 1104–1119. [CrossRef]
28. Eyre, M.; Wetherelt, A.; Coggan, J. Evaluation of automated underground mapping solutions for mining and civil engineering applications. *J. Appl. Remote Sens.* **2016**, *10*, 20–39. [CrossRef]
29. Cabo, C.; Del Pozo, S.; Rodríguez-Gonzálvez, P.; Ordóñez, C.; González-Aguilera, D. Comparing Terrestrial Laser Scanning (TLS) and Wearable Laser Scanning (WLS) for Individual Tree Modeling at Plot Level. *Remote Sens.* **2018**, *10*, 540. [CrossRef]
30. Besl, P.; McKay, M. A method for registration of 3-D shapes. *IEEE Trans. Pattern Anal. Mach. Intell.* **1992**, *14*, 239–256. [CrossRef]
31. Wujanz, D.; Barazzetti, L.; Previtali, M.; Scaioni, M. A Comparative Study among Three Registration Algorithms: Performance, Quality Assurance and Accuracy. *Int. Arch. Photogramm. Remote Sens. Spat. Inf. Sci.* **2019**, 779–786. [CrossRef]
32. Schnabel, R.; Wahl, R.; Klein, R. Efficient RANSAC for Point-Cloud Shape Detection. *Comput. Graph. Forum* **2007**, *26*, 214–226. [CrossRef]
33. Cloud Compare. GPL Software (Version 2.9.1). Available online: www.danielgm.net/cc/ (accessed on 7 August 2019).
34. Rodríguez-Martín, M.; Rodríguez-Gonzálvez, P.; Gonzalez-Aguilera, D.; Fernandez-Hernandez, J. Feasibility study of a structured light system applied to welding inspection based on articulated coordinate measure machine data. *IEEE Sens. J.* **2017**, *17*, 4217–4224. [CrossRef]
35. Rodríguez-Gonzálvez, P.; Rodríguez-Martín, M. Weld Bead Detection Based on 3D Geometric Features and Machine Learning Approaches. *IEEE Access* **2019**, *7*, 14714–14727. [CrossRef]
36. Höhle, J.; Höhle, M. Accuracy assessment of digital elevation models by means of robust statistical method. *ISPRS J. Photogramm. Remote Sens.* **2009**, *64*, 398–406. [CrossRef]
37. Hasan, A.; Pilesjö, P.; Persson, A. The use of LIDAR as a data source for digital elevation models–a study of the relationship between the accuracy of digital elevation models and topographical attributes in northern peatlands. *Hydrol. Earth Syst. Sci. Discuss.* **2011**, *8*, 5497–5522. [CrossRef]
38. Rodríguez-Gonzálvez, P.; Garcia-Gago, J.; Gomez-Lahoz, J.; González-Aguilera, D. Confronting Passive and Active Sensors with Non-Gaussian Statistics. *Sensors* **2014**, *14*, 13759–13777. [CrossRef]
39. Herrero-Huerta, M.; Lindenbergh, R.; Rodríguez-Gonzálvez, P. Automatic tree parameter extraction by a Mobile LiDAR System in an urban context. *PLoS ONE* **2018**, *13*, e0196004. [CrossRef] [PubMed]

Technical Note

Use of a Wearable Mobile Laser System in Seamless Indoor 3D Mapping of a Complex Historical Site

Andrea di Filippo [1], Luis Javier Sánchez-Aparicio [2], Salvatore Barba [1], José Antonio Martín-Jiménez [2], Rocío Mora [2] and Diego González Aguilera [2,*]

[1] Department of Civil Engineering, University of Salerno, Via Giovanni Paolo II, 132, 84084 Fisciano (SA), Italy; andrew89.adf@gmail.com (A.d.F.), sbarba@unisa.it (S.B.)

[2] Department of Land and Cartographic Engineering. University of Salamanca, Higher Polytechnic School of Avila, Calle de los Hornos Caleros, 50, 05003 Avila, Spain; luisj@usal.es (L.J.S.-A.); joseabula@usal.es (J.A.M.-J.); rociomora@usal.es (R.M.)

* Correspondence: daguilera@usal.es

Received: 4 November 2018; Accepted: 26 November 2018; Published: 28 November 2018

Abstract: This paper presents an efficient solution, based on a wearable mobile laser system (WMLS), for the digitalization and modelling of a complex cultural heritage building. A procedural pipeline is formalized for the data acquisition, processing and generation of cartographic products over a XV century palace located in Segovia, Spain. The complexity, represented by an intricate interior space and by the presence of important structural problems, prevents the use of standard protocols such as those based on terrestrial photogrammetry or terrestrial laser scanning, making the WMLS the most suitable and powerful solution for the design of restoration actions. The results obtained corroborate with the robustness and accuracy of the digitalization strategy, allowing for the generation of 3D models and 2D cartographic products with the required level of quality and time needed to digitalize the area by a terrestrial laser scanner.

Keywords: cultural heritage; restoration; indoor mapping; laser scanning; wearable mobile laser system; 3D digitalization; SLAM

---

1. Introduction

The guidelines for the conservation and enhancement of cultural heritage, codified in the Athens Charter and repeatedly reiterated by subsequent documents up to the most recent Krakow Charter [1], underline the importance of multidisciplinary and scientific approaches for the management of interventions in cultural heritage sites [2].

Currently, the use of new technologies for the data acquisition in the architectural field has reached widespread diffusion, mainly due to the ability to digitalize artifacts with great precision and to the possibility of generating informative models useful for the analysis, simulation, and interpretation phases [3]. The most popular techniques, which have now become a reference standard, are modern photogrammetry [4] and laser scanning [5]. Photogrammetry acquires two-dimensional images that require mathematical processing to derive 3D information. Through precise formulations based on projective or perspective geometry [6], it transforms the data extracted from the images into three-dimensional metric coordinates and colors [7]. For its part, laser scanning is able to directly obtain the 3D point spatial position [8,9] with high accuracy and without lighting conditions, especially over homogeneous surfaces where photogrammetry cannot provide reliable results.

The main products obtained from both 3D point clouds and 2D orthoimages techniques have been used for the virtual reconstruction of cultural heritage sites [10], the analysis of rock-art paintings [11,12], the creation of accurate numerical simulations [13], or even the analysis of pathological processes [14,15], among others.

Besides the wide range of advantages that these solutions can offer, the digitalization of large and complex areas, especially indoor scenarios, generally entails the use of a large amount of images (in the case of photogrammetry) or scan stations (in the case of the laser scanner), deriving in a time consuming fieldwork and thus in an important error propagation [16,17]. Hybrid solutions, such as mobile mapping systems (MMSs), have emerged with great capabilities and possibilities in the last few years, allowing the management of different sensors and the possibility to operate in complex outdoor and indoor scenarios [18–21], minimizing error propagation.

Since their early development in the late 1980s, MMSs have been progressively improved in order to provide increasingly more precise and denser data, acquired in a shorter amount of time [22]. Besides the progresses in optical sensors, one of the key advances in MMS is related to spatial referencing technology. While the very early applications were restricted to environments where the sensor positions were computed using ground control, advantages in satellite and inertial technology make spatial referencing possible in previously unknown and undiscovered places [23,24]. Furthermore, the miniaturization and cost reduction of components have played a fundamental role in the spread of MMS, allowing for more flexible, portable, and low-cost systems [22]. This attribute, within the capacity of generating 3D point clouds by means of a spatial referencing in previously unknown environments, has allowed the application of this technology in Unmanned Aerial Vehicles (UAV) [23,25,26], Unmanned Ground Vehicles (UGV) [27–29], or equipped in backs (e.g., the Leica Pegasus back-pack, the Heron MS-2 back-pack or the Kaarta Stencil) [22,24,30]. Meanwhile, the use of the two first platforms could reduce the problems associated with travelling speed during the data acquisition, as well as improving the time efficiency during the survey. Moreover, their application in indoor and narrow spaces (common in Cultural Heritage) could entail some problems. These limitations place the wearable mobile laser systems (WMLS) as a potential solution for mapping indoor environments, as it is possible to obtain a 3D point cloud of the environment with a centimeter's accuracy [22,24,30]. However, this accuracy could be strongly affected by the characteristics of the trajectory, such as the travelling speed or the path followed [22,31].

Under these assumptions, this paper evaluates the suitability of a wearable mobile laser system for the digitalization of a complex indoor environment belonging to a cultural heritage building, as well as for the generation of cartographic products required for its conservation and restoration. This wearable system combines laser scanning technology and an inertial measurement unit (IMU) in portable equipment that can be handled by an operator while walking through the cultural heritage site. This sensor acquires point clouds on the move, thanks to the Simultaneous Localization and Mapping algorithms (SLAM) [32,33], without needing the support of a global navigation satellite system (GNSS). During this evaluation, we took into account the different parameters that could influence the final quality of the 3D point cloud. These parameters are: (i) the identification of critical areas; (ii) the prevision for closing loops; (iii) the traveling speed; (iv) the time spent to obtain the 3D point cloud; and (v) the density of the point cloud.

2. Materials and Methods

2.1. Equipment

The WMLS tested in the case study was the ZEB-REVO, commercialized by GeoSLAM (Figure 1) [34], which consisted of a 2D time-of-flight laser scanner (Hokuyo UTM-30LX-F from Hokuyo Automatic Co., Osaka, Japan) rigidly coupled to an IMU mounted on a rotary engine. The motion of the scanning head on the motor drive was stored in a processing unit located in a small backpack and provided the third dimension to generate 3D information. This computer was equipped with batteries that fed the hand-held laser scanner through a special connection cable. A 3D SLAM algorithm was used to combine the 2D laser scan data with the IMU data, in order to return accurate 3D point clouds, following the full SLAM approach implemented in the robotic operative system (ROS) library [35]. With a 360° vertical field of view and 30 m of range in ideal indoor conditions (which was reduced to 15–20 m in real working circumstances), the operator moved through the indoor environment capturing more than 43,000 points per second. Regarding accuracy, the manufacturer declared that its value is 1–3 cm in relative terms and 1–30 cm in absolute positioning for a 10-min scan, with the closing of a single loop [34]. Table 1 shows further technical specifications.

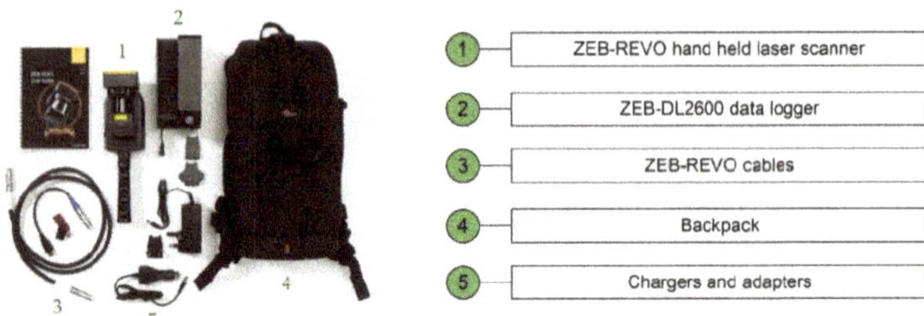
1
2
3
4
5
ZEB-REVO hand held laser scanner
ZEB-DL2600 data logger
ZEB-REVO cables
Backpack
Chargers and adapters

Figure 1. Main components of the wearable mobile laser systems (WMLS) used for data acquisition.

Table 1. Technical specification of the Geo Simultaneous Localization and Mapping (SLAM) ZEB-REVO device.

Parameter	Value	Parameter	Value
Total device dimension (mm)	220 × 180 × 470	Laser measuring principle	Time of flight
Scanner dimension (mm)	86 × 112 × 287	Scanner resolution	0.625° H × 1.8° V
Total device weight (kg)	4.10	Wavelength (nm)	905
Scanner weight (kg)	1.00	Orientation system	MEMS IMU
Head rotation speed (Hz)	0.5	Camera	GoPro
Operating time (h)	4	Scan rate	100 lines/s 43,200 points/s
Field of view	270° (H) × 360° (V)	Points per scan line	432 (0.325° int)

2.2. Methodology

The formalization of a schematic procedural pipeline for data acquisition and management represents a fundamental step to test the effective possibility of using WMLS for tracking complex indoor environments.

In this regard, it is possible to identify a succession of methodological phases that characterize an inspection with this approach: (i) the survey design (planning of the path); (ii) the data acquisition (protocol and basic rules); (iii) the post-processing (SLAM algorithm to compute the sensor trajectory and map the environment); and (iv) the cartographic product generation (three-dimensional and two-dimensional digital models). Figure 2 outlines the main steps of the applied methodology.

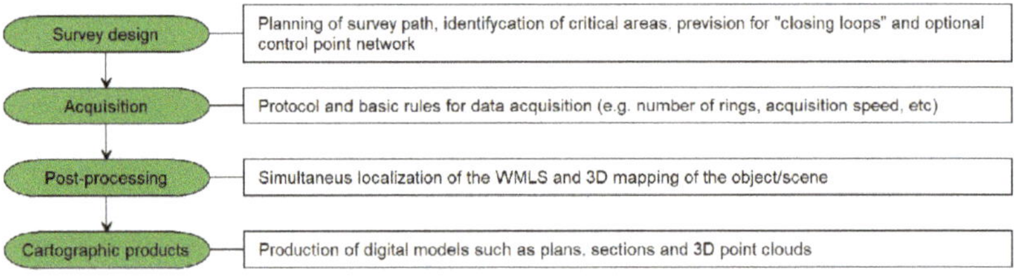
Survey design
Planning of survey path, identifycation of critical areas, prevision for "closing loops" and optional control point network
Acquisition
Protocol and basic rules for data acquisition (e.g. number of rings, acquisition speed, etc)
Post-processing
Simultaneus localization of the WMLS and 3D mapping of the object/scene
Cartographic products
Production of digital models such as plans, sections and 3D point clouds

Figure 2. Methodological phases of the procedural pipeline for surveys with a WMLS.

2.2.1. Survey Design

Before conducting data acquisition, the user should plan the proposed survey path in order to identify potential problem areas, such as doorway transitions, stairwells, open spaces, and smooth walled passageways, generally with poor geometrical features. It should be noted that features are significant if the ratio of their size and their range is approximately 1:10 (e.g., a feature must be textgreater0.5 m in size for a distance of 5 m). In addition, if there are not sufficient features along the direction of travel, the SLAM algorithm cannot correctly determine forward motion. In these cases, the user can proceed in the following ways:

- Improve the background with additional references (e.g., boxes in a corridor or a parked vehicle in an open field);
- ensure that those limited features are scanned repeatedly as you move through the scene by pointing the WMLS in their direction. As a result, more measurement points will define the element, assuring its use as reference during the post-processing;
- avoid acquiring moving objects (e.g., passing pedestrians or vehicles) since the SLAM algorithm may support on them as static features.

The planning should also consider "closing loops" wherever possible. The approach used to transform the raw scan data into a point cloud uses a method analogous to the close traverse technique applied for surveys [36], in that a previously known position is used to determine its current location. The re-surveying of a known area allows the spreading of the compounded error around the loop and the improvement of the accuracy of the resulting point cloud. As a minimum, it is mandatory that the operator starts and ends the survey in the same position to ensure at least one loop closure (Figure 3).

In general, it is better to do circular loops rather than "there and back" loops where the path simply doubles back on itself [31]. This applies to horizontal and vertical rings—i.e., if possible enter and exit through different doors and move between floors via distinct stairwells. It is important to scan the closed loop regions carefully to ensure that the key features are scanned from a similar perspective. It may be necessary to turn around to return to a region from another direction. This is a crucial feature in poor environments.

Figure 3. Example of a survey design with the WMLS. The compensation of the accumulated error is possible thanks to the closure of the path following different track and acquiring the internal cloister area from different points of the route.

2.2.2. Acquisition

The process of scanning using WMLS is an important step since the collected data will inevitably influence the level of quality of the cartographic products. For these reasons, it is useful to define the following operative rules:

- Inspect the site of interest in advance in order to identify critical areas not detected during planning and remove any obstacles along the way;
- make easily accessible all the connections between different rooms and floors, such as doorways or stairs, so as to ensure safe passage of the operator and avoid moving objects during the scanning;
- walk slowly in order to have a good coverage and a high-resolution data. If the forward movement is too fast there may not be enough repeated features for the SLAM algorithm to transform the raw laser data into a point cloud;
- pay attention to the transition areas and tight curves that must be travelled slowly, guaranteeing a period when the scanner can display features on both sides. The same care is necessary when we change from a closed space (feature rich) to an open environment (feature poor);
- split large surveys into more than one scan mission. This is to avoid big file sizes as well as to reduce any drift effect and thus error propagation that might be created in the SLAM data. It is recommended that each scan mission is limited to 30 min [34].

2.2.3. Post-Processing: The Generation of the 3D Point Cloud

The post-processing approach adopted for the case study was the Simultaneous Localization and Mapping (SLAM) algorithm, which addressed the problem of positioning a mobile system in an unknown environment and provided its 3D mapping. Access to SLAM was justified in two ways: (i) placing the system within a space or environment (pose estimation or trajectory computation); (ii) 3D modelling of the environment (mapping or reconstruction).

A large variety of SLAM solutions are available; they can be classified either as filtering or smoothing [32]. Filtering approaches model the problem as an online state estimation, where the state of the system consists in the current instrument position and the map. The estimation is augmented and refined by incorporating new measurements as they become available. To highlight their incremental nature, filtering approaches are usually referred as online SLAM methods [37]. Conversely, smoothing approaches estimate the full trajectory of the instrument from the full set of measurements. They address the so-called full SLAM problem and typically rely on least-square error minimization techniques.

GeoSLAM algorithm is able to perform both an open-loop incremental solution for online SLAM and a closed-loop global registration for full SLAM (as in the case study). However, it is appropriate to introduce the general characteristics of the algorithm in order to understand its performance. For GeoSLAM formulation, the trajectory can describe the position of the sensor during data acquisition and can project raw laser measurements (2D laser profiles or segments) into a registered 3D point cloud when necessary. Data processing is an incremental (the segments are registered one-by-one) and iterative procedure following a framework similar to the iterative closest point (ICP) algorithm:

- The first step identified corresponding surface patches from the laser point cloud. The patches were determined by spatially decomposing the scene into a multiresolution voxel grid, controlled by the "voxel density" parameter; increasing it caused the algorithm to use smaller voxels. Clusters of laser points that were both spatially and temporally proximal were identified and used to compute surface properties based on the centralized second-order matrix of the point coordinates [33]. The surface normal was obtained from the eigenvector corresponding to the minimum eigenvalue of the second-order matrix. The surface planarity, computed from the ratio of the eigenvalues, was used to discard elements that were not approximately planar. These properties were used to establish a first registration of the segments. During this stage, the following constrains were applied: (i) a filter for retaining only reciprocal correspondences and (ii) several boundary conditions that ensured continuity with the previous segment [33].

- In the second step, the estimated trajectory between two proximal surfaces was refined, minimizing the errors between matching surfaces and deviations from the measured IMU accelerations and rotational velocities. During this process, the following elements were considered into a cost-function (function to be minimized): (i) surface element match errors, (ii) IMU measurement deviations, and (iii) initial condition constraints. It is worth mentioning that the consideration of the IMU measurements ensured the estimated trajectory was smooth. The above terms of the cost-function were non-linear with respect to the rotational correction. Taking this into consideration, the algorithm used a linearization of the system by means of the Taylor expansion.

On the first iteration, the previously unprocessed trajectory segment was initialized by integrating the accelerometer and gyro measurements from the IMU. Since the processing window was advanced by a fraction of its length, the first section of the trajectory segment was already estimated from the previous time step; thus, the IMU data were only required to propagate the trajectory for the remainder of the window. As new data were acquired, the algorithm proceeded by processing a segment of the trajectory whose extremes were represented by positions occupied by the system in two well-defined moments. Next, it advanced the window by a fraction of its length from the previous time step. The dimension of segments was defined by the "window size" parameter. The number of iterations was controlled by the "convergence threshold" parameter; increasing it raised the maximum number of iterations for each processing step and reduced the convergence threshold during the online registration phase.

Considering that the previous process was an incremental procedure (also called open-loop solution) in which each segment was registered with respect to the previous segment, an error accumulation could be produced. In order to minimize the error accumulation, and considering that the data acquisition followed a close-loop path (Figure 3), the GeoSLAM algorithm applied a global registration procedure in which the close-loop restriction was taken into account. During this stage, the algorithm operated along the entire trajectory with one large window, instead of considering the trajectory in small segments (open-loop solution). Eventually, it was possible to give priority to flat surfaces in the search for feature correspondence. Moreover, it is worth mentioning that the laser scanner used by the tested WMLS was a line scanner (Figure 1) (Table 1). This laser was complemented with a rotational engine that allowed us to obtain a 2D profile in each period of time (segment of the scene). All of this was carried out by an operator. The movement of the operator along the scene guaranteed the acquisition of successive segments characterized by a certain overlap that the SLAM algorithm registered, allowing for the creation of the 3D point cloud.

2.2.4. Cartographic Products Generation

The generation of plants and sections required a vectorization of the 3D point cloud obtained from the fusion of the different paths. Thus, the following approach was carried out: (i) the extraction of sections from the 3D point cloud and (ii) the vectorization of the sections.

Regarding vectorization, the most efficient solutions are those that allow semi-automatic feature extraction from point clouds. The least-cost-path algorithms are particularly interesting [38] for our case study, since the automatic solutions are not able to recognize (and possibly not represent) the different objects in the scene and are influenced by the characteristic noise of the point clouds returned with a SLAM approach. Conceptually, a least-cost-path algorithm can be divided into two steps. In the first step, data points are linked with their nearest neighbors using a spherical search radius slightly larger than the point cloud resolution to produce a neighborhood network. A cost function, which represent the effort of moving along points in this network (hereafter referred to as "edges") is estimated. The second step derives the least-cost path between user-defined network points, providing the estimated trace. Once a trace has been estimated, manual adjustments can be easily applied by adding intermediate waypoints and recalculating the relevant least-cost paths.

3. Results

3.1. Case Study: An XV Century Palace

With the aim of evaluating the potentialities and limitations of the WMLS technology, we selected a gothic palace in ruins, located in the urban center of Segovia (Castile and León, Spain), as a case study (Figure 4). The palace was erected in the 15th century as a consequence of the economic expansion experimented by the city, following the precepts of the civil plateresque architecture [39]. Its fronts are made of sandstone masonry and brick masonry. They stand out for their horizontality, showing a ratio height/width of 1/2. With respect to the inner composition, the construction shows the characteristic appearance of the palaces erected during this epoch, with two annular traces in the two first plans and an inner cloister in the center of the construction [39]. These elements are integrated on five levels: (i) basement; (ii) ground floor; (iii) mezzanine; (iv) first floor and; (v) gallery (maybe added during the XIX century), filling an area of about 3000 m^2 and holding more than 50 rooms. The internal courtyard is characterized by a front porch that transfers part of its weight to the masonry and partly to stone columns with granite base (Figure 5).

The construction is mainly made up of sandstone coming from the local quarries. In the lower part (foundations and basement) a masonry in stones connected with mortar prevails, to which a brick masonry and a half-timbered work are replaced, proceeding upwards (Figure 5).

Recent restoration measures have added structural elements in reinforced concrete, such as beams and pillars in order to avoid the collapse of the structure.

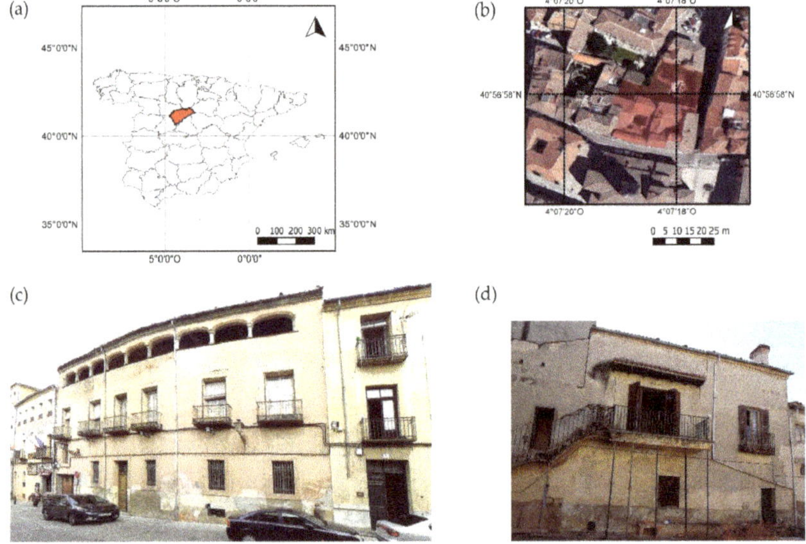

Figure 4. (**a**) Location of the province of Segovia; (**b**) the palace in Segovia; (**c**) main front facade of the palace; (**d**) back front facade of the palace.

Figure 5. Details of the internal cloister.

Timber slabs and open-node (Figure 6) trusses complement these masonry elements. The wooden floors are present both in the simple form, with single beams, and in the form composed of double row of beams and overlapping joists (Figure 7). The secondary frame is made up of a wooden plank in which a conglomerate jet has been made; it is composed of fine aggregates and mortar with a watering function and a thickness of a few centimeters. The finish provides the substrate and the relative flooring.

Figure 6. Gallery; detail of an open-node truss.

Figure 7. Internal cloister; detail of a wooden floor with a double row of beams and overlapping joists.

Nowadays, the state of conservation of the structure is deficient. The infiltration of water and the acid attack promoted by bird excrement have produced the rotting of the wooden elements with the consequent collapse of some floors (Figure 8). Rainfall has also caused the deterioration of sandstone masonry and the detachment of plasters in many environments. The capillary rising of moisture from the foundations has favored saline efflorescence and the appearance of mildew, phenomena accentuated by rainwater. These factors, along with the dust accumulated in the palace's many rooms, make the use of traditional digitalization techniques such as terrestrial photogrammetry or terrestrial laser scanning nearly impossible. These characteristics place the WMLS as the most suitable recording tool due to its flexibility, low weight, and capacity of recording large indoor areas without the support of any position global navigational satellite system (GNSS) [40].

Figure 8. Main cloister; detail of a collapsed wooden floor due to the rotting of the beams.

3.2. Mission Planning

Based on the guidelines previously defined (Sections 2.2.1 and 2.2.2), the surveyed palace was divided into four acquisition paths, each one with a different casuistic found during the digitalization of the heritage buildings (Figure 9):

- path 1 includes interior rooms, the main cloister, and a minimal part of the garden;
- path 2 includes interior rooms, the main cloister, and a part of the garden and a linear gallery;
- path 3 includes interior rooms, the main cloister, and the garden and the street front;
- path 4 includes interior rooms, the main cloister, and a smaller courtyard and the entire garden.

With these premises, the digitization of the internal areas took about 65 min (Table 2). Furthermore, using a terrestrial laser scanner (Faro Focus 3D) with the resolution at 1/5 and the quality at 3x, every scan of the interior took about two and a half minutes. Considering the time needed for the setup of the different stations, a total of 350 min was required for a survey of 70 stations (number compatible with the dimensions of the building), which was six times the period required with the WMLS.

Table 2. General information about the WMLS paths.

Path	Covered Floors	Walked Distance (m)	Acquisition Time (min)	Average Speed (m/s)
1	1	339.80	18.39	0.31
2	4	328.35	17.11	0.32
3	3	426.89	18.47	0.39
4	1	226.70	9.94	0.38

Figure 9. Plan view of the case study.

3.3. Data Processing and Registration

The GeoSLAM algorithm introduced in Section 2.2.3 was controlled by three fundamental parameters:

- The "convergence threshold," which increased or decreased the number of iterations for each processing step during the local and global registration phases. For the case study, a low value of the parameter was chosen, guarantying a quality of data that did not require a large number of iterations;
- the "window size," which defined the size of data samples processed by the algorithm. This helped to encompass the errors that occurred during the local registration phase. An intermediate value for this parameter was a perfect compromise between model quality and required hardware resources;
- the option "prioritize planar surfaces" was used for the case study and considered with very planar surfaces during the global registration phase. This helped to improve the global registration of very large data sets, common in the case of indoor mapping.

This setting guaranteed a processing time of the single path similar to the time required for the detection of the same (Table 3).

Table 3. Processing time of raw data with default setting of full SLAM parameters.

Default Algorithm Configuration		
Path	**Processing Time (min)**	**Difference between Processing and Acquisition time (%)**
1	23.18	≅20
2	23.76	≅21
3	25.31	≅25
4	12.42	≅22

It took approximately 83 min (about five times less time that was required to record the seventy stations needed by the terrestrial laser scanner) to solve the full SLAM problem in the four routes.

In order to perform the alignments between the different paths, a network of artificial targets (spheres) wa distributed around the cloister and on the internal and external facades (Figure 10)

(Figure 11). The use of spheres was not mandatory for the execution of the alignment, but this approach made it possible to analyze the registration error between the different paths. The centroids of homologous spheres, extracted by means of the RANSAC Shape Detector algorithm [41] were used as reference pairs. The error associated with the pairs was quantified through the discrepancy in the overall coordinate system between the spatial coordinates of the two homologous centroids (located in two different paths). The root mean square error (RMSE) of the whole registration was just over 3 cm for all the alignment processes achieved (a value perfectly compatible with the WMLS accuracy).

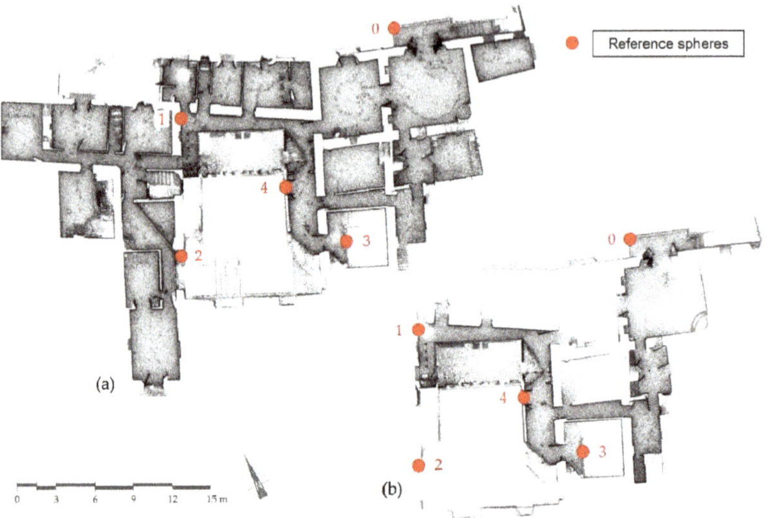

Figure 10. Reference sphere distribution on the first floor for the path 1 (**a**) and path 2 (**b**).

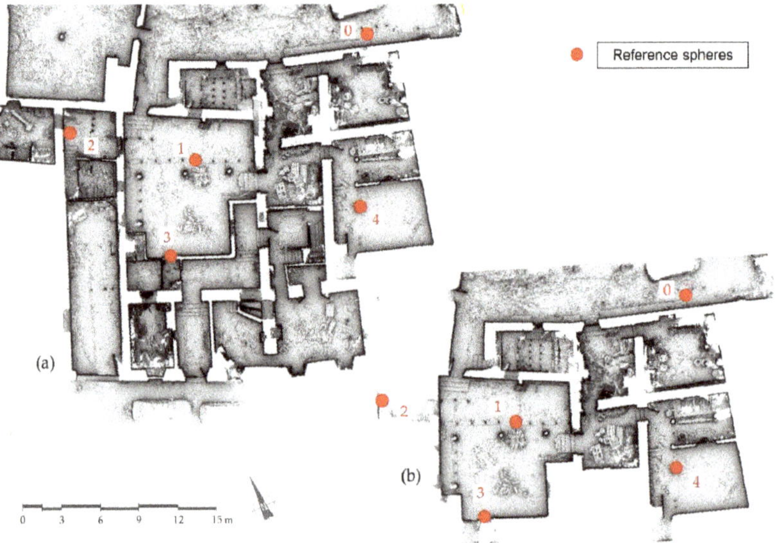

Figure 11. Reference sphere distribution on the ground floor for the path 3 (**a**) and path 4 (**b**).

3.4. System Validation

A further quality control of the acquired data was performed through a comparison with an outdoor network surveyed with a terrestrial laser scanner (Faro Focus 3D, employed for the detection of the front facade, the garden, and the internal cloister). The network consisted of 11 stations acquired with a resolution of 1/2 and a quality of 2x (Figure 12). The scans were aligned using the spherical targets and then an ICP algorithm, in order to create a ground truth.

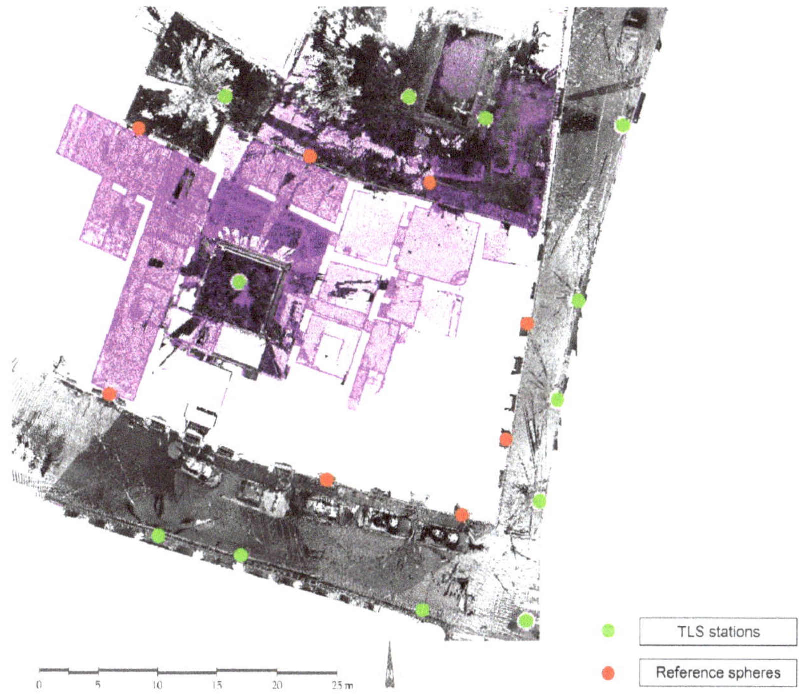

Figure 12. Reference sphere distribution on the first floor for the path 2 (red circles) and the outdoor TLS network (green circles). Scans coming from the terrestrial laser scanner are outlined in the grey color, whereas scans coming from the WMLS for the path 2 are represented in the purple color.

Considering the implemented cartographic products, the following validations were carried out: (i) a local validation and (ii) a global validation. The local validation was a comparison based on fitting flat geometric primitives on homologous point clouds. Some distance measurements were extracted, where each value was the mean of five distance measurements between two planes located on the opposite walls or between the floor and ceiling (Figure 13) (Figure 14). Starting from this data, the system was validated trough the indexes introduced by Nocerino et al. [22] (1) (2):

$$\mathrm{RLME} = \left(\frac{\mathrm{D_{Zm}} - \mathrm{D_{Fm}}}{\mathrm{D_{Fm}}}\right) \times 100 \tag{1}$$

$$\mathrm{RLMA} = 1:\mathrm{ROUND}\left(\left|\frac{100}{\mathrm{RLME}}\right|\right) = 1:\mathrm{ROUND}\left(\left|\frac{\mathrm{D_{Fm}}}{\mathrm{D_{Zm}} - \mathrm{D_{Fm}}}\right|\right) \tag{2}$$

The relative measurement error (RLME) was computed as the relative difference between measured distance $\mathrm{D_Z}$ for the GeoSLAM ZEB-REVO and the distance for the Faro Focus 3D, assumed as reference length D_F. The relative length measurement accuracy (RLMA) was defined as the rounded absolute reciprocal value of the RLME times 100. Table 4 reports the results.

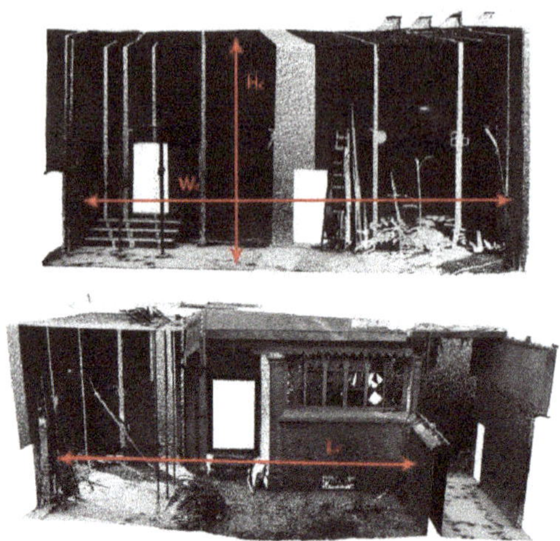

Figure 13. Distances measured for the accuracy evaluation.

Figure 14. Distances measured for the accuracy evaluation.

Table 4. Measured distances with standard deviations (σ), relative length measurement errors (RMLE) and accuracies (RMLA).

		Faro Focus 3D		GeoSLAM ZEB-REVO			
		D_{Fm} (m)	σ (m)	D_{Zm} (m)	σ (m)	RLME	RLMA
Main cloister	L_1	10.279	0.008	10.268	0.010	−0.107	≈1:1000
	W_1	10.096	0.013	10.085	0.022	−0.109	≈1:600
	H_1	4.869	0.014	4.869	0.012	0.001	≈1:7000
Garden	L_2	22.730	0.012	22.756	0.011	0.114	≈1:800
	W_2	18.766	0.023	18.787	0.025	0.112	≈1:900

In the case of the main garden, the RLME assumed a slightly greater value than the main cloister. This was due to two factors: in the case of a SLAM system, the greater probability of error accumulation over longer distances (L_2 and W_2) and the decline in performance of the latter with outdoor acquisitions (being designed for indoor surveys). The global validation was carried out by comparing the centroids of the sphere network captured by both systems (Figure 12), with an approach similar to the one defined in Section 3.3. The RMSE obtained was around 3 cm, which was in line with the accuracy of the proposed method. In both the validation approaches, the accuracy of the WMLS was between 1 and 3 cm, which was compatible with the data provided by the manufacturer [34].

3.5. Plan and Section Restitution

The post-processing returns 3D point clouds, characterized by the number of points (and therefore of bytes required for its archiving), was not suited for the vectorization of two-dimensional products. Moreover, their density was not uniform, since it was related to the traveling speed along the paths and the overlapping of some parts of the clouds in the registration phase. Table 5 provides a schematic summary of the magnitude and mean surface density values for each point cloud. As can be seen from the comparison with Table 2, an increase in average speed corresponded to a decrease in the mean surface density.

Table 5. Point cloud features; the surface density is estimated by counting for each point the number of neighbors inside a sphere of three centimeters radius (R) and dividing this value by the sphere max section (πR^2).

Path	Number of Points	Mean Surface Density (points/m^2)	Number of Points after 1 cm Subsampling	Mean Surface Density after 1 cm Subsampling (points/m^2)
1	32,955,139	29,648	14,056,546	6140
2	30,059,713	28,399	14,032,348	5594
3	33,839,213	20,428	16,766,690	5348
4	16,720,282	18,488	9,210,536	4634

In order to generate plans and sections, a subsampling strategy based on distance (1 cm) was applied without producing any loss of detail that would compromise the quality of the products. As a result, a 3D point cloud with more than a 100 million points was obtained for the whole historical palace (Figure 15).

The final step involved tracing the two-dimensional products. To do this, several sections along the different floors of the building were extracted. Previously, in the application of the least-cost-path algorithm, a low pass filter was applied [42]. This strategy locally fit a plane around each point of the cloud and then removed those points far from the fitted plane (Figure 16a). For the present case study, the following parameters were used: (i) a neighbored search radius of 0.03 m and (ii) a relative error of 1 sigma for the exclusion of the points. Next, we applied the filtered point cloud using the least-cost-path algorithm, which allowed for the automatic vectorization of the plant (Figure 16b). Finally, some manual adjustments were conducted on the obtained vectorization data, allowing for the creation of plans in a quick and accurate way (Figure 17).

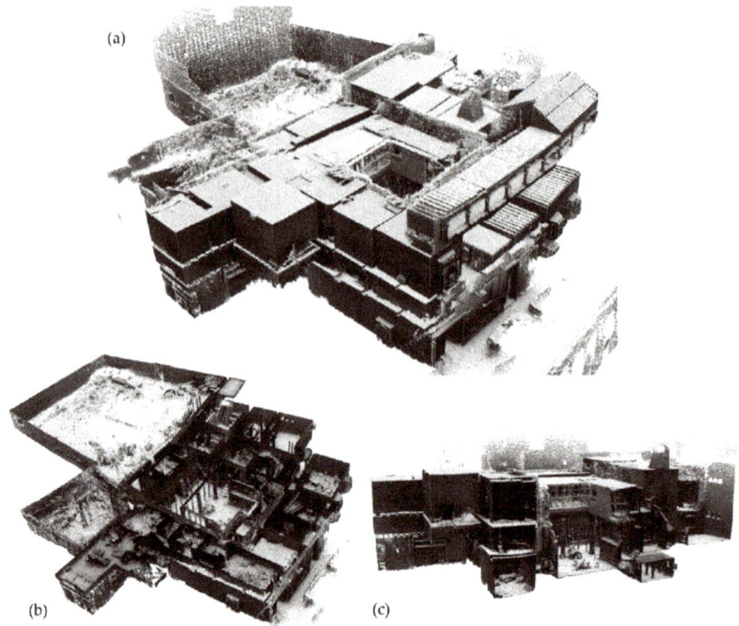

Figure 15. Final point cloud with more than a 100 million points: (**a**) perspective view; (**b**) horizontal section; (**c**) vertical section.

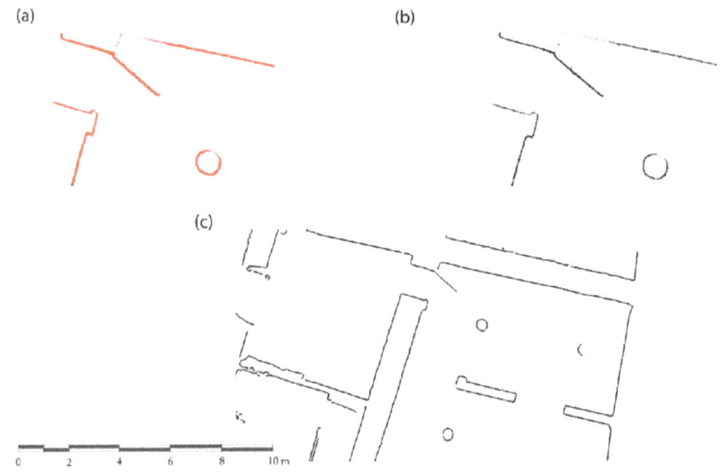

Figure 16. Detail of the vectorization process: (**a**) section extracted from the point cloud; (**b**) section filtered; (**c**) application of the least-cost-algorithm over the filtered section.

Figure 17. Ground floor plan and first floor plan.

4. Discussion

Based on the obtained results, as well as its proven its efficiency, accuracy, portability, and weight, it can be seen that the WMLS tested in our case study offers a potential solution for mapping complex cultural heritage sites.

Regarding the efficiency, the WMLS needed just 63 min to capture the data (distributed in a total of four paths) and 83 min to solve the SLAM problem. In both cases, the system outperformed the time estimated for a terrestrial laser scanner to digitalize the same construction (around 70 scan stations). In comparison with other state-of-the-art MMS, such as the Leica Pegasus back-pack or the Heron MS-2 back-pack, the proposed WMLS solution had a lower data acquisition rate (43,000 points per second against 600,000 points per second captured by the Leica Pegasus back-pack and 700,000 points per second captured by the Heron system). The WMLS capture rate is suitable in terms of its density, capacity to detect geometrical features, and for its ability to map indoor cultural heritage environments.

Regarding the accuracy, the point cloud returned by the WMLS guaranteed an accuracy within a centimeter in relation to the point cloud obtained by a terrestrial laser scanner. These values were similar to those obtained by Nocerino et al. [22] for indoor environments and those obtained by Cabo et al. [43] in an outdoor environment with many geometrical features. This accuracy seems to be linked with the number of geometrical features present in the scene, as well as the planning of the data acquisition using several close-loops.

Concerning the portability, the tested sensor had approximately 4 kg of weight and small dimensions (220 × 180 × 470 mm) (Table 1) in comparison with other MMS, such as the Leica Pegasus backpack, with an estimated weight of 12 kg and dimensions of 310 mm × 270 mm × 730 mm, the Heron MS-2 back-pack mapping solution, with a total weight of 11 kg, and the NavVIS 3D system, which requires a cart to support its sensors [22,24,44]. These characteristics are especially relevant for the digitalization of the palace, since this building shows narrow areas (Figure 8) in which the other MMS would have problems.

5. Conclusions

In this paper, a wearable mobile laser system (WMLS) has been presented and tested for the digitalization of a complex cultural heritage building. This device highlights for its lightweight, flexibility in comparison with other traditional techniques, such as terrestrial photogrammetry or terrestrial laser scanning. The combination of this instrument, mainly composed by a 2D Hokuyo laser scanner and an Inertial Measurement Unit, together with the Simultaneous Location and Mapping approach, allows the acquisition of indoor environments dynamically without the necessity of stations or the use of global navigation satellite systems.

Investigation results estimate that the WMLS requires the complete digitalization of the entire structure, with around 3000 m^2, in about 65 min. A quick comparison demonstrates how the acquisition time of the wearable system is approximately equal to the sixth part of the time required by the terrestrial laser scanner (Faro Focus 3D), providing the correct accuracy and density of data for the creation of sections and plans for restoration projects. The post-processing phase itself is around five times shorter than the corresponding registration of the laser scans.

The process of registering the different paths, which uses spherical target centroids as control points, returns a RMSE of about 3 cm, compatible with the accuracy of the analyzed system, and is able to offer excellent results in complex environments, such as the one described in this paper. These potentialities are also confirmed by the comparison with an external network generated by a terrestrial laser scanner. During this stage, the accuracy of the system has been evaluated at two different scales: (i) at the local scale using the relative length measurement error (RLME) and the relative length measurement accuracy (RLMA), and (ii) at the global scale, where the root mean square error (RMSE) between the centroids detected by each system, WMLS and laser scanner, have been confronted. In both cases the accuracy of the system is estimated between 1 and 3 cm.

Future investigations could concern the integration of the point clouds generated by the WMLS into building information modelling (BIM) systems. Another future development could concern the process of coloring automatically the point cloud of WMLS by synchronizing the acquisition path with video captured by GoPro and the subsequent projection of the frames on the cloud itself.

Author Contributions: All authors conceived and designed the experimental campaign; A.d.F., L.J.S.-A., and D.G.A. performed the experimental campaign; A.d.F., L.J.S.-A., J.A.M.-J., and R.M. performed the pre-processing and post-processing steps; A.d.F., L.J.S.-A., and S.B. wrote the article and all authors read and approved the final version.

Funding: The authors wish to thank HERGONSA S.L. for the access and the economic support to develop this case study.

Acknowledgments: Authors wish to thank the V SUDOE INTERREG for providing the framework of the HeritageCARE project, Ref. SOE1/P5/P0258. This work has also been framed in the research project Patrimonio 5.0 by Junta of Castilla y León, Ref SA075P17.

Conflicts of Interest: the authors declare no conflict of interest.

References

1. Krakow Charter 2000: Principles for Conservation and Restoration of Built Heritage. Available online: http://hdl.handle.net/1854/LU-128776 (accessed on 20 September 2018).
2. Torres-Martínez, J.A.; Seddaiu, M.; Rodríguez-Gonzálvez, P.; Hernández-López, D.; González-Aguilera, D. A multi-data source and multi-sensor approach for the 3D reconstruction and web visualization of a complex archaelogical site: The case study of "Tolmo De Minateda". *Remote Sens.* **2016**, *8*, 550. [CrossRef]
3. Garcia-Gago, J.; Gomez-Lahoz, J.; Rodríguez-Méndez, J.; González-Aguilera, D. Historical single image-based modeling: The case of Gobierna Tower, Zamora (Spain). *Remote Sens.* **2014**, *6*, 1085–1101. [CrossRef]
4. Waldhäusl, P.; Ogleby, C. 3 × 3 rules for simple photogrammetric documentation of architecture. In *International Archives of Photogrammetry and Remote Sensing, Proceedings of the Close Range Techniques and Machine Vision, Melbourne, Australia, 1–4 March 1994*; Australian Photogrammetric and Remote Sensing Society: Melbourne, VIC, Australia, 1994.
5. Barber, D.; Mills, J. *3D Laser Scanning for Heritage: Advice and Guidance to Users on Laser Scanning in Archaeology and Architecture*; Historic England: Swindon, UK, 2007, ISBN 978-1848025219.
6. Kraus, K.; Waldhäusl, P. *Photogrammetry: Fundamentals and Standard Processes*; Dümmler Köln: Munich, Germany, 1993, ISBN 978-3427786849.

7. Fabio, R.; Sabry, E.H. Image-based 3D Modelling: A Review. *Photogramm. Rec.* **2006**, *21*, 269–291. [CrossRef]
8. Blais, F. Review of 20 years of range sensor development. *J. Electron. Imaging* **2004**, *1*, 13. [CrossRef]
9. Guidi, G.; Russo, M.; Beraldin, J.-A. *Acquisizione 3D e Modellazione Poligonale*; McGraw-Hill: New York, NY, USA, 2010, ISBN 978-8838665318.
10. Kalay, Y.; Kvan, T.; Affleck, J. *New Heritage: New Media and Cultural Heritage*, 1st ed.; Routledge: London, UK, 2007, ISBN 978-0415773560.
11. Torres-Martínez, J.A.; Sánchez-Aparicio, L.J.; Hernández-López, D.; González-Aguilera, D. Combining geometrical and radiometrical features in the evaluation of rock art paintings. *Digital Appl. Archaeol. Cult. Herit.* **2017**, *5*, 10–20. [CrossRef]
12. Gonzalez-Aguilera, D.; Muñoz-Nieto, A.; Rodriguez-Gonzalvez, P.; Menéndez, M. New tools for rock art modelling: Automated sensor integration in Pindal Cave. *J. Archaeol. Sci.* **2011**, *38*, 120–128. [CrossRef]
13. Sánchez-Aparicio, L.J.; Riveiro, B.; Gonzalez-Aguilera, D.; Ramos, L.F. The combination of geomatic approaches and operational modal analysis to improve calibration of finite element models: A case of study in Saint Torcato Church (Guimarães, Portugal). *Constr. Build. Mater.* **2014**, *70*, 118–129. [CrossRef]
14. Sánchez-Aparicio, L.J.; Del Pozo, S.; Ramos, L.F.; Arce, A.; Fernandes, F.M. Heritage site preservation with combined radiometric and geometric analysis of TLS data. *Autom. Constr.* **2018**, *85*, 24–39. [CrossRef]
15. Del Pozo, S.; Herrero-Pascual, J.; Felipe-García, B.; Hernández-López, D.; Rodríguez-Gonzálvez, P.; González-Aguilera, D. Multispectral radiometric analysis of façades to detect pathologies from active and passive remote sensing. *Remote Sens.* **2016**, *8*, 80. [CrossRef]
16. Guidi, G.; Beraldin, J.-A.; Ciofi, S.; Atzeni, C. Fusion of range camera and photogrammetry: A systematic procedure for improving 3-D models metric accuracy. *IEEE Trans. Syst. Man Cybern. B* **2003**, *33*, 667–676. [CrossRef] [PubMed]
17. Stumpfel, J.; Tchou, C.; Yun, N.; Martinez, P.; Hawkins, T.; Jones, A.; Emerson, B.; Debevec, P.E. Digital Reunification of the Parthenon and its Sculptures. In Proceedings of the VAST, Brighton, UK, 5–7 November 2003; pp. 41–50.
18. Paparoditis, N.; Papelard, J.-P.; Cannelle, B.; Devaux, A.; Soheilian, B.; David, N.; Houzay, E. Stereopolis II: A multi-purpose and multi-sensor 3D mobile mapping system for street visualisation and 3D metrology. *Rev. Fr. Photogramm. Teledetec.* **2012**, *200*, 69–79.
19. Remondino, F.; Toschi, I.; Orlandini, S. Mobyle Mapping Systems: Recenti sviluppi e caso applicativo. *GEOmedia* **2015**, *19*, 6–10.
20. Al-Hamad, A.; El-Sheimy, N. Smartphones based mobile mapping systems. *Int. Arch. Photogramm. Remote Sens. Spat. Inf. Sci.* **2014**, *40*, 29. [CrossRef]
21. Piras, M.; Di Pietra, V.; Visintini, D. 3D modeling of industrial heritage building using COTSs system: Test, limits and performances. *Int. Arch. Photogramm. Remote Sens. Spat. Inf. Sci.* **2017**, *42*, 281. [CrossRef]
22. Nocerino, E.; Menna, F.; Remondino, F.; Toschi, I.; Rodríguez-Gonzálvez, P. Investigation of indoor and outdoor performance of two portable mobile mapping systems. In *Videometrics, Range Imaging, and Applications XIV*; International Society for Optics and Photonics: Munich, Germany, 2017.
23. Kumar, G.A.; Patil, A.K.; Patil, R.; Park, S.S.; Chai, Y.H. A lidar and imu integrated indoor navigation system for uavs and its application in real-time pipeline classification. *Sensors* **2017**, *17*, 1268. [CrossRef] [PubMed]
24. Lehtola, V.V.; Kaartinen, H.; Nüchter, A.; Kaijaluoto, R.; Kukko, A.; Litkey, P.; Honkavaara, E.; Rosnell, T.; Vaaja, M.T.; Virtanen, J.-P. Comparison of the selected state-of-the-art 3d indoor scanning and point cloud generation methods. *Remote Sens.* **2017**, *9*, 796. [CrossRef]
25. Flener, C.; Vaaja, M.; Jaakkola, A.; Krooks, A.; Kaartinen, H.; Kukko, A.; Kasvi, E.; Hyyppä, H.; Hyyppä, J.; Alho, P. Seamless mapping of river channels at high resolution using mobile lidar and uav-photography. *Remote Sens.* **2013**, *5*, 6382–6407. [CrossRef]
26. Opromolla, R.; Fasano, G.; Rufino, G.; Grassi, M.; Savvaris, A. LiDAR-inertial integration of UAV localization and mapping in complex enviroments. In Proceedings of the 2016 International Conference on Unmanned Aircraft Systems (ICUAS), Arlington, VA, USA, 7–10 June 2016.
27. Niu, X.; Yu, T.; Tang, J.; Chang, L. An online solution of lidar scan matching aided inertial navigation system for indoor mobile mapping. *Mob. Inf. Syst.* **2017**, *2017*, 4802159. [CrossRef]
28. Shamseldin, T.; Manerikar, A.; Elbahnasawy, M.; Habib, A. SLAM-based Pseudo-GNSS/INS Localization System for Indoor LiDAR Mobile Mapping Systems. In Proceedings of the IEEE/OIN PLANS 2018, Monterey, CA, USA, 23–26 April 2018.

29. Pierzchała, M.; Giguère, P.; Astrup, R. Mapping forests using an unmanned ground vehicle with 3D LIDAR and graph-SLAM. *Comput. Electron. Agric.* **2018**, *145*, 217–225. [CrossRef]
30. Lagüela, S.; Dorado, I.; Gesto, M.; Arias, P.; González-Aguilera, D.; Lorenzo, H. Behavior analysis of novel wearable indoor mapping system based on 3d-slam. *Sensors* **2018**, *18*, 766. [CrossRef] [PubMed]
31. Farella, E. 3d mapping of underground environments with a hand-held laser scanner. In Proceedings of the SIFET Annual Conference, Lecce, Italy, 8–10 June 2016.
32. Thrun, S. Simultaneous localization and mapping. In *Robotics and Cognitive Approaches to Spatial Mapping*; Jefferies, M.E., Yeap, W., Eds.; Springer: Luxembourg, 2008; pp. 13–41, ISBN 978-3-540-75388-9.
33. Bosse, M.; Zlot, R.; Flick, P. Zebedee: Design of a spring-mounted 3-d range sensor with application to mobile mapping. *IEEE Trans. Robot.* **2012**, *28*, 1104–1119. [CrossRef]
34. GeoSLAM Technology, ZEB-REVO Solution Brochure. Available online: https://gpserv.com/wp-content/uploads/2017/01/ZEB-REVO-Brochure-v1.0.3.pdf (accessed on 15 August 2018).
35. Quigley, M.; Conley, K.; Gerkey, B.; Faust, J.; Foote, T.; Leibs, J.; Wheeler, R.; Ng, A.Y. ROS: An open-source Robot Operating System. In Proceedings of the ICRA Workshop on Open Source Software, Kobe, Japan, 12–17 May 2009; p. 5.
36. Kavanagh, B.F. *Surveying Principles and Applications*, 7th ed.; Pearson Education: Upper Saddle River, NJ, USA, 2006, ISBN 978-0137009404.
37. Grisetti, G.; Kummerle, R.; Stachniss, C.; Burgard, W. A tutorial on graph-based SLAM. *IEEE Intell. Transp. Syst. Mag.* **2010**, *2*, 31–43. [CrossRef]
38. Thiele, S.T.; Grose, L.; Samsu, A.; Micklethwaite, S.; Vollgger, S.A.; Cruden, A.R. Rapid, semi-automatic fracture and contact mapping for point clouds, images and geophysical data. *Solid Earth* **2017**, *8*, 1241. [CrossRef]
39. Navarro, P.C. The Royal Country Estates around the Monastery of El Escorial: Medieval Tradition and Flemish Influence. *Rev. EGA* **2014**, 46–53. [CrossRef]
40. Misra, P.; Enge, P. *Global Positioning System: Signals, Measurements and Performance*, 2nd ed.; Ganga-Jamuna Press: Lincoln, MA, USA, 2010, ISBN 9780970954428.
41. Schnabel, R.; Wahl, R.; Klein, R. Efficient RANSAC for point-cloud shape detection. In *Computer Graphics Forum*; Blackwell Publishing Ltd.: Oxford, UK, 2007; pp. 214–226.
42. Han, X.-F.; Jin, J.S.; Wang, M.-J.; Jiang, W.; Gao, L.; Xiao, L. A review of algorithms for filtering the 3D point cloud. *Signal Process. Image Commun.* **2017**, *57*, 103–112. [CrossRef]
43. Cabo, C.; Del Pozo, S.; Rodríguez-Gonzálvez, P.; Ordóñez, C.; González-Aguilera, D. Comparing Terrestrial Laser Scanning (TLS) and Wearable Laser Scanning (WLS) for individual Tree Modeling at Plot Level. *Remote Sens.* **2018**, *10*, 540. [CrossRef]
44. Heron Wearable Mobile Mapping. Available online: https://gexcel.it/en/solutions/heron-mobile-mapping (accessed on 18 November 2018).

Article

Geometric Characterization of Vines from 3D Point Clouds Obtained with Laser Scanner Systems

Ana del-Campo-Sanchez *, Miguel Moreno, Rocio Ballesteros and David Hernandez-Lopez

AgroForestry and Cartographic Precision Research Group. University of Castilla—La Mancha, Regional Development Institute, Campus Universitario s/n, 02071 Albacete, Spain; miguelangel.moreno@uclm.es (M.M.); rocio.ballesteros@uclm.es (R.B.); david.hernandez@uclm.es (D.H.-L.)

* Correspondence: ana.delcampo@uclm.es

Received: 30 August 2019; Accepted: 10 October 2019; Published: 12 October 2019

Abstract: The 3D digital characterization of vegetation is a growing practice in the agronomy sector. Precision agriculture is sustained, among other methods, by variables that remote sensing techniques can digitize. At present, laser scanners make it possible to digitize three-dimensional crop geometry in the form of point clouds. In this work, we developed several methods for calculating the volume of vine wood, with the final intention of using these values as indicators of vegetative vigor on a thematic map. For this, we used a static terrestrial laser scanner (TLS), a mobile scanning system (MMS), and six algorithms that were implemented and adapted to the data captured and to the proposed objective. The results show that, with TLS equipment and the algorithm called convex hull cluster, the volumes of a vine trunk can be obtained with a relative error lower than 7%. Although the accuracy and detail of the cloud obtained with TLS are very high, the cost per unit for the scanned area limits the application of this system for large areas. In contrast to the inoperability of the TLS in large areas of terrain, the MMS and the algorithm based on the L_1-medial skeleton and the modelling of cylinders of a certain height and diameter have solved the estimation of volumes with a relative error better than 3%. To conclude, the vigor map elaborated represents the estimated volume of each vine by this method.

Keywords: *Vitis vinifera*; terrestrial laser scanning; plant vigor; mobile mapping; precision agriculture; vine size

---

1. Introduction

Precision agriculture strategies that apply remote sensing techniques are widely used, particularly in viticulture [1]. Information obtained from satellites, airborne cameras, and ground-based sensors (among others) over the earth's surface is a trend in research and innovation activities that has been applied to precision agriculture. With these techniques, not only can we obtain the spectral response of the surface of crop from a determined point of view (aerial or ground-based), but we can also obtain the approximate crop geometry [2].

Canopies drive the main vegetal processes, such as photosynthesis, gas interchange, and evapotranspiration. These processes are directly related to sunlight interception and the microclimate generated by the plants [3]. Efforts to measure the spatial parameters in canopies have been made with simplified geometrical models, as proposed in [4], through parameters like the leaf area index (LAI), "point quadrat" [5,6], leaf area density (LAD) [7], tree area index (TAI) [8–10], ground canopy cover (GCC) [11], tree row LiDAR volume (TRLV) [12], surface area density (SAD) [3], and photosynthetically active radiation (PAR) [13], among many others. Canopy characterization and monitoring help improve crop management through the estimation of water stress, the affection by pests and weeds, nutritional requirements, and final yield. This monitoring could be performed with a network of ground sensors [14,15] and/or remote sensing techniques at any scale.

When applying satellite or airborne-based remote sensing techniques, users receive the data captured by sensors as a set of images (bidimensional data) or as a set of isolated measurements. However, less attention has been paid to other types of information that can contribute significantly to the geometry characterization of the canopy, such as 3D point clouds. It is possible to obtain an accurate 3D model of a crop from aerial imagery using photogrammetry techniques [2,16–18], but the point of view of these images does not build a true 3D model. Instead, it builds what is called a 2.5D model. These flights obtain images from a nadir perspective, so all objects are projected on a horizontal plane. The lower part of the canopy structure of the plant is hidden from the sensor, and therefore, it is ignored in the data acquisition process. To solve this limitation, laser scanning is becoming a promising technology in precision agriculture. These systems can be mounted on static tripods [19], aircraft [20], land vehicles [8], or be used as hand-held systems [21,22], so scanning can be done from several perspectives. Further, laser scanning systems can be mounted on drones, which facilitates data acquisition in the process of biomass mapping [23]. However, the autonomy of these systems can limit the applicability of these types of systems.

Point clouds taken by active sensors, such as laser scanners, are generated via the return of light pulses that are emitted and received by the sensor. A light pulse is emitted from a known point in space with a specific direction. This pulse travels in a straight line through the air until it intercepts the object's surface where the pulse is reflected. The distance between the sensor and the scanned surface is measured by receiving the return with three possible methods, depending on the construction of the equipment: (1) the time of flight (TOF); (2) the phase shift; or (3) optic triangulation. The location of the scanned point can be estimated by comparing it with the base. These sensors are commonly accompanied by a rotating mirror converting a single scan direction to a full plane of scan directions (perpendicular to the rotated axis). The platform where the equipment is mounted is used to determine if the scanning is performed from a fixed point (static terrestrial laser scanner) or a trajectory (mobile mapping, hand-held, or aircraft). Mobile systems can measure the objects from several perspectives. However, to obtain a complete and accurate digitalization there should be as many perspectives as the number of the object's faces. Crops are an especially difficult object to measure due to the irregular shape of their canopies [19,24].

Due to the capture process, static laser scanning equipment generally registers a higher point density and with higher accuracy than mobile integrated laser scanning systems [25]. However, they are less operational because, to cover a wide area and avoid occlusions, the number of scanning stations is very high, and therefore, the time to acquire the information is also high. However, the cost of acquisition is often lower than that for mobile equipment because the integrated sensors required for both types of equipment are more sophisticated for mobile systems. The postprocessing of the captured information is more laborious on static platforms because the static equipment must solve the joint of each single scan and its georeferencing. However, for the mobile and aircraft systems, the trajectory of the capture is measured, and the integration of all sensors is solved, which facilitates the matching of all the point cloud in an automatic manner. Hand-held portables also need subsequent georeferencing.

The system can integrate other sensors to collect the spectral response of the scanned object. The spectral response of the object can then be integrated in the built 3D point cloud. The final point cloud jointly provides the geometry and spectral response of the scanned surfaces [26]. It can, for instance, account for the bidirectional reflectance distribution function (BDRF), which is a main issue in high-resolution remote sensing techniques in vegetation [27]. The integration of global positioning systems (GPS) has also achieved great success in geolocating and scaling geomatic products [28,29]. Greater integration has been done with an inertial measurement unit [30]. Clearly, integration of software and hardware devices expands the scanned variables and improves accuracy, although it makes processing more difficult and tedious.

The use of laser sensors to digitize the 3D components of crops (particularly in viticulture) is new but promising. One of the first attempts to use laser scanning in viticulture was the studied in [31], which calculated the light interception of each vegetal organ with a laser beam mounted on a structure with an arc shape. The use of LiDAR on vineyards has increased since then. The total canopy volume can be characterized, but other agronomic parameters of interest can also be directly estimated with this technology, such as canopy height and fruit position, among many others, or indirectly estimated, such as LAI, canopy porosity, and others, through the generation of the relationships between these agronomic variables with the measured geometrical characteristics of the canopy. In addition, knowing the spatial disposition of certain plant organs could optimize some treatments of a localized nature. For example, the autonomous detection of fruits can determine the yield [32–34] or automate its harvest. Moreover, a spray application of any phytosanitary material [10,35–38], grapevine sucker detection [26], weeding, quantification of biomass storage [19,39–41], pest prevention [42], and any other treatment that may be necessary to achieve sustainable, desired results can now be accurately applied, and even automated, with current technology.

In precision viticulture, canopy characterization is directly related to the quantitative and qualitative production potential of a vineyard [34,43–46]. The canopy's structure, position, and orientation (among others) are what defines vegetal performance [46] because light interception and canopy microclimates are driving factors for energy and gas interchange and evapotranspiration. Further, in viticulture, the correct balance between vegetative growth (shoot and leaf "production") and reproductive development (grape production) is the key to optimizing grape production and quality [47]. Several parameters are defined in viticulture with this aim, such as vine capacity, vine vigor, crop load, and crop level [48]. Monitoring all these parameters could be a benefit of the use of 3D characterization using remote sensing techniques (and particularly laser scanning systems). The quantification of the total biomass produced (vine capacity) is also crucial for estimating carbon sequestration by vineyards [49].

This work is focused on the development of a new methodology, software, and procedure for data acquisition to determine the volume occupied by the trunks of vines from 3D point clouds taken by laser scanning equipment. A comparison between a static and mobile laser scanner was also performed. After calibration, the procedure was applied to a real case study to determine maps for the vine trunk volume as a measure of vine capacity (or plant vigor). The difficulties, weaknesses, and future requirements of this technology are fully applied, and the objective of characterizing vine capacity is also analyzed and discussed.

2. Materials and Methods

2.1. Proposed Procedure

Figure 1 shows the workflow of the proposed methodology. Due to the complexity of the shape of the vines and the non-destructive condition of this study, this study starts with an accurate volume calculation of a vine-shaped artificial object (VSAO) to obtain the real volume data of an object of similar geometry to validate the proposed methodology. The VSAO is composed of two PVC pipes (5 cm in diameter) arranged in a "T" shape, resembling the trunk and the arms of the vineyards driven on a trellis (Figure 2). The choice of the diameter of the pipes used was made according to the average diameter observed in the vines of the vineyard. Likewise, the dimensions and shape of the artificial object were generated by simulating the geometries of the vines in this vineyard. To calculate the volume of the VSAO, the diameter and length of both cylinders were measured.

Field data were acquired and processed for a test area with two laser scanning systems: static terrestrial laser scanner (TLS) and mobile mapping system (MMS). Each piece of equipment produced a colored 3D point cloud with accurate geolocation and high density. These point clouds were the input data used to calculate the volume of the VSAO with specifically developed algorithms. These algorithms, which will be fully described in this manuscript, return the volume of the VSAO from the point clouds obtained with the different systems. After comparing the calculated volume with the actual volume of the VSAO, a volume accuracy can be determined for each system. This process is called the calibration process. Once calibrated, the methodology and the best algorithm will be applied to a real case study on a vineyard located in the southeast of Spain.

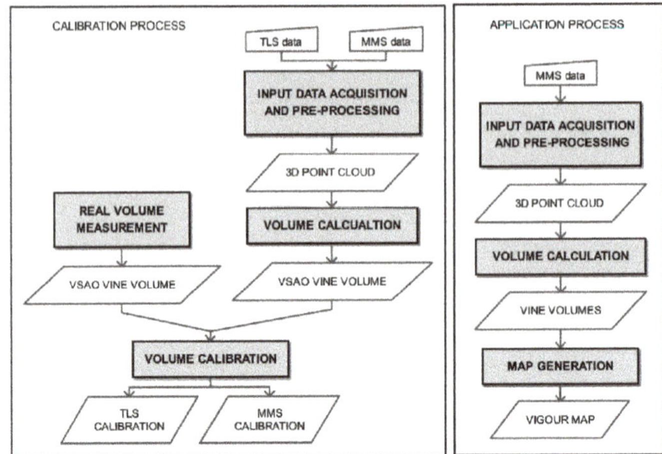

Figure 1. General procedure workflow. TLS: terrestrial laser scanner. MMS: mobile mapping system. VSAO: vine-shaped artificial object.

Figure 2. Vine-shaped artificial objects (VSAOs) in the test area of the scan.

2.2. *Study Areas for Calibration Process and Application of the Proposed Methodology*

A calibration area where the VSAOs were scanned is located in a practice field at the University of Castilla, La Mancha, Albacete (Spain). This vineyard has experimental and teaching purposes, so its state and morphology are highly heterogeneous. However, it covers the typical physical characteristics of trellis systems with a drip irrigation system, where possible occlusions, slope changes, vegetation height, etc., occur. Two identical VSAOs were located in places were vines were missing in positions similar to those of actual vines. Figure 2 shows the location of the two VSAOs in the scanned area. Since the objective was to determine the volume occupied by the vine's trunk, measurements were obtained without leaves and prunes. This state is the most appropriate for scanning the evidence of crop vigor, because the accumulated vigor appears in the perennial parts of the plant and not in the deciduous parts, such as leaves or prunes.

A real application of this methodology was implemented in a vineyard located in the southeast of Spain (38.728928°, −1.470696° EPSG:3857, Figure 3). The study area comprises an area inside of a 0.58 ha vineyard. Trellis driving is separated 3 m between strips and 1.5 m between vines. In this plot, different irrigation treatments have been performed since 2016, as described in Figure 3. These different treatments will, in the future, drive new developments of this canopy. These features make this place an interesting application area due to their high variability. These treatments have been applied only during the last two years, so they have not yet resulted in noticeable differences in the trunk diameter. However, determining plant vigor using the proposed methodology can provide useful information about nutritional and irrigation requirements in the decision making process. Vines with higher vigor would demand more nutrients and water than those with lower vigor. Also of interest is the determination of carbon sequestration by plants, which only accounts for perennial wood and not for shoots or leaves that are removed every year. Thus, with the proposed methodology and data acquisition procedure, vigor maps can be obtained to help farmers better manage their vineyards.

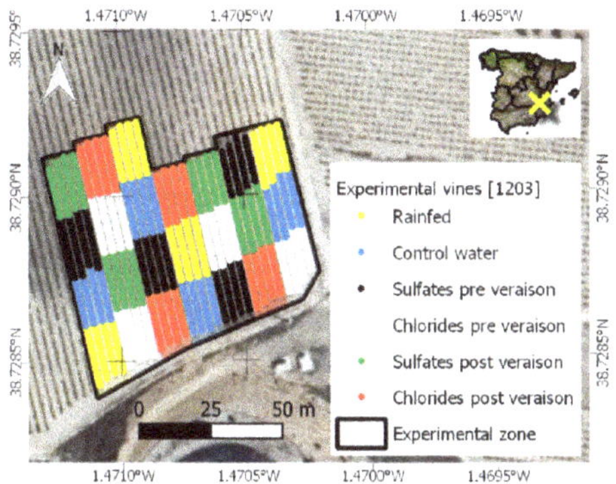

Figure 3. Situation map (EPSG:3857).

2.3. Equipment

The TLS equipment was a FARO Focus3D X 330 (FARO Technologies, Inc., Lake Mary, Florida) (Figure 4a), which utilizes phase shift technology to read the distance to an object. This reader was mounted on a Manfrotto Super Pro Mk2B tripod and Manfrotto 3D Super Pro head (Manfrotto, Cassola, Italy) (Figure 4a) for each single scan station. This equipment's field of view is almost complete, with a 360° view on a horizontal plane and 300° on a vertical plane because of its gyratory base and rotation mirror. The scan resolution was configured to 6 mm to 10 m, with a beam divergence of 0.19 mrad (1 cm to 25 m) and a ranging error of ±2 mm (10 to 25 m). It also contains an RGB camera, GPS receiver, electronic compass, clinometer, and altimeter (electronic barometer) to approximately correlate the individual scans in postprocessing. For the accurate joining of different point cloud-calibrated white spheres, an ATS SRS Medium (ATS Advanced Technical Solutions AB, Mölndal, Sweden) was used (Figure 4a). The information captured with this equipment was processed by the software SCENE 6.2 (FARO Technologies Inc., Lake Mary, United States), resulting in a unique georeferenced and colored 3D point cloud.

MMS is a Topcon IP-S2 Compact+ (Topcon Corporation, Tokyo, Japan) (Figure 4b). This is a system that integrates five laser scanners, a 360° spherical digital camera with six optics, an IMU (Inertial Measurement Unit), a dual frequency GNSS receiver, and a wheel encoder. The laser scanners are all SICK LMS511-10100S01 (SICK AG, Waldkirch, Germany), and the spherical camera is a FLIR LadyBug 5+ (FLIR Integrated Imaging Solutions Inc., Richmond, Canada). The system was mounted in a regular 4 × 4 car, with additional batteries and a control system (a high performance rack system, with an i7 processor, 32 GB RAM, and redundant SSD with industrial USB 3.0). The capture software was Topcon Spatial Collect 4.2.0 (Topcon Corporation, Tokyo, Japan). The postprocessing software for the data collected by this equipment was the Topcon Geoclean Workstation 4.1.4.1 (Topcon Corporation, Tokyo, Japan). The resulting product was a georeferenced and colored 3D point cloud. Considering the mobile condition of the equipment and the integration of the sensors, the five possible returns for each laser scanner were filtered to the highest intensity to ensure false observations (noise points).

(a)

(b)

Figure 4. Laser scanner equipment. (**a**) A TLS mounted on a tripod with a head and calibrated sphere. (**b**) The MMS operating on a 4 × 4 vehicle.

For the acquisition of geolocation data, we used a GPS-RTK (global positioning system—real time kinematic) with Topcon HiPer V (Topcon Corporation, Tokyo, Japan) receptors and postprocessing software MAGNET Tools 5.1.0 (Topcon Corporation, Tokyo, Japan) in the reference point measurement, using the MMS trajectory solution with centimetric precision.

The main characteristics of both laser scanners are reviewed and compared in Table 1.

Table 1. The laser scanner's technical characteristics.

Characteristic	TLS	MMS
Brand and model	FARO Focus^{3D} X 330	Topcon IP-S2 Compact+
Laser principle	Phase shift	Time of Flight
Number of evaluated echoes	1	5
Wavelength	1550 nm	905 nm
Beam divergence	0.19 mrad	11.9 mrad
Maximum scan rate	976,000 points/s	150,000 points/s
Range	0.6 to 330 m	0.7 to 80 m

2.4. Data Acquisition

For the calibration process, VSAOs were ubicated on two points where vines were missing, as can be seen in Figure 2. Scanning with TLS and MMS was performed with the spatial configuration shown in Figure 5. Six TLS stations were used, at 1.60 m from terrain, around the two VSAOs. The MMS trajectory was three rows on each side of the VSAO. On the chosen date (February 2019), the crop was pruned, and the sprouting had not yet started to avoid occlusions.

On 11 May 2018, data were acquired for the real case, with the sprouting just having started, so there was no occlusion of leaves. The MMS was driven by all rows of the experimental zone (Figure 3) and the two contiguous rows to each side of the perimeter, to cover all the delimited areas of interest with enough overlap.

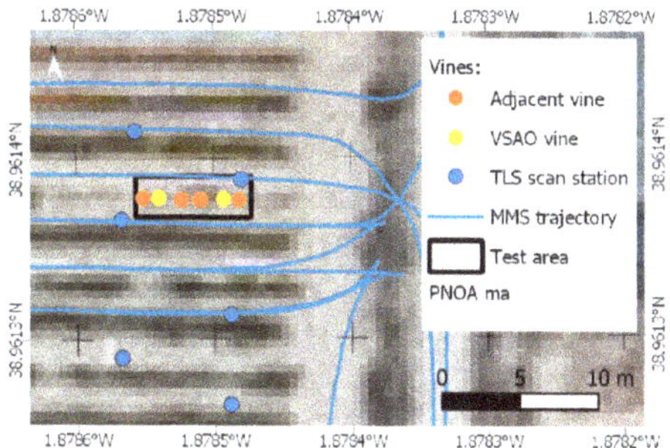

Figure 5. Test area, location of the VSAOs, TLS stations, and MMS trajectory (EPSG:3857).

2.5. *Algorithms for Volume Calculation*

Several algorithms that help in the process of volume calculation from point clouds have already been developed [50,51]. However, these are general algorithms that require adaptation and calibration for different object shapes, as well as sparse and noisy point clouds, such as in the case of vine trunk volume calculation. Other algorithms, such as the L_1-medial skeleton [52], can help develop new algorithms for volume calculation, which is one of the main contributions of this paper. The shape of the vine trunks is highly irregular; these trunks are located in an adverse environment for data acquisition, which demands point cloud treatment, algorithm evaluation, calibration, and adaptation. This process should be incorporated in a tool that performs these tasks in an automatic manner. With the methodology and tool developed in this manuscript, these requirements are fulfilled. The proposed methodology includes the development, adaptation, and implementation of a set of algorithms developed in the C++ language and a classification algorithm implemented in MATLAB (Mathworks Inc., Massachusetts, USA); all of these algorithms have been integrated into a unique piece of software.

The imported information includes:

- A text file with the approximate coordinates for each vine base, which can be obtained with a GNSS-RTK or a high resolution orthoimage, among others.
- A text file describing the main parameters of the project: the name, approximate dimension of the searched figure (Figure 6), input and output file paths, formats, coordinate reference systems, etc.
- A point cloud in the LAS file format [53] or compressed LAZ.
- A text file with the position of each single scan performed by TLS.

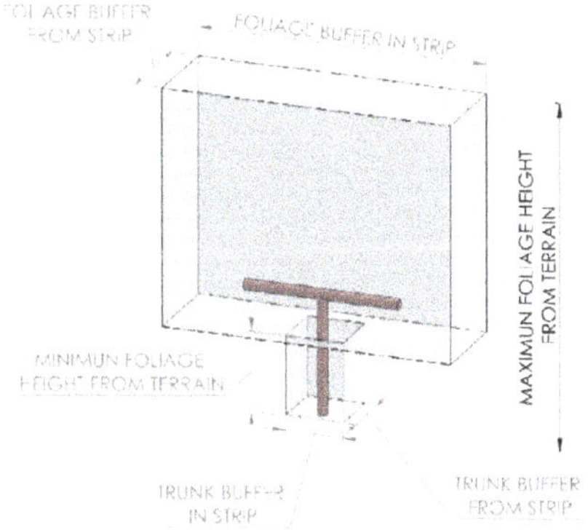

Figure 6. Approximate squared envelope of a vine.

Three strategies have been implemented and evaluated to calculate the volume occupied by the vine: (1) OctoMap [50], which is an algorithm to generate volumetric 3D environmental models based on voxelization of the occupied space; (2) a convex hull cluster (CHC) [51] that closes the convex envelope of previously clustered sets of points according to geometric and radiometric criteria; and (3) volume calculation from the trunk skeleton (VCTS), which obtains the volume of an object from the distance between each point of the cloud and the internal structure of the object, generated with the L_1-medial skeleton [52] algorithm. These three algorithms will be described below. These are some volume calculation strategies that we have adapted to the characteristics of the point clouds captured by our TLS and MMS. However, none of these strategies are ready to be applied to the specific case of calculating the volume of vine trunks. In this paper, we describe the new developments and adaptations required for the case study of vine volume calculation. This is especially crucial in the case of MMS, where point cloud data are sparse and noisy, but whose applicability is higher due to the wider areas covered. Automated point cloud classification based on RGB values, point selection based on trunk shape, and the development, adaptation, and calibration of algorithms for volume calculation are the main contributions of this work.

Before applying any of these three algorithms, the acquired point cloud should be preprocessed to produce a point cloud with a high quality and three-dimensional definition of the scanned object. If automated clipping, classification, and debugging processes are not enough to define the vine shapes, possible manual editing of the resulting point cloud can be performed. The latter is a step that should be avoided to ensure a highly automatic process.

It should also be noted that the tools for each algorithm (i.e., 3D viewers) have also been incorporated in the implementation of the algorithms used in this work, since most of the modelling libraries from the point clouds include them for their determination and use in different workflows.

The processing steps are summarized in Figure 7. The algorithm implements six different and independent processes that require parameter definitions that are adequate for each datum and can be applied to each vine separately. Intensive work has been performed to determine the parameters that best apply to this case study; these parameters will be shown in results section.

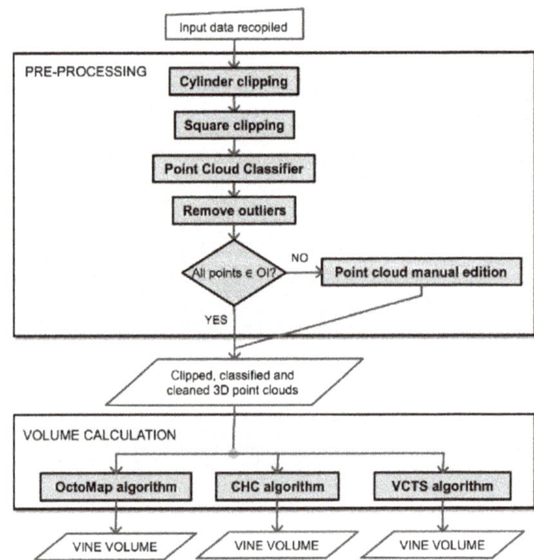

Figure 7. Software flowchart. OI: object of interest. CHC: convex hull cluster. VCTS: volume calculation from the trunk skeleton.

2.5.1. Point Cloud Preprocessing

The raw data collected by TLS and MMS are processed with the software supplied with the equipment—FARO SCENE for TLS and the Topcon Geoclean Workstation for MMS—integrating the information from the different sensors that each system incorporates (Section 2.3). As a result, these software packages return two georeferenced and colored 3D point clouds. However, before applying the algorithms to volume calculation, it is necessary to preprocess the georeferenced and colored point cloud to (1) obtain the point cloud relative to each individual vine; (2) eliminate any point cloud belonging to leaves; and (3) remove any outliers that appear because of the adverse environment in which the measurements were obtained. To perform these preprocesses, software was developed that permits the automated application of this preprocess.

(1) Cylinder and square clipping subprocesses

With the cylinder clipping algorithm, the input point clouds are segmented for each vine as a cylinder with two possible criteria from which to choose the radio: the ROI (region of interest) buffer (found at a half distance between the contiguous vines in its strip) or the fixed distance between vines in the same row. The cylinder centers are determined by the coordinates of the vine base collected by centimetric GPS-RTK measurement.

The square clipping step algorithm permits one to approximate cropping to a composed figure of two superposed straight parallelepipeds, one for trunk definition and one for arm definition (Figure 6), depending on the type of pruning performed. In the case study, the scanned vines were pruned with Guyot, so only the trunk was characterized. This process helps to improve the point cloud depuration and removing noise and other elements.

A review of the editable parameters of these two steps is listed in Table 2.

Table 2. List of the editable parameters of the cylinder and square clipping processes.

Name	Description	Possible Values
ROI [1] buffer	Method to segment first cylinder	fix@distance, computed@halfMeanDistance
Trunk buffer from strip	Length in meters. See Figure 6	0.050 to 0.500
Trunk buffer in strip	Length in meters. See Figure 6	0.050 to 0.500
Minimum foliage height from terrain	Length in meters. See Figure 6	0.100 to 1.000
Maximum foliage height from terrain	Length in meters. See Figure 6	0.400 to 2.000
Foliage buffer from strip	Length in meters. See Figure 6	0.100 to 1.000
Foliage buffer in strip	Length in meters. See Figure 6	ROI buffer, distance (0.500 to 2.000)
Sensor type	Sensor type choice	TLS [2], MMS [3]

[1] Region of interest. [2] Terrestrial laser scanner. [3] Mobile mapping system.

(2) Point cloud classifier subprocess

This process consists of classification according to standardized classes [53]. This allows one to segment the points that define the woody part of the plant in case the canopy is developed. The implemented algorithm is a semi-automatic segmentation of points using only their color. This approach is an application of computational vision techniques based on an artificial neural network (ANN) capable of clustering points with similar radiometric responses. This process has two subprocesses that are clearly differentiated: training a neural network and applying it.

This algorithm is a further development of the leaf area index calculation software (LAIC) [54], which is applied to point clouds. For training, only one vine has to be used. In the input point cloud loading, the RGB color space is transformed into a CIE-Lab color space (Commission Internationale de l'Eclairage (Lab)), where L is lightness, a is the green to red scale, and b is the blue to yellow scale. In this way, we transform the color space of the three components (R, G, and B) to two components (a and b). Then, a cluster segmentation (k-means) is performed on a determined number of clusters (2 to 10), considering this bi-dimensional variable (coordinates a and b). The user then identifies, in a supervised process, which cluster of points represents the woody vine part. With this selected cluster, an ANN is trained. A minimum percentage of successfully classified points with the ANN-calibrated model in the supervised process should be reached (usually 95%). Afterward, the trained ANN is applied to all vines, assigning a class to the points that represent the woody parts of each vine.

The processing parameters for this tool are listed in Table 3.

Table 3. List of the editable parameters of the point cloud classifier.

Name	Description	Possible Values
Calibrate or apply	Calibrate (1) or apply calibration (0)	1 or 0
Class for trunk	Code classifier as trunk [53]	13 to 31
Hidden nodes	Number of hidden nodes	2 to 15
Input nodes	Number of input nodes	3 to 3
Iterations	Number of iterations	10 to 2000
Minimum calibration accuracy	Minimum percentage of successfully classified points with the ANN [1] calibrated model in the supervised process, %	50 to 100
Output nodes	Number of output nodes	1 to 1

[1] Artificial neural network.

(3) Remove outliers subprocess

It is possible that previous processes were not able to accurately segment the vine and required an automatic outlier detection process. This process is parameterized according to the density and disposition of the points expected in the segmented figure. This program implements two different algorithms that are executed consecutively. Both are classes from the Point Cloud Library (PCL) [55], and both are filters of outlier points. The first is the statistical outlier removal [56] algorithm, which detect outliers based on a threshold calculated as the standard deviation of the distance for each point to a certain number of neighboring points. The second algorithm is the radius outlier removal [57]. This filter considers a point as an outlier if the point does not have a given number of neighbors within a specific radius from their location. Detected outliers are classified as a noise point class (Class 7 [53]) in both processes. A list of processing parameters is given in Table 4.

Table 4. List of editable parameters for the remove outliers process.

Name	Description	Possible Values
Class to use	LiDAR class [53] to use	−1 to 31 [1]
Statistical sample neighbors for SOR [2] algorithm	Number of sample neighbors to compute mean distance	10 to 1000
Statistical std threshold for SOR algorithm	Threshold of standard deviation of computed mean distance	0.1000 to 10.0000
Radius minimum neighbors for ROR [3] algorithm	Number of minimum neighbors	1 to 1000
Radius search for ROR algorithm	Radius search (meters)	0.0010 to 1.0000

[1] −1 for all classes. [2] Statistical outlier removal. [3] Radius outlier removal.

2.5.2. Volume Calculation with the OctoMap Algorithm

OctoMap is an algorithm, programmed as an open-source C++ library, to generate volumetric 3D environment models [50]. 3D maps are created by taking the 3D range measurements afflicted with underlying uncertainty. Multiple uncertain measurements are fused into a robust estimate of the true state of the environment as a probabilistic occupancy estimation. The OctoMap mapping framework is based on octrees. Octrees are hierarchical data structures for spatial subdivisions in 3D [58,59]. Space is segmented in cubic volumes (usually called voxels), which represent each node of the octree. Cubic volumes are recursively subdivided into eight subvolumes until the given minimum voxel size is reached (Figure 2 in [50] or Figure 8c). The resolution of the octree is determined by this minimum voxel size (Figure 8). The tree can be cut at any level to obtain a coarser subdivision if the inner nodes are maintained accordingly [50].

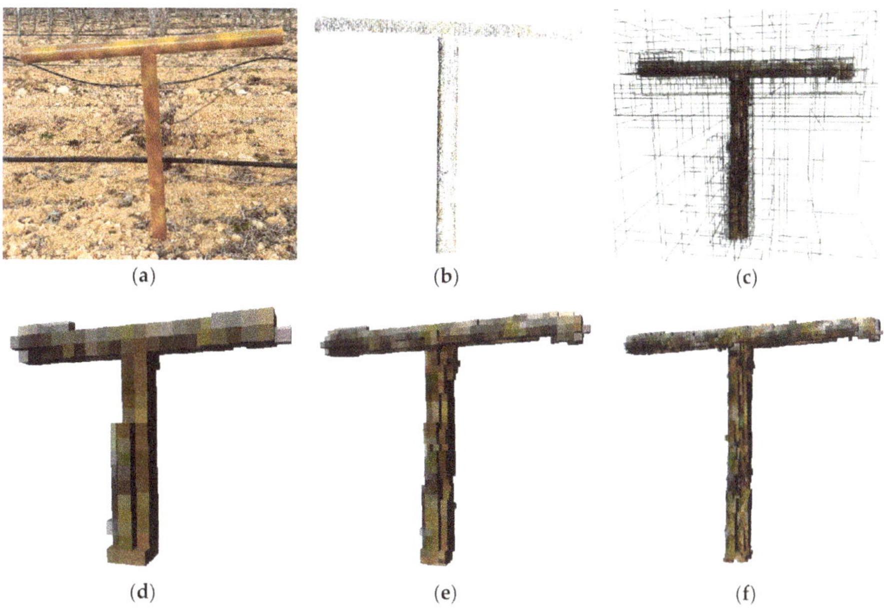

Figure 8. Example of VSAOs at voxel resolutions. (**a**) Real VSAO. (**b**) 3D point cloud from TLS. (**c**) Voxelization. Voxel resolutions of (**d**) 0.04; (**e**) 0.02; and (**f**) 0.01 m.

Voxels are treated as Boolean data, where initialized voxels are measured as an occupied space (1), and null (0) voxels are free or unknown spaces. It should be noted that we only measured the face of the object shown from the position of the equipment, with existing occlusions. Therefore, each measurement establishes free voxels between the observer and the detected surface (occupied voxel), and all those behind are defined as unknown voxels.

OctoMap creates maps with low memory consumption and fast access time. This contribution offers an efficient way of scanning and the possibility of achieving multiple measurements that can be fused in an accurate 3D scanned environment. In contrast to other expeditious approaches focused on the 3D segmentation of single measurements, OctoMap is able to integrate several measurements into a model of the environment.

Taking advantage of the flexibility of writing data, this framework ensures the updatability of the mapped area, as well as its resolution, and copes with the sensor noise. The state of a voxel (occupied, free, or unknown) can be redefined if the number of observations with different states is higher than the times it was previously observed with its initial state.

Furthermore, the appropriate formulas in the algorithm control the possibility of a voxel to be changed based on its neighbors and the number of times that it has been modified. Thus, the quantity of the data is reduced to the number of voxels that must be maintained. This clamping method is lossless because its thresholds avoid the losses of full probabilities.

The subprocess called OctoMap is an adaptation of the OctoMap algorithm [50] for the purpose of estimating the volume of vines. The editable parameters for the processing point clouds with our adapted algorithm are listed in Table 5.

Table 5. List of the editable parameters of OctoMap.

Name	Description	Possible Values
Class to use	LiDAR class [53] to use	−1 to 31 [1]
Compute free voxels	Compute free voxels	True, False
Voxel resolution	Voxel linear resolution in meters	0.0010 to 1.0000

[1] −1 for all classes.

2.5.3. Volume Calculation with the CHC Algorithm

The volume calculation algorithm called convex hull cluster (CHC) is the result of the integration of an algorithm from PCL (Point Cloud Library) [52] implemented into our software. It uses the method of voxel cloud connectivity segmentation (VCCS) [51], which generates volumetric over-segmentations of 3D point cloud data, known as supervoxels. These elements are searched as variant regions of k-means clusters through the point cloud (considered a voxel octree structure). They are evenly distributed across 3D space and are spatially connected to each other (Figure 9). Thus, each supervoxel maintains 26 adjacency relations (6 faces, 12 edges, and 8 vertices) in voxelated 3D space with its adjacent neighbors.

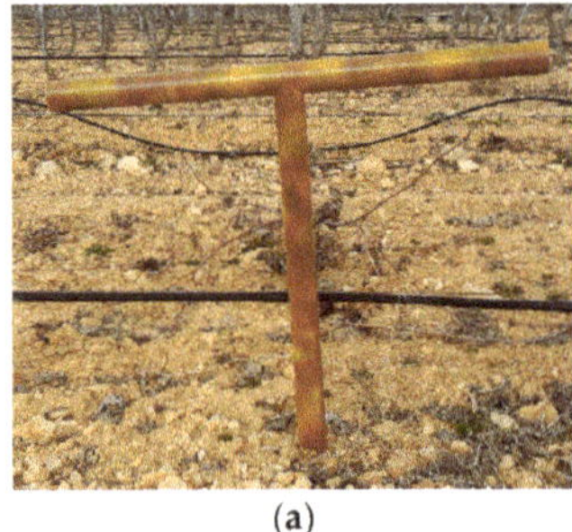
(a)

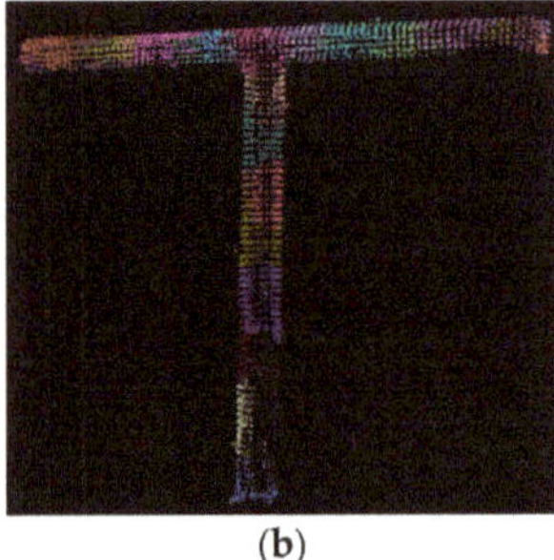
(b)

Figure 9. (a) VSAO; (b) example of a clustering TLS point cloud with the convex hull cluster algorithm.

The process starts from a set of seed points distributed evenly in space on a 3D grid with an established resolution (R_{seed}) where the point cloud is located. The voxel resolution (R_{voxel}) is the established size of the voxel's edge. The seed voxels begin to grow into supervoxels until they reach the minimum distance from the occupied voxels. If there are no occupied voxels near any point of the cloud among the grown supervoxels, and there are no connected voxels among their neighbors, the isolated seed voxel is deleted.

The seed points are expanded by a distance measure calculated in a feature space consisting of spatial extent (normalized by the seeding resolution), color (the Euclidean distance in normalized RGB space), and normals (the angle between surface normal vectors).

The supervoxels' growth is an iterative process that uses local k-means clustering.

In this process, confirmed voxels are ignored. In this way, processing is sped up, and the amount of information that needs to be taken into account is reduced. The iterations end when all supervoxels have been confirmed or rejected, and, therefore, all points in the cloud belong to a specific cluster. The editable parameters of the supervoxel clustering are listed in Table 6.

Table 6. List of the editable parameters of the convex hull cluster.

Name	Description	Possible values
Class to use	LiDAR class [53] to use	−1 to 31 [1]
Color weight	Weight of color variable	0.000 to 1.000
Normal weight	Weight of normal variable	0.000 to 1.000
Spatial weight	Weight of spatial variable	0.000 to 1.000
Seed resolution	Seed linear resolution (meters)	0.0200 to 2.0000
Voxel resolution	Voxel linear resolution (meters)	0.0010 to 1.0000

[1] −1 for all classes.

2.5.4. Volume Calculation with the VCTS Algorithm

The L_1-medial skeleton is an algorithm that generates a curved skeletal representation of scanned objects as 3D point clouds. This curved skeleton defines a simplified inner abstraction of the 3D shape of the object, which facilitates analysis of that shape.

This skeleton consists of nodes and segments linked together, as shown in Figure 10. A line string is formed by all segments whose nodes are up to two segments long. The nodes belonging to three (or more) segments define the end of a line string and the beginning of two different line strings.

Figure 10. (**a**) An MMS point cloud imported in the L_1-medial skeleton viewer. (**b**) Skeleton results (grey points are the input point cloud, green points are the nodes, and red lines are segments).

Although this algorithm is not conditioned by previous assumptions of the geometric shape of the object, we start from the premise that the vine trunk can be modelled as the sum of the volumes enclosed by the cylinders defined by each segment of the skeleton or by the cylinders defined by each line string. Knowing the skeleton and its segments, all the points of the cloud are clustered according to the segment to which they belong. This clustering is based on the proximity of the point to the segment as the minimum (orthogonal) distance between them.

In the first case, the height of each cylinder is taken as the length of each segment. For the estimation of the radius, the parameters of the mean and median centralization of the minimum distances that exist between each point and its segment were used. In the second case, the height of the cylinder is considered to be the sum of the lengths of the segments that comprise it, and the radius as the mean and median of the minimum distances between the points and the segments.

These four volume estimation strategies have been called "VCTS segment mean", "VCTS segment median", "VCTS line string mean", and "VCTS line string median". These strategies are designed to solve the problem that, for a segment, each point provides a different radius, which can be due to real changes or noise in the point cloud. The success of the algorithm depends on defining a suitable strategy to estimate a single radius value that represents the segment of the object.

To this point, it is necessary to highlight that in the extremes of the vines, the skeleton strategy can fail, because there are few points near the base for the presence of soil, stones, vegetation, and the upper section due to the obfuscation caused by the aperture of the arms. Again, it is necessary to find a strategy to estimate the dimensions of the cylinders at the extremes of the figure. The problems that arise in the clustering of points and the assignation of each segment or string are solved in the following ways:

- The extreme points that are not assigned to any segment or line string (because they are not enclosed between planar sections fixed by nodes and segments) are added to the nearest segment or line string in each case, and extend until they reach the same conditions of belonging as the rest of the points assigned to this segment or line string. This, by default, avoids errors when quantifying the total volume of the vine.
- The assignment to segments or line strings is unique to each point, so all the points are assigned to a single segment or line string, which reduces errors by excess in the zones of insertion between elements (segments or line strings).
- For the mean and median of the L_1-medial skeleton segment, when the segments are given without any assigned point (because the density is not great enough), the radius of the cylinder is considered to be the minimum found in the segments of its line string.
- Because of the low density of the points, their quality and the probability of missing scanned faces, for the VCTS algorithm's mean and median, if the radius of a cylinder is lower than the mean (or median) radius of the line string to which it belongs in by as many units as the threshold establishes, its radius is considered to be the mean (or median) radius of the line string.

The editable parameters of these algorithms are listed in Table 7.

Table 7. List of the editable parameters of the VCTS algorithms.

Name	Description	Possible Values
Class to use	LiDAR class [53] to use	−1 to 31 [1]
Algorithm	Chosen strategy to set radius and height of cylinders	Segment mean, segment median, line string mean, line string median
Minimum outlier threshold	Threshold for elimination of rough errors in meters	0.0010 to 1.0000

[1] −1 for all classes.

2.6. Validation Analysis

For the validation of the methods, the absolute and relative errors made in the estimates of the calculation of the volume of both VSAOs carried out with the six proposed strategies were calculated. However, other factors have also been taken into account in order to determine the true possibilities of each sensor and volume calculation algorithm, which will be fully analyzed in the results and discussion. The real value of the volume of each VSAO has been obtained thanks to the simplified form of the pipes that form it. In order to calculate the volume of these artificial objects, the diameter and length of both cylinders that compose them were measured. Afterwards, the equation of the cylinder volume (the circular area of the base multiplied by its height) was applied.

2.7. Generation of Vine Size Maps

Crop vigor maps were elaborated with the GIS (geographical information system) QGIS 3.4.3 Madeira (QGIS Development Team) through volumes calculated by the developed software in relation to the geolocation of each vine. The output data of the implemented algorithms were written to an ESRI Shapefile (Environmental Systems Research Institute, Inc., Redlands, USA) with a geometry type point. Each point feature represented a vine in the vineyard, which included a field with the values of the estimated volumes for each vine. This vector layer was represented with an appropriate graduated color ramp to show, with 7 classes, how vigorous each vine was (its volume). A 2 cm orthoimage of the ground sample distance (GSD) was used as the background layer (the product of the solution of a photogrammetric flight block taken via an airborne RGB camera in an unmanned aerial vehicle (UAV) at a later date). As an aid for the delimitation of the experimental area of this vineyard, the extent of each treatment, and its replications, was also represented. Due to its semi-automatic character, this calculation was applied only to a random selection of 10% of the scanned vines.

3. Results

3.1. 3D Point Cloud Acquisition and Preprocessing

The input data acquisition performed in the test area with the TLS equipment resulted in a colored and georeferenced point cloud. It was performed via six scan stations (Figure 5) with five calibrated spheres as targets for scan joining and georeferencing. The mean target distance error was 9.98 mm and 0.021°, with a deviation of 1.57 mm and 0.009°. The configuration parameters were as follows: a horizontal resolution of 10,240 points, a vertical resolution of 4267 points, a horizontal angular area of 0° to 360°, a vertical angular area of 90° to −60°, and 4× quality.

The GPS positions registered by this equipment have a precision with a 5 m error. Each scan station takes 11.15 minutes.

The same area covered by the MMS collected a colored and georeferenced point cloud. The start and stop angles of its scanners were 65° to 185°, 65° to 185°, −5° to 185°, −5° to 115°, and −5° to 115°. These angles are suitable according to the assembly and position of the scanners in the platform and the occlusions that define the rest of the equipment of the system, including the vehicle itself. The assembly of the scanners can be seen in Figure 4b. The scan frequency was set to 100 Hz, and the pulse repetition frequency (PRF) for the system configuration was 134,000 points per second. The contamination level was set to level 2 for all of them. Level 2, for this equipment, is a high level of contamination that fits the requirement of the agricultural environment in which the data were acquired. The spherical camera captured images in 5 m intervals. The recording of the GPS observations was kept at 10 Hz. The capture lasted 27 minutes and drove 1.7 km due to the maneuvers and the obstacles that the driver of the vehicle had to avoid. The car's advance speed was set to approximately 1 m/s to increase the cloud point density.

Both clouds were cut with the same defined area of 18.15 m^2, which included six vines (represented in Figure 5). The TLS cloud had 5,613,180 points, while the one captured with the MMS had 188,984 points.

For data acquisition from the real vineyard, the same capture configuration was used as in the test area. The scanning work was divided into four independent captures of 34, 50, 54, and 35 minutes, covering a total of 27.2 km. It should be noted that despite the size of the scanned area, the vehicle had to find accesses to the strips and have a clear path for kinematic alignment at the beginning of each shot. Thus, the distance travelled far exceeds the actual scanning distance. The environmental acquisition conditions (reflection, dust, etc.) and the characteristics of the equipment's techniques generated several outliers, which were cleaned with a filter using points based on an intensity value between 225 and 826 and a minimum threshold of four neighbors in a cube with a 0.09 m sized side.

The final point cloud clipped to the experimental zone (0.58 ha; represented in Figure 3) produced 38,833,904 points.

Point cloud preprocessing was done with the parameters reviewed in Table 8. These parameters were determined by the conditions of the capture. Each situation (sensor, meteorology, crop status, etc.) has specific characteristics to which these parameters have been adapted. The election of dimensional parameters was established considering such aspects as the point density of the clouds, the plantation framework, and the vine size or its prune. For outlier removal, we considered ambient conditions (i.e., dust presence); the skill of the equipment operator and vehicle driver, as well as ground conditions (for constant-speed scanning); and the equipment's functional requirements and limitations.

Table 8. List of the utilized parameters in preprocessing processes.

Clipping Parameters	TLS [2] Test Area	MMS [3] Test Area	MMS Real Case
ROI [1] buffer	computed@half MeanDistance	computed@half MeanDistance	computed@half MeanDistance
Trunk buffer from strip	0.3 m	0.3 m	0.3 m
Trunk buffer in strip	0.3 m	0.3 m	0.3 m
Minimum foliage height from terrain	0.4 m	0.4 m	0.4 m
Maximum foliage height from terrain	1.2 m	1.2 m	1.2 m
Foliage buffer from strip	0.6 m	0.6 m	0.6 m
Foliage buffer in strip	ROI buffer	ROI buffer	ROI buffer
Sensor type	TLS	MMS	MMS
Point Cloud Classifier Parameters	**TLS Test Area**	**MMS Test Area**	**MMS Real Case**
Class for trunk	13	13	13
Hidden nodes	5	5	5
Input nodes	3	3	3
Iterations	60	60	60
Minimum calibration accuracy	95	95	95
Output nodes	1	1	1
Remove Outliers Parameters	**TLS Test Area**	**MMS Test Area**	**MMS Real Case**
Class to use	13	13	13
Statistical sample neighbors for SOR [4] algorithm	50	50	10
Statistical std threshold for SOR algorithm	2	0.5	0.5
Radius minimum neighbors for ROR [5] algorithm	10	2	6
Radius search for ROR algorithm	0.05	0.2	0.09

[1] Region of interest. [2] Terrestrial laser scanner. [3] Mobile mapping system. [4] Statistical outlier removal. [5] Radius outlier removal.

Manual editing of the point cloud was done in the segmented vines when the results were anomalous.

3.2. Volume Calculation Results

The processing parameters for volume calculations with OctoMap, CHC, and VCTS are reviewed in Table 9.

These parameters were chosen in an iterative, manual, and supervised process of selection based on the improvement of the results. The visual interpretation of the three-dimensional modelling of each algorithm (whether in the form of voxels, the clustering of points, or simulation of internal structures), and the approximation of the estimated volume to the actual volume of the VSAO, were the criteria for optimizing these values.

The OctoMap algorithm needs to know the precise location of the sensor with respect to each scanned point in order to determine if the resulting voxels represent occupied or undefined space, and, therefore, makes it impossible for us to calculate the volume enclosed by the clouds of points taken with MMS (with this algorithm), as in [60]. In addition, the density of the points attained with MMS does not allow us to calculate the volume of the vines with the CHC algorithm, since the dimensions of some parts (i.e., the trunk's diameter) exceed the mean density of the cloud; moreover, its precision is low and it lacks the geometric definition of the figure due to occlusions or a lack of perspective when scanning, as in [60]. The results of these volume calculations are shown in Table 10.

The results obtained follow the four strategies proposed for the estimation of the height and radius of the identified cylinders, as shown in Table 10.

Table 9. List of the utilized parameters in the volume calculation algorithms.

OctoMap Parameters	TLS [1] Test Area	MMS [2] Test Area	MMS Real Case
Class to use	13	-	-
Compute free voxels	False	-	-
Voxel resolution	0.01	-	-
CHC [3] Parameters	**TLS Test Area**	**MMS Test Area**	**MMS Real Case**
Class to use	13	-	-
Color weight	0	-	-
Normal weight	1	-	-
Spatial weight	1	-	-
Seed resolution	0.15	-	-
Voxel resolution	0.01	-	-
VCTS [4] Parameters	**TLS Test Area**	**MMS Test Area**	**MMS Real Case**
Class to use	-	13	13
Minimum outlier threshold	-	0.01	0.01

[1] Terrestrial laser scanner. [2] Mobile mapping system. [3] Convex hull cluster. [4] Volume calculation from the trunk skeleton.

Table 10. Volume calculation results and errors committed.

Scanner	Volume Calculation Algorithm	VSAO [1] A (dm^3)	VSAO B (dm^3)	Absolute Error (dm^3)	Relative Error (%)
TLS [2]	OctoMap	2.329	1.666	0.793	28.41
TLS	Convex hull cluster	2.807	2.450	0.179	6.40
MMS [3]	VCTS [4] segment median	2.868	2.341	0.264	9.44
MMS	VCTS segment mean	2.863	2.665	0.099	3.55
MMS	VCTS line string median	2.843	2.128	0.358	12.81
MMS	VCTS line string mean	2.916	2.779	0.069	2.46
	Real volume	2.790	2.790		

[1] Vine-shaped artificial object. [2] Terrestrial laser scanner. [3] Mobile mapping system. [4] Volume calculation from the trunk skeleton.

Considering these results, the maximum error is reached with the OctoMap algorithm for the cloud scanned with TLS. This is a default error of 28.41%, which may be mainly due to the lack of faces on the scanned object.

The convex hull cluster algorithm improves the volume estimation from this cloud (with a 6.40% error). However, better approximations are given from MMS clouds with strategies based on the calculation of the mean radius per segment (3.55%) and the line string (2.46%).

Figure 8 shows the VSAO (Figure 8a), and the 3D point cloud taken by its TLS (Figure 8b). One can see the lack of scan points where the irrigation pipe is near the VSAO trunk, as well as under its arms. These occluded parts are also visible in the result of the OctoMap algorithm in Figure 8f, where there are no voxels. The same lack of occlusions is visible in the CHC algorithm results (Figure 9b), where it can be observed that the cluster segmented in the arms is thinner than the cluster segmented in the trunk. In both cases, the TLS point cloud underestimates the exterior surface of the VSAO and, consequently, the volume that these two algorithms estimate.

In the same manner, the occlusion of the lower arms is also noticed in the point cloud captured with the MMS equipment (Figure 10a), affecting both the position of the skeleton (not centered) and the determination of the radius of the segmented cylinders (lower than the real value).

3.3. Generation of Vine Size Maps

Based on the results obtained in the validation process, the methodology that achieves the lowest error from the equipment and makes large-scale data collection (MMS) feasible in a real case study was applied. This methodology applies the VCTS line string mean algorithm to 10% of the total scanned strains (120 vines of 1203 vines), which were randomly selected. In the final generated map (Figure 11), values between 1.0 and 8.0 dm^3 are observed. This variability may be due to various factors such as new vines planted to replace vines with problems, the lack of consistency in pruning methods, increased occlusions, uneven scanning speed, and other reasons for managing a vineyard or the technical limitations of the scanning method with this equipment in this type of scenario. No differences between the treatments were found (as expected), because these differentiated treatments started only two years ago. We expect to find these differences in the volume of the canopy but not in the trunk volume.

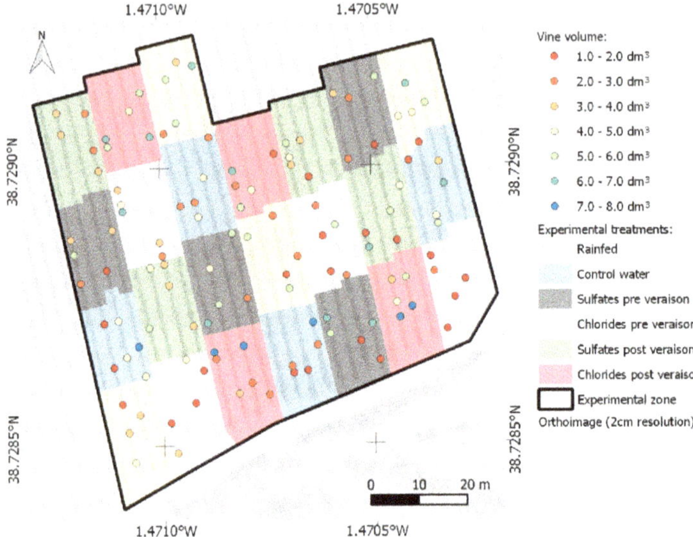

Figure 11. Calculated vine size map (scale 1:1000, EPSG:3857).

It should be noted that the operational character of the MMS equipment allows a complete mapping of the vineyard with an acceptable density of points, as can be seen in Figure 12.

Figure 12. (**a**) Real vine. (**b**) Preprocessed MMS point cloud taken from the same vine.

4. Discussion

It should be noted that the technical limitations of each piece of equipment have guided this work in its two subobjectives: to test the possibility of calculating the volume occupied by the trunks of vines in a vineyard using clouds from points taken with TLS equipment, and to extrapolate the best possibility to a real case study scanned with MMS, where it would be feasible to elaborate a vine size map based on the volume of each vine.

Firstly, the TLS point clouds have digitalized, with high detail and precision, the areas of the vines that were within their reach. Nevertheless, the areas occluded behind the equipment itself or its intermediate elements were not captured, thus making the three-dimensional definition of the scanned object incomplete (in this case, the vines of the vineyard), similar to the problems found in [61]. Secondly, the mobile capture system of the MMS solves this deficiency, as the number of views taken of the object covers most of its faces. This result could be achieved by increasing the number of TLS scan stations, but, considering the large number of faces these objects have, this process would be too costly for the intended purpose. However, the point density and low-quality cause other problems (also treated in [21]). Thus, the approach to treat and evaluate the obtained data should be different. In fact, the different algorithms implemented obtained different results depending on the scanning systems utilized because of the differences in the types of information acquired.

Taking advantage of the TLS (detail and precision) and considering their limitations (occlusions), the two proposed algorithms (OctoMap and CHC) can estimate the volume occupied by the scanned vines with the proposed methodology. OctoMap does not need to know the entire figure if the occlusions are smaller than the calculated voxel size [50], which makes this method appropriate for TLS, where many faces of the object are occluded. Indeed, we identified a defect error in the results due to the occlusion of the lower face of the arms, which could be solved by placing the scan stations at a lower height from the ground. The CHC algorithm is not as strongly affected by this lack of scanned faces since in the case of figures with simple geometries, such as cylinders, the closing of the convex envelopes of the point groupings obtained by this algorithm is accurate and resolves the occlusions suffered by the cloud.

Nevertheless, the capture performance of the TLS makes it unfeasible to survey large extensions of land, such as those covered by agricultural crops. However, the application of the tested algorithms to MMS point clouds is not possible because they have lost the precision and definition conditions required by OctoMap and CHC. In addition, those two algorithms also resulted in lower accuracy in the determination of the VSAO volume for TLS. Thus, it can be concluded that OctoMap and CHC are not the most appropriate algorithms for this case study. We recommend increasing future efforts in developing strategies for the skeleton algorithm.

The proposed change of strategy that focuses on modelling algorithms based on the internal structure of the objects (L_1-medial skeleton [52]), and not on determining the closure of their surfaces (OctoMap [50] and CHC [51]), has made it possible to estimate the volumes of individualized vines scanned with MMS point clouds, as the results show. Even the estimation of volume with this strategy improves, in some cases, the values obtained with TLS clouds (3.55% and 2.46% errors for the VCTS segment mean and line string mean, respectively, compared with a 6.40% error with the CHC algorithm and TLS point clouds). However, in the case of a TLS in which two faces of the trunk are perfectly defined but there are many occlusions because of the lack of perspective, the skeleton algorithm returned many errors that require further development to be robust and usable.

The extrapolation of this methodology to a real case study has identified several alterations that make it difficult to obtain the individualized volumes, which should be overcome in future work. On the one hand, the segmented point clouds of some vines do not define their shape due to the poor quantity and quality of the scanned points. This makes manual additions to the cloud subjective, in order to clean outliers that have not been identified, which is also seen in previous automatic processes, and contributes to the generation of incoherent skeletons. Therefore, the volumes obtained in these cases are inaccurate due to poor quality data acquisition, which can be solved by a better vehicle to transport the MMS, avoiding the generation of dust, decreasing the speed (to increase point density), and maintaining constant speed, among many other factors. On the other hand, strategies based on the VCTS algorithm are semi-supervised and require visual inspection of the generated skeleton before applying the volume calculation. Consequently, this methodology is time-consuming and materially cost-intensive. The increase in the costs that this methodology would require is also affected by the number of vines that require manual editing because of anomalies. Thus, the more variable the scanned noise is, the less the cleaning can be automated, and the more manual editing is needed. More effort towards the complete automation of this process should be developed, because it is the most promising algorithm with this objective.

As probable areas of work based on this experience, improvements will be developed in the automation and handling of algorithms. Further, it will be necessary to improve the conditions during data acquisition, taking into account the generation of dust, the constant speed of the vehicle during data acquisition, and other factors. In addition, since these are parameterized algorithms, their evaluation and optimization for each case study is necessary, so we intend to develop methods that consolidate an appropriate choice based on the improvements that occur in the acquisition of point clouds. Of course, more algorithms that allow the estimation of volumes will be tested, as in [62]. Another challenge to address is the determination of the volume occupied by the canopy (not only the trunk), which will require other algorithms and software development.

Thus, the proposed methodology and developed software are the first step towards promising technology to characterize the geometry of woody crops in order to help decision-making in crop management.

5. Conclusions

In this work, different strategies for calculating vine volumes from point clouds captured with static and mobile terrestrial laser scanners were developed in order to elaborate maps of the vegetative vigor of crops, particularly vine size. The proposed methodology makes use of laser scanning systems in precision agriculture, a promising technology; however, the experience has left several improvements to be solved to improve the obtained results, such as (1) improving the data acquisition; (2) increasing the automation of the result generation to avoid current manual data treatments; and (3) refining the algorithms to better determine the volume.

The results have revealed that the calculation of volumes from different scanning systems requires different algorithms because of the variability in the density of point cloud, noise, and occlusions. TLS point clouds are more accurate using the CHC [51] algorithm (with a 6.40% relative error), while the most complete and accurate results are obtained from MMS point clouds using the VCTS with the L_1-medial skeleton [52] line string mean algorithm (2.46% relative error). The VCTS could not be applied to TLS because of the occlusions that appear with this system, but considering the results using MMS, it is an interesting algorithm to apply to these systems after its adaptation.

The potential of laser scanning equipment has been demonstrated in agronomic challenges, as well as the application of three-dimensional point clouds to the three-dimensional digital characterization of vegetation. However, in this first approach, an intensive manual editing process is required, which should be solved in future developments. Nevertheless, these are the first experiments with this technology, and outstanding results were obtained by this working group, so future prospects are positive.

Author Contributions: A.d.-C.-S. is the principal investigator and corresponding author. She led and supervised the overall research and field data collection. She also wrote the paper with the support of the remainder authors. M.M. structured the paper in collaboration with R.B. D.H.-L. developed the software and helped to interpret the final results and limitations. M.M. developed the algorithm for point cloud classification. R.B. helped with the agronomic interpretation of the results and the design of the data acquisition process. All authors were key in the interpretation of the results.

Funding: This research was funded by Junta de Comunidades de Castilla-La Mancha, grant number SBPLY/17/180501/000251.

Acknowledgments: The authors of this work would also like to thank the technical, administrative, etc. collaboration provided by the rest of the members of the AgroForestry and Cartographic Precision Research Group (PAFyC-UCLM). We would also like to thank the Centro de Edafología y Biología Aplicada del Segura of the Consejo Superior de Investigaciones Científicas (CEBAS-CSIC) for the management, care and monitoring of the vineyard that we have used as a real case study.

Conflicts of Interest: The authors declare no conflict of interest.

References

1. Njoroge, B.M.; Fei, T.K.; Thiruchelvam, V. A Research Review of Precision Farming Techniques and Technology. *J. Appl. Technol. Innov.* **2018**, *2*, 9.
2. Ballesteros, R.; Ortega, J.F.; Hernández, D.; Moreno, M.Á. Characterization of Vitis vinifera L. Canopy Using Unmanned Aerial Vehicle-Based Remote Sensing and Photogrammetry Techniques. *Am. J. Enol. Vitic.* **2015**, *66*, 120–129. [CrossRef]
3. Shultz, H.R. Grape canopy structure, light microclimate and photosynthesis. A two-dimensional model of the spatial distribution of surface area densities and leaf ages in two canopy systems. *J. Grapevine Res.* **1995**, *34*, 211–215.
4. Watson, D.J. Comparative Physiological Studies on the Growth of Field Crops: I. Variation in Net Assimilation Rate and Leaf Area between Species and Varieties, and within and between Years. *Ann. Bot.* **1947**, *11*, 41–76. [CrossRef]
5. Smart, R.E.; Shaulis, N.J.; Lemon, E.R. The Effect of Concord Vineyard Microclimate on Yield. I. The Effects of Pruning, Training, and Shoot Positioning on Radiation Microclimate. *Am. J. Enol. Vitic.* **1982**, *33*, 99–108.
6. Smart, R.E.; Shaulis, N.J.; Lemon, E.R. The Effect of Concord Vineyard Microclimate on Yield. II. The Interrelations between Microclimate and Yield Expression. *Am. J. Enol. Vitic.* **1982**, *33*, 109–116.
7. Steduto, P.; Hsiao, T.C.; Fereres, E.; Raes, D. *Crop Yield Response to Water*; Steduto, P., Ed.; FAO Irrigation and Drainage Paper; Food and Agriculture Organization of the United Nations: Rome, Italy, 2012; ISBN 978-92-5-107274-5.
8. Rosell Polo, J.R.; Sanz, R.; Llorens, J.; Arnó, J.; Escolà, A.; Ribes-Dasi, M.; Masip, J.; Camp, F.; Gràcia, F.; Solanelles, F.; et al. A tractor-mounted scanning LIDAR for the non-destructive measurement of vegetative volume and surface area of tree-row plantations: A comparison with conventional destructive measurements. *Biosyst. Eng.* **2009**, *102*, 128–134. [CrossRef]
9. Arnó, J.; Escolà, A.; Vallès, J.M.; Llorens, J.; Sanz, R.; Masip, J.; Palacín, J.; Rosell-Polo, J.R. Leaf area index estimation in vineyards using a ground-based LiDAR scanner. *Precis. Agric.* **2013**, *14*, 290–306. [CrossRef]
10. Walklate, P.J.; Cross, J.V.; Richardson, G.M.; Murray, R.A.; Baker, D.E. Comparison of Different Spray Volume Deposition Models Using LIDAR Measurements of Apple Orchards. *Biosyst. Eng.* **2002**, *82*, 253–267. [CrossRef]
11. Steduto, P.; Hsiao, T.C.; Raes, D.; Fereres, E. AquaCrop—The FAO Crop Model to Simulate Yield Response to Water: I. Concepts and Underlying Principles. *Agron. J.* **2009**, *101*, 426–437. [CrossRef]
12. Sanz, R.; Rosell, J.R.; Llorens, J.; Gil, E.; Planas, S. Relationship between tree row LIDAR-volume and leaf area density for fruit orchards and vineyards obtained with a LIDAR 3D Dynamic Measurement System. *Agric. For. Meteorol.* **2013**, *171*, 153–162. [CrossRef]
13. García de Cortazar, V.; Acevedo, E.; Nobel, P.S. Modeling of par interception and productivity by Opuntia ficus-indica. *Agric. For. Meteorol.* **1985**, *34*, 145–162. [CrossRef]
14. Burrell, J.; Brooke, T.; Beckwith, R. Vineyard computing: Sensor networks in agricultural production. *IEEE Pervasive Comput.* **2004**, *3*, 38–45. [CrossRef]
15. Matese, A.; Vaccari, F.; Tomasi, D.; Di Gennaro, S.; Primicerio, J.; Sabatini, F.; Guidoni, S.; Matese, A.; Vaccari, F.P.; Tomasi, D.; et al. CrossVit: Enhancing Canopy Monitoring Management Practices in Viticulture. *Sensors* **2013**, *13*, 7652–7667. [CrossRef] [PubMed]
16. Mathews, A.; Jensen, J.; Mathews, A.J.; Jensen, J.L.R. Visualizing and Quantifying Vineyard Canopy LAI Using an Unmanned Aerial Vehicle (UAV) Collected High Density Structure from Motion Point Cloud. *Remote Sens.* **2013**, *5*, 2164–2183. [CrossRef]
17. Pichon, L.; Ducanchez, A.; Fonta, H.; Tisseyre, B. Quality of Digital Elevation Models obtained from Unmanned Aerial Vehicles for Precision Viticulture. *OENO One* **2016**, *50*. [CrossRef]

18. Weiss, M.; Baret, F.; Weiss, M.; Baret, F. Using 3D Point Clouds Derived from UAV RGB Imagery to Describe Vineyard 3D Macro-Structure. *Remote Sens.* **2017**, *9*, 111. [CrossRef]
19. Keightley, K.E.; Bawden, G.W. 3D volumetric modeling of grapevine biomass using Tripod LiDAR. *Comput. Electron. Agric.* **2010**, *74*, 305–312. [CrossRef]
20. Tarolli, P.; Sofia, G.; Calligaro, S.; Prosdocimi, M.; Preti, F.; Fontana, G.D. Vineyards in Terraced Landscapes: New Opportunities from Lidar Data. *Land Degrad. Dev.* **2015**, *26*, 92–102. [CrossRef]
21. Cabo, C.; Del Pozo, S.; Rodríguez-Gonzálvez, P.; Ordóñez, C.; González-Aguilera, D.; Cabo, C.; Del Pozo, S.; Rodríguez-Gonzálvez, P.; Ordóñez, C.; González-Aguilera, D. Comparing Terrestrial Laser Scanning (TLS) and Wearable Laser Scanning (WLS) for Individual Tree Modeling at Plot Level. *Remote Sens.* **2018**, *10*, 540. [CrossRef]
22. Bauwens, S.; Bartholomeus, H.; Calders, K.; Lejeune, P. Forest inventory with terrestrial LiDAR: A comparison of static and hand-held mobile laser scanning. *Forests* **2016**, *7*, 127. [CrossRef]
23. Brede, B.; Calders, K.; Lau, A.; Raumonen, P.; Bartholomeus, H.M.; Herold, M.; Kooistra, L. Non-destructive tree volume estimation through quantitative structure modelling: Comparing UAV laser scanning with terrestrial LIDAR. *Remote Sens. Environ.* **2019**, *233*, 111355. [CrossRef]
24. Wahabzada, M.; Paulus, S.; Kersting, K.; Mahlein, A.K. Automated interpretation of 3D laserscanned point clouds for plant organ segmentation. *BMC Bioinf.* **2015**, *16*, 248. [CrossRef] [PubMed]
25. Lin, Y.; Jaakkola, A.; Hyyppä, J.; Kaartinen, H. From TLS to VLS: Biomass estimation at individual tree level. *Remote Sens.* **2010**, *2*, 1864–1879. [CrossRef]
26. Yaxiong, W.; Shasha, X.; Wenbin, L.; Feng, K.; Yongjun, Z. Identification and location of grapevine sucker based on information fusion of 2D laser scanner and machine vision. *Int. J. Agric. Biol. Eng.* **2017**, *10*, 84–93.
27. Walthall, C.L.; Norman, J.M.; Welles, J.M.; Campbell, G.; Blad, B.L. Simple equation to approximate the bidirectional reflectance from vegetative canopies and bare soil surfaces. *Appl. Opt.* **1985**, *24*, 383–387. [CrossRef]
28. Escolà, A.; Martínez-Casasnovas, J.A.; Rufat, J.; Arnó, J.; Arbonés, A.; Sebé, F.; Pascual, M.; Gregorio, E.; Rosell-Polo, J.R. Mobile terrestrial laser scanner applications in precision fruticulture/horticulture and tools to extract information from canopy point clouds. *Precis. Agric.* **2017**, *18*, 111–132. [CrossRef]
29. Llorens, J.; Gil, E.; Llop, J.; Queraltó, M.; Llorens, J.; Gil, E.; Llop, J.; Queraltó, M. Georeferenced LiDAR 3D Vine Plantation Map Generation. *Sensors* **2011**, *11*, 6237–6256. [CrossRef]
30. del-Moral-Martínez, I.; Rosell-Polo, J.; Company, J.; Sanz, R.; Escolà, A.; Masip, J.; Martínez-Casasnovas, J.; Arnó, J.; del-Moral-Martínez, I.; Rosell-Polo, J.R.; et al. Mapping Vineyard Leaf Area Using Mobile Terrestrial Laser Scanners: Should Rows be Scanned On-the-Go or Discontinuously Sampled? *Sensors* **2016**, *16*, 119. [CrossRef]
31. Poni, S.; Lakso, A.N.; Intrieri, C.; Rebucci, B.; Filipetti, I. Laser scanning estimation of relative light interception by canopy components in different grapevine training systems. *J. Grapevine Res.* **1996**, *35*, 177–182.
32. Grocholsky, B.; Nuske, S.; Aasted, M.; Achar, S.; Bates, T. A Camera and Laser System for Automatic Vine Balance Assessment. In *2011 ASABE Annual International Meeting Sponsored by ASABE*; American Society of Agricultural and Biological Engineers: Louisville, KY, USA, 7–10 August 2011.
33. Herrero-Huerta, M.; González-Aguilera, D.; Rodriguez-Gonzalvez, P.; Hernández-López, D. Vineyard yield estimation by automatic 3D bunch modelling in field conditions. *Comput. Electron. Agric.* **2015**, *110*, 17–26. [CrossRef]
34. Smart, R.E. Principles of Grapevine Canopy Microclimate Manipulation with Implications for Yield and Quality. A Review. *Am. J. Enol. Vitic.* **1985**, *36*, 230–239.
35. Gil, E.; Escolà, A.; Rosell, J.R.; Planas, S.; Val, L. Variable rate application of plant protection products in vineyard using ultrasonic sensors. *Crop Prot.* **2007**, *26*, 1287–1297. [CrossRef]
36. Kang, F.; Pierce, F.J.; Walsh, D.B.; Zhang, Q.; Wang, S. An Automated Trailer Sprayer System for Targeted Control of Cutworm in Vineyards. *Trans. ASABE* **2011**, *54*, 1511–1519. [CrossRef]
37. Llorens, J.; Gil, E.; Llop, J.; Escolà, A.; Llorens, J.; Gil, E.; Llop, J.; Escolà, A. Ultrasonic and LIDAR Sensors for Electronic Canopy Characterization in Vineyards: Advances to Improve Pesticide Application Methods. *Sensors* **2011**, *11*, 2177–2194. [CrossRef] [PubMed]

38. Walklate, P.J.; Richardson, G.M.; Baker, D.E.; Richards, P.A.; Cross, J.V. Short-range lidar measurement of top fruit tree canopies for pesticide applications research in the United Kingdom. In *Advances in Laser Remote Sensing for Terrestrial and Oceanographic Applications*; International Society for Optics and Photonics: Orlando, FL, USA, 1997; Volume 3059, pp. 143–152.
39. Castelan-Estrada, M.; Vivin, P.; GAUDILLIÈRE, J.P. Allometric Relationships to Estimate Seasonal Above-ground Vegetative and Reproductive Biomass of *Vitis vinifera* L. *Ann. Bot.* **2002**, *89*, 401–408. [CrossRef] [PubMed]
40. Chatzinikos, A.; Gemtos, T.A.; Fountas, S. The use of a laser scanner for measuring crop properties in three different crops in Central Greece. In *Precision agriculture'13*; Stafford, J.V., Ed.; Wageningen Academic Publishers: Wageningen, The Netherlands, 2013; pp. 129–136.
41. Keightley, K.E. Applying New Methods for Estimating in vivo Vineyard Carbon Storage. *Am. J. Enol. Vitic.* **2011**, *62*, 214–218. [CrossRef]
42. English, J.T. Microclimates of Grapevine Canopies Associated with Leaf Removal and Control of Botrytis Bunch Rot. *Phytopathology* **1989**, *79*, 395. [CrossRef]
43. Carbonneau, A. Recherche sur les Systèmes de Conduite de la Vigne: Essai de Maitrise du Microclimat et de la Plante Entière Pour Produire Économiquement du Raisin de Qualité. Ph.D. Thesis, Université de Bordeaux 2 (FRA), Bordeaux, France, 1980.
44. Mabrouk, H.; Carbonneau, A.; Sinoquet, H. Canopy structure and radiation regime in grapevine. 1. Spatial and angular distribution of leaf area in two canopy systems. *J. Grapevine Res.* **1997**, *36*, 119–123.
45. Mabrouk, H.; Sinoquet, H.; Carbonneau, A. Canopy structure and radiation regime in grapevine. 2. Modeling radiation interception and distribution inside the canopy. *J. Grapevine Res.* **1997**, *36*, 125–132.
46. Ross, J. *The Radiation Regime and Architecture of Plant Stands*; Springer Netherlands: Dordrecht, The Netherlands, 1981; ISBN 978-94-009-8649-7.
47. Dry, P.R.; Loveys, B.R. Factors influencing grapevine vigor and the potential for control with partial rootzone drying. *Aust. J. Grape Wine Res.* **1998**, *4*, 140–148. [CrossRef]
48. Steyn, J.; Aleixandre Tudo, J.; Aleixandre Benavent, J.L. Grapevine vigor and within vineyard variability: A review. *Int. J. Sci. Eng. Res.* **2016**, *7*, 1056–1065.
49. Morandé, J.A.; Stockert, C.M.; Liles, G.C.; Williams, J.N.; Smart, D.R.; Viers, J.H. From berries to blocks: Carbon stock quantification of a California vineyard. *Carbon Balance Manag.* **2017**, *12*, 5. [CrossRef] [PubMed]
50. Hornung, A.; Wurm, K.M.; Bennewitz, M.; Stachniss, C.; Burgard, W. OctoMap: An efficient probabilistic 3D mapping framework based on octrees. *Auton Robot* **2013**, *34*, 189–206. [CrossRef]
51. Papon, J.; Abramov, A.; Schoeler, M.; Wörgötter, F. Voxel Cloud Connectivity Segmentation—Supervoxels for Point Clouds. In Proceedings of the 2013 IEEE Conference on Computer Vision and Pattern Recognition, Portland, Oregon, USA, 23–28 June 2013; pp. 2027–2034.
52. Huang, H.; Wu, S.; Cohen-Or, D.; Gong, M.; Zhang, H.; Li, G.; Chen, B. L1-medial skeleton of point cloud. *ACM Trans. Graph.* **2013**, *32*, 1. [CrossRef]
53. American Society for Photogrammetry and Remote Sensing (ASPRS). LAS SPECIFICATION VERSION 1.4—R13. Available online: https://www.asprs.org/wp-content/uploads/2010/12/LAS_1_4_r13.pdf (accessed on 10 May 2019).
54. Córcoles, J.I.; Ortega, J.F.; Hernández, D.; Moreno, M.A. Estimation of leaf area index in onion (Allium cepa L.) using an unmanned aerial vehicle. *Biosyst. Eng.* **2013**, *115*, 31–42. [CrossRef]
55. Rusu, R.B.; Cousins, S. 3D is here: Point Cloud Library (PCL). In Proceedings of the 2011 IEEE International Conference on Robotics and Automation, Shanghai, China, 9–13 May 2011; pp. 1–4.
56. PCL Class Statistical Outlier Removal. Available online: http://www.pointclouds.org/documentation/tutorials/statistical_outlier.php (accessed on 10 May 2019).
57. PCL Class Radius Outlier Removal. Available online: http://pointclouds.org/documentation/tutorials/radius_outlier_removal.php (accessed on 10 May 2019).
58. Meagher, D. Geometric modeling using octree encoding. *Comput. Graph. Image Process.* **1982**, *19*, 129–147. [CrossRef]
59. Wilhelms, J.; Van Gelder, A. Octrees for Faster Isosurface Generation. *ACM Trans. Graph.* **1992**, *11*, 201–227. [CrossRef]

60. del Campo Sánchez, A.; Moreno Hidalgo, M.Á.; Hernández López, D. Determinación del vigor del viñedo mediante caracterización tridimensional basada en tecnología láser escáner. In *Libro de Actas del I Congreso de Jóvenes Investigadores en Ciencias Agroalimentarias*; CIAIMBITAL (Centro de Investigación en Agrosistemas Intensivos Mediterráneos y Biotecnología Agroalimentaria. Universidad de Almería): Almeria, Spain, 2018; pp. 8–13.
61. López-Lozano, R.; Baret, F.; García de Cortázar-Atauri, I.; Bertrand, N.; Casterad, M.A. Optimal geometric configuration and algorithms for LAI indirect estimates under row canopies: The case of vineyards. *Agric. For. Meteorol.* **2009**, *149*, 1307–1316. [CrossRef]
62. Mei, J.; Zhang, L.; Wu, S.; Wang, Z.; Zhang, L. 3D tree modeling from incomplete point clouds via optimization and L1-MST. *Int. J. Geogr. Inf. Sci.* **2017**, *31*, 999–1021. [CrossRef]

MDPI
St. Alban-Anlage 66
4052 Basel
Switzerland
Tel. +41 61 683 77 34
Fax +41 61 302 89 18
www.mdpi.com

Remote Sensing Editorial Office
E-mail: remotesensing@mdpi.com
www.mdpi.com/journal/remotesensing

www.ingramcontent.com/pod-product-compliance
Lightning Source LLC
Chambersburg PA
CBHW051713210326
41597CB00032B/5467

* 9 7 8 3 0 3 9 2 8 0 1 8 6 *